European Commission

Proceedings of the International Symposium

The learning society and the water-environment

La société cognitive et les problèmes de l'eau

Paris, 2-4 June 1999

Edited by

A. Van der Beken, M. Mihailescu, P. Hubert and J. Bogardi

Organised within the framework of

European Thematic Network of Education and Training
ETNET, ENVIRONMENT WATER
SOCRATES Programme of the European Commission

INTERNATIONAL HYDROLOGICAL PROGRAMME IHP
UNESCO, United Nations Educational, Scientific and Cultural Organisation

Directorate-General
Research

A great deal of additional information on the European Union is available on the Internet. It can be accessed through the Europa server (http://europa.eu.int).
De nombreuses autres informations sur l'Union européenne sont disponibles sur Internet via le serveur Europa (http://europa.eu.int)

Cataloguing data can be found at the end of this publication.
Une fiche bibliographique figure à la fin de l'ouvrage.

Luxembourg: Office for Official Publications of the European Communities, 2000
Luxembourg: Office des publications officielles des Communautés européennes, 2000

ISBN 92-828-8308-6

Printed in Belgium

PRINTED ON WHITE CHLORINE-FREE PAPER
IMPRIMÉ SUR PAPIER BLANCHI SANS CHLORE

Preface

Since time immemorial education and training have been selective methods for humanity in its search for knowledge and skills. It was somehow always understood that the well being of people and the economic wealth were dependent upon understanding of nature and managing the natural resources.

Educational systems were tried out and they became, during the last two centuries, gradually an essential action-tool of governments to ensure that the whole of the population of a country would participate in its economic development.

Today this "**learning society**" has been defined as the merging of the educational world and the economic world. But at the end of the millennium, the globalization of both information and economy challenges our educational systems and methods. Advanced learning technologies have been introduced and the Internet makes old and new education and training tools available to a multitude of users, un-hoped for ten years ago. Needs analysis, planning, cost-effective designs, implementation, delivery and updating, quality assessment and control, validation or accreditation and, in general, the sustainability and socio-economic and cultural values of methods and tools are questioned. This International Symposium aimed at highlighting all these issues from the point of view of the water-environment.

The **International Hydrological Programme (IHP) of UNESCO** and its theme "Transfer of Knowledge Information and Technology" is the appropriate worldwide framework. The **European Thematic Network of Education and Training (ETNET) for Environment-Water,** Project 25899-CP-3-9831-BE-ERASMUS-ETN funded by the SOCRATES/ERASMUS Programme of the Directorate-General XXII for Education, Training and Youth of the European Commission (EC), is the forum building on existing partnerships between the educational world and the economic world of the water-related sectors, including the many professional and scientific associations.

Objectives of the Symposium were:

- to provide a dialogue for presentations in an efficient way on topics of mutual interest and on results of past and on-going projects;
- to explore the needs for and to foster new developments of programmes and projects;
- to promote intensive interdisciplinary co-operation;
- to improve understanding and broaden the knowledge in all matters related to the concept of the Learning Society and the Water-Environment issues;
- to offer an exchange of views among the providers of education and training, the software and multimedia developers and the end-users;
- to stimulate international perspectives and networking for enhancing the quality and pooling of expertise;
- to promote international co-operation, more specifically with UN-agencies, EU-programmes, IGOs and NGOs.

The Proceedings of the Symposium follow the organisation into seven THEMES for which key-note lectures were invited and 52 submitted papers were selected. A rapporteur for each THEME has written a short summary, which is published at the beginning of each THEME. The Editors have added 13 poster-contributions for which extended abstracts were available. These offer additional information about on-going projects. At the end of the Proceedings is given a list of CD-ROMS, Internet papers and Internet courses with contact-addresses or URLs, extracted from the papers or posters presented at this Symposium.

The editors would like to thank: Mr. G. Neveu and his OIE-team who have been expedient for some financial aspects, the members of the Scientific Committee of the Symposium who helped to select the papers and took up the role of chair-person, rapporteur or keynote lecturer, Mr. W. Gilbrich who has contributed in many ways to this Symposium.

The Symposium Secretariat was in the excellent hands of Dr M. Mihailescu with the support of the Laboratoire de Géologie Appliquée, Université Pierre et Marie Curie. The smooth running of the Symposium at UNESCO's headquarters was the result of the very efficient support given by Mrs. Ouraghlia of the UNESCO IHP-secretariat.

Last but not least, the Editors gratefully acknowledge the important financial support for the organizing of the Symposium offered by UNESCO (venue, simultaneous translation, logistics, travel and subsistence costs), WMO and UNDP (travel and subsistence costs) and the European Commission through its project ETNET.ENVIRONMENT-WATER and related projects managed by TECHnology for WAter REsources (TECHWARE).

In addition, these Proceedings are made available thanks again to an extra-ordinary support of the Thematic Programme Energy, Environment and Sustainable Development, 5th Framework Programme for Research, Technological Development and Demonstration of the European Commission. In this way, one of the objectives of the Symposium "promotion of international co-operation, more specifically with UN-Agencies, EU-programmes, IGOs and NGOs" has been reached. The Editors hope that this objective and the many other objectives will continuously be in the mind and serve as guidelines for action among the readers of these Proceedings.

September 30, 1999

The Editors:

A. Van der Beken, Brussels
M. Mihailescu, Bucharest
P. Hubert, Paris

Foreword

Since 1987 many students and teaching staff have benefited from the education and training programmes of the European Commission. These programmes, known under their acronyms ERASMUS, SOCRATES, COMETT, LEONARDO, TEMPUS are now well known not only throughout the whole of Europe but also worldwide. They have created much interest in non-European countries and serve as models for transnational and international co-operation programmes.

The new initiative of UNESCO, called "GOUTTE of Water" (Global Organisation of Universities for Teaching, Training and Ethics), is based on experience gained within the European context.

Within the European Commission these programmes span the whole formal and non-formal education and training sector, and it has been recognised that research and education always go together. Therefore, the Framework Programme for Research and Technological Development and Dissemination is also funding accompanying measures, which include training activities. In the field of water, training grants for researchers and education activities in the framework of Advanced Study Course have been particularly supported within the past and ongoing environmental RTD programmes.

The SOCRATES/ERASMUS Thematic Network Projects are a new activity launched in 1996. Their aim is to establish a European-wide forum of academia and non-academia, by discipline or other linking theme, which can investigate the state of the European co-operation in the field, test out new ideas for education and training, argue the future discussions for the curriculum and relate academia to the needs of the outside world and the profession, all by means of specific projects.

The European Thematic Network of Education and Training (ETNET) for ENVIRONMENT-WATER, funded under this SOCRATES/ERASMUS scheme, develops successfully several of these specific projects: they provide further analysis (the specific projects on "Continuing Profession Development Programmes for Water" and "European Paradigm of Integrated Water Management"), further information gathering (the specific project on "European Postgraduate Degrees in Hydrology and Water Management"), or a test phase to see if a particular approach really works (the specific projects "European Engineering Graduate School of Environment-Water" and "ODL for Water").

Obviously, all these specific projects of ETNET have a strong linkage with research, technological development and dissemination and more specifically with the specific theme "Energy, Environment and Sustainable Development" of the 5th Framework Programme of the European Commission. For these reasons, the initiative taken by ETNET and UNESCO's International Hydrology Programme " to organize the International Symposium "The Learning Society and the Water Environment", was very appropriate and timely.

The European Commission therefore decided to publish the Proceedings of this International Symposium so underpinning the efforts made by the participants to improve the quality of education and training and show the mechanisms for a continuous dynamic of practice and evaluation, innovation and implementation.These proceedings are an important source for future initiatives of ETNET and other organisations or networks.

I. V. MITCHELL
Head of Unit ERASMUS
European Commission
DG XXII – Education, Training and Youth

Table of contents *Table*

Theme A

Theme B

Theme C

Theme D

Theme E

Theme F

Theme G

Conclusion

Posters

Ladies and Gentlemen,

The key to sustainable, self-reliant development is education - Education that reaches out to all members of society...
Education that provides genuine lifelong learning opportunities for all.
We must be ready, in all countries, to reshape education so as to promote attitudes and behaviour conducive to a culture of sustainability.
The emergence of educational networks is one of the backbones of a learning society.
Networking in education, as a dual experience of teaching and learning together, helps build the society of the future.
The European Union provides one of the best examples in the world today of large-scale educational exchange and research networks.

UNESCO is utilising the potential of educational networking in its programmes too.
At university level, it has launched its UNESCO Chair and university partnership programme UNITWIN.
This facilitates the exchange of knowledge and ideas on a global scale, involving and strengthening institutions of higher learning in the developing countries.
University chairs in water or water-related disciplines have been, or are being, created in Jordan, Morocco, Malawi, Sudan, Lesotho, South Africa, Russia, Ukraine and France.
As you can see, most of these UNESCO chairs are in countries and serving regions facing the most severe water shortages or environmental problems.
As a further step, these chairs are expected to become the nucleus of wide-spread university partnerships.

I am happy and proud to announce that five weeks ago the first water-oriented UNITWIN Network: "Réseau méditerranéen de l'eau", a Mediterranean Water Network, was launched in Cannes, France.
Its 13 initial partners from universities, and municipalities are lead by the UNESCO Chairholder, Professor Raoul Caruba at the University of Nice Sophia-Antipolis.

In order to make networking an efficient vehicle of the Learning Society, knowledge transfer and communication within networks must also be part of early learning.
Therefore, UNESCO has also created school networks within the UNESCO Associated Schools Project.
Water and the Environment are among its central themes.
One of the "Flagship" projects is the Blue Danube project, linking together schools in the countries, which share the Danube.
UNESCO's 2000+ programme on science and technology education at high school level has developed a water-related teaching module.

When we talk about learning, we should not only think about formal education.
What is more, water-education, both formal and informal, should not only target the professional community.
In a Learning Society, education penetrates all daily life.
We should explore all avenues to raise public awareness in particular among young people.

Messieurs les Présidents,
Excellences,
Mesdames, Messieurs,

S'agissant des moyens d'informer et de sensibiliser le grand public, l'UNESCO s'est associée au projet de la société française STRASS Productions intitulé "L'or bleu". Cette encyclopédie interactive sur CD-ROM figure sur la liste des candidatures pour le Prix des Princes, qui couronne les meilleurs matériels éducatifs. Cette association entre l'UNESCO et une entreprise du secteur privé a reçu le soutien de la FAO, du Comité français pour l'UNICEF, de l'Union européenne et des autorités françaises. Cet exemple de partenariat, que j'ai choisi parce qu'il est le plus récent, est loin d'être le seul. Preuve que la société d'apprentissage (je préfère personnellement cette expression à celle de "société cognitive", car elle souligne le caractère continu du processus) est déjà bel et bien une réalité.

Mesdames, Messieurs,

Les trois jours qui viennent s'annoncent passionnants, car ce colloque est le premier à explorer les rapports entre la société d'apprentissage et la problématique de l'eau. Les analyses et les pistes contenues d'une part dans le "rapport Delors", issu des travaux de la Commission mondiale sur l'éducation pour le XXIe siècle, et d'autre part dans le livre blanc de la Commission européenne, intitulé "Enseigner et apprendre - Vers la société cognitive" pourront être appliquées à ce domaine d'une actualité vitale que constituent l'eau et l'environnement.

A considérer les thèmes que vous allez aborder - développement des ressources humaines, sensibilisation du public, analyse des besoins de formation, évaluation et accréditation, formation continue et à la carte, enseignement à distance, formation par la recherche, mobilité et internationalisation, etc. -, on peut être certain que les débats feront le tour de la question. Je me bornerai donc à conclure par trois questions, qui devraient à mon sens guider en permanence notre réflexion commune.

Première question: comment mettre en valeur le rôle de l'éducation dans l'élaboration de solutions durables aux problèmes de l'eau? Je me suis déjà implicitement référé au parallèle entre l'éducation, clé d'un avenir viable, et le projet Vision mondiale de l'eau. Etant donné qu'il vise à définir les moyens d'assurer une bonne gestion de l'eau pour un environnement mondial durable, ce projet doit inclure un volet "éducation" qui lui permette d'atteindre ses objectifs. J'ai la conviction que les débats et conclusions de ce colloque pourront fournir la base d'un processus de consultation sectorielle sur "l'eau et l'éducation" aux fins et dans le cadre de la Vision mondiale de l'eau.

Deuxième question: comment intégrer la "nouvelle éthique de l'eau" à l'éducation ? Il semble que cette expression, "nouvelle éthique de l'eau", que j'avais employée à Marrakech lors du premier Forum mondial de l'eau, ait fait florès. Mais les mots ne suffisent pas. Encore faut-il que des mesures soient prises pour commencer à les concrétiser pour guider les comportements quotidiens. C'est, encore une fois, par l'éducation que le contenu de cette nouvelle éthique peut être transmis, tant aux responsables d'aujourd'hui qu'aux jeunes générations. Je compte sur vous tous pour y contribuer, dans vos différentes sphères d'activité, et je compte aussi sur les titulaires des chaires UNESCO pour garder cette nécessité à l'esprit. La "nouvelle éthique de l'eau", loin de se figer en un dogme universel, doit se traduire en modalités concrètes chaque fois adaptées aux différents environnements climatiques, socio-économiques et culturels.

Troisième et dernière question: comment utiliser au mieux les nouveaux médias à des fins d'éducation ? Activité humaine fondamentale, l'éducation crée entre les groupes comme entre les individus des rapports riches et profonds. Ce processus peut être renforcé, facilité ou amélioré par les nouveaux médias, mais non remplacé par eux. Les nouvelles technologies de l'information nous permettent de rassembler, d'organiser et de transmettre des données très rapidement et quelle que soit la distance. Il faut donc que nous définissions de nouvelles façons d'enseigner, d'apprendre et d'interagir. La forme sous laquelle circule l'information ne doit pas se substituer au contenu; le transfert de l'information et du savoir ne doit pas être une fin en soi. N'oublions pas que, à l'ère de l'information, la capacité humaine d'assimiler les connaissances n'a pas changé; seules les capacités techniques ont évolué.

L'"homo virtualis", avec toutes ses potentialités, ne remplace pas l'"homo sapiens". Faisons en sorte qu'il puisse comprendre et résoudre les problèmes de l'eau pour assurer aux générations futures un environnement durable.

Tous mes vœux vous accompagnent dans vos débats.

Theme A

The learning society. Human resources development and capacity building. Interdisciplinarity. Public awareness

La société cognitive. Développement des ressources humaines et des compétences. Interdisciplinarité. Sensibilisation du public.

Thème A: La société cognitive. Développement des ressources humaines et des compétences. Interdisciplinarité. Sensibilisation du public.

Rapporteur: R. A. Lopardo (Argentine)

La scéance du Thème A a été présidé par M. Besbes (Tunisie) avec un sage contrôle du temps et un agil developpement des discussions.

Pendant la scéance on a eu l'exposé introductif de Mme. E. Almeida-Teixeira (Union Européenne) et six des présentations orales acceptées. Pour des raisons imprévues le travail du Pays Bas (M P. De Laat et M. A. Ramsuidersingh) a été reprogrammé pour une autre scéance et les travaux du Portugal (M. F. Nunes Correia) et de la France (M. W. Gilbrich) n'ont pas été exposés.

Dans l'exposé introductif, Mme. Almeida-Teixeira a presenté une vue générale et très compréhensive du régard de l'Union Européenne sur la Société Cognitive par rapport à l'eau et l'environnement.

L'éductation constitue un facteur très impportant pour une société très complexe, et elle est etroitement liée au problème de l'emploi, face aux changements sociaux et à la présence des marchés concurrentieles et du marché du travail. Dans ce cadre, on a proposé une "Europe du Savoir", qui doit faire face à cinq objectifs: nouveaux savoirs, école-emploi, combat de l'exclusion, compétence en trois langues et partage des investissements.

L'Union Européenne doit faire une contribution à l'éducation de qualité, en respectant l'indépendance de chaque état membre, mais avec la coopération des universités de ces pays. Les programmes SOCRATES et ERASMUS en sont d' excellents exemples. Ils permettront de améliorer le dialogue entre les professeurs et les chercheurs des divers pays. D'autre part, il est déjà en place le programme ETNET, dans le domaine de l'eau et de l'environnement. On propose de parvenir à un modèle multimédia européen et peut être universel qui devra tenir compte du marché du travail.

On a aussi exposé le sujet de l'évaluation et le problème de l'autoévaluation de l'apprenant. Finalement, malgré que la nouvelle technologie permettra d'acquérir des connaissances, elle n'est qu'un complément de l'education traditionnelle, car elle ne pourra jamais remplacer le contenu. La technologie permet de passer le massage mais n'est pas le message lui-même.

M. Besbes a demandé la possibilité d'élargir les programmes de l'Union Européenne vers d'autres pays.

M. Roux a présenté une expérience très attirante sur l'information et l'éducation dans le domaine de l'eau et de l'environnement dans le cadre des agences franVaises dans ce domaine. Avec les principes de "l'éducation des jeunes et par les jeunes", "prise de conscience collective" et "bonne gouvernance", les auteurs proposent des "classes d'eau" dans le système éducatif et l'élaboration d'une charte sociale de l'eau, qui va être soumise à la Conférence de La Haye l'année prochaine. L'expérience a permis de mettre les jeunes en contact avec les problèmes de l'eau et de l'environnement. Le financement du programme est assuré parce qu'il est basé sur le système educatif conventionnel, plus un effort économique (modeste) du programme lui-même, financé par les usagers de l'eau, à travers des agences. Lorsqu'on a demandé s'il y avait des rapports avec des programmes d'autres pays, les auteurs mentionnent que jusqu'à présent il n'y a que des contacts entre professeurs franVais et italiens.

M. Lopardo a présenté des idées sur la formation de l'ingénieur en sciences de l'eau avec une vision intégrale de l'environnement, en tenant compte du besoin d'une forte formation technique sur les domaines de base et d'une formation interdisciplinaire au niveau post-universitaire. L'auteur veut décourager la formation de grade du type trop généraliste, car ces professionnels ne seront pas utiles pour l'intégration des équipes transdisciplinaires dont on a besoin por les études d'évaluation d'impacts sur l'environnement.

M. Loiskandl a présenté un excellent témoignage d'une expérience européenne sur l'éducation dans les domaines de l'eau et de l'environnement avec l'intégration d'un plan d'études autrichien à une dimension européenne. Pendant la présentation l'auteur a posé les problèmes de la mobilité des étudiants et de la révision de la formation de l'université d'origine, et il a proposé la réduction de la durée moyenne des études et l'introduction de modules. Cette adaptation contribue à l'enseignement continu au niveau post-universitaire.

M. Feng a proposé un programme d'éducation du type "pyramidal" sur tois niveaux (éducation publique, éducation vocationnelle et éducation professionnelle), pour être appliqué dans le domaine de l'eau et de l'environnement dans les pays en voies de développement, avec la collaboration et l'aide économique des pays developpés et des organisations internationalles.

M. Wind a présenté une démostration de l'utilisation des modèles mathématiques de systèmes interdisciplinaires de gestion d'eau comme méthode d'éducation universitaire et aussi pour l'éducation du public en général. L'auteur met en rélief les problèmes de sélection des échelles appropriées du temps et de l'espace, mais aussi comment on peut mesurer les incertitudes et mettre en évidence comment quantifier les problèmes.

M. Khapdekar a basé son exposé sur deux programmes developpés en Inde sur la participation des citoyens dans le contrôle de la qualité des eaux. L'auteur a constaté que le controle de la qualité de l'eau peut contribuer à l'éducation conventionnelle, à travers des expériences des programmes de "Prise de Conscience de la Pollution du Gange" et "Jalapareekshana".

En conclusion, on peut resumer les propositions, experiences et sujets traités pendant la scéance du Thème A en trois axes fondamentaux (idées forces):

a) La coopération comme moyen d'assurer une éducation à vie sur l'eau et l'environnement à tous les niveaux.

   Expérience des "classes d'eau" avec la coopération entre les agences et le système educatif. Expérience de compatibilisation interuniversitaire européenne, avec la mobilité des étudiants et l'inclusion de modules. Expérience sur le contrôle de la qualité des eaux dans l'éducation des enfants. Programmes ERASMUS et SOCRATES. ETNET domaine de l'eau.

b) La définition d'un cadre approprié pour la formation des professionnels sur l'eau et l'environnement

Comment introduire la vision intégrale de l'environnement dans la formation des professionnels des sciences de l'eau sans perdre la spécialisation et pour les préparer au travail dans des équipes transdisciplinaires. Programme du type pyramidal pour l'éducation publique, vocationnelle et professionnelle.

c) L'utilisation de nouvelles technologies pour faciliter l'éducation sur la gestion de l'eau et de l'environnement

On a constaté l'application de modèles mathématiques sur les aspects integrales de l'environnement, comme exemple de systèmes interdisciplinaires de la gestion de l'eau. Ce type de nouveaux moyens peut beaucoup aider.

Keynote lecture

Europe of Knowledge

M.-E. Almeida-Teixeira

European Commission
Directorate General XXII
Education, Training and Youth
Brussels, Belgium

It is with great pleasure that I am here, today, on behalf of the Directorate General for Education, Training and Youth, (DGXXII) of the European Commission, particularly the Unit responsible for Higher Education, in order to participate in the International Symposium organised by the UNESCO International Hydrological Programme and ETNET – Water - Environment – the European Thematic Network of Education and Training in the area of Water-Environment.

I have the difficult task to be the first to address you, particularly in an area that is of utmost importance for the present and future of our society – The Learning Society, the need for quality education and training, and public awareness, namely on what concerns the Water-environment.

I will give you an overview on how, from the point of view of the European Union, we are working towards building the « Europe of Knowledge ». Some of the initiatives that were undertaken will be described.

The Learning Society

To give you an idea on how we (at the European Commission) did start setting the foundations of the Learning Society, let me go back as early as 1996, which was declared, by the Commission, the European Year of Long Life Learning .

It is an evidence that the European society is experiencing multiple and complex changes. Three are particularly important:

- the information society,
- the globalisation,
- the scientific and technical civilisation.

Both Member States and the European Community, as a whole, have dedicated great reflection to these.

Education and Training are at the crossroads of two main axis of the European society's development:

- because , in our culture, they are considered as one of the key factors of personal achievement as well as of social promotion,
- but also, because they determine our economic competitiveness and, consequently, the level and quality of our employment.

The correlation between education and employment becomes increasingly closer. In 1994, 11% of the working force without secondary education was unemployed, while 8% of those having attained such level, or 5% of those holding a higher education diploma were unemployed.

In addition, Continuing Learning and Training are increasingly important, but Continuing Education is insufficiently developed in Europe.

Only 7% of the European workforce benefit from a training action annually. Not withstanding, Continuing Education is essentially restricted to big corporations and to workforce already better qualified from the start.

At the end of our century, the main objective, both to individuals and organisations, is to develop a constant capacity for (rapidly) acquiring new knowledge.

This important and unavoidable change implies that both objectives and contents of the Education and Training systems should be reconsidered in order to respond to the needs of the European economy competitiveness and the quality of employment.

In 1995, the Commission published the White Paper on "Teaching and Learning: Towards the Learning Society" with the objective of placing education and training at the very centre of our society.

The White Paper set out five general objectives as follows :

Objective 1. Encourage the acquisiton of new knowledge,

2. Bringing the school and the business sector closer together,

3. Combat exclusion : offer a second chance through school,

4. Proficiency in three Community languages,

5. Treat material investment and investment in training on an equal basis.

The multiple debates, at Member-State level, with social partners, employers, universities, non-governmental organisations, among other, contributed to the foundations of the new concept of a "Europe of knowledge".

"Europe of knowledge" is simultaneously a reality under construction and an objective to meet.

It situates itself at the interception of two processes:

- the emergence of the "knowledge society", which will profoundly mark the next century, and
- the European construction, one of the most significant historical processes of the last 40 years.

Besides, the European Commission proposed, in the Agenda 2000, to make the policies which are the driving forces of the « Society of Knowledge » – innovation, research, education and training – one of the four fundamental axis of the Internal Policies of the Union.

In this context the European Summit, in November 1997, in Luxembourg, devised a coordinated European strategy for the employment, putting particular emphasis on Education.

Also , in November 1997, the Communication from the Commission, entitled "Towards a Europe of knowledge" presented the guidelines for the future Community actions in the fields of Education, Training and Youth. These were positioned at the centre of gradual construction of a European education area, in relation to three dimensions: knowledge, citizenship and competence.

On the one hand, it is necessary to give all individuals the possibility of further acquiring and deepening their knowledge in a lifelong perspective.

On the other hand, it will facilitate an enhancement of citizenship through the sharing of common values, and the development of a sense of belonging to a common social and cultural area, and mutual understanding of our cultures.

In addition, this European area should be a space of promoting the acquisition of competencies, developing employability, made necessary through changes in work and its organisation.

This means that it is increasingly necessary to promote creativity, flexibility, adaptability, the ability to "learn to learn" and to solve problems, on a lifelong basis.

The European Union's action in the fields of Education and Training has also considered the fundamental role played by the new Information and Communication Technologies (ICT).

Information and communication technologies are strongly impacting in our economical and social lives. They are already having a considerable impact in the areas of Education and Training. Everyone's access to information and knowledge has improved, and ICTs enable new ways and complementary possibilities of teaching and learning, more flexible and personalised.

The Amsterdam Treaty which came into force on 1st May of this year (1999) (after being ractified by the Member- States of the European Union) sets out in its Article 149 (ex- article 126 of the Treaty of Maastricht), the contribution that the European Union should make to the development of quality Education in the respect of national and cultural diversity.

Particularly, its point 1 states that : « The Community contributes to the development of quality education by encouraging cooperation between Member-States and by supporting and supplementing their actions while fully respecting the responsibilities of the Member-States for the content of teaching and the organisation of education systems and their cultural and linguistic diversity ».

Actions undertaken

I am confident that most of you have heard about the programme ERASMUS, and particularly the student and teacher mobility activities that take place in its framework. Since, 1996, with the approval of the SOCRATES programme, the European Community Programme in the field of Education that the activities in the field of Higher Education (Chapter still called ERASMUS) are part of SOCRATES, whose scope is significantly larger than that of the former ERASMUS Programme.

Interuniversity Cooperation initiated with the ERASMUS programme, is, since 1996, part of the European (or international) cooperation strategic activities of the Higher Education Institutions themselses, thus confering some struturation to the various activities of the universities towards education.

Besides, the SOCRATES Programme introduced a new action entitled 'University cooperation in areas of common interest', which is often refered as Thematic Network Projects.

ERASMUS Thematic Network Projects are projects which aim at confering a European dimension to a given discipline or area of study, by means of cooperation primarily amongst European Higher Education Institutions which are concerned with the area of study.

Besides, Thematic Networks aim at assessing curriculum innovation, improving teaching methods, development of teaching materials and fostering the development of joint programmes and specialised courses, as well as improving the dialogue between academics and other partners, be they socio-economic partners, non-governmental organisations, professional associations, to mention a few examples.

Provided the objectives are adhered to, Thematic Networks partnerships have a significant freedom to interpret the action in terms of activities.

The work undertaken may take the form of background reflection on issues of major relevancy to the area at play, the formulation of recommendations and prospective views leading to progress in the field of study, addressed by the Thematic Network, as well as the development of specific actions or projects.

Cooperation in Thematic Networks should count with the participation of all countries and that of the institutions that are concerned with the study area at stake.

Indeed the SOCRATES programme is open to the participation of the Member-States of the European Union and those of the European Economic Area as well as to a number of Central and Eastern European countries and Cyprus (at present, a total of 11 associated countries) which have concluded negotiation agreements with the European Commission in order to cooperate in the Community action programme in the field of education.

Besides cooperation is open to the participation of non-university partners, particularly that of entreprises, non-governmental organisations, professional associations, etc, where relevant.

Most of you are aware of the fact that the ETNET – Water - Environment is one of the Thematic Network projects that the European Commission supports since September 1996, in the framework of the SOCRATES programme.

ETNET – Water-Environment Thematic Network addresses issues related to water and particularly how the teaching of this area, of utmost importance, should be improved either via Curriculum Development, including specific course modules, or by producing teaching materials which may be used at different levels of education.

Particular attention is often not given to water, unless it is at the center of natural disasters (like, flooding, landslides, etc) or industrial disaters (like, extensive contamination often man-driven).

However, scientists have drawn the attention to the importance of water and to the severe consequences that an uncontrolled disruption of its cycle may have to everyone's life, be it personnal or in society, including in the economic sphere.

Between the organisation of the World Water Conference in Mar del Plata, Argentina, in 1977 and the 1998 UNESCO's Conference « L'eau : une crise imminente ? », that a multitude of meetings were organised either at scientific or more political level.

In the aftermath of the 1977 Conference in Argentina, 1981-1990 was declared the 'Water Decade'. The objective was that of enabling all humans beings to access drinkable water by the year 2000 : unfortunately, we are far from meeting such an objective.

The 1982 Rio de Janeiro Conference on 'Sustainable Development' was part of the Decade, and since then, that a World Water Day is celebrated every year on 22nd March.

Different studies available draw the attention to the overall scarcity of water, particularly to the main factors which are responsible for its 'crisis' or even what is refered as the 'water bomb'.

These are :

1. the inequality of water resources repartition, be it worlwide or inside a given country,

2. the wastage of water - and its non efficient management

3. pollution

4. increase in human population, particularly in world areas where the shortage of this natural resource is already rather significative.

Water, as many other aspects linked to our environment, are often neglected when an apparent more immediate economic situation is at play, and governments opt for short term solutions, instead of more sustainable decisions.

That may also explain the lack of a stronger willingness as to address the water problem worlwide, particularly by integrating scientific findings as well as supporting water usage or management by adequate legislation.
It was in the context of Thematic Network Projects, that at the end of 1996, a Call for proposals was launched by the European Commission aiming at implementing the Objective 1 (Encourage the Acquisition of New Knowledge) of the White paper on 'Teaching and Learning : Towards the Learning Society'

As such, the activity should :

- Enable a comparison of existing courses or modules, and contribute to the establishment of European-wide levels of 'core' knowledge.

- Enable a reflection, at European level, of course contents contributing to their increased quality or further development.

as well as

- Provide for increased synergies between the 'formal', including continuing education and adult education, and the 'informal' educational systems.

TEST-EAU, which will be demostrated during this Symposium aims at obtaining a test for self-assessment and validation of knowledge acquired in the area of Water-Environment.

The ETNET Thematic Network provided for the right forum where a comparative analysis of relevant curricula could take place as to determine the 'core contents' of a multimedia product of quality, which can be recognisable Europe-wide, and possibly outside European borders.

The fact that Thematic Network Projects are 'fora' open to the participation of non-university partners, provides for the taking into account of new requirements of the labour-market and/or societal needs in their activities, particularly those related to curriculum development and production of teaching materials.

The underlying principle of the above experiences, carried out since 1996, lead to its general acceptance by the Members States of the European Union, which have also given a favourable opinion towards integrating it in the second phase of the SOCRATES Programme.

Alternative Educational Pathways are together with Adult Education the two main pillars of the Grundtvig chapter of the poposal for a second phase of SOCRATES, to be valid from 1st of January 2000.

Besides training and professional organisations have also welcome the positive contributions that tests for the self-assessment of individuals may give to the building up of the Europe of knowledge.

The general participation of universities and the quality assurance they provide to the process, together with the gradual demonstration that synergies between the diploma ("formal education") and alternative complementary ways of acquiring knowledge (often via an "informal system") are not only possible, but contribute to a mutual enrichment and further university curriculum development.

Sensitisation of professional organisations and their adoption of quality products having gathered an European calibration and acceptance consensus should be further improved.

Issues that are still being addressed relate to authors' rights of multimedia products or products to be accessed via the World Wide Web. That is complex and, particularly so, when the innovative aspects are important, difficult to foresee at the outset of a project, which itself, also receives the contribution of several partners.

Besides, any useful educational material or tool requires regular update and calibration, otherwise the strong congregation of efforts may come next to nil in a increasingly shorter time (both from the context viewpoint and the technological support). Such regular upgrading constitutes a task that the team that developed such products should address and resolve in a favourable way.

I wish to thank the organisers of this International Symposium for having held it, in an area of such importance, where, research and development, on the one hand, and education (and training), update and awareness raising, on the other, are constantly needed.

Information et éducation à l'eau et à l'environnement. Une nécessité aujourd'hui pour une bonne gestion

Roux L. - A.[1]**, Roche P. - A.**[2]
[1] Président du Conseil d'administration de l'Agence de l'eau Rhône-Méditerranée-Corse
[2] Directeur de l'Agence de l'eau Seine-Normandie

Résumé
Les organismes de l'eau, et particulièrement les agences de l'eau françaises, ont été amenés à s'intéresser progressivement à l'information, à la communication puis à l'éducation sur l'eau, d'abord avec les jeunes, puis vers tous les publics. En effet on sait aujourd'hui qu'une bonne gestion de l'eau doit être faite avec les usagers et les bénéficiaires, comme y incite notamment une Charte sociale de l'eau qui va être soumise à la Conférence de La Haye de mars 2000.
Cette communication décrit les réalisation des six agences françaises dans ce domaine et, plus succinctement, celle d'une dizaine de pays. Elle annonce aussi le lancement par l'Académie de l'eau française d'une enquête sur une vingtaine de cas qui devrait déboucher sur un Réseau d'échange sur Internet. Elle propose enfin l'organisation d'une première rencontre entre experts en liaison avec l'UNESCO et les PHI, fin 2000.

Mots clés: Information sur eau et environnement - Communication - Formation - Classes d'eau - Public concerné - Pédagogie.

1. Introduction

Un peu partout, les gestionnaires de l'eau mesurent progressivement l'importance d'une éducation à l'eau pour les citoyens en constatant le rôle majeur des usagers et des bénéficiaires de la gestion de l'eau pour contribuer à l'élaboration des politiques publiques et apporter leur soutien à leur mise en oeuvre.. C'est le cas par exemple de l'impact du gaspillage de l'eau ou de rejets inconsidérés de détritus divers dans les collecteurs considérés comme le vecteur universel (le « tout à l'égout ») ou encore de beaucoup d'autres comportements individuels et collectifs très perturbants et irréfléchis.

C'est pourquoi depuis plus d'une dizaine d'années, ont été développées des actions de sensibilisation, d'explication ou même de formation et d'éducation tournées vers divers publics, adultes et jeunes.

Dans cette stratégie, la place réservée aux jeunes, tant des élèves des écoles, des lycées que des étudiants, est forte car ils sont les acteurs de demain et constituent un relais de choix vers leurs parents.

La présente communication se propose d'exposer d'abord les actions d'information, d'éducation et de formation entreprises par les agences françaises de l'eau et leur évolution.

On présente ensuite des actions analogues menées dans une dizaine de pays étrangers. Enfin, on évoque l'inventaire plus large, engagé par l'Académie de l'eau française, afin de mettre en commun des expériences diverses au profit de tous.

2. L'action des agences françaises de l'eau

2.1. Pourquoi les agences de l'eau s'intéressent-elles à l'éducation et à la formation ?

Les six agences de l'eau françaises sont l'exécutif des comités de bassin qui tiennent un rôle clé dans la planification de l'eau et dans le financement des ouvrages d'intérêt commun. Ces comités réunissent des représentants de trois catégories : les maîtres d'ouvrage publics et privés, l es administrations et les usagers et bénéficiaires de l'action (industriels, agriculteurs, consommateurs domestiques, pêcheurs, associations, etc.).

Les agences, qui bénéficient aujourd'hui d'une expérience de plus de trente ans, ont été très tôt confrontées à un étrange paradoxe. Elles ont pour mission d'accélérer l'équipement correct du pays par une programmation rationnelle et concertée, assortie d'aides financières significatives provenant de recettes assurées, mais n'ont pas de lien direct avec le public alors qu'elles sont en contact quotidien avec beaucoup de ses représentants.

Ainsi rapidement, les agences de l'eau, après avoir limité leurs actions de formation et d'information au public très restreint de leurs partenaires directs, ont été conduites à élargir ces actions à des publics de plus en plus importants (réf. 1).

Ces actions d'éducation et de formation, engagées depuis très longtemps par les agences, ont pris des formes diverses selon les agences, respectant en cela les enjeux propres à chaque bassin. On trouvera, ci-après, d'abord un résumé de ces principales actions, agence par agence, puis en conclusion quelques remarques générales sur l'évolution de leurs actions.

2.2. Agence de l'eau Adour-Garonne

Cette agence a mis au point une mallette pédagogique avec un manuel d'utilisation destiné aux enseignants du primaire, avec des fiches pédagogiques pour les élèves.

Dans l'enseignement secondaire, des dossiers pédagogiques sont adressés aux établissements d'enseignement qui en font la demande. Ces dossiers pédagogiques, créés par l'agence de l'eau Loire-Bretagne, ont été adaptés au cas de l'agence Adour-Garonne.

Dans les lycées, 500 conférences sont organisées chaque année et un CD-Rom portant sur un jeu de rôle ayant pour thème les acteurs de l'eau a été reproduit en 5 000 exemplaires et est adressé aux classes des lycées qui en font la demande. Dans les lycées agricoles, une exposition spécifique est mise à disposition par cette agence.

Enfin, une lettre d'information publiée à raison de 4 numéros par an est diffusée aux enseignants de tous les niveaux.

2.3. Agence de l'eau Artois-Picardie

Une documentation adaptée est adressée par l'agence aux enseignants du primaire qui en font la demande. Des sessions d'information sont organisée pour les instituteurs et sont réalisées par le chargé des affaires scolaires de l'agence.

Dans les collèges également, une documentation adaptée est adressée aux enseignants. Certains documents font l'objet d'un envoi systématique à tous les établissements scolaires.

Des conférences en milieu scolaire sont confiées à un organisme spécialisé. Dans un premier temps, 200 conférences ont été réalisées dans un département pilote.

2.4. Agence de l'eau Loire-Bretagne

L'agence à mis en place une politique d'actions scolaires en 1992 comprenant :

La production de supports éducatifs diffusés très largement, comme le document intitulé « Le grand voyage de Perle d'eau » pour les élèves et 4 cahiers du maître pour les enseignants du primaire ainsi que des fiches pédagogiques, comme un dépliant sur le cycle de l'eau « Ma planète bleue ».

Des actions partenariales avec des relais et des publics cibles, notamment « 1000 défis pour ma planète », la « Journée mondiale de l'eau », en partenariat avec les autres agences. En 1999, un livret intitulé « Raconte-moi l'eau » a été également diffusé dans les collèges, enfin une action menée avec des villes pilotes sur le thème « Economies d'eau ».

Des actions avec des réseaux et des interlocuteurs spécialisés comme « Ecole et nature ».

Des actions avec des relais du monde scolaire, notamment :
des stages pour les étudiants en IUFM et interventions dans leur formation même
une convention triennale pour le développement de projets pédagogiques sur l'eau avec le Rectorat de Poitiers, l'Institut de formation en éducation à l'environnement et un partenariat avec l'agence Adour-Garonne.

2.5. Agence de l'eau Rhin-Meuse

L'agence Rhin-Meuse a mis au point une campagne de sensibilisation en milieu scolaire : « Vive l'eau, les jeunes se mobilisent », qui s'adresse aux enfants du primaire (cycle 3) et aux élèves du secondaire et du supérieur.

Après 8 années de cette campagne et 8 000 journées de l'eau, ce sont 163 000 jeunes qui se sont engagés dans la croisade organisée par l'agence au service de l'environnement. Les enfants qui participent à cette action sont rendus le plus actif possible. A cette fin, l'agence leur fournit des documents et des animations. Après la journée de l'eau, la classe part en visite sur le terrain : station d'épuration, captages, production d'eau potable, démonstration de pêche, analyses en laboratoire, etc.). Les élèves sont ensuite invités à faire le compte-rendu de leur expérience sous forme de poèmes, contes, dessins, expositions, films, spectacles.

Dans les établissements secondaires, techniques et supérieurs, des conférences-débats sont organisées. D'une durée de 2 heures, ces conférences-débats sont adaptées de manière à sensibiliser leur jeunes en formation professionnelle sur les enjeux, les questions et les préoccupations liés à l'eau dans leur vie domestique et professionnelle, pour le futur. Ainsi 142 conférences ont permis de sensibiliser 4 026 futurs professionnels (école d'infirmières, de sages-femmes, métiers de l'automobile, secteur paramédical).

2.6. Agence de l'eau Rhône-Méditerranée-Corse

Les cibles prioritaires de cette agence sont les 16 800 établissements scolaires (de la maternelle au lycée). de Rhône-Méditerranée-Corse.

Dans le primaire, l'agence a élaboré de nombreux documents spécialement adaptés :
3 livrets abondamment illustrés constituant les 3 premiers numéros d'une série intitulée « River Jack » (l'eau potable, la station d'épuration, la vie de la rivière)
« Ma planète bleue »
« Connaître l'eau, c'est connaître la vie »
Au total, environ 100 000 documents ont été adressés dans les écoles primaires.

Dans les collèges, l'agence organise des conférences de 2 heures, complétées par des visites sur le terrain. L'ensemble est financée par l'agence à raison de 1 000 conférences par an confiées par convention à un organisme spécialisé. De plus une documentation appropriée est distribuée dans les collèges.

Une nouvelle action très prometteuse a été lancée en 1999 dans l'ensembles des collèges du bassin, liant eau et environnement. Elle s'appuie sur un dossier pédagogiques très complet intitulé « La rivière m'a dit », préparé par la Fédération Rhône-Alpes de protection de la nature (FRAPNA) (réf. 2), avec l'appui de Elf et Unigestion Holding, de Alp Action et de l'agence. Ce kit, qui s'inspire de réalisations anglaises, comporte un certain nombre de livrets pédagogiques sur l'eau, les rivières, la vie en rivière, les berges, les vallées, complété par quelques instruments d'observation (loupes, thermomètres, etc.). Il a pour objet de faire comprendre les liens étroits entre l'eau, la vie aquatique, la flore le long des rivières et la faune, tant par des fiches très suggestives que par la vérification par les élèves de certains de ces aspects sur le terrain.

2.7. Agence de l'eau Seine-Normandie

L'agence de l'eau Seine-Normandie offre deux actions en faveur de l'éducation et de la formation au domaine de l'eau.

Un envoi de documentation sur demande des enseignants provenant de diverses origines (autres agences de l'eau, fédération de pêche, divers administrations -santé, environnement, agriculture, etc.- ou bien conçue par elle-même, Itinéraires pédagogiques)

Un module de formation, appelé « classe d'eau », dont la conception originale est due à Claude SALVETTI, ingénieur à l'Agence de l'eau Seine-Normandie, adapté à tous les niveaux scolaires de la maternelle à la terminale, à de nombreux niveaux universitaires et aux adultes concernés par l'eau ou la pédagogie.

Mis à part le niveau de maternelle, le programme d'une classe d'eau se déroule en milieu scolaire, sur 5 jours consécutifs, selon une organisation en 3 parties :
exposés en classe, adaptés aux niveaux d'âge, par des responsables d'usage ou de fonction (un maire, un agriculteur, un pêcheur, un fonctionnaire de la police des eaux, un industriel, etc.)
travail en atelier sur le thème de l'eau (l'eau en français, en mathématiques, en chimie, en physique, en sciences de la vie et de la terre, en musique, en art plastique, géographie, culture physique, philosophie, etc.)
visites de terrains d'ouvrages ou de sites relatifs à l'eau (station d'épuration, de traitement d'eau, captage, château d'eau, écluse, égouts, promenade en bateau, etc.).
Chaque élève a un livre de bord comprenant une documentation générale sur l'eau.

Chaque classe d'eau réalise une production à l'issue de la classe d'eau qui se termine par une séance de clôture en présence des parents d'élève, des intervenants, des élus locaux, de la presse, etc., au gré des organisateurs qui sont les enseignants eux-mêmes.

L'agence leur adresse un modèle de programme de la classe d'eau correspondant au niveau qui les concerne et un livre de bord type dont ils s'inspirent pour élaborer leur propre classe d'eau. Au reçu du programme adressé par l'enseignant, l'agence verse à la coopérative de l'école une subvention forfaitaire de 3 600 F par classe d'eau.

Le graphique ci-après indique la répartition des classes d'eau depuis la création du module en 1987 jusqu'à 1998 et montre que plus de 4 000 classes d'eau ont été réalisées.

Le modèle a été adapté aux adultes en enseignement supérieur (Agro, ISAB, DESS pollution chimique et environnement, Faculté de pharmacie de Paris, DEUST énergie et environnement, IUFM, etc.) mais également pour des publics spécifiques (infirmières en milieu scolaire et universitaire, élus départementaux, élus municipaux, chefs de PME-PMI, cadres de la Caisse française de développement, etc.

Chaque type de classe d'eau possède son modèle de programme et son livre de bord qui sont adressés à la demande aux interlocuteurs de l'agence qui sont intéressés.

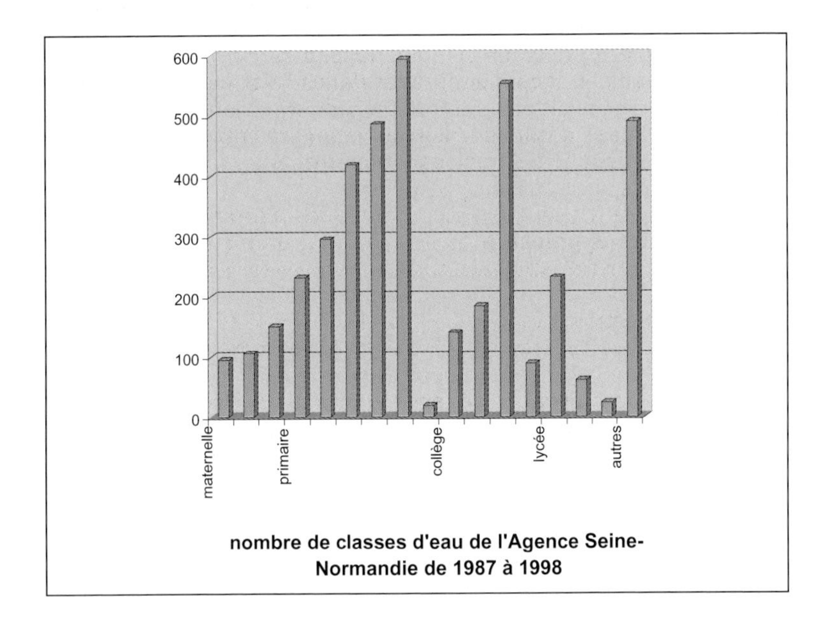

nombre de classes d'eau de l'Agence Seine-Normandie de 1987 à 1998

2.8. Quelques remarques sur ces actions et sur leur évolution

Au début ces actions se sont appuyées sur les enseignants et les instituteurs agissant comme relais vers les jeunes du primaires puis du secondaire grâce à des documents pédagogiques concernant l'eau., l'action des pouvoirs publics en France, mais bien sûr à partir d'exemples dans chaque bassin.

Ces documents très largement distribués, comportant souvent certains destinés aux maîtres et d'autres aux jeunes, ont été complétés par des conférences de 2 heures, des visites sur le terrain sous forme d'une journée permettant une bonne information.

Certaines agences, notamment Seine-Normandie, sont passées à une véritable formation à l'eau avec des classes d'eau d'une semaine adaptées à chaque niveau.

Progressivement des extensions se sont développées en direction d'un public plus large, d'abord vers les adolescents puis vers des spécialistes de différents domaines et en direc-

tion de la formation des maîtres. Trois autres tendances peuvent être notées :
le développement des démarches interagences et l'échange d'expérience entre celles-ci,
un partenariat avec d'autres organismes intéressés par les actions venant s'ajouter à celles entre agences,
un élargissement en direction de l'environnement, comme le montre la nouvelle action de l'agence Rhône-Méditerrarnée-Corse.

Cette ouverture vers d'autres expériences augure bien des retombées que doit avoir l'enquête sur les réalisations d'autres pays développées ci-dessous.

3. Quelques expériences internationales connues de l'académie de l'eau

Les motivations des agences de l'eau pour les actions d'information et d'éducation soulignées au début de cette communication sont également celles de beaucoup d'organismes chargés de la gestion de l'eau ou de l'environnement un peu partout dans le monde. La montée en puissance des principes de « bonne Gouvernance » donne des responsabilités nouvelles aux citoyens et nécessite en effet de les informer et même de les éduquer aux problèmes de l'eau. Un projet de « Charte sociale de l'eau » (réf. 3) devrait venir conforter cette démarche. On sait aussi que la préservation de la santé vis-à-vis de l'eau passe par une éducation à l'hygiène et à l'environnement (réf. 4).

La priorité donnée à la gestion de l'eau par la demande depuis la Conférence de Dublin a conduit à mieux informer et préparer les usagers aux problèmes, ce qui a provoqué de nombreuses opérations intéressantes de formation.

L'existence de ces expériences multiples ont conduit l'Académie de l'eau, avec l'appui des agences françaises de l'eau, à mettre sur pied un groupe de travail sur l'eau et l'éducation, présidé par
M. Pierre HUBERT, pour organiser des échanges entre les organismes qui se sont engagés dans cette tâche. L'objectif premier est de bâtir progressivement un « Réseau internet d'échanges » à partir d'un noyau d'une dizaine de pays et d'autant d'organismes les plus concernés ou les plus motivés.

3.1. Des actions conduites par des organismes impliqués dans la gestion de l'eau

On peut citer, dans ce domaine, le **Massachusetts Water Resources Authority-MWRA** à Boston USA (réf. 5) qui organise depuis une dizaine d'années des classes d'eau dans les écoles de la métropole urbaine ou l'Office de l'environnement (AMA) de la **Communauté autonome de Madrid** (réf. 6, 7) qui a créé des classes de nature avec le « Canal Isabel II » un peu analogue à « L'eau m'a conté » de l'agence de l'eau Rhône-Méditerranée-Corse. Elles s'appuient sur des écomusées dans plusieurs des « parcs régionaux » madrilènes, chacun avec un thème local qui est repris pour la classe en ville.

La **Préfecture d'Osaka** (réf. 8) s'est appuyée au maximum, pour la mise en place de son « Nouveau plan », sur la coopération avec la population en favorisant la création de « Community improvments projects » de façon à focaliser les bonnes volontés. Ces communautés d'individus sont des participants actifs pour organiser des symposiums, conduire des recherches spécifiques, bâtir et réaliser un projet d'amélioration d'un lieu de vie Pour les inciter à agir dans de bonnes directions, la Préfecture distribue des prix récompensant les meilleures initiatives, des brochures et posters, organise des concours et des compétitions. Enfin un « mois des communautés » est l'occasion d'encourager les habitants à participer au nettoyage des abords des routes, des rivières et des parcs ou encore à « jardiner » dans des centres où chacun peut trouver des conseils et des directives.

A **Londres** (réf. 7), avec l'appui de l'Environment Agency, Thames Water mène une politique active d'information sur les problèmes d'eau et notamment sur ses liens avec l'environnement, en particulier le long de la Tamise et de l'estuaire. Il en est de même à **Amsterdam** (réf. 7) où la politique d'information de la ville soulevée par les pouvoir publics est élargie aux problèmes d'environnement.

Au **Maroc** (réf. 9), le support de l'action est l'Office national de l'eau potable (ONEP), responsable de la mobilisation des ressources en eau. Formation et sensibilisation s'adressent aux enfants, aux éducateurs, aux usagers de l'eau, aux médias, aux femmes et aux associations. Les messages et le contenu de la formation sont adaptés à chacun des publics, ainsi qu'aux secteurs urbains et ruraux. Son objectif est essentiellement d'inciter à économiser et à ne pas gaspiller l'eau qu'il faut aller chercher loin. La figure ci-jointe détaille les modalités d'action de cette formation à l'eau au Maroc, à Casablanca.

3.2. Des initiatives « citoyennes »

Les trois exemples suivant ont été conçus et organisés par des Universités et des organismes non liés à l'eau mais aux valeurs de société, comme le « Rotary club ».

L'**Université de Padoue** (réf. 10) cherche à développer les liens et les connaissances entre l'eau et le territoire, à partir du cas de la Province de Padoue, très fortement urbanisée et industrialisée qui a néanmoins donné lieu à une civilisation d'eau pour le biais des multiples utilisations agricoles, artisanales et urbaines de l'eau et qui a su maintenir un paysage aquatique protégé. Le projet s'appuie sur une dizaine de modules dispensés par les moyens d'éducation à distance à partir de liaisons télématiques utilisant un Software de communication et des questionnaires de monitoring et de vérification.

En **Belgique** (réf. 11), l'opération « Sources » est dispensée dans des écoles volontaires pour des élèves de 5e et 6e, avec l'appui de l'Université d'Anvers et du WWF (World Wide Fund for Nature) et la sensibilisation à la qualité de l'eau douce en 4e et 5e avec l'appui du Rotary Club. Ces deux actions visent à développer les valeurs environnementales s'appuyant sur l'aspiration vers le meilleur.

En **Algérie** (réf. 12), il y a un grand désir d'agir dans ce domaine. Sans doute les actions dans les pays de l'Islam pourraient s'appuyer sur la valeur privilégiée que le Coran a donné à l'eau (réf. 13).

4. Le projet d'enquête et de réseau de l'Académie de l'eau

La première étape comporte :

L'établissement d'un questionnaire utilisant les données des 9 cas qu'on vient de citer, s'ajoutant au cas des 6 agences de l'eau et à une dizaine de cas choisis parmi les expériences concrètes reçues lors de l'Appel à communication de la Charte sociale (réf. 3), portant totalement ou partiellement sur la communication, l'information et l'éducation. On utilisera aussi des données de l'UNESCO (réf. 14) et celles du Symposium sur l'eau, la ville et l'urbanisme (réf. 15). Il comportera les 6 thèmes suivants :

Quelle cible ? Enfants, femmes, éducateurs, spécialistes, journalistes, abonnés, associations...
Quels moyens ? Média, enquêtes, plaquettes, écoles...
Quelle pédagogie ?
Quel financement ?
Quels résultats ?

L'envoi du questionnaire à la vingtaine des cas utilisés sera fait par courrier et par internet. Il y sera joint une note d'une dizaine de pages sur l'expérience des agences répondant aux 6 thèmes ci-dessus.
L'analyse des réponses et l'édition d'un rapport d'étape qui sera disponible en décembre 1999.

La seconde étape, dont le financement sera recherché en liaison avec l'UNESCO et les PHI des pays concernés, comportera :

La mise sur internet des documents synthétiques obtenus et la constitution d'un réseau d'échanges

L'organisation d'une réunion de travail d'une quarantaines de spécialistes issus notamment des PHI des pays étudiés et d'experts internationaux. Son objectif sera d'élargir progressivement l'échantillon à partir de la diffusion des résultats de l'étude et des recommandations issues de cette réunion.

ACTIONS D'EDUCATION DIRECTES SELON LA CIBLE

Justification du choix	Messages	Le comment?
• Une éducation efficace commence dès le jeune âge • Représentent plus de 50 % de la population • Les problèmes de ressources s'accentueront à l'avenir (besoins en hausse, qualité en baisse) • Ce sont des prescripteurs (ils peuvent influencer les adultes)	• Découverte de l'eau • Respect de l'eau - Économie - Non-pollution • Relation eau potable/santé - Qualité de l'eau potable - Éviter l'eau non contrôlée	• Ecoles (25.000 enfants/an): exposés, expositions, affichages • Colonies de vacances (300.000 enfants/an) • Salon de l'enfant (200.000 visiteurs/an dont une majorité d'enfants) • Visites d'installations ONEP 1/ Toutes les actions au profit des jeunes sont organisées en collaboration avec les ministères de l'Education nationale, et de la Jeunesse et des sports, les autorités locales, associations... 2/ L'Inauguration du programme des jeunes par montgolfière 3/ Affiches sur la base des dessins d'enfants
• Contact permanent avec les jeunes • Servent de relais privilégiés étant donné leur mission de formateurs	• Importance de l'eau • Rôle des jeunes pour l'avenir • Rôle socio-économique de l'eau • Relation eau/santé	• Mettre à la disposition des supports adéquats • Visites d'installations • Semaines d'information
• Clients (droit à l'information) • Pouvoir d'influences familiales (chefs de famille) • Directement concernée par la facture (payeurs)	• Qualité de l'eau • Economie de l'eau • Eviter la pollution • Etapes de production (efforts) • Tarifs • Assainissement	• Envoi de mailing (dépliants) • Information sur a facturation • Lettre de vœux (plus conseils) • Lettre sur les problèmes particuliers (qualité...)
• Vecteurs importants de transmission de messages • Ont une influence sur la décision du lecteur • Evitent fausses informations et rumeurs	• Rareté des ressources • Les nouveaux projets • Les problèmes de qualité • Les coûts des projets • Les tarifs de l'eau	• Organisation de journées de presse • Visites des installations d'AEP • Interviews particulières • Dossiers de presse • Informations systématiques
• Educatrices des générations futures • Premières utilisatrices de l'eau • Pouvoir d'influence dans la famille • Payeurs (chefs de famille)	• Economie • Pollution • Tarifs • Utilisations abusives • Conseils de bonne utilisation	• Foyers féminins (exposés, débats, séminaires) • Associations • Visites d'installations
• Spécialités de leurs actions	• Valeur de l'eau • Qualité de l'eau	• Mettre à la disposition des supports adéquats • Visites d'installations

Références

(1) Salvetti C. La pédagogie et la formation des utilisateurs d'eau. Colloque L'eau et la vie des hommes au XXIe siècle) in : Les cahiers du MURS? numéro 34, 4e trimestre 1997

(2) Fédération Rhône-Alpes de protection de la nature. La rivière m'a dit. Accompagnement pédagogique de l'Education Nationale

(3) Académie de l'eau. La charte sociale de l'eau. Conférence-débat au CNAM, 25 mars 1999, avec le Président I. Seragelin, Vice-Président de la Banque mondiale

(4) Professeur Dausset J., Prix Nobel de médecine, Président de l'Académie de l'eau. Les liens eau-santé in : Après demain, décembre 1998

(5) Quang Trac N'Guyen, Valiron F. MWRA-MAPC Monographie de Boston pour l'Académie de l'eau, mai 1996 (Symposium L'eau, la ville et l'urbanisme, UNESCO Paris, 10/11 avril 1997)

(6) Garcia C., Valiron F. Monographie réalisée pour l'Académie de l'eau, novembre 1996. (Symposium L'eau, la ville et l'urbanisme, UNESCO Paris, 10/11 avril 1997)

(7 Bizet B., Professeur à EAPLD. Politique métropolitaine d'environnement : Amsterdam, Berlin, Londres, Madrid et Milan. In : Metropolis, janvier 1994

(8) Kovacs Y. Pacific Consultants. Monographie d'Osaka

(9) Regragui M. L'ONEP (Maroc) : l'éducation et la formation dans le domaine de l'eau. Colloque L'eau et la vie des hommes au XXIe siècle in : Les Cahiers du MURS, n° 34, 4e trimestre 1997

(10) Université de Padoue. « Adottiamo l'acqua » : Pojet d'apprentissage coopération télématique pour l'éducation à l'eau in : Congrès international de Kaslik, Liban, juin 1998

(11) De Backer L.W. Université catholique de Louvain. L'eau et le respect du politique in : Congrès international de Kaslik, Liban, juin 1998

(12) Kerdoun A. Université de Constantine, Algérie. Education et sensibilisation en vue d'une protection de l'eau en Algérie in : Congrès international de Kaslik, Liban, juin 1998

(13) Dr Amery Hussein A. Colorado School of Mines, USA. Culturally-sensitive education for water sharing and protection : an islamic perspective in : Congrès international de Kaslik, Liban, juin 1998

(14) UNESCO-UNEP. Freshwater Resources : Education environmental module. International Environmental programme. Paris 1995

(15) Verdeil V., Valiron F. Académie de l'eau. Colloque L'eau, la ville et l'urbanisme. UNESCO Paris, 10/11 avril 1997 in : Les cahiers IAURIF, avril 1997

Water Sector Capacity Building Strategies, Methods and Instruments in a World of Knowledge Networks

Stratégies, méthodes et instruments pour le développement de capacité dans le secteur de l'eau, dans un monde de réseaux de connaissance.

P.J.M. de Laat & A.S. Ramsundersingh
International Institute for Infrastructural, Hydraulic and Environmental Engineering (IHE)
P.O.B. 3015
2601 DA Delft
The Netherlands

Abstract
Since the international consensus on human resources development for the water sector in 1965, only a few efforts have been made to establish local or international providers of trainers. In developing countries the enrolment of students in higher education and the national expenditures in R&D are very low. At the same time, the water sector is threatened, because of the lack of rational management, while the need for more and better quality of water is growing. Training of individuals has been one of the strategies to improve the water sector performance. In more recent years, the development of indigenous education, training and research capacity, has been started up. Through this capacity better knowledge and more professionals can be made available to the sector. In the immediate future the use of regional and international knowledge networks will increase the efficiency in knowledge base and human resources development.

Résumé
Depuis le consensus obtenu en 1965 sur la nécessité de développer les ressources humaines dans le secteur de l'eau, seulement de minimes efforts ont été fait pour établir des centres éducatif locaux ou régionaux. Dans les pays en voie de développement, la participation dans l'éducation supérieure et les dépenses dans le domaine de la recherche sont très faibles. Ceci malgré le fait que le secteur de l'eau est menace: il y a un manque de gestion rationnel, tandis que les besoins d'eau - en termes de quantité et qualité - sont croissants. L'éducation intensive d'individus a été une des stratégies employées pour augmenter la performance du secteur de l'eau. Ces dernières années, certaines activités se sont concentrées sur le développement de capacités locales sur le terrain de l'éducation et de la recherche scientifique. Le résultat est une plus grande connaissance locale et un plus grand nombre de professionnels locaux au service du secteur de l'eau. Dans le future immédiat, les réseaux de connaissance régionaux et internationaux vont permettre une augmentation de l'efficacité dans le développement de centres de connaissances dans les pays du Tiers Monde, et vont ainsi accroître la performance du développement de ressources humaines.

Introduction

The importance and need for reinforcing higher education, particularly in developing countries, has been spelled out in various documents of the World Conference on Higher Education, organised by UNESCO in October 1998 (e.g. Towards an Agenda 21 for Higher Education). With regard to the developing world most universities have not been able to keep pace with the spectacular technological development that took place during the past two decades. There are various reasons, though they may not apply to the same degree for all of the countries:

- The dramatic growth in the number of students in tertiary education in the second half of this Century had a negative impact on the quality of education in most developing countries. World-wide student enrolments increased six-fold from 13 million in 1960 to 82 million in 1995.
  Many universities could not cope with these large numbers of students, and the simultaneous decline in entrance level, without lowering the scientific standard.

- For universities established in the colonial past, traditional areas of education (law, humanities, Social science) often remain to be most important (Castells, 1993). The higher educational system was, and sometimes still is, seen as the provider of the political elite to serve the administration after independence. The political orientation of students may cause ideological manifestations that could well paralyse the functioning of the entire university. Moreover, the traditional educational areas are less expensive and easier to teach than technological studies for which experiments (laboratories) and specific materials are required. Attempts to increase the level of education in science and technology are further hampered by lack of staff trained in the latest technology. Only 25 percent of the total enrolment in higher education programmes are attributed to science and technological studies. Most of these enter education programmes in the field of agriculture, biology and ecology, while a low number enters engineering studies. Also, the existing science and technology (engineering) studies do not aim at producing professionals who can deal with present and future challenges. These programmes only cater for training in pure science or engineering. The curricula of most universities and training institutions in the world do not provide future professionals with knowledge and skill on technical management, multi-disciplinary, participative and communicative methods and tools, financing, privatisation and decentralisation issues, etc.

- During the last decade most developing countries experienced economic stagnation. In most cases this has led to a substantial reduction in the financial resources available for higher education. Most universities lack, therefore, adequate training and research facilities, such as audio and visual teaching aids, text books, journals, reference materials, laboratory and computational facilities, Internet connections, instruments and equipment.

- Most of the developing countries only spend (on average) 0.5 percent of their Gross National Product (GNP) on R&D, while the developed countries spend (on average) 2.5 percent of their GNP. Therefore the development of a national or indigenous techno-scientific infrastructure will and can not easily take place. At the same time know-how must be imported, for which there is no capacity to absorb and/or to adjust to local conditions.

The critical role of science and technology in the ability of a country to be competitive in the present information-based economy is generally understood. The absence of highly qualified human resources potential and lack of techno-scientific infrastructure in many

The environmental revolution suggests to include basic environmental topics in the curricula, as proposed by a team of authors of IAHR and UNESCO (Kobus et Al, 1994) in a very interesting paper. Nevertheless, it seems to be not enough.

The environmental concept has to be present in all steps of any hydraulic project, as an additional requirement. Students have to incorporate gradually this vision, along all the undergraduate chairs. That can't be reach through the incorporation of isolated subjects, like "ecology introduction" or "environmental engineering". New engineers should associate their points of view about "structural stability" and "economic feasibility" (which are usually obtained through all technical subjects during their studies) with "environmental stability" and "environmental feasibility). Those themes must be present during the whole formation of all engineers, independently their later specialization on environmental impact assessment aspects.
If wishes to train an specialist on "water and environment engineering", able to participate in transdisciplinary teams of environmental impact studies, is inexorable the necessity of postgraduate studies, in interdisciplinary teams, in which is possible to incorporate the concepts about subjects that engineering can't reach, to allow the interaction with another professionals of different disciplines.

5. Conclusions and recommendations

It is proposed to introduce the environmental subjects into the existing topics in all engineering chairs. It needs the education of professors and assistants by means of teams of "formation for educators", not easy to be accepted.

The environmental concept has to be present in all steps of any hydraulic project, as an additional requirement, and the students have to incorporate this vision
For interdisciplinary formation, the post-graduate frame is proposed. It is particularly recommended the cooperation of engineers with ecologists and biologists in research and development projects. Postgraduate courses can help to the interaction of all professionals related to water and environment. In some countries, several of these initiatives have already been implemented. In Latin America, engineers have not a good education in environmental topics and they are not forced to take postgraduate courses for an interdisciplinary formation.
If a comprehensive discussion is established at universities around these initiatives, the new ideas will be finally incorporated and the international cooperation spreads them, the hydraulic engineer will be able to face the challenges of XXI Century with success.

References

De Lío, J.C., Menéndez, A.N. & Loschacoff, S.C. (1989) Un ejemplo de la complementación entre modelación física y matemática (An example of physical and mathematical complementation), X congreso Nacional del Agua, Córdoba, Argentina.
Hjort, P., Kobus, H., Nachtnebel, H.P., Nottage, A. & Robarts, R. (1991) Relating hydraulics and ecological processes, Journal of Hydraulic Research, Vol 29, extra, pp. 8-19.
Kobus, H., Shen, H.W., Plate, e. & Szollosi-Nagy, A. (1994) Education of hydraulic engineers, Journal of Hydraulic Research, Vol. 32, N° 2, pp. 163-181.
Lopardo, R.A., De Lío, J.C. and Vernet, G.F. (1987) The role of pressure fluctuations in the design of hydraulic structures, in "Design of Hydraulic Structures", edited by R. Kia y M.L. Albertson, Colorado State University, Fort Collins, U.S.A., pp. 161-175.

Integrating the Curriculum of Land & Water Management and Engineering to European Water Sciences

Intégration de la Formation en Aménagement et Régimes de l'Eau vers des "Sciences de l'Eau Européennes"

W. Loiskandl, H.P. Nachtnebel, M. Calderón

Universität für Bodenkultur Wien, BOKU
(University of Agricultural Sciences Vienna)
Institute for Hydraulics and Rural Water Management, Institute for Water Management, Hydrology and Hydraulic Engineering; both Muthgasse 18, 1190 Vienna, Austria
Centre for International Relations, Gregor Mendel Strasse 33, 1180 Vienna, Austria

Abstract

The objective of this paper is to present the impacts of integration into the European dimension to a local curriculum. In the past, both international activities and student mobility had to be arranged individually, accompanied sometimes by discouraging obstacles. By joining the European learning society and encouraging students to include a semester abroad in their curriculum, not only the number of "mobile students" increased but also the need for reviewing the curriculum at the home institution arose. One first result was the reorganization of the course of study to reduce the average study duration and to increase the flexibility of the curriculum by introducing modules. An introduction of a bachelor degree is now under discussion. To achieve a better compatibility with partner universities emphasize has to be given on learning agreements and the introduction of European modules. The ongoing adaptations are a contribution to continuous education on an European postgraduate level which already became a key role in development of society.

Résumé

L'objectif de cet exposé est de présenter les effets de l'intégration d'un plan d'études local à une dimension Européenne. Dans le passé les activités internationales ainsi que la mobilité des étudiants devaient être réglées au cas par cas, parfois accompagnées par des obstacles décourageants. Associer la société cognitive Européenne et encourager les étudiants à suivre un semestre à l'étranger a permis l'augmentation du nombre "d'étudiants mobiles", et également demandé une révision de la formation à l'université d'origine. Un premier résultat concerne la réorganisation des cours dans le but de réduire la durée moyenne des études et de permettre une plus grande flexibilité de la formation par l'introduction de modules. La création d'un diplôme à baccalauréat est actuellement en discussion. Pour disposer d'une meilleure compatibilité avec les universités partenaires, un effort particulier doit concerner les conventions d'études et l'introduction de modules Européens. L'adaptation continue contribue à l'enseignement continué au niveau post-diplôme européen qui joue désormais un rôle clé dans le développement de la société.

Introduction

Becoming an associated member to European study exchange programmes in 1991 had a drastic impact on Austria's education at university level in general. In this paper particularly the impacts on the study branch land & water management and engineering at the university of Agricultural Sciences, Vienna (Universität für Bodenkultur Wien, BOKU) will be discussed. Encouraging students to include a semester abroad in their curriculum did not only increase the number of "mobile students" but also initiated a review of the curriculum at the home institutions. Our liberal study policy – for example, students are free to decide when to take the exams, either at the end of the respective course or at a later examination date offered by the institute - became questionable and restrictions or regulations are in discussion. The present curriculum in engineering, which is presented in the following chapter, is set up in a central European style. More recent developments like postgraduate and short courses and international education collaboration extend the scope of the universities.

The "Universitäts-Organisationsgesetz UOG-1993" (University Organization Act) promoted the decentralization of university-level education (Study in Austria, 1995). The new concepts are rather based on modules (specialization course blocks) to allow more flexibility for the single student. Emphasize is given to specialization which is directly related to water and environmental science. An implementation of a bachelor degree is under discussion. Still more work has to be done to achieve a better compatibility between the various educational systems. To reach this goal, learning agreements and the introduction of European modules are needed. Compatibility may not be confused with homogenization, i.e. individual characteristics of each university should be preserved for the benefit of all other partners.

Austria's education system in engineering

Education related to engineering education (basically provided at technical universities and since short time at the "Fachhochschulen ") in Austria is offered at university level in form of regular studies and Ph.D.-programmes. The minimum regular study period is fixed with five years to obtain a diploma. As there are very few restrictions for a single student on how to organize the study, the average study time exceeds the regular duration considerably. To illustrate the background of a student and the diversification in engineering training, the education system of Austria is summarized in Fig.1 (Years indicated are the required minimum age). The study requirement is basically the secondary school leaving-certificate (obtained at a grammar-school, i.e. gymnasium) or an equivalent authorized study admission certificate. Recently the education system was extended by the so called "Fachhochschule".

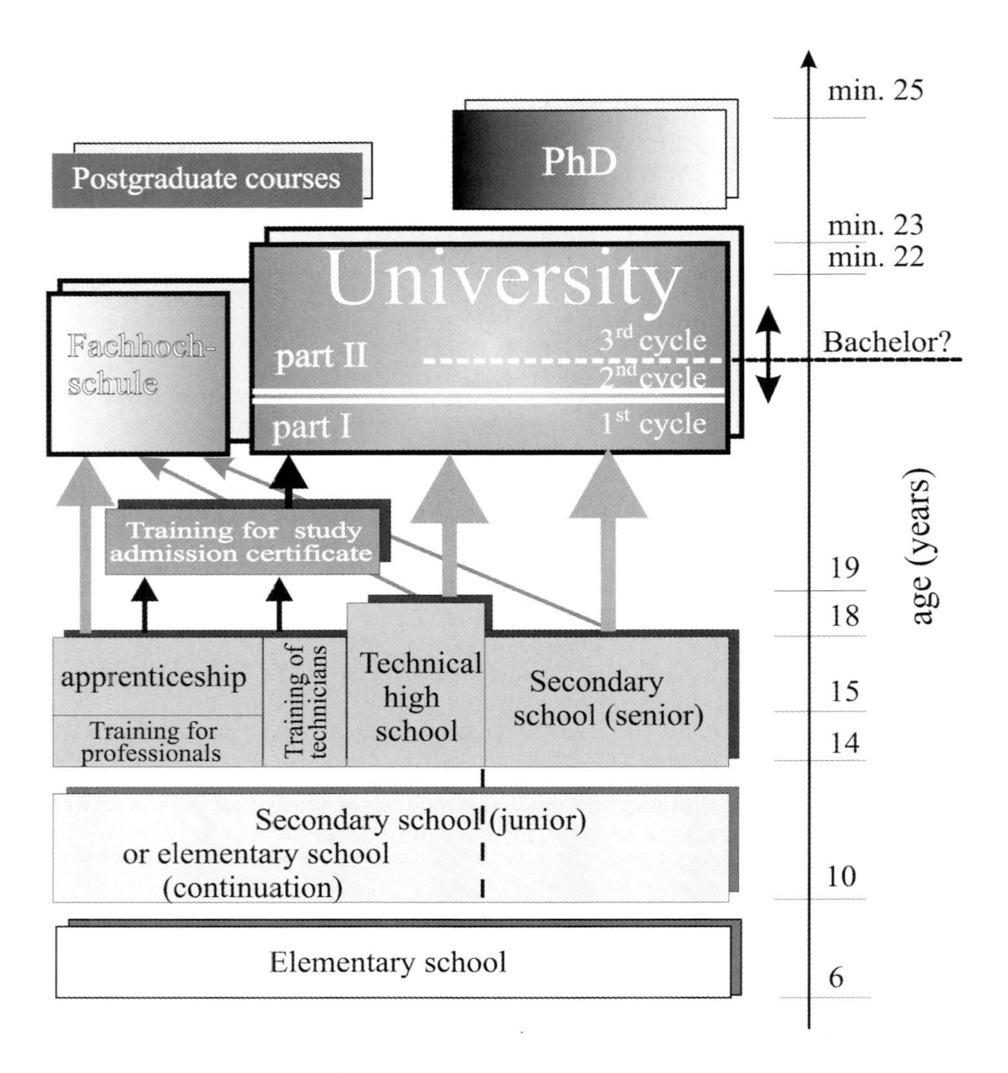

Figure 1: Education system in Austria related to engineering education and training

Apart from the technical universities, the "Universität für Bodenkultur Wien (BOKU)" is the only Austrian university which combines theoretical knowledge with the practical application of technology, economics and the study of nature on a scientific basis.

Regular degree programme of land & water management and engineering

The course of study is divided into two parts, intended as a once-through study. An academic year consists of two semesters, winter and summer term. By introducing a bachelor degree the two parts could be converted into cycles (Van der Beken, 1998/1) leading to a better compatibility with other European curricula. In 1996, the new study concept was approved by the Austrian government (BMWV, 1997). The major changes resulted in a decrease of the weekly workload and an increase in flexibility for setting up an individual curriculum (Fig.2), hence including courses or modules from outside the BOKU is much easier now.

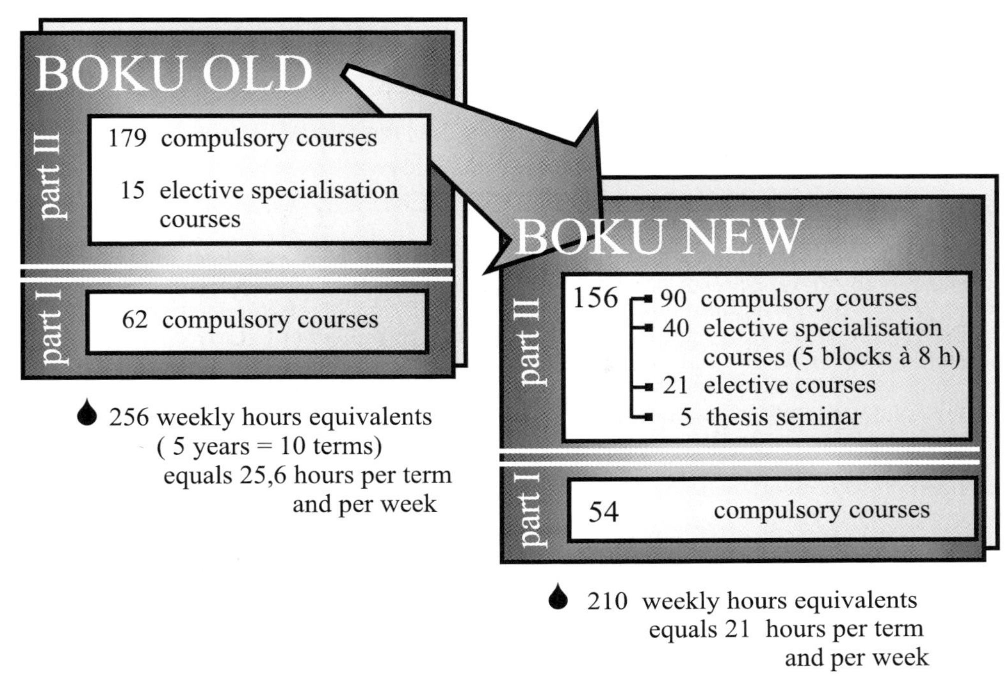

Figure 2: Overview of the new curriculum of land & water management and engineering

The following explanations are related to the new study plan. During Part I (comprising the first four semesters) the following compulsory subjects are offered: mathematics, botany, geology, soil science, chemistry, physics, geodesy, etc.
Part II consists of compulsory subjects - in which students get a training in: economy and law; civil engineering; geotechnical engineering and applied geology; hydrology, water management and hydraulic structures; rural water management; sanitary engineering, water provision and industrial water management; waste management and waste treatment; transport and highway engineering; rural planning and development; water ecology and water-ecological planning; surveying, remote sensing and geo-information systems - and of optional courses again divided in: a) specialization topics and b) a smaller share of individual courses not restricted by any thematic regulations. Students have to choose five optional course-blocks related to the institutes of the study branch or other Socrates partner institutes offering additional specialization in a particular field of land & water management and engineering. The remaining optional courses are completely independent of the regular study program and are not bound by any means to the home university. A minimum number of courses has to be taken in a foreign language. Writing the final thesis ("Diplomarbeit") in a foreign language contributes to that requirement. Obligatory in part II are ten weeks of practical training. The obtained degree is "Diplom Ingenieur".

PhD-programme

In 1906, the BOKU officially was granted the right to award the academic degree of "Dr.nat.tech". The admission requirement is the completed regular study (Dipl.-Ing. degree) or equivalent prevailing studies for foreign students. Doctorate programme students have to enrol as regular students and attend doctoral courses during a minimum period of four semesters. Courses are selected individually by the students in collaboration with the assigned supervisor and approved by the head of the commission for the Ph.D. programme. The supervisor must have the "venia docendi". The topic of the Ph.D. thesis has to be related to a study branch of the BOKU. The Ph.D. thesis is first evaluated by two experts and then defended in a final examination (rigorosum). Although the minimum time requirement for a Ph.D. study is two years, the average study duration is 3 to 4 years in reality.

Postgraduate Study Programmes

Study Programme Technical Environmental Protection: At present this study programme is offered in co-operation with the Technological University of Vienna. Students qualifying for this programme are graduates of the TU's, the Montan University of Leoben, the BOKU, as well as graduates in the field of architecture and of equivalent study courses of other universities in Austria and abroad. The study period comprises a minimum of four semesters, whereas the current average is six semesters. The focus on practical work involves interdisciplinary projects (such as remediation of soils, environmental compatibility studies, waste management concepts, etc.). The demand for graduates is growing, among others, due to more frequent performance of environmental compatibility studies. Recently it was decided to convert this postgraduate programme into an independent self-financing interuniversity programme.

Continuous education: International Courses and short courses are especially tailored for professionals who are not able to stay off work for a longer period. Courses were given in Hydraulic Computation, Geostatistics, Geographical Information Systems (GIS) and topics related to water, soil and environmental sciences. Especially continuous education and training will be probably a fast growing sector, hence the biggest challenge for all education institutions in future. The blue print of CET-WARE (Continuous Education and Training for Water Resources Engineering) included the following statement in relation to water environment: "With the world-wide recognition of the paramount importance of the human resources capital for sustainable and long-term development, CET-WARE activities are essential in all countries" (Bogardi et al., 1995).

Student mobility

Perhaps the strongest influence on creating "international" human capital via student mobility arose from the ERASMUS programme. For a newcomer to an existing network the first task was the creation of a functional organization to deal with the new international relations. Fortunately the government of Austria was aware of that need and supported the establishment of centres for international relations. At the university local coordinators were appointed to organize the technical part. From the beginning one difficulty was the so called central European curriculum, which has no provision for a bachelor degree and is therefore not congruent to many other European education systems. Since the first steps into "Europe's Learning Society", many alternations of the curricula - especially at technical universities - have been forced by the new learning environment as it was previously discussed.
To quantify BOKU's student mobility development within the framework of ERASMUS/SOCRATES, the number of incoming students is compared with outgoing students (Table 1).
Furthermore, the first column in table 1 shows the number of first year students enrolled for land & water management in the academic years 1992/93 to 1998/99. This number is taken as an indicator for the total amount of students enrolled. As one goal of BOKU's internationalization strategy is to send out about 15% of the students, the correlation between the number of first year and outgoing students proves that this aim is more or less achieved in the field of land & water management and engineering. However, at present, we are still sending more students abroad than we are receiving at our institution,

although this imbalance is constantly being reduced and reciprocity was improved during the last two years. On the contrary, the time BOKU students spend at foreign universities could still be increased: the average duration of the ERASMUS/SOCRATES study period abroad is slightly higher for incoming than for outgoing students.

Table 1: Socrates - Student mobility flow from 1992/93 to 1998/99

	BOKU	Incoming			Outgoing		
year	1. year Students	Students	Months	Average Months/ Student	Students	Months	Average Months/ Student
92/93	215	0	0	0	2	11	5,50
93/94	230	3	17	5,67	7	36	5,14
94/95	261	0	0	0	14	71	5,07
95/96	262	5	37	7,40	23	114	4,96
96/97	250	5	31	6,20	25	189	7,56
97/98	175	19	121,5	6,39	20	112	5,6
98/99	220	22	146	6,64	35	214	6,11
S		54	352,5	6,53	126	747	5,93

It is also interesting that the most favourite countries of destination for BOKU students (namely Great Britain, Spain, France, Italy and The Netherlands –, Figure 3) do not tally with the countries of origin of the majority of incoming students (Belgium, Sweden, and Great Britain –, Figure 4). Consequently, efforts are made to reach a more even distribution and a tendency of spreading can be observed.

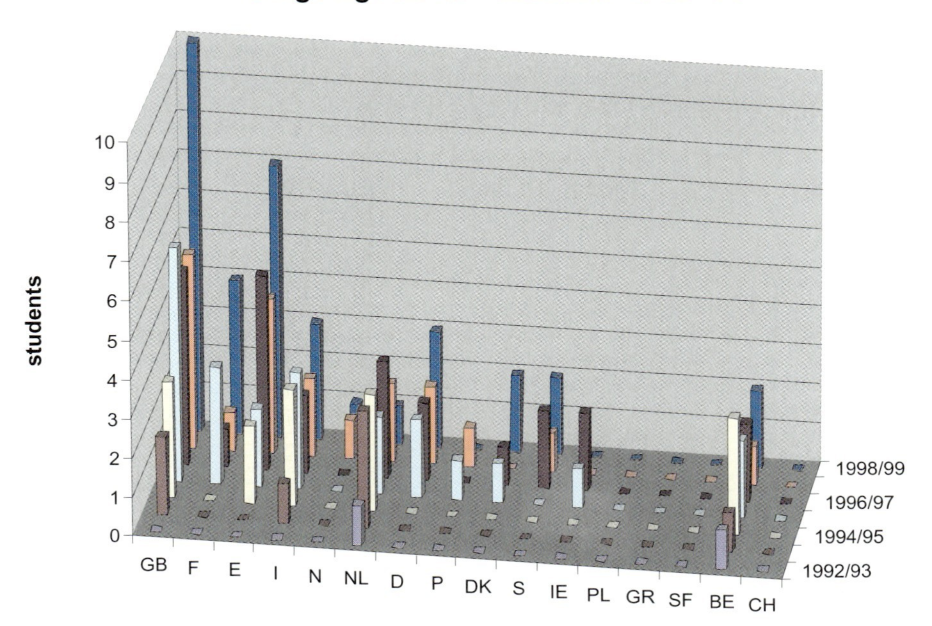

Figure 3: Outgoing Socrates students (1992/93 to 1998/99) by countries of Europe

Country of Origin of Incoming KTWW 1992 - 1999

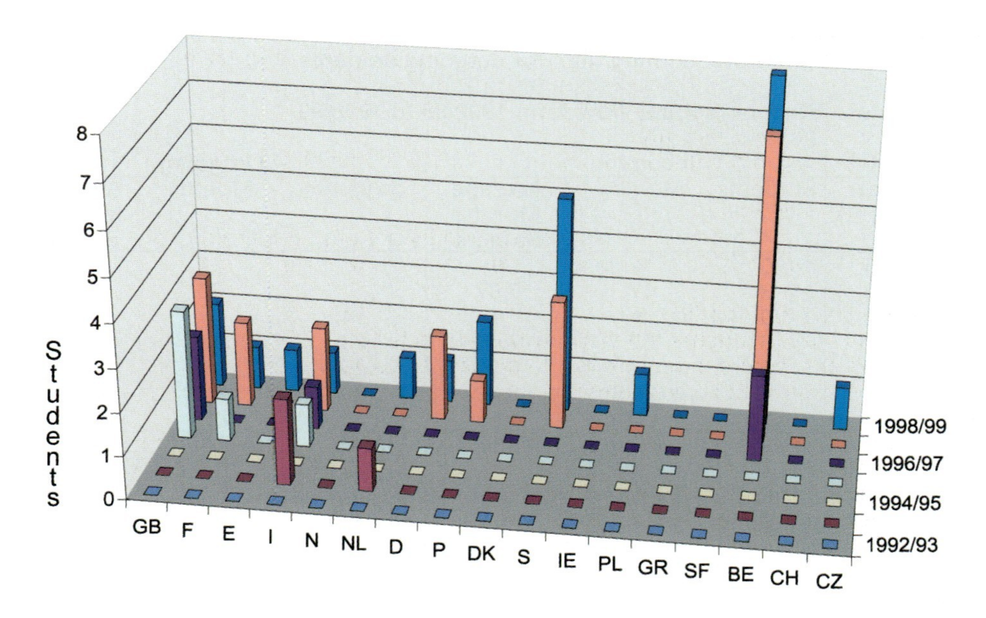

Figure 4: Incoming Socrates students (1992/93 to 1998/99) by countries of origin

Besides ERASMUS/SOCRATES, other educational activities are related to TEMPUS and LEONARDO. Main partners of TEMPUS activities are from Poland (PANSED and STEEM), Romania (WATERMOST) and Hungary (ICER; Table 2).

Table 2: TEMPUS - Student mobility flow from 1996 to 1999

Year	Students	Incoming students from Months/Student	Country	Project name
1996	3	6	Hungary	ICER
1996/97	1	9	Poland	PANSED
1997	4	6	Hungary	ICER
1997	1	3	Poland	PANSED
1997	3	4	Romania	WATERMOST
1998	1	6	Poland	PANSED
1998	4	3	Romania	WATERMOST
1999	1	3	Poland	STEEM
1999	1	3	Romania	WATERMOST
		Outgoing students to		
196/97	1	6	Poland	PANSED
1997	1	2	Romania	WATERMOST

Within TEMPUS the incoming students are dominating. In future TEMPUS activities are continued under Socrates since the two programmes are merging, this should result also in a better reciprocity with this partners. Bilateral treaties with other countries complete the scale of international co-operations.

Integration - Continuing

Substantial achievement was reported, but education and training is a continuous process, hence efforts which may be divided in immediate and long term tasks are still needed.

Immediate tasks of study policy are:

- Stabilization of outgoing students at about 15 – 20 % of the student number of an academic year
- Improvement of reciprocity
- Harmonization of work loads at home and abroad by setting up learning agreements
- Full implementation of ECTS (European Credit Transfer System)

Further goals of the study policies are:

- As only a part of all students (15-20% proposed) include a study abroad at a host university in their curriculum, attention has to be paid to those staying at home. Therefore, in order to promote a European dimension at each university staff mobility is a necessity.
- Participation in the development of European modules, as started within the thematic Socrates-network ETNET.ENVIRONMENT-WATER (European Thematic Network of Education and Training for ENVIRONMENT-WATER, Van der Beken, 1998/2), where five specific projects (European Postgraduate Degree in Hydrology and Water Management, European Graduate School of Hydraulics, Forum for Open and Distance Learning (ODL) for Water and European Paradigm for Integrated Water Management) are comprised.
- Deepening of co-operation of all learning and training institutions in Europe related to Water-Environment society.
- Extension of co-operations to overseas.

Finally, and this could have been stated at the beginning as well, the question "who is a student ?" needs a definition. Since professional development and lifelong learning became a new paradigm no one can exclude himself of being a student in one way or another. With the recognition of continuous training and education new demands continuously will be brought forward by the learning society to all educational institutions.

References

Austrian Rectors´ Conference and the Austrian Academic Exchange Service (1995) STUDY IN AUSTRIA, Vienna, Austria

Bogardi, J., Gilbrich, W.H., Van der Beken, A. & Wooldridge, R. (1995) Blueprint Cet-ware: Continuing Education and Training for Water Resources Engineering, Brussels, Belgium.

Bundesministeriums für Wissenschaft und Verkehr, (1997) Erlaß GZ. 68.716/23-I/A/3/97 vom 10. Juli 1997.

Van der Beken, A. ed. (1998/1) Guide to educational terminology, ETNET.ENVIRON-MENT-WATER (European Thematic Network of Education and Training for ENVIRONMENT-WATER), Brussels, Belgium.

Van der Beken, A. ed. (1998/2) General report for the second year of activity 1997-1998 ETNET.ENVIRONMENT-WATER (European Thematic Network of Education and Training for ENVIRONMENT-WATER), Brussels, Belgium.

A Pyramidal Education Program for Sustainable Water-Environment in Developing Countries

Guozhang Feng

Northwestern Agricultural University, CHINA

Abstract

The importance of water-environmental education in developing countries is briefly mentioned. The structures of both knowledge and education are recognized as fuzzy and fractal pyramids. A three-level education program called pyramidal education program (PEP) is proposed for water-environmental education in developing countries. The three levels are public education, vocational education and professional education, respectively. Some achievements of China's water-environmental education in the three levels are briefly introduced. The role of different sectors of the society at different levels of water-environmental education is discussed. The collaboration and support of international organizations and developed countries are emphasized. The PEP may be suitable to any other education.

Key words: water-environmental education, pyramidal education program, fuzzy and fractal pyramid, public education, vocational education, professional education, developing countries, and China.

1. Introduction

Water-environment is one of the most essential parts of human living surroundings and the nature and plays a very important role in sustainable development of the world. However, undesirable water-environmental changes have become critical restrictions to long-term continuous progress of the world and are impacting on social progress, economic growth, human health and environmental security of the globe, especially on those of developing countries. A large number of developing countries which once had less water-environmental pressure are now suffering from severe water-environmental pollution, deterioration and even water-environmental-related disasters.

The water-environmental issues in developing countries are mainly caused by human activities and in chief water-related human activities. They are strongly concerned with non-appropriate exploitation or over-withdraw of water resources without effective measures for water-environmental protection, non-rational allocation of the water to different sectors of the society and the nature, hazardous pollution of water bodies or water-environments including conscious and unconscious disposition of a large amount of solid wastes and polluted water into rivers, streams, lakes, aquifers and so on, and lack of highly qualified capacity of water-environmental protection. All of these are directly and/or indirectly concerned with scientific, technologic, administrative, legislative and financial aspects.

Rapid population growth, urbanization and industrialization, large-scale farmland reclamation and deforestation, over-exploitation and consumption of natural resources except water resources, and persistent droughts and catastrophic floods, are some other major human or natural related causes of water-environmental deterioration.

For sustainable development in the forthcoming new millenium, developing countries have to resolve water-environmental issues, together with other environmental issues as a whole. To protect water-environment from further damage, improve water-environmental quality and mitigate water-environmental disasters need advanced science and technology, sufficient investment, appropriate management and the public's awareness. All of these need effective and efficient water-environmental education. In fact, as an archaic Chinese proverb said that "it might take a decade to grow forest, but a hundred to cultivate people", water-environmental education is the most important and underlying means in resolving water-environmental issues. It is imperative to carry out wide-rang water-environmental education in developing countries, especially in those suffering from severe water-environmental damage.

Water-environmental education is complicated system engineering concerning in all sectors of the society and difficult to carry out for most developing countries. It needs to seek for a suitable way for the education with feasible and efficient manners. In this paper a water-environmental education program is proposed within a general framework based on the experience of China's environmental education and other actions in environmental protection. The program is called pyramidal education program (PEP).

2. Structures of knowledge and educations

2.1. Knowledge structure

It is widely recognized that knowledge structure is a typical pyramid. In a knowledge pyramid, a large amount of elementary knowledge is located at the base of the pyramid. New knowledge is rooted of the base. Higher-level knowledge grows from lower-level knowledge. Different elements or kinds of knowledge at a same level produce different kinds of higher-level knowledge. The higher the level is, the more academic the knowledge is. Therefore, the knowledge varies from elementary or general knowledge at the bottom to highly academic knowledge or advanced science and technology at the top of the pyramid.

Similarly, The knowledge structure of water-environment is also a pyramid. It is composed of different levels of water-environmental knowledge ranging from general knowledge to highly advanced science and technology of water-environment.

Like water-environment itself, which is a major component or subsystem of the complicated global or regional environmental systems, water-environmental knowledge is a major component of the complicated environmental knowledge system as well. It is concerned with many kinds of knowledge in natural and social sciences. It means that the structure of water-environmental knowledge may not be a simple pyramid but a complicated one.

The complicated knowledge structure of water-environment may be further expressed by some fractal structure. The basic structure of the knowledge as the basic component of the fractal structure may be simplified as a triangle, in which higher-level knowledge is born of a combination of two kinds of lower-level knowledge. This basic structure is shown in Figure 1a where *a* and *b* stand for the two different kinds of lower-level knowledge respectively, and c the higher-level knowledge developed from *a* and *b*. Accordingly, a number of the basic structures may be produced by the manner of two-two combined development of the knowledge, such as the triangles *6abe*, *6bcf*, *6cdg* and so on as shown in Figure 1b. Furthermore, some triangles based on the higher-level may be obtained, such as triangles *6efh* and *6fgi* in Figure 1b. By doing this in turn, the highest-level knowledge would be produced; e.g. the j indicated top of the triangle *6hij* in Figure 1b. It is evident that Figure 1b is somewhat like a Sierpinski gasket that is a very popular fractal structure (Barnsley, 1988; Zhang, 1996).

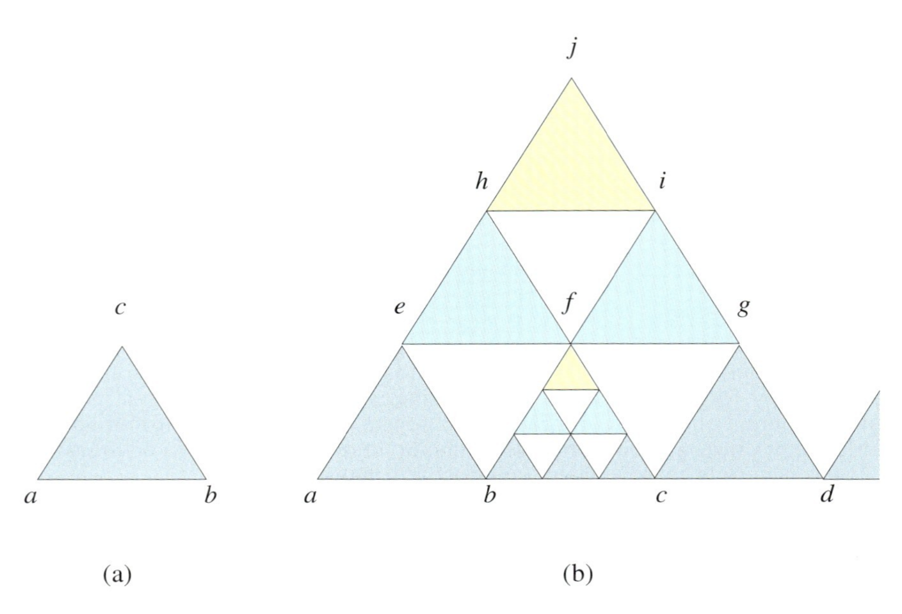

Figure 1. The pyramidal structure of knowledge expressed by triangles in one-dimensional plane: (a) basic structure; (b) fractal structure

In addition, knowledge structure has typical fuzzy properties. In a knowledge structure, any basic structure may be divided into several smaller one and even infinitive according to the scales of the differentiability of knowledge components. Correspondingly, the knowledge may also be divided into infinitive levels or hierarchies. The differentiability of knowledge is actually fuzzy because there is no clear demarcation line between any pair of adjacent knowledge levels.

Accordingly, knowledge structure may finally be recognized as a fuzzy and fractal structure, in which the development of knowledge is infinitive but with gaps in the structure. It is undoubted that the knowledge of water-environment is of course a fuzzy and fractal structured pyramid.

It should be pointed out that the development of knowledge is essentially bi-directional: vertical and horizontal. The results of the two-two combinations of different knowledge may be a kind of new knowledge on either a higher-level or the same level. It means that real knowledge structure is more complicated than the one mentioned above. However, for practical purpose in investigation of different levels of knowledge, the unidirectional (vertical) development of knowledge is acceptable.

2.2. Education structure

Education is the learning or transferring process of knowledge (Gilbrich, 1997). Undoubtedly, it is also a typical pyramid in structure with fuzzy and fractal properties. Like the knowledge of water-environment, which is a major part of the environmental science dealing with complicated environmental system, water-environmental education is also a major part of environmental educational system. In fact, water-environmental education itself is indeed a complicated education system. The structure of water-environmental education can also be regarded as a pyramid. It is composed of all patterns, methods and means of education, all the educators and educatees and almost all sectors of the society, as well as the knowledge of water-environment. However, for the purpose of practice, it may be appropriate to simplify the education into a three-level pyramid (Figure 2). As an education program, it is hereafter called pyramidal education program (PEP).

3. The pyramidal education program

As mentioned above, in the PEP water-environmental education is divided into three general levels, namely, Level A, Level B and Level C, locating in the education pyramid from the bottom to the top in alphabet as illustrated in figure 2, where the pyramid is simplified as a triangle. In this education program, Level A, Level B and Level C represent public education, vocational education and professional education, respectively; and each level represents quite a wide range of knowledge. However, for any developing country, regardless of any level of the PEP, the common objective is to enhance its capacity of protection, restoration, reconstruction and improvement of the damaged water-environment so that it is able to keep water-environment and all the environments as a whole in safe state for sustainable development of the developing world. Meanwhile, regardless of how the education is divided it should be considered as a complete system and be comprehensively developed.

3.1. Level A: public education

Level A stands for public education or social education. The purpose of this level of water-environmental education is to raise the public's awareness of water-environment and its protection. The primary task of this level of education is to lead every member of the society to understand the basic relationship between the human and the environment, to set up correct consciousness of water-environment, and to participate in water-environmental protection and related work self-consciously. It needs to lead the public to understand the scarcity of water available for sustainable supply and frailty of water-environment, as well as the importance of a safe water-environment to sustainable development of their homelands, their countries and the world. Therefore, this level of water-environmental education is basically a kind of general knowledge education or water-environmental education. It should be fully knowledgeable but not highly academic.

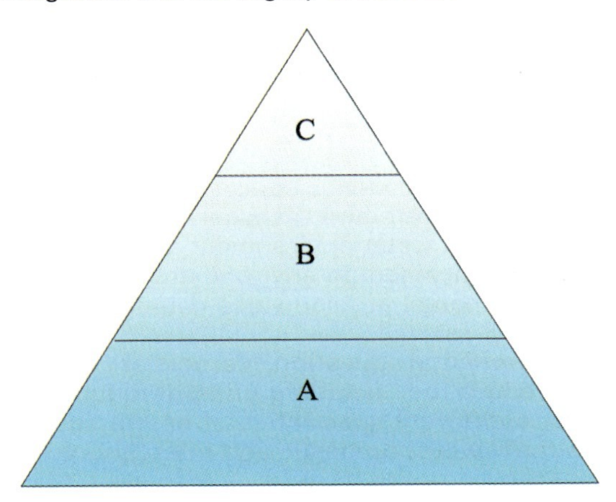

Figure 2. The pyramidal education program (PEP) of water-environmental education (A, B and C stand for three different education levels, i.e. Level A, Level B and Level C, which represent public education, vocational education and professional education, respectively)

Practically, this level of water-environmental education is one aspect of capacity building and a reflection of civilization of whole the society. It is nearly concerned with the education of human morality, ideology, behaviour and personality. It should be a kind of rational education in stead of political propaganda.

Obviously, in this level of water-environmental education, each social member himself is either an educatee or an educator in general meaning. It is the largest-scale action in water-environmental education and needs completely cooperation and collaboration among governments, governmental and non-governmental academic and non-academic organizations, the media, and almost all other agencies as well as the public themselves in the society.

China, the largest developing country in the world, has made considerable achievements in its economic development and social progress in last two decades. However, with socio-economic development, accompanied by rapid industrialization and urbanization, China encountered severe environmental deterioration. Particularly, water-environment disasters have become and still are countrywide environmental problems (The State Environment Protection Administration (SEPA[1]) of the People's Republic of China, 1998). Facing the environmental problems which are severely impacting on the country's continuing development, China has made enormous endeavor in its environmental protection, especially water-environmental protection, in which education serves as a forthgoer (Information Office of the State Council, 1996; National Environmental Protection Agency et al. 1997; Wu, 1997). From pre-schools to universities, from individuals to communities, and from single social sector to whole the society, water-environmental education has become a very important part of the nation's education. All types of education such as formal education, non-formal education, informal education, are playing important roles in the public education of water-environment. Take Shanghai, the largest city of China, as an example, public environmental education now takes an important position in elementarily education, in which the pre-school and school environmental education has made considerable achievements (Deng et al. 1997; Wan, 1997; Zhang, 1997; Hu, 1997). This kind of environmental education has been successfully popularized to all over the country (NEPA et al. 1997; SEPA of the People's Republic of China, 1998).

In this kind of water-environmental education, the roles of governments from the center to local, the public and the media such as newspapers, radio, television and any means being capable of bringing the elementary environmental knowledge into publicity are very important. In particular, women's contribution is considerably beneficial.

3.2. Level C: professional education

Level C stands for professional or academic education. The purpose of this education level is to foster highly qualified scientists and managers who should be capable of scientific research and decision-making in water-environment and relevant fields of environmental science. It represents the advancement in environmental science and technology and the capacity of resolving environmental problems of a country.

In this kind of water-environmental education, learners are educated and trained through regular or formal higher education, including undergraduates, graduates and postgraduates at different levels in different specialities of environmental science. They can be cultivated through two main ways: domestic and international education.

([1]) SEPA stands for the State Environment Protection Administration, which is China's highest (ministerial) administrative authority in environmental management. It was translated into the National Environmental protection Agency (NEPA) in some documents, especially in those prior to 1998.

China has made considerable progress in the first kind of education. By the end of 1995, there were 206 specialities on bachelor's level in environmental science in 140 universities and colleges, in which some of the specialities were the same in different universities. At the same time, up to 223 institutions, including most of the universities and colleges in bachelor's education and a certain number of research institutions such as the Chinese Academy of Sciences, had the authorities of awarding master's degrees in 51 different specialities, 77 for doctorates in 39 different specialities, and several places for post-doctorates in environmental and environmental-related sciences (NEPA et al. 1997). In these environmental-related specialities, some of them are specified water-environmental specialities, some are water-environmental-leading ones, and most of others offer a certain number of courses in water-environment. These numbers have increased since then. On-the-job training in this kind of water-environmental education is also properly developed. Quite a large number of young people have been sent to developed countries to study environmental science. However, the domestic education is the major part of the education.

In this kind of water-environmental education, the roles of universities and related academic agencies are absolutely important and imperative. Particularly, the higher education system in water science plays a key role in this kind of water-environmental education. Developing countries should pay more attention to develop domestic education and corresponding education agencies and to foster their own scientists and managers.

3.3. Level B: vocational education

Level B covers all other aspects of the PEP excepting for those involved in level A and level C. This educational level of water-environment may be probably regarded as vocational education and further divided into two classes: skill training and reeducation.

The skill training means that in this kind of water-environmental education the major aim is to foster so called middle-level specialists, commonly, technicians and related persons with certain skills in water-environment and related fields. It can be realized through formal and non-formal education in vocational or special schools and other education agencies such as universities, by any other patterns and means of education. In the formal education the trainees are basically the regular school students while in the non-formal education the trainees are mainly working people who usually take courses on-the-job in part time.

The reeducation is typical continuing education. It includes knowledge renew, skill improvement, continuous professional education of the persons who are working in water-environment and related fields. This kind of education should be well planned and rationally arranged in accordance with the needs of the society and backgrounds of the educatees or trainees.

In China, most of the education agencies engaging in professional environmental education also play a key role in this kind of water-environmental education. Most of on-the-job training is carried out by these agencies in the forms of short-term training and long-distance education. At present, it is carried mainly through corresponding teaching as present. Meanwhile, hundreds of vocational or special schools offer environmental courses. Particularly, through the continuing education and so-called specialized education to administrators who are working in governmental agencies dealing with environmental issues, on-the-job training has made significant achievements. In the past decade, more than 10,000 training classes were held and over 400,000 persons were well trained (NEPA, et al. 1997).

In this kind of water-environmental education the role of both the governments and education agencies involved are equivalently important. Also, the roles of professional associations, especially those in water science and environmental science are very important. All the associations involved are capable of making significant contributions to the education. Meanwhile, the contribution of international collaboration and support is important and should not be neglectable. In quite a lot of programs or projects on environmental education in developing countries, international support plays an important role.

4. Need of international collaboration and support

It is undoubted that every developing country plays a key role in its water-environmental education and development of water-environmental science and technology. So does in all the environments. However, due to the well-known reasons of relatively undeveloped education characterized by lack of educational infrastructure, lack of qualified professional educators and lack of sufficient investments in education, some developing countries could not make visible progress in water-environmental education even though they may make obvious endeavor. A common challenge to developing countries is rapid population growth and less GDP (gross domestic products). They have to increase their food production and industrial products to cover their basic requirements. It inevitably affects on environmental education and leads to undesirable environmental deterioration, especially water-environmental deterioration in water-short countries in arid and semiarid regions.

Facing with persistent water-environmental damage, most developing countries may be able to make clear decision on water-environmental education but difficult to put the decision into practice for some of the developing countries. To do this in perfection, the collaboration and support of international organizations and developed countries are completely necessary. International organizations and developed countries should make active effort as usual and even better in water-environmental education of developing countries. They should provide more help to developing countries in educators' exchange, educational facilities and finance.

The globe is our common homeland. To protect it from any damage is our common responsibility. All members of the globe should work together for keeping our common homeland to be in safe water-environment and sustainable development in the hopeful new millenium.

5. Conclusions

The education of water-environment is complicated system engineering. Its structure can be recognized as a fuzzy and fractal pyramid. For practice, a pyramidal education program (PEP) for water-environmental education is proposed, in which the education is divided into three general levels: public education, vocational education and professional education.

To developing countries, all the three levels of water-environmental education are very important and should be comprehensively developed. However, the emphases should be reasonably redressed in different developing stages according to the state of environment and the development of socio-economy.

The roles of governments and professional organizations are important at all levels of the education. Women's contributions are particularly important, particularly in public education, and should be brought into full play.

The collaboration and support of international organizations and developed countries in water-environmental education of developing countries are completely necessary and important, and should be well continued in the future.

As a 'program', the PEP only made some general description within the framework of the pyramidal education structure without providing concrete proposals on water-environmental education. Any detailed program may be produced within the framework.

There is no universal educational model for all the world. Each developing country has its special socio-economic and environmental state and needs special model for its water-environmental education. All actions in water-environmental education of developing countries should be focused on their special water-environmental situation whether or not self-organized and international supported.

The concept and theory may be suitable to any other education and should be further studied in deep.

References

Barnsley, M.F. (1988). Fractals Everywhere. Boston: Academic Press, Inc.
Deng, W.J., Yu, Y.S. and Hu, C.C. (1997). The Role of the Coordination Committee of Environmental Education in Primary and Secondary Schools in Shanghai.
Gilbrich, W.H. (1997). Guide to Educational Terminology. ETNET.ENVIRONMENT-WATER, Brussels.
Hu, C.C. (1997). The Benefits of Environmental Education in the Community.
Information Office of the State Council (1996). Environmental Protection in China (white paper).
National Environmental Protection Agency, Propaganda Department of the CPC Central Committee, State Education Commission (1997). The National Action Program for Environmental Publicity and Education (1996-2010).
The State Environment Protection Administration (SEPA) of the People's Republic of China. (1998). The State of the Environment China 1997.
Wan, J.Y. (1997). Strengthening Environmental Education of in Kindergarten Using Various Methods.
Wessel, J. (1999). A Guide to the Needs of Education and Training in the Water Sector. ETNET.ENVIRONMENT-WATER, Brussels.
Wu, C.H. (1997). China Environment Report. Published by the Professional Association for China's Environment.
Zhang, J.Z. (1996). Fractals, Tsinghua University Press, Beijing.
Zhang, M.C. (1997). Environmental Protection and Personality Education.

Learning Society?

Une société cognitive?

Gilbrich W. H.

France

Abstract

The philosophy of the Symposium assumes a learning society. After an excursion through selected cultures in antiquity and in the more recent past the paper questions whether, at present, society is well utilising its technology and learning techniques in order to constitute a mew learning society. The paper relates creative learning with the general societal attitudes and the stresses that the present society in the industrialised countries needs the generation of new ethic and spiritual approach in order to become a learning one in contrast to utilising its possibilities for entertainment, as a consume-oriented man. The paper raises the question of purpose and aims of learning, whether the learning society strives at higher material levels of mankind or whether a higher culture, intelectual and spiritual form of life is the desired outcome.

Résumé

La philosophie du symposium admet l'existence d'une société cognitive. Après un voyage dans un certain nombre de cultures de l'antiquité ainsi que dans celles d'un passé plus récent, la contribution pose la question de savoir si, à présent, la société utilise correctement sa technologie et ses techniques d'apprentisage pour constituer une nouvelles société cognitive. Elle fait un lien entre l'apprentissage crétif et les attitudes générales de la société, puis insiste sur le fait que la société actuelle dans des pays industrialisés a besoin de voir apparaître une nouvelle approche éthique et spirituelle afin de devinir une société de connaissance à but de divertissement de consommation de masse uniquement. Le document soulève la question des objectifs de l'apprentisage. Il se demande si la société cognitive a pour seul but d'élever le niveau matérial de l'homme ou bien si elle vise à lui apporter une forme de vie culturelle, intelectuelle et spirituelle plus élevée.

The philosophy of the Symposium assumes a learning society and implicitly also assumes the existence of a non-learning society, at least in the past. But the title obviously also is based on the idea that the learning society originates from or distinguishes itself from a non-learning society because the term "learning" means action. The end-product would be a "learned" society, status rather than action. In fact, the idea is fascinating to consider whether today's society really is a learning or a learned one or whether this concerns only a fraction of the society, in the worst case a minority. It appears worthwhile to review both past and presence whether learning, learned societies have existed and whether also the opposite is true. If a non-learning society exists and if it is quantitatively significant the symposium title merits an interrogation mark.

Philosophers from antiquity onwards have argued whether mankind in its entirety will reach higher levels in full harmony and equality and almost all religions include a message of a paradise. On the other hand, particularly Asian schools and religions (China, India,

Persia) have developed a dualistic approach and the Christian religion and there again much more some of its developments is based on dualism. Good and bad, heaven and hell are the most common expressions. However philosophers during the past three centuries have speculated whether the modern mankind really is subject to dualism and today's political streams and some systems are based on the idea of equality. Political models are being discussed and sometimes have been tried out in practice that everybody is absolutely equal in a human community in which all people have the same function, value, income - and education. If put into reality, will this be a learning or even a learned society?

Doubts must be expressed that the entirety ever will constitute a learning and learned society but that only a fraction will participate in the learning process and the question is the order of magnitude, is it a minority only or a general mass movement. If it is a minority only, one should be aware of the political implications : support to elite groups is suspect in a number of political and sociological systems. But even if a broad mass movement could be observed the question must be posed about quality, contents and nature of the knowledge transferred in the learning process. Interesting as the question is it cannot be followed here, namely to discuss the amount, nature but also the philosophical and ethic background of such knowledge, skills included. What makes an educated person, who is learned? What is the ideal, the specialist, the generalist, what is the role of a studium generale (not needed or essential), does learned mean academic? When propagating a learning society we should not forget to define the desired results. Will we be able to offer a facet in which science, ethics and the hidden desires of human beings are harmonically balanced because knowledge alone soon will lead to disappointment. Hence, "learning" is not only an academic exercise but needs a spiritual environment. The few examples of former cultures will underline this.

A few years ago an American fundamentalist ecologist wrote that the spiritual decline of mankind started with the invention and use of tools. Human development, in fact, did not follow this ideal of the absolutely not-learning society, but quite in contrary, was based on the couple capability of learning and the wish to learn and apply. However, intensity and goals of learning have been quite different through out the past, and the struggle for survival was by no means the only parameter. Although we may admire the end products such as the stone - carved relicts of Egyptian culture we have problems to identify ourselves with the former learning or learned societies.

The antiquity in general was illiterate. In many cases our perception is based on erroneous assumptions. Particularly the eighteenth and nineteenth century believed the classic Greek to be a people of philosophers. Particularly during the Romantics stories and pictures showed the Agora of Athens crowded with wise men, engaged in profound discussion and the writers would record the findings to be studied elsewhere. Old Greece an intellectual paradise? No word about the army of uneducated people, workers, women and slaves. In reality, the percentage of learned people was a tiny fraction from a privileged population group, quantitatively close to zero. The cradle of philosophy a learning society - certainly not. The Greek mixture of socio-economic circumstances and education today neither is understandable nor can it serve as a model.

We are deeply impressed by the cultural heritage of old Egypt. From reading their texts we have an idea of the high level of their knowledge and of their thoughts. However, neither the Pharao nor the High Priests could write, whether or how much they could read will remain a secret, probably they were able to identify the hieroglyphics. There was a perfect distribution of functions. A special profession of writers, stone-carvers translated the ideas of the thinkers into readable signs. Thanks to their "technology" and to the climatic conditions we are able to trace their ideas and to appreciate the depth of their ideas. What has been written down originates from either the high aristocracy of Pharao's administration or from the priests. Since religion, state, knowledge largely formed a unit it is fairly difficult to describe the learned society. The learning process was strictly controlled, the knowledge-holders were selective in choosing their disciples and the transfer of knowledge was oral instruction. It is sure that the oral transfer much exceeded what is written on stone because secrets were the power of the knowledge-holders. Quantitatively, the number of educated people (say administrators and priests) was relatively large. Occasional revolts occurred when the labourers no longer could feed this clan. It is important to note that knowledge was reserved to a closed society which was most eager to defend its rights and privileges. A learning or learned society? One can affirm this question only partially. The percentage within the population was large (on the eve of a revolt probably in the order of a third) but access was limited and controlled. Judging from today's perception another observation is strange to us. A person would not learn for himself but for fulfilling his role within the religion - state oriented distribution of functions, the given place in society.

One more observation must be made. Today we are used to seeing writing, reading and thinking to possibly forming a unit. In Egypt these functions were separated and this fully corresponds to the role given by the Gods to everybody, from the Pharao downwards to the writer. From this follows that the amount of learned people depended on the number and role of persons engaged in the temples or in the services of the Pharao and the high aristocracy. As a consequence one may speak of a demand - defined learned society which cannot serve as a model for today. Also the political idea behind forbids its application. The clan of priests and administrators ruled the state. They were not interested in power - sharing. In controlling learning the conservative Egyptian society could preserve its privileges, and they were considerable: exemption from labour on the construction sites, from field work but also from taxes as opposed to holding the power. As impressive the products of the Egyptian learned society are they cannot be applied today as the socio-political system is not compatible with our one and for us it is surprising enough that the Egyptian system kept stable over some two thousand years. This stability means that the contents of learning has not much changed since the Thinites period or Pharao Djoser, only little efforts have been made to expand knowledge. Learning has to be understood as reaching a tradition-fixed quantum and not as going beyond. Our understanding of "learning" is dynamic and not conservative.

The Roman Empire should be briefly looked at. Its literature still today finds high attention. However, illiteracy was prevailing as by its very nature the Roman Empire was not intellectual. When nevertheless rich, free men were fond of art and literature and when the Romans had a high sense for reporting and recording the bulk of written and educators were Greek slaves and it was the Greek slaves who transferred the Greek philosophy to Rome and in this way only it reached the Occident.

It is most interesting to look into the initial period of Christianity. It is reported that Jesus himself had a profound knowledge of what we consider the Old Testament. There is no evidence that he was able to read it, oral communication from the Synagogue must be assumed. The fact that once he wrote a word into the sand probably should not be over-interpreted. His disciples probably were illiterate. Except most of the letters of Paulus the origin of the New Testament is heavily disputed by researchers and secondary sources must be supposed.

However despite these deficiencies a book religion evolved and the role of written word was decisive for the development of the two millennia to follow as it created new forms of learning societies. - As a curiosity it should be mentioned that in the Middle-ages the legend came up that the mother of Jesus was highly educated, could read and write, and she even advanced to be the patron of numerous universities. - It must be stressed that today's culture and education is based on the Christian tradition and that this new knowledge, new contents of learning programmes filled the gap following the fall of the Roman Empire; the new Church filled the intellectual vacuum.

While the Roman senatorial class enjoyed a high cultural standard and it formed the learned (not necessarily learning) society the Germanic successor Kingdoms formed a completely new social order. The ruling class were warriors, powerful and illiterate. It took some time until they discovered and then protected the cultural and educational heritage present in the Church but centuries passed until Charles-the-Great launched a programme designed for a general raise of the educational level. He failed. Much later, Emperor Otto was able to sign only after somebody (a monk or a priest) had pre-drawn the "O" and the "T". Also the knights were illiterate throughout, some of their wives had rudimentary education. It is amusing to read that one of the knights, Hartmann von der Aue, in a poem emphasizes that he was able to read books. Hence, the ruling class, the broad nobility, did not form a learning society; their role was kept by the Church and more precisely, not by priests but by monks. If the Roman art of writing has survived it is thanks to cloisters and abbeys. Research has shown that the level of education was horrifyingly low and the tenth century generally is called the dark on. The learned society of that period today probably would not be counted learned but it kept the connection with antiquity and after the tenth century gradually evolved towards the European system of learning. With the growing importance of literacy the uneducated warrior class could only decline, unable and unwilling to change the sword against the feather.

A learning age came up with the enlargement of the number of learners and with higher level of teaching and the increase of number of subjects. Firstly only in abbeys or episcopal administration schools then in training foundations, the up-rising concerned both quantity and quality and new types of teaching institutions emerged, the pre-runners of the universities. Originally pure institutions of the Church they gradually developed a certain intellectual independence (today hard to believe: the sorbonne was particularly loyal to the Church over a long time) and finally became practically autonomous, financed by the crown.

This intellectual up-rising produced a new societal group, the academic teachers, researchers, university staff. Not rich in terms of money, not powerful in military or political terms but a really learned society, small in numbers but dynamic and the influence was enormous: in many political disputes no longer the sword but the word was decisive. Knowledge became power. This up-rising was highly supported by the invention of printing. The costs for written information fell dramatically and the new techniques also permitted to widely diffuse the pamphlets and books. A new age had begun, open for almost all groups of the society, changing completely the life in general but also the intellectual pattern of all societal groups. The learning society was no longer class-oriented or group-restricted but evolved towards a popular movement with the ultimate goal (which was not intended at that time) that everybody would have access to education. Hence, the conditions for a learning society were set, also a democratisation of learning and of professional chances, but these were not fully foreseeable some four hundred years ago. Also, it must be admitted that the way was very long until education was fully liberated and the obstacles partially came from the old institutions defending their power but largely also in the form of costs. As long as the State did not cover the costs even for a minimal education programme and as long as children of any age had to help their parents, education was the privilege of those who could afford the costs.

The movement to increase learning covered all of Europe, of course with some centres. As opposed to the Egyptian case learning now meant an explosion of knowledge. The ideal was not to know a well qualified knowledge well by heart but to know more. Again unlike Egypt, learning was an international effort; in terms of that time one could speak of a globalisation and this was much helped by the fact that Latin universally was the language of universities. With national languages pushing Latin aside some countries experienced stagnation or even a fall - back and it is interesting enough that today the global learning process is based on the wide usage of the English language.

It cannot be the purpose of the paper to historically follow the societal development throughout the past centuries although it would be of high interest to see the (relative) decline of formerly leading institutions or societal groups as well as the rise of new ones. Renaissance, the enlightenment, the romantics, the industrial revolution, they all contributed to the increase of knowledge, they all drew new population groups into the learning exercise and the period until the First World War can be considered a peak in the philosophical understanding of learning. To learn and to be learned was the aim of the so-called Bildungsbürgertum and probably never before and never after this time the prestige of a professor was as high. In fact, there was a learning society although a limited one. The limiting factor was money because learning was costly. Hence, the idea is justified to believe that with increased government funding of learning the ideal could and can be reached that everybody participates in learning and that the Twentieth Century would constitute the golden age of a learning society, no war, no poverty but a high cultural, ethic and educational level. No doubt, this dream is reflected in almost all programmes of political parties in our days.

Before reflecting on present problems, one has to be aware of the fact that the high educational level referred to above was largely limited to Europe and North America. Illiteracy in other continents, in colonies, even was considered an advantage. Today's vision however is global.

Until some time after World War Two the assumption in general was (a) illiteracy in the industrialised countries would go down to zero, (b) illiteracy in the developing countries would dramatically decrease and the educational level hopefully one day would reach the level of the industrialised countries, and (c) that the general societal attitude would inspire everybody to happily participate in the learning process so that the educational and cultural level visibly would reach never experienced hights, the golden age of intellectual culture.

Such intellectual culture cannot be defined, in its extreme it would mean that everybody reaches almost academic level on a broad basis of a "studium generale", hence something similar to the "Bildungsbürgertum". No doubt, the condition is literacy, although the examples drawn from antiquity demonstrated that knowledge and literacy at that time normally were separated. Today, this separation is impossible. There are many definitions for illiteracy and they all are vague, arbitrary. The threshold determines the number of illiterates and thus the problem is open to statistical manipulation. The intended political aim determines the count.

According to an OECD study about 14% of the German population (making some 11 million people) is not able to correctly read, write, calculate. The figure obviously depends on the interpretation of "correctly" and other counts restrict the number to some four million (or 5%). France reports similar figures and they are probably representative for most of the European countries.

There are abundant statistics for the developing countries and they all suffer from the counting procedure applied. However, many countries admit a rate exceeding 50%. According to Unesco sources the number of illiterates is almost constant but not the percentage : the alphabetisation campaigns to not cope with the increase of population.

The reasons for illiteracy in developing countries are well known, and the list is long: economic poverty, political priorities, natural catastrophes, civil wars, cultural and religious barriers, education-hostile fundamentalism, exclusion of ethnic groups, linguistic problems, mostly a mix of reasons. The number of countries affected is high and no improvement can be seen. Where educational programmes are based on external funding the economic situation and the political attitude of the donors is decisive and thus out of the reach of the recipient country.

Even the existence of public schooling programmes is no guarantee for the results. We need not go to the developing countries but we can study the misery at home : the state prescribes the number of schooling years but not the target. A pupil will be released after the compulsory number of years irrespective of the knowledge gained. Hence, one can postulate a desirable level but there are no means that it will be reached by all pupils. At a time of high unemployment these young people have almost no chance on the labour market. The efforts made by many governments to focus special programmes on this group of pupils unfortunately do not yield the desired results and the problems probably must be searched for in the society itself and not necessarily in the schooling policy. The paper will address some of these societal problems in the industrialised countries although it could be understood unjust not to discuss the developing countries, particularly those areas suffering from natural disasters, wars and refugee problems.

In the industrialised countries schooling is obligatory and free-of-charge. Governments are investing considerable funds, sometimes the highest item in the budget besides defense and subventions. Nevertheless, the high number of illiterate or semi-illiterates is alarming.

Apart from medically-mentally ill persons (for whom more money is being spent than the public is aware of) there is a large number of pupils who practically are standing outside of the schooling system, even if they are registered. It is not surprising that many of them belong to unstable population groups, victims of migrations: immigrants, street-children, inhabitants in "quartiers difficiles", illegal inhabitants. The number of these marginal groups can only be estimated and much depend on the political standpoint, even the governments are lacking precision. French sources estimate figures much exceeding one million; official counts are much lower.

It would be a great mistake to relate the problem only to immigrants. Even highly advanced social systems could not prevent a surprisingly large number of native marginal groups, of step-outs and their descendents, of people for whom the public social welfare does not suffice but there is also a significant group of people who detest the present culture and its labour - and income and consumer philosophy and who no-matter-how disappear from public statistics. One only knows that the number is higher than tolerable. While our society often is inclined to give up the adults it should not forget that the main victims are the children.

Drugs are responsible for a large number of children to step out and for many of them it is a way of no return, for both dealer and victim. The public although aware of the drug problem apparently is powerless.

If a part of our society is sick it is doubtful whether police and educational systems can remedy the situation. Many children live in a criminal environment, the age for the first criminal act is dramatically decreasing - Kids of ten years are among them. If our society wishes to remedy the sick part then it has to deal with the complicated interactive network of criminality, drugs, prostitution, over-population in difficult areas, no-hope attitude = easy to say but nobody has a solution. Hence one has to state that a significant part of the population is and will remain outside of the learning society. As long as there is the conviction that the application of criminal energy leads to higher income than education nothing will change.

While there are many material reasons for the marginal groups to exist one should not underestimate the personal will. Whether it fits into a political belief or not, there is a certain fraction of population which neither wishes to nor is willing to learn and to work. It is the spirit of a time which determines the order of magnitude of this social group. This group under no circumstances will become a part of the learning society, one only can hope that it is a small fraction, Much more decisive is the attitude and behaviour of the so-called middle-class and their attitude towards learning, also their order of magnitude as the majority of society, will determine whether they belong to a learning society or not. If the majority of people cannot be counted "learning" we would be forced to state that only a minority forms the learning society with the result that the entirety of the society is not a learning one. The availability of tools for learning is not decisive. What counts is the usage by at least so many persons that the expression "learning society" is justified. If the share is small the expression is an illusion.

The general behaviour of adults does not support a learning society although means for learning are readily available as never before. Sales figures speak a clear language. The flood of journals, pocket books, comics, computers contain elements suitable for learning. The bulk however serves for entertainment. Billions are being spent but they do not leave traces in an increase in education. On the contrary, the prevailing TV programmes and printed matter have only limited cultural or educational value and do not improve learning. High-level TV programmes and printed products often benefit from subventions earned through mass products of lower level. Pseudo - scientific programmes suppress critical thinking and their contents often is considered the truth which satisfies superficially. Unless there are sensations there will be little interest. Tests have been made to measure the audience/spectator rate for scientific emissions : people are switching off with the exception of a very small nucleus. Regretfully, one has to state that the large variety and the low cost of educational material or presentations does not enjoy the acceptance it could have. The present generation has unique chances for learning but only a minority makes use. The beneficiaries are not the carriers of scientific knowledge but the entertainment sector. It may be symbolic that these days a seventeen year old girl has been discovered to be the most beautiful girl of France : She immediately broke off her school education in favour of a nicer life with more money.

While many people are willing to recognise the low level and the intellectual non-value of entertainment as presented in the mass media the computer and its in-house application as PC enjoys a bonus. The PC can serve for learning and there is a lot of software on the market. Again, sales figures reveal that the bulk of sales concerns entertainment and games. Some people advocate that computer games inspire fantasy and creative thinking; other people detest it. In fact, the danger cannot be neglected that people content themselves in computer language and picture, that they forget the art of formulating sentences but only think along a sequence of pictorial sequences. The program seems to deliver all information desired by the user but in reality it limits the information. The high sophistication of programs risks to reduce the creativeness for own innovation. While the value of the computer for offices cannot be overestimated its value as a private means for learning remains doubtful at large population scale. This is not the fault of the computer or the program but that of the user. The easy access to information is an advantage but it may spoil the user so that he tends to avoid complications. However, higher learning never was without great intellectual efforts and the easy way the computer offers may make our minds sleepy. The computer is able to deliver rational solutions for almost everything but it will not promote ethics and cultural aspects. One may deplore that the PC at home largely serves for games (the sales figures are fantastic) and that learning plays only a minor role, but this societal attitude reflects what many people fear : entertainment and games are dominating in our society in which the mass media daily show how easy much money can be gained in this sector. If the brilliant technical devices in our hand are not used for a break-through to higher cultural, educational levels, if ethics and spiritual values play only a limited role, is it the fault of the learning society that it does not successfully advocate its own case, that it cannot attract the masses to join higher human aims.

A learning society? Yes, it is technically possible, all pre-conditions are fulfilled. However, in order to conserve the interrogation mark into an exclamation mark much conviction work needs to be done. The learning society is not the automatic consequence of our technology but it will exist only after a spiritual, ethic conversion of our society away from the low plains of simple consumer - attitude of criticless application of our technology to the steep peaks of higher understanding. The learning society conditions a societal change which gives more space to intellect, spiritual higher values rather than to commercial instincts and entertainment. Members of the learning society must not give up the fight.

It is not the purpose of this contribution but it would be more than interesting to discuss the aims of learning, the philosophical framework, the purpose, whether to be able to earn more money or whether to become recognissant for human values. Material or spiritual aims? Accumulation of knowledge or are there higher issues?
Rome started to loose when this most powerful society changed from hard work to "panem et circensem".

Wadbos: a case study for the design of interdisciplinary water systems

Wadbos: étude de cas pour l'enseignement interdisciplinaire de gestion d'eau

Wind H. G., Van Lieshout G. M. G. M.
University of Twente, Department of Civil Engineering,
The Netherlands

Abstract
One of the topics in the design of mathematical models of interdisciplinary water systems is the selection of the appropriate time and space scales. A didactical approach for this problem is presented, together with some simple exercises.

Résumé
L'un des objectives dans la conception des modèles mathématiques de systèmes interdisciplinaires de gestion d'eau vise à sélectionner les échelles appropriées du temps et de l'espace. Une approche de détection de ce problème a été présentée et complétée d'exercices simples.

1. Introduction

The Northern border of the Netherlands is formed by an estuary: the Waddensea. This estuary is used for many functions such as: nature reserve, recreation, sand mining etc. Hence a combination of economical and ecological aspects. In order to be able to formulate a sustainable policy, all these aspects and their mutual interactions have to be taken into account. In WadBOS, the acronym for Waddensea Decision Support System, such an integrated system for policy formulation is built. In addition this system is also used as a teaching instrument to teach students to understand and design integrated systems. In this note one aspect of integrated systems is focussed: the aggregation level or the spatial and temporal resolution.

Decreasing of the aggregation level often leads to more detail. Initially students often have the impression, that accuracy increases with increasing detail. This may be true if only one part of the DSS is investigated, e.g. the geographical data. However geographical information is only part of an impact analysis in integrated systems, as shown in Figure 1.

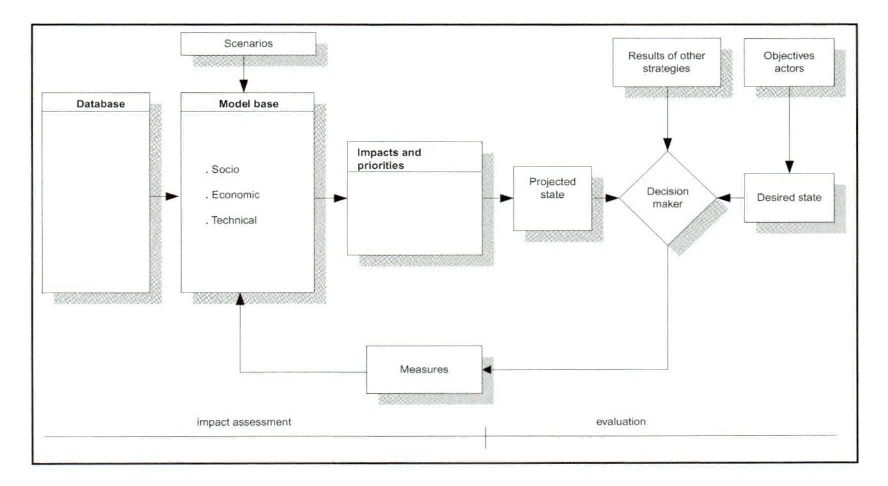

Figure 1. Model prediction and uncertainty sources

Figure 1 shows that the output of the impact analysis of the water system is the result of the modelbase, database, scenario's and measures. The accuracy of the projected state depends on the same four sources. Hence the accuracy of the data base, or the spatial resolution, should be comparable with the accuracy with which the physical, social and economic processes are represented. A similar argument can be used for the temporal resolution. This design perspective is elaborated in this paper.
From a practical point of view it is important to design models of water systems at the appropriate aggregation level of temporal and spatial resolution, as more detail requires more data, more processes and more interactions. As the temporal and spatial scale often are related the number of interaction increases with n3 or more. This implies an increase in costs and time.

The teaching objective of this project is to teach students that integrated water systems, such as shown in Figure 2, can be described on various aggregation levels and that an aggregation level has to be selected: too much detail takes too much time and yields too much results. In the next part the didactical approach will be explained followed by some simple examples. This note ends with a brief discussion.

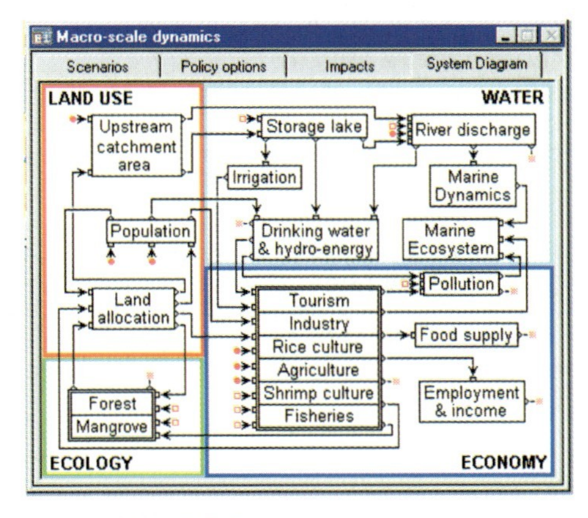

Figure 2. System diagram of WadBOS

2. Didactical approach

As a starting point for a global outline of the didactical approach in the course, we analysed the educational objectives into more detail.

To describe the objectives, we take a simple model in which two variables, A and B are coupled to provide an output-variable Z = f(A,B) see Figure 3. Both for variable A and for B the uncertainty depends upon the aggregation level at which it is measured: the lower the aggregation level, the smaller ΔA or ΔB.

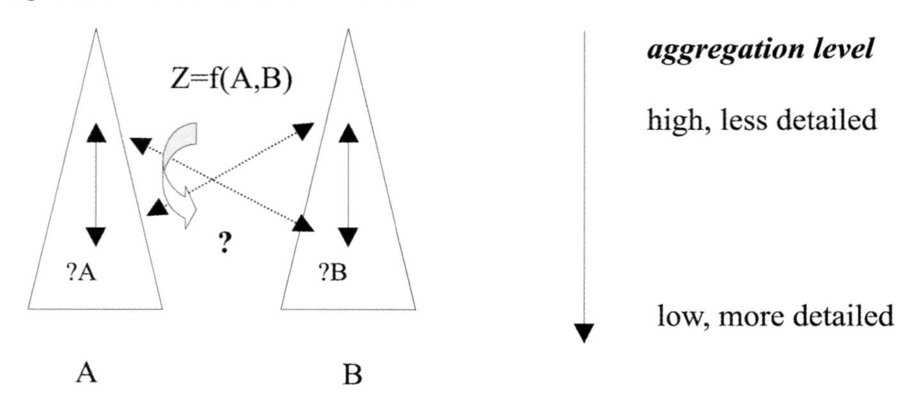

Figure 3. Didactical principles

The uncertainty in Z is a function of the uncertainties in A and B: $\Delta Z = f(\Delta A, \Delta B)$.
The problem to be solved is: "How to choose ΔA and ΔB (or: at which aggregation level A and B should be determined) to provide a proper value of ΔZ?" Or, in terms of the picture above: "What should be the "slope" of the dotted arrow?"
To solve this kind of problems, students should:
a) understand the meaning of the picture above
b) be able to determine ΔA as a function of the aggregation level of A
c) be able to determine ΔZ for given values of ΔA and ΔB.
These objectives can be characterised as " applied mathematical analysis".
In a realistic situation, in which a client needs the value of Z as a basis for technical or political decisions, the student should also be able to:
d) identify criteria and norms for the range of ΔZ which is appropriate for the user of the output variable of the model;
e) make an appropriate choice of the aggregation level of A and B, to get an output Z with the appropriate value of ΔZ.
These objectives are more of the "design" type.

The learning process of the student should pass through three stages: *orientation (on contents and methods), exercises and feed-back on straightforward cases and exercises and feed-back on more complex cases* . The table 1 presents the three stages together with their typical learning activities, teaching activities and learning outcomes.

Table 1. Learning and teaching activities for the design of integrated water systems

Stage	Learning activity	Teaching activity	Learning outcome: to be able to:
Orientation	study oral and written presentations analyse situations; discover principles imitate problem solving methods	provide written or oral presentation and explanation provide cases from which principles ca be derived describe and demonstrate problem solving methods	define and describe concepts and principles describe problem solving approach and methods
Exercise Straightforward Cases	solve presented exercises	provide exercises which can be solved by straightforward application of standard methods provide feed-back on output and approach of students exercises provide a "help-desk" function	solve straightforward problems (not complex, well known domain)
Exercise Complex cases	analogous to the former phase, but now on problems of a more complex character or in another domain as used in the former phases (transfer).		

This didactical scheme has been applied both to teach the more "analytical" objectives and for the more "design-like" objectives. In the latter case, a further splitting up is required: in a first phase the constituent design processes are covered separately and in the final phase the process as a whole is practised.

In the next section some examples will be presented of tasks and exercises which are typical for the various phases of the teaching-learning process as depicted above. The relation between increasing the aggregation level Z and the exercise is shown in Table 2.

Table 2. Relation between exercises and level of aggregation

	Exercises		
Increasing Level Of Aggregation ↑	Number of strips of the musselbank Exercise 1,2	Flood damage Assessment	System reduction
	= 2	Innundated area exercise 5	Individual contribution of sources exercise 7
	n = 4	Number of averaged house exercise 4	Contribution of sources and uncertainty exercise 8
	n = 8		
	n = 17	Individual house exercise 3	

3. Examples with a single variable

Introduction.

The first example deals with the representation of the cross-section of a mussel bank. In this example the variable Z, introduced in the didactical part, is the area of the cross section, while the related variables A and B are the width of the strip ΔX and the number of strips n. Furthermore this example is restricted to the data base in Figure 1.The first exercise points towards the importance of a norm in order to select an appropriate representation. In the second exercise a number of representations of the cross sections is obtained and together with a norm one of the representations is selected.

Exercise 1: representation of a the area section of a musselbank.

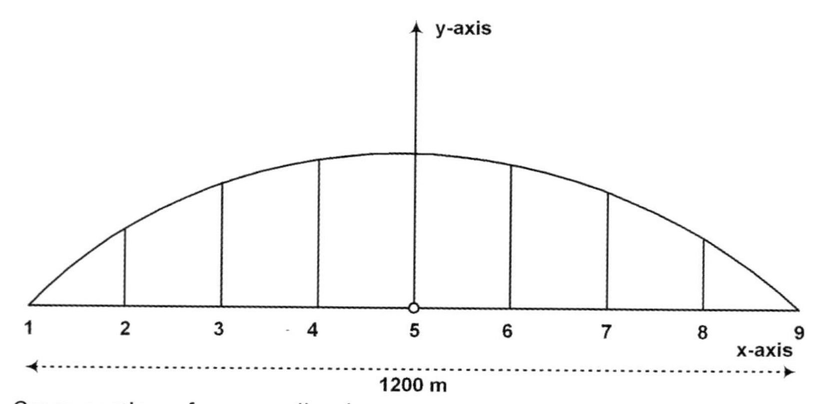

Figure 4. Cross section of a musselbank

In Figure 4 the vertical section of a mussel bank is presented. The width of the bank is 1200m and the height is 2m. The question is: which spatial resolution should be chosen? Suppose one would select the spatial resolution in such a way, that the shape of the section is clearly recognisable or distinguishable. The conclusion than would be: a selection of an appropriate representation from Table 3 cannot be made, because the terms "recognisable" or distinguishable" are not defined.

Table 3. Number of strips and spatial resolution

Number of strips n	Width of the sections ΔX.
2	600 m
4	300 m
8	150 m
17	75 m

Exercise 2: representation of the area of a section of a musselbank, with a given accuracy.

In order to be able to calculate the area of the musselbank in Figure 4, the cross section of the musselbank will be described by $y=b(a^2-x^2)$ where a and b are constants.
The exact value of the area O of the cross section is O = 4 ba^3 /3. If the mussel bank is represented by a number of strips of equal width Δ x, then the spatial resolution Δx is related to the number of points n and the width 2a by Δx=2a/(n-1). The area A_n of n strips can be expressed by $A_n = \Delta x(0{,}5\, f_0 + f_1 + f_2 + f_3 + \ldots + f_{n-1} + 0.5 f_n)$, the trapezium rule. The relative error Δ in the representation of the area follows from
$\sigma n = \{(O-A_n)/O\}100\%$. Exercise 2 is an example where a wide variety of spatial representations of the musselbank is obtained, each with a different Δx. Based on the required accuracy , the appropriate representation can be obtained from Table 4.

Table 4. Relative error in the estimate of the area

Number of points		
	Area m^2	Relative error (%)
3	1200	25.
5	1500	6.25
9	1575	6.25
18	1594	0.3

Lesson: for the selection of an appropriate spatial resolution a norm is required.

4. Examples with more than one variable

Introduction.

The focus in these examples is on the determination of the spatial resolution in case of flood damage assessment. Flood damages are assessed by various types of objects, like houses, farms, industries etc. For each of these objects a depth-damage relation is required. The flood damage per object is calculated form the depth damage curve and the local inundation depth. Uncertainty sources are: the level of the foundation of the building relative to the river and the water level in the river. A source of inherent uncertainty is the uncertainty in the damage due to differences in contents and type of building. The flood damage in the following examples is assessed on three levels of aggregation of the village: the individual houses, the "averaged" houses and the inundated area.

The variable Z, in chapter 2, is in this case the flood damage S. The flood damage is, after the introductory exercises 3 and 4, affected by more than one variable: the number of objects and the damage per object. The first variable relates to the data base in Figure 1, while the second variable is part of the systems description. Exercise 5 deals with flood damage assessment of a village. The aim of the exercises is to teach students that if one of the variables is uncertain, reduction of the uncertainty in one of the other variables does not lead significantly lead to a reduction of the uncertainty in S. This notion is used in exercise six to formulate a required degree of accuracy for the water level, given the uncertainty in the depth-damage relation. In the exercises 7 and 8 these concepts are used for system reduction and for the related spatial representation.

Exercise 3: assess the flood damage S by estimating the damages s_i of the individual houses.

The flood damage S follows in this case from: $S = \sum_{i=1}^{n} s_i$. Some disadvantages of this method are that it cannot be used for estimating flood damages in advance and the method is time consuming.

Exercise 4: estimate the flood damage S based on the number of houses n and a depth-damage curve.

The flood damage follows in this case from the product of the number of houses n and the average flood damage per house s : $S = ns$. If the uncertainty σ in the flood damages are uncorrelated, then the uncertainty in the flood damage S equals $\sigma_{tot} = \sigma\sqrt{n}$, while, if the uncertainty σ is fully correlated, the uncertainty in S is $\sigma_{tot} = \sigma n$. The real uncertainty will be between these two estimates. The approach of estimating the flood damage S in exercise six is acceptable if the uncertainty σ_{tot} in the damage S is acceptable.

Exercise 5: estimate the flood damage S based on the inundated area and a depth-damage curve.

In this example the inundated area is known. Furthermore the average value of the flood damage and standard deviation of the area per house including gardens, streets etc. are known. This information allows for an estimate of the number of houses n. The flood damage follows from *S=nsi*. If s and n are uncorrelated then the uncertainty in the flood damage can be expressed as:

$$\frac{\sigma^2_{Stot}}{S^2_{tot}} = \frac{\sigma^2_n}{n^2} + \frac{\sigma^2_{si}}{s^2_i} \qquad 1$$

Lesson: the uncertainty in the flood damage S is determined by the uncertainty in the damage σ_s and the uncertainty σ_n in the number of houses n. If for example the relative uncertainty in the flood damage is much larger than the uncertainty in the number of houses, then equation 1 shows that the uncertainty in the flood damage does not reduce if more accurate methods are used to determine the number of houses. A simple method by using the flooded area and the average area per house is then sufficient in this approach.

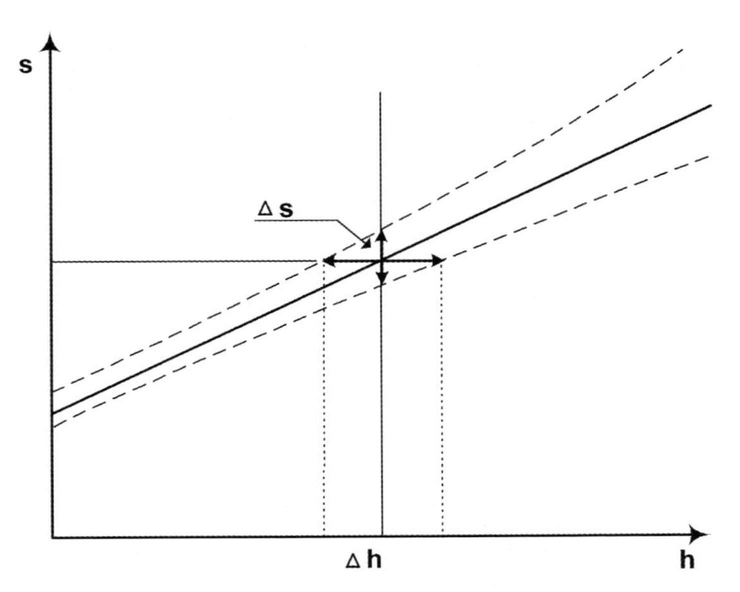

Exercise 6: Relation between uncertainty in flood damage and the accuracy of the water level.
Figure 5. Water level accuracy and flood damage uncertainty

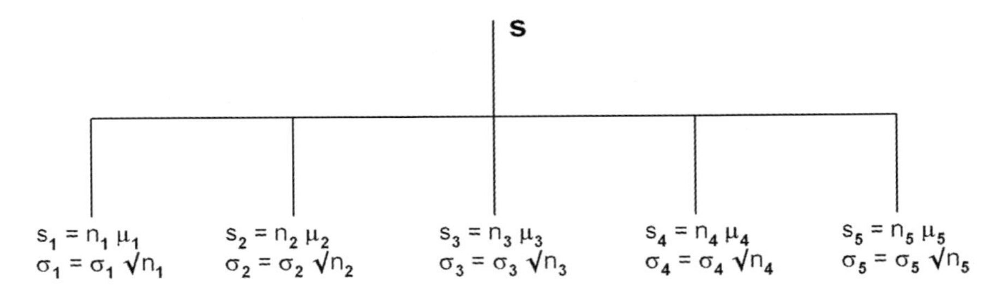

Figure 6. Damage tree

In Figure 5 the depth-damage relation is shown, including the uncertainty band as a function of the water depth h. Estimate the accuracy Δh for the water level h, given the uncertainty ΔS.

Lesson: if the uncertainty in one variable is known, then this information can be used to select an appropriate spatial representation for another variable.

Exercise 7: reduction of the number of contributing sources, based on their individual contribution.

The average flood damage S for n sources equals $S = \sum_{i=1}^{n} n_i s_i$. Suppose that there are ten cities and towns. The relative contribution of these sources to the flood damage ranges from one to ten. Do you have to take all cities and towns into account estimating S or could some cities be left out? Find the arguments.

Exercise 8: reduction of the number of contributing sources, based on their individual contribution to the absolute value and the uncertainty.

Following the previous exercise, the uncertainty in the contribution of each sources is 10%. However, from one of the intermediate sources the uncertainty is 40%.
It is assumed that the uncertainty in the damage between the cities are uncorrelated. The uncertainty in S then equals $\sigma^2_{Stot} = \sum_{i=1}^{n} \left[\sigma^2(n_i) s_i^2 + \sigma^2(s_i) n_i^2 \right]$. Do you have to take all cities an towns into account inassessing the flood damage S or could some cities be left out? Find the arguments.

Lesson: knowledge of the contribution of the individual sources can be used for reduction of the system and the related spatial representation.

4. Discussion

In this paper a contribution is made to the selection of the appropriate spatial resolution of integrated river and estuarine systems. This notion is also applied for system reduction. The tools presented in this paper are rather simple and can easily be extended towards a course on statistics. The problem for students is the application of these tools for system design. This becomes clear in the selection of an appropriate spatial resolution:
Representation of the cross section of a musselbank (1, 2)
Flood damage assessment per house, for n houses or a flooded area (3, 4, 5)
Selection of an appropriate level of uncertainty of a variable based on the uncertainty of the remaining variables is the topic of exercise 6. Students find this approach difficult because they are not used to formulate first the required uncertainty and use this information to select an appropriate model concept. Usually students are asked to apply just one model concept and they are not aware that many model concepts are possible.
System reduction in exercises 7 and 8 is difficult for students as they are tempted to include all elements in the system. This process of system reduction often takes place in the problem formulation phase, where reality is reduced to a model which is fit for analysis. Examples in flood damage assessment are the neglect of the spatial development in time or non monetary values. Both aspects are often important but uncertain contributions to flood damage.

In this paper no attention has been paid to the sensitivity analysis, uncertainty analysis etc. The reason is that these techniques are used in the validation phase of integrated systems. In this paper the focus is on the design of these systems, where the selection of the spatial and temporal resolution takes place.

The main attention has been focussed on the spatial resolution. It is expected that a similar approach can be followed for the temporal resolution.

The step towards more realistic exercises learns that the gap between the theory in the first exercises and reality is still very large. Maybe more attention should be paid to the didactical difficulties in the design phase of water systems.

Water quality monitoring for environment education and citizen empowerment: experiences from India

Contrôle de la qualité de l'eau pour l'éducation environnementale et la participation des citoyens: expériences de l'Inde.

Khadpekar V. S.
Centre for Environment Education (CEE), India]

Abstract
Water Quality Monitoring (WQM) can promote awareness and action for environmental quality and public health. It can reinforce children's learning at school, and empower citizens lacking access to safe drinking water to redress this lacuna. The paper reports experiences from two programmes in which CEE separately addressed these two contexts of WQM. In the Ganga Pollution Awareness Programme schoolchildren monitored water quality along the river and interpreted the results for their environmental implications. In Jalapareekshana NGOs in rural areas tested drinking water sources for potability, using WQM as a basis to forge partnerships of citizens and public health authorities. The methods adopted, constraints faced and efforts to overcome them are described. The strengths and limitations of WQM kits, a key component of both programmes, are discussed. Finally the systemic changes needed to maximise the benefits of such WQM activities are mooted.

Résumé
Le contrôle de la qualité de l'eau (CQE) peut promouvoir la connaissance et l'action sur la qualité de l'environnement et de la santé publique. Il peut contribuer à l'enseignement des enfants à l'école, et à l'empouvoirment des citoyens privés d'accès à l'eau potable, pour combler cette lacune. Cette étude fait état de deux programmes dans lesquels la CEE a abordé de manière séparée ces deux cas de CQE. Dans le programme Prise de Conscience de la Pollution du Gange (GPAP) les écoliers ont vérifié la qualité de l'eau le long de la rivière et ont interprété leurs résultats environnementales. Dans le programme Jalapareekshana, les ONG en zones rurales on testé les sources d'eau potable en utilisant le CQE comme la base d'un partenariat entre les citoyens et les autorités responsables de la santé publique. On y décrit les méthodes qui ont été adoptées, les contraintes auxquelles il a fallu faire face et les efforts consentis pour les vaincre. Les points forts et les limites des boîtes d'outils pour CQE, éléments - clés des deux programmes, sont abordés. Enfin les changements systémiques indispensables pour optimiser les effets bénéfiques des actions du CQE font l'objet de débats.

1. Experience 1: Ganga Pollution Awareness Programme (GPAP)

Fieldwork is an effective way to learn about the environment. Its context-specificity makes learning more intimate, creative and enduring than is possible through textbooks, lectures, or labwork. It adds substance to the theoretical framework supplied by the other methods.

Conventional school labwork in control conditions yields largely predictive and prescriptive results, which fieldwork precludes. Its findings, including errors, represent reality which must be explained by the experimenters, compelling them to progress from ritual to research. This reinforces learning.
Water quality monitoring (WQM) holds several possibilities for environmental education (EE). Water is a basic support of life and of the environment, on whose quality as a whole water quality has direct bearing. It can be studied within the average scholar's limits of theoretical knowledge and access to equipment. A school-based WQM programme is described below.

Project Background: In 1987 the Government of India, Ganga Project Directorate (GPD) asked CEE to develop a programme to involve science students in secondary and higher secondary schools along the Ganga, a major Indian river, in monitoring its water quality. The students would work in field (riverside) conditions, supervised by their teachers, with support by the education authorities of the states through which the Ganga flows, Uttar Pradesh (UP), Bihar and West Bengal (WB).

For CEE this was an opportunity to promote its philosophy of `learning by doing', bringing students to apply to a live situation at their doorsteps the science learnt in the classroom and to interpret the environmental implications of the findings. We also envisaged a network for exchanging the information thus collected from different locations, along nearly 2000 km of the river, to strengthen the programme through shared concern among discrete student groups working over a large geographical area.

Planning the Programme: GPD recommended taking up the programme in about 100 schools, restricting participation exclusively to students from either Std. IX or Std. XI. There was to be no mix of students from different levels, and in each school the entire class named had to participate. CEE, in consultation with the state education authorities, identified 50 schools in UP, 25 in Bihar and 25 in WB. Largely non-élite, these were distributed over rural as well as urban areas.

In deciding the activities to be included, CEE took into account the limitations of the students' science knowledge, the hazards of fieldwork on a major river, and the infrastructural limitations of the schools. Accordingly, nine physical and chemical parameters were shortlisted:

1. temperature
2. colour
3. odour
4. turbidity
5. hardness
6. pH
7. nonfilterable solids
8. presence of ammonia
9. dissolved oxygen (DO)

BOD and COD, though important, were excluded as their tests demanded rigor beyond the students' abilities. Riverine life-forms were excluded owing to the risks inherent in a group of young people working on a major river with a massive flow.

We initially thought that tests of volatile parameters like temperature and odour, and fixation of the DO sample, could be done on site and the rest, including DO measurement, in school labs. But many schools did not have labs. So a field kit with reagents and equipment to perform tests was developed. To avoid expensive meters and to highlight the connection of the tests to the science learnt at school, simple time-tested methods were adopted, with some modifications to allow them to be done in the field without risk of mishap.

The kit had many innovations. For example, the standard lab thermometer is graduated 0-100°C; its length, 40 cm, makes it fragile; its mercury column is difficult to read in outdoor conditions. The actual temperature range needed in the kit was 0-40°C. An alternative thermometer from the market, graduated 0-50°C, less than 20 cm long, and having red alcohol instead of mercury, was used. The volumes of reagent (1-10 ml) needed in the tests obviated the need for fragile and unwieldy burettes and pipettes. Plastic syringes of 2 and 10 ml, each labelled for use with a unique reagent to prevent contamination, were used instead. To the greatest extent possible, plastic replaced glassware. Thus the kit was compact and sturdy, low in accident risks, and suited for use by children in the field.

A teacher's manual accompanying the kit explained how to perform each test, its scientific basis, the chemical reactions involved, how to record the findings and interpret their environmental significance, and how to prepare reagents for replacement. It gave guidelines for field discipline and the precautions to be taken while doing the tests. Each school got a set of three kits (to ensure that they were actually used and that every student got an opportunity to do at least one test), reagent stock for a year's work, and the manual.

Operationalization: Workshops were held to orient education administrators and teachers to such work and to give them hands-on experience of the kits. The teachers, who had the necessary science knowledge, found it a novelty to perform experiments outside a lab. They saw that the kits were portable labs, and were eager to use them with students. Two adept teachers from each state were made resource persons for the respective states.

The next phase initiated students and the teachers who would conduct the programme, at workshops by CEE scientists and state resource persons. GPAP was launched in about 80 schools. Sampling rounds were done at approximately two-month intervals. Results were encouraging in some cases and disappointing in others. Student interest was uniformly high, but from teachers and school managements it was mixed.

Some teachers imaginatively went beyond minimal programme requirements, giving students a creative learning experience. Included by clerical error, a school with no science teaching programme produced top quality work. A slum school, with a mathematics teacher in charge of science, not only did competent monitoring but also mapped the stretch of river, recording pollution sources. One teacher voluntarily expanded the scope of activities, drew up a plan to set up a lab and study centre outside the school system where anyone interested could work, and tried to mobilise local funding to implement his plan.

Elsewhere, some teachers found co-curricular work, to be done outside regular school hours, an imposition and turned in disappointing results. In some cases teachers were enthusiastic but not the school managements. They participated only because of instructions from the education authorities.

CEE compiled the results for circulation to the participant schools to start what was hoped would evolve as a network for exchange of programme information among them. GPD showed the findings to the Prime Minister, whose appreciative comments on the work were communicated to all the participants.

Problems and solutions: The problems faced were academic and administrative. Academically, the emphasis on quantitative exactness typical of science education in schools led to some work of doubtful value. For example, the test for turbidity, done with the Secchi disk, required an average of two readings, yielding results to the second decimal place of a cm. Teachers were told that the disk was a crude device, and that precision exceeding the nearest round figure in cm was insignificant. They could not accept that a 'scientific' method could be 'crude', or grasp the idea of statistical significance.

Interpreting and correlating results presented other kinds of problem. The manual dealt with this, but few teachers paid it serious attention. Thus temperature, DO and turbidity were individually measured, but attempts to correlate them were few. The work was not done in a spirit of scientific inquiry.

Administratively, late start of activities, interruption by rain and flood, examination and vacation schedules, allowed only two or three sampling rounds in the year instead of the proposed six. Incidental expenses on transport and food for the students were also a problem. Gaps between rounds varied from school to school. At the end of the year students graduated out of the programme. Few serious efforts were made to initiate the new entrants who replaced them.

The academic problems were resolved, with moderate success, in dialogue with the teachers. The administrative problems were systemic, and could not be resolved within the time frame of the pilot phase of the programme.

Learnings: The enthusiasm of students and interested teachers was heartening, suggesting a potential for EE through scientific inquiry. Administratively, co-curricular work, especially in high schools, seems to have limited scope. The system at this level is oriented almost entirely to examinations and academic results. It has no room for activities which do not match this orientation. An administrator in UP, highlighting this dilemma, said that he appreciated the educational value of the programme, but could be sustain it only if it was implemented in all the state's schools and with an exam orientation, which would make it vulnerable to all the stifling influences of the system.

Such activities have a potential in alternative forums like science clubs and eco-clubs, which exist in some schools. The scheme of Socially Useful Productive Work (SUPW), prevalent in many school, also holds possibilities, including a modest budget for incidental expenses, which were not provided for under GPAP.

CEE's evaluation concluded that such alternatives must be explored to sustain the programme. Students should not enrol by academic level alone. Only those actually interested should be involved. Restricting participation by academic level is in fact a serious obstacle. That the science competence needed for such work obtains only at certain levels is debatable logic. Child-to-child learning can complement curriculum- and teacher-centred learning. If a mix of students from different levels works together, the juniors will not only be equipped for basic activities (such as recording temperature and odour), they can also learn from the seniors some advanced activities for which they are not formally trained. Such mixing can ensure that if some students graduate from the group, others will remain in it who are competent to continue the work and to orient newcomers. This will ensure continuity of the programme and, over time, help add dimensions and activities. The idea of a separate lab and study centre to bring together interested people from beyond the participating schools is worth considering for integrating strategic options. Other lessons from GPAP radically alter its basic objective of promoting WQM for EE, and take it from the realm of education to action. Such a possibility arose in the programme discussed next.

2. Experience 2: Jalapareekshana

GPAP attracted NGOs working with communities lacking access to safe drinking water. They felt that testing the potability of water would be a valuable activity to empower these communities for better control over their environmental health conditions. WWF-India held a training workshop for this, using the GPAP kit. One of its results was the modification of the kit to make it amenable to such use.

Project Background: A breakthrough came when, in 1988, the National Drinking Water Mission (NDWM) of the Government of India asked CEE to develop a kit-based programme for NGOs and community groups to test drinking water and report its quality regularly to public health authorities, enabling timely action to prevent waterborne disease. The Jalapareekshana programme resulted from NDWM's request. CEE was required to develop it around a kit designed by Defence Laboratory Jodhpur (DLJ) for use by soldiers to select suitable camping sites. It could test some parameters which the GPAP kit had excluded, like fluoride, iron and, most importantly, faecal contamination - a significant cause of water-related morbidity in India. It was an elaborate kit, with a battery-run incubator and proprietary reagents (the GPAP kit had no trade secrets).

Planning the programme: CEE discussed the details of the work envisaged with NDWM and CAPART (Council for Advancement of People's Action and Rural Technology). The latter suggested a many NGOs from different parts of India as potential partners. For operational convenience it was decided to initially focus on the southern and western parts of the country. Twenty NGOs were contacted to obtain a picture of the field conditions in which the project was to be implemented. Fifteen of them sent 27 representatives to an orientation workshop jointly organized by CEE, DLJ and Gujarat Jalseva Taleem Institute (GJTI), on water quality parameters and their health implications (conducted by GJTI), use of the kit (DLJ) and planning, reporting and support communication (CEE). At the end of five days the participants returned to the field, each carrying a kit and a work plan developed during the workshop.

Operationalization and learnings: The programme was implemented for a year as a pilot. With experienced and committed NGOs in charge, CEE did not get closely involved in monitoring the field activities, carried out in geoclimatic zones ranging from arid (less than 100 mm rain) to heavy rainfall (more than 2000 mm) areas. Most of the NGOs regularly reported their activities and experiences. The reports were educative in terms of how potent ideas can run up against obstacles which seem trivial but can make or break a programme.

The most common feedback was that the DLJ kit was not right for the kind of activity being done. In most areas just one or two parameters of drinking water quality were critical. The multi-parameter capabilities of the kit could not be optimally utilised. Secondly, maintaining the kit (which needed special skills) and replacing reagents (which had to be ordered by post) proved major problems. Thirdly, the monitoring activity often conflicted with other priorities of the NGOs (one problem in this regard pertained to kit design. In the original DLJ design the incubator was meant to be run by dry battery cells. However soldiers were found using the cells in transistor radios. So the incubator was redesigned to run off lead-acid batteries of jeeps. For many NGOs this meant decommissioning their solitary jeep for the 24 hours needed to complete the test for faecal contamination). Fourthly, many NGOs felt that just establishing the unpotability of water through tests was not enough. Simple methods of making it potable, manageable by the people, were essential. Such methods were not available for all the critical parameters, and in a crisis situation, the health authorities or other specialists could not be relied upon to rush into remote localities to help.

More wide-ranging and fundamental issues were also raised. Many experienced NGOs felt that, in water scarcity situations, people attach great importance to the availability of adequate quantities of water. Only if this basic need was fulfilled would it be meaningful to think about water quality. The idea of citizens doing potability tests and reporting their results to public health authorities was good in principle, but unless the authorities were responsive it was a futile exercise. With their trained scientists and laboratory infrastructure, they tended to be skeptical about reports from 'barefoot scientists' working in community groups. While some organizations did have sympathetic individuals, they could help only in their private capacities, not as members of the organizations.

3. Conclusions

The experiences reported above are from projects which, owing to systemic constraints, could not proceed beyond the pilot phase. The sponsoring governmental agencies lost interest in them after the individuals (in both cases senior bureaucrats) who inspired them had moved on. The work initiated still continues in a few places, in a disjointed and uncoordinated manner. The critical mass and momentum so necessary to make a significant difference are missing. On the positive side, the projects triggered some similar participatory experiments in various parts of India by agencies who were not partners. This is heartening. There is still room to hope that a coalition of such initiatives will emerge to gain the necessary momentum.

For CEE, GPAP was a precious learning experience in making EE locale-specific while relating it to the contents of formal science curricula. A decade after the pilot, requests are still received to provide inputs to WQM-based educational programmes. One important realization is that a kit is at best an aid. Its effectiveness depends entirely on how imaginatively it is used. Good educational aids do not automatically mean good education.

Specifically on the subject of kits, while multi-parameter kits are useful for EE, for practical functions such as testing water potability, single-parameter kits seem more appropriate. In any given locale, only a few parameters are critical, and the need is for reliable kits to measure them singly. Multi-testing capacities cannot be optimally used and are a wasted investment.

Finally, an important lesson from both the experiences is that educating educators (the GPAP case) and executives (the *Jalapareekshana* case) is essential if such experiments in education or empowerment are to make any meaningful difference in the way systems work. This emerges as the greatest challenge of a learning society in which knowledge and skills increasingly need to be harnessed for sharing between people committed to making the world a better place to live in.

The sustainability principle in water resources management requires education and training as missing link between problems and solutions (examples from South America)

Bundschuh J.
Darmstadt University of Technology, Institute of Geology and Palaeontology, Germany

Abstract
Water as central part for sustainable development is strongly interconnected with numerous social-economic, financial, environmental, scientific-technical and political-institutional aspects. To solve the related problems, which were studied for the eight largest cities of Argentine and which are similar to those found in all over South America, requires a comprehensive interdisciplinary and inter-institutional approach, where experts of all fields must work together considering all aspects in a well-balanced form. The absence of such experts and the additionally required environmental awareness of the entire society, make the installation of corresponding facilities for education, training and information as well as an international co-operation indispensable to solve or to prevent water related problems and water crises by establishing and executing the therefor necessary action plans. Some possibilities are presented for the case study of Argentina.

1. Introduction

Due to the high water availability in the investigated case studies of Argentine and many other developing countries, water was considered in the past as unlimited available, as free and hence of low value. The rapid increase of the water demand within the last decade requires that more and more water, especially more and more groundwater, must be exploited. Primary reasons are the increasing direct and indirect water consumption due to the growing urban population and the excessive water demand (from user and supplier side). On the other side decreases the water availability continuously. Reasons are the increasing contamination, the overexploitation and the decreasing recharge e.g. due to climate change, etc. Both processes are related to numerous social-economic, political-institutional, environmental and scientific-technical aspects which are interconnected in a very complex way.

The fundamental water related parameters and necessary data for their description were analysed for the largest cities of Argentine (Buenos Aires, Rosario, Cordoba, La Plata, Tucumán, Santa Fe, Salta, Mar del Plata). Aim was to investigate the social-economic, financial, environmental, scientific-technical and political-institutional water related problems, their reasons, parameters and data to describe them and their complex interrelations (Fig. 1).

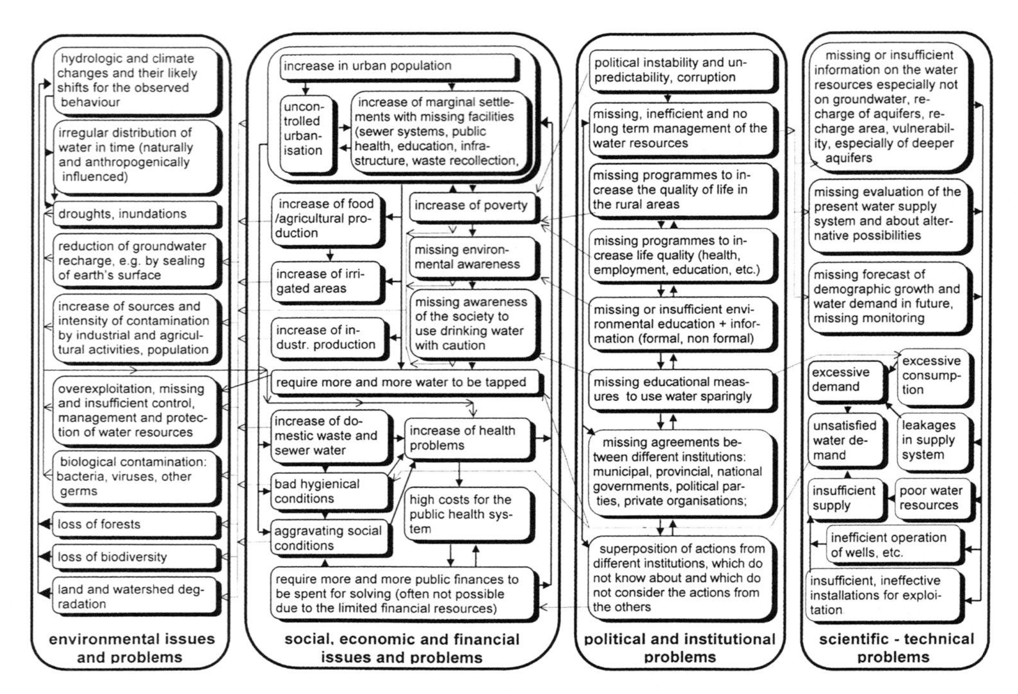

fig. 1: Interrelations between the social-economic, financial, environmental, scientific-technical and political-institutional water related problems and parameters

With different extent, all investigated cities show many of the worst symptoms of the regions underdevelopment: vast areas, great numbers of poor people, high concentrations of contamination and traffic congestion. Many of these problems affect the environment and the water resources within and outside of the urban areas (e.g. by increasing sources and intensities of contamination and by increasing quantities of exploited water which are due to increases of industrial production, of sewage and of domestic waste). The uncontrolled increase in population and the directly related decrease of living conditions cause an increase of environmental impacts which are directly related to health problems. They, as do so the aggravating social conditions require more and more public finances to be spent, a requirement that often can not be met due to the limited financial resources. Living conditions in urban areas are interrelated with political and institutional problems, e.g. missing programmes to increase the social conditions by corresponding programs due to: political instability and unpredictability, missing programmes to improve the quality of life in rural areas (to decrease the migration into the cities), missing programmes to increase living standard, missing agreements between different institutions, and missing interdisciplinary and inter-institutional co-operation. Missing or insufficient programs for environmental education and information (formal and non-formal) lead to an increase of environmental contamination and results in an excessive water consumption. Excessive water consumption together with leakages in the water supply system cause an excessive water demand. Excessive demand and insufficient supply result in an unsatisfied water demand.

2. Education and training as missing link between problems and solutions

Rising questions: What is necessary to solve the described problems? How can necessary action plans been formulated? What can the policy and decision makers do? How can they act in the appropriate form if they do not have the necessary background information and knowledge about available resources, contamination hazards, existing sources of contamination, the interrelation between the social-economic conditions? How can they formulate environmental laws without including important parts like the polluter pays principle as one of the most powerful tools to prevent or minimise pollution at its source? How can they establish programmes to diminish contamination, if they do not know about the priorities of the cases, if they do not know to which extent the particular contamination sources affect the water resources, if they do not know which contaminated site must be completely sanitated and which ones can be treated by cheaper measures, e.g. by sealing and if they do not know about effective and the most economic treatment methods e.g. of industrial effluents? How can they decide whether a polluting industrial plant must be closed immediately or if it is sufficient to oblige them special protection measures which they have to introduce within a certain time (proper treatment plants, other cleaner production methods, relocation to more appropriate places)? How can they plan the future demand for the capacity of the water supply, the capacity of water treatment plants and municipal waste deposits if they do not have demographic studies and prognoses of the demand in future? How can they know up to which volume of water they may exploit without overexploiting the aquifers if they do not have hydrogeological information? How can they know which measures are necessary to protect the water resources if they do not have corresponding studies? How can they optimise the water supply system if they are not metering the water, if they have no knowledge about performance indicators and benchmarking? How can they decide which are the most needed short term actions and which can remain for middle and long term consideration. How can they prevent or minimise disasters if the do not consider warnings from international warning systems? Why for example did they not consider with enough strength the information by the National Weather Service of the United States which successfully predicted the start and scope of the "El Niño" in 1998 half a year before it started?

This selection of some questions shows that water-related problems can only be solved by interdisciplinary and inter-institutional co-operative work of high-level professionals and a participation of the society. This requires that all professionals have an actualised high standard of fundamental and applied knowledge. Only so the natural scientists, the social scientists, the economists, the engineers, the technicians and the society can give consulting, information, pressure, public force, etc. to the decision makers (from industry, politics, administrations, public and private institutions and agencies, agronomy, etc.) to force them and to give them the necessary knowledge as background for action plans. A claim which is generally far from reality in developing countries. Hence corresponding applied training and educational programmes for professionals and an increase of the academic education quality for more qualitative professionals in future generations as well as environmental education for the whole society must be provided (Fig. 2). As example therefore, INASLA, a centre for education, training and information on water resources which was founded at the university of Salta in 1995 with the support of the German Academic Exchange Service (DAAD) and the German Agency for Technical Co-operation (GTZ), is presented.

Forming environmental awareness of the society: Missing awareness of the society, as such is fundamental for public participation in policy, is one of the main reasons for the increase of environmental contamination, the water misuse and the explosively increasing volumes of domestic waste, etc. On the other side all individuals of the society are directly concerned with water related problems. In the industrialised countries, growing awareness of environmental problems among the populations has forced government authorities to introduce laws and regulations protecting natural resources like air, soil and water against pollution. In developing countries such awareness and such pressure on the authorities is hardly to be found. Hence there exists no corresponding force on governmental authorities by the population.

Since the environmental problems appear, develop and expand with such a high velocity, it is not enough to offer the environmental education and training only to a limited part of the population like pupils or students or only to a limited part of the world like to the industrialised countries. The velocity of environmental education and training must be adapted to the velocity of the increase of environmental problems and must be offered to the whole world. Each individual of the society must receive an environmental responsibility.

The academic freedom of the universities gives them liberty in their teaching. Hence they can adapt themselves quickly to the changes and to the new requirements of the society. The presented problems indicate the urgent need of the universities to offer a service of environmental education and training for all adults by public information, co-operation with institutions, organisations and - very important - the mass media. Possibilities are: courses for adults (adapted to their level of knowledge), public speeches or discourses (e.g. in cultural centres, parish halls, etc.), articles and advertisements in newspapers and journals, flyers, exhibitions, programmes in television/radio, special channels in the television especially designed for education purposes, information by the internet (esp. for continued education and continuation training), edition and publication of education materials, etc.

INASLA is particularly involved in the following three programmes: (1.) Co-operation with the mass media for awareness forming, (2.) forming an awareness of the society to reduce the quantity of domestic waste, and (3.) forming an awareness of the society to use drinking water with caution.

Education, training and international co-operation for professionals: Missing environmental education and training and the actual deficiency of qualified engineers and scientists specialised in the field of the environment and water resources, as well as professionals

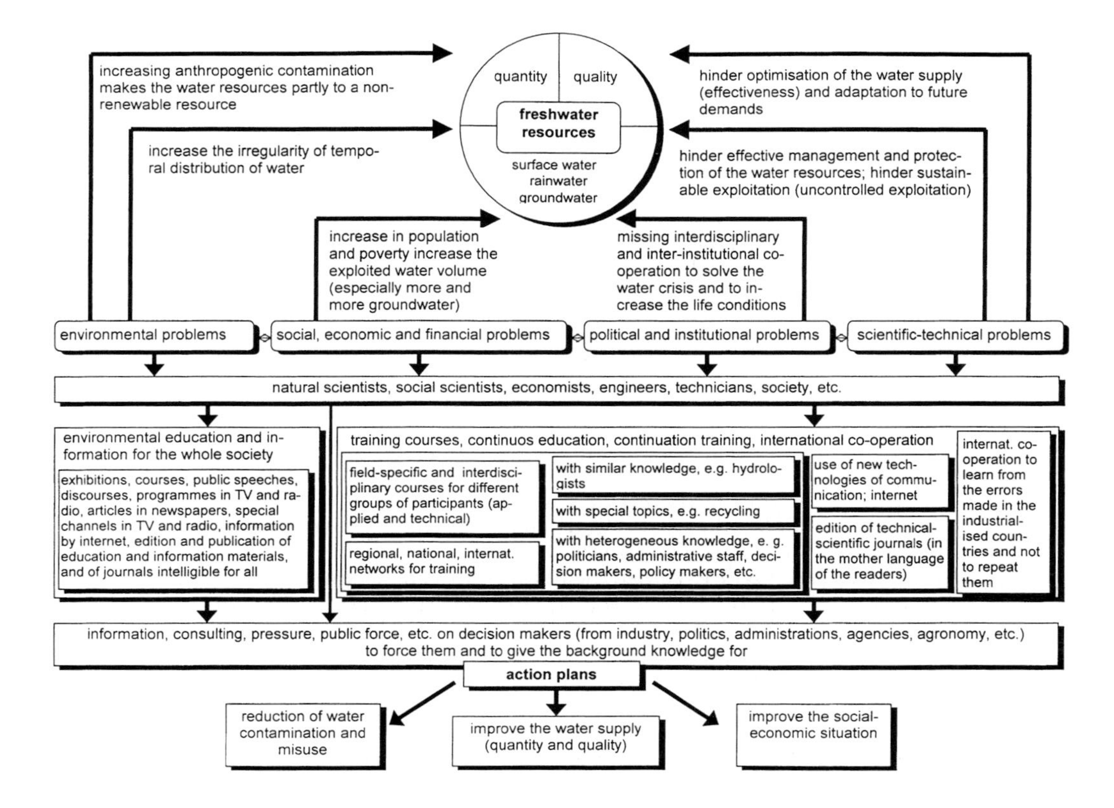

fig. 2: Connection between the water related problems from Fig. 1, their impacts on the water resources and the different possibilities of training, edaction and information as prerequisite for the formulation of action plans to solve them.

from other disciplines involved in the water-related problems, which can be recognised daily within the whole Latin America and the absolute necessity of such professionals, the installation of adequate training capabilities is of essential importance. The universities, as places of the highest level of knowledge and independent institutions (academic freedom) have an especial role and responsibility to offer a service of environmental education and training on a high level. Some possibilities which are provided by the case study INASLA are:

- To offer practically oriented postgraduate courses, continued education and continuation training. The contents are related especially to the problems and the requirements of South America.
- To create regional, national and international networks for environmental education and training.
- To edit scientific/technical journals (in the mother language of the reader).
- To introduce new technologies of communication like the internet.
- To expand the bilateral exchange of students, postgraduates and scientists with other argentine, Latin-American or German universities and institutions.
- To advise about possibilities of international scientific contacts, exchange programmes and sources of international information and to initiate these.
- To carry out scientific investigations and to supervise diploma works and Ph.D. thesis from local and German students and to do research work. Co-operation in projects with other universities and in interdisciplinary projects.
- To organise seminars, congresses and other scientific or technical meetings.
- To sign contracts with governments, organisms, institutions related to the water resources and the environment to exchange actions, experiences and to develop together tasks (e.g. continued education and continuation training).
- International co-operation in environment and public health (learning from the errors of the industrialised countries). INASLA is involved en the execution of the following applications: (1.) deposition of waste, reduction of waste (municipal, industrial), (2.) treatment of existing contaminated sites (abandoned industrial sites), (3.) treatment of contaminated sites of actual industries and (4.) environmental legislation.

3. Conclusions

Water-related problems can only be solved by interdisciplinary and inter-institutional co-operative work of high-level professionals and a participation of the society. This requires that all professionals have an actualised high standard of fundamental and applied knowledge. Only so the natural scientists, the social scientists, the economists, the engineers, the technicians and the society can give consulting, information, pressure, public force, etc. to the decision makers (from industry, politics, public and private administrations and services, agencies, agronomy, etc.) to force them and to give them the necessary knowledge as background for decisions and actions.

This requires the installation of adequate training capabilities for regular students, employees, staff, professionals, etc. which are involved in one or the other form in water related problems, and a environmental education for the society. Only so the base for multidisciplinary action plans to reduce contamination, to improve the water supply and to improve the social-economic conditions can be provided which allow a sustainable use of the water resources and which guarantees a sustainable future of the world society.

Well-trained professionals and a well-informed society are primordial for developing solutions and the therefore necessary decisions and actions in water resources management which must be defined always:

- Considering the interdisciplinary connections between water supply, hydrologic, environmental social, economic and health aspects
- Understanding that the protection of water resources and environment do not neces-

sarily mean additional costs for the public system but that it means controversy prevention of high costs and decreasing life quality which will rise in future if protection is neglected.

- Being aware of the need to form multi-municipal, multi-provincial, multi-national and international commissions
- Realising that development and improvement of the whole education system is a principal possibility, to decrease the environmental impacts and to improve health problems (hence decreasing costs for the public health), to lower the rate of birth as well as the urbanisation process.
- Including scientists and technical experts as consultants in political decisions.

Les mathématiques et l'ingénieur de l'eau au XXIème siècle

Mathematics and the water engineer at the 21th century

Thirriot C.

Institut National Polytechnique de Toulouse, France

Résumé
De nombreux codes (MATLAB, MATHEMATIKA, MACTCAD) permettent déjà de réaliser des calculs analytiques par ordinateur. Cette possibilité va aller s'accélérant. En conséquence l'effort de formation en mathématiques des ingénieurs hydrologues ou hydrauliciens risque de s'amenuiser. Que conserver de la culture mathématique et comment la conserver ?

Abstract
Various computer programms as MATLAB, MATHEMATIKA, MATCAD are already used to perform analytical calculations. This ability will go increasing. Consequently the effort to train in mathematics for hydrologists and hydrulician engineers would be decreasing. What to preserve from the present mathematical culture and how to do it ?

1. Introduction

Que restera-t-il de l'exercice mathématique après que MATLAB, MATHEMATIKA et MATCAD aient engendré des enfants encore plus puissants qui dispensent de l'effort de réflexion individuelle. Sera-ce pour les mathématiques de notre adolescence comme pour le latin au début du siècle. Les gens cultivés des siècles passés pensaient ne pas pouvoir vivre intelligemment sans latin qui explique la génèse de la langue. Eh bien dans les parkings, les pressings, les supermarkets, la vie continue ! Dans vingt ans (ou moins) les automates mathématiques rendront inutile le calcul analytique.

Cette communication voudrait être une incitation à la réflexion critique communautaire pour savoir s'il faut laisser aller les choses ou bien protéger de manière prospective une culture mathématique avant qu'elle n'aille rejoindre les trésors Mayas.

Cette discussion pourra se faire en trois étapes :
- l'état des lieux
- l'évolution imminente
- le fonds de commerce de l'approche mathématique pour l'ingénieur hydrologue.

2. L'état des lieux

L'analyse portera essentiellement sur le cas français. Des récents événements dans les lycées, il ressort que les programmes vont être allégés après avoir été bien bouleversés pendant les trente dernières années après l'émergence trop hâtive des mathématiques dites modernes à la maternelle.

Dans les Ecoles d'Ingénieurs, depuis plusieurs années, on saute allègrement les démonstrations sans que les élèves se soient plaints douloureusement. Les cours magistraux sont mieux orientés vers les applications. Mais le système pédagogique est différent de la méthode américaine que je découvrais avec étonnement et admiration dans le livre de CRANDALL: "Engineering Analysis" en 1958, où l'on partait de l'exemple du système hydraulique oscillant galerie-cheminée d'équilibre pour aboutir aux espaces vectoriels et aux systèmes à n équations différentielles.
Y aura-t-il un Euro des mathématiques pour un enrichissement mutuel de l'enseignement de l'algèbre et de la géométrie dans l'Europe sans frontières ?

Les français, très souvent contents d'eux-mêmes disent que toutes les nations nous envient notre système d'enseignement des mathématiques mais sommes-nous si sûrs de l'excellence de notre pédagogie ? Combien d'élèves au sortir du baccalauréat ont-ils ressenti concrètement que les mathématiques étaient un instrument privilégié de la raison critique, pas uniquement fondé sur la déduction, armature outrancièrement apparente, mais que l'intuition y puisait aussi ses racines.

En hydrologie, tout commence par l'arithmétique des bilans qui devrait être un réflexe issu de la culture mathématique. Et ce bilan ne doit pas être un religion sectaire du chiffre ou plutôt des chiffres qui sortent en rangs serrés de l'ordinateur mais une vision réaliste et sereine qui acclimate l'incertain. Beau paradoxe de la mathématique considérée comme science de la certitude et rend lucide et rationnelle l'appréhension de l'incertitude. Elle intervient dans la théorie des erreurs, indiquant déjà au premier degré par les différentielles les relations qui peuvent s'accrocher entre les incertitudes sur les divers paramètres caractéristiques d'un modèle de crue ou d'infiltration.

Plus ambitieuse, elle passe ensuite du recensement statistique à la probabilité pour proposer des silhouettes de fonction de répartition. Bien sûr, l'hydrologue n'est pas le seul à faire ce saut de l'échantillon particulier à la description synthétique d'une population. Cela avait déjà commencé avec d'Alembert dans les sciences qui deviendront plus tard sociales. Mais l'hydrologue d'aujourd'hui qui a biberonné peut être contre son gré, dans sa jeunesse à l'algèbre moderne sait que ces fonctions de répartition sont un cadre flou de la réalité. Et il en retiendra que plusieurs modèles sont possibles à condition qu'ils rendent compte convenablement des situations asymptotiques. Il n'est que de rappeler le cas de la formule de Gumbel décrivant rationnellement les états extrêmes qui sont par exemple les débits de crue maximaux annuels.

La probabilité cumulée F(Q) d'une valeur de débit inférieure ou égale à Q étant

$$F(Q) = e^{-e^{-\frac{Q-Q_o}{Q^*}}}$$

le comportement asymptotique pour les très grandes valeurs a la simplicité de l'évolution exponentielle

$$F(Q) = 1- e^{-\frac{Q-Q_o}{Q^*}}$$

Et ce comportement exponentiel fait rejaillir toute une culture !

Exemple naïf et banal de l'intégration des mathématiques dans la manière de penser et de réagir des gens des générations actuelles mais qui risque de se dissiper peu à peu dans une civilisation presse-bouton informatique.

Mais il ne s'agit pas d'opposer de manière stérile boulimie informatique et calcul à la plume. Le calcul littéral présente, c'est sûr, le très grand avantage de la généralité mais seulement lorsqu'on sait visualiser à l'intérieur de la conscience l'image de la fonction manipulée. Un cosinus, une exponentielle provoquent une image bien nette dans le cerveau mais déjà les fonctions de Bessel qui pour beaucoup de problèmes radiocirculaires sont l'avatar des sinus et cosinus trouvés en géométrie rectangulaire, ces fonctions de Bessel paraissent bien ésotériques au profane et ne prendront vraiment corps qu'avec la tabulation numérique et encore mieux la représentation graphique autrefois compulsée dans les livres et maintenant instantanément exhibée sur l'écran du microordinateur.

Mais qui se soucie encore des fonctions de Bessel ou plus largement des fonctions spéciales qui dans ma jeunesse étaient l'argument d'un cours annuel sur les Méthodes Mathématiques de la Physique ?

Et pourtant elles surgissent dans des problèmes qui préoccupent l'ingénieur de l'eau comme la pérégrination des pesticides dans le sol vers la nappe phréatique ou dans l'expansion d'une crue vers les zones humides latérales à la rivière.

A titre d'exemple reprenons brièvement les équations dans le cas de l'entraînement uniquement vertical des pesticides en solution.

2.1. Les conditions hydrodynamiques

On considère que l'écoulement de l'eau est permanent, saturé ou non saturé, mais uniforme dans un milieu poreux supposé homogène. Le seul paramètre est alors la vitesse moyenne de pore u.

2.2. Les échanges avec la paroi et l'effet du puits

Les pesticides sont en concentration C dans l'eau en écoulement. Ils sont en concentration équivalente C' dans l'eau figée, dans les zones mortes et sur les parois des grains de sol. C'est cette partie C' qui va être active en particulier dans la lutte microbienne.
Il y a échange entre la concentration mobile C et la concentration immobile C' suivant une cinétique linéaire. Le flux de matière de la partie mobile vers la partie immobile est supposé représenté par :

$$\phi 1 = h(C - C') \quad (1)$$

Le flux de retrait de matière (assimilation par les plantes, les microbes, etc.) est pris en compte par un effet de puits caractérisé par un flux f2 tel que :

$$\phi_2 = - h'C' \quad (2)$$

La variation de la population C' peut finalement être représentée par l'équation de continuité :

$$\alpha \frac{\partial C'}{\partial t} = \phi_1 + \phi_2 \quad (3)$$

le coefficient a assurant la correspondance entre la concentration adsorbée S comptée habituellement par rapport à la masse de sol sec et la concentration C' homogène à la concentration C comptée dans l'eau vive.

2.3. L'équation de transport

L'eau vive transporte les pesticides comme un soluté. Sans échange et sans diffusion longitudinale chimique ou hydrodynamique, le bilan de matière conduit à l'équation d'évolution de la concentration C :

$$\frac{\partial C}{\partial t} + u\frac{\partial C}{\partial c} = 0 \quad (4)$$

En tenant compte de l'échange avec les zones de stockage inertes, il vient :

$$\frac{\partial C}{\partial t} + u\frac{\partial C}{\partial c} = -\phi 1 = -h\,(C - C') \quad (5)$$

2.4. Le modèle retenu

Il est composé des deux équations gérant les évolutions conjointes des concentrations C et C'

$$\frac{\partial C}{\partial t} + u\frac{\partial C}{\partial x} + h(C - C') = 0 \quad (6)$$

$$a\frac{\partial C'}{\partial t} + h\,(C' - C) + h'\,C' = 0 \quad (7)$$

La coordonnée spatiale x est comptée suivant la verticale descendante. Le milieu est supposé semi-infini, sans borne en profondeur.

Jusqu'ici, dira-t-on, il n'y a pas de crainte à avoir pour les générations futures. L'apprentissage universitaire les conduira vers la réflexion critique de la modélisation mathématique. Cela nous paraît évident aujourd'hui et pourtant ce n'est pas assuré définitivement. Et pourtant l'élaboration de ce système d'équations avec ses hypothèses soupesées, est un exercice de réflexion critique sur la réalité qu'il serait dommage de perdre. D'un point de vue pédagogique, déjà avec ces équations, par l'analyse inspectionelle (comme l'appelait BIRKHOF) qui est de l'analyse dimensionnelle sur équations, on peut obtenir des informations intéressantes : h et h' sont des inverses de temps qu'on pourrait dire de relaxation ; a coefficient de stockage temporaire va intervenir pour modifier les temps caractéristiques du phénomène ; après choix d'une longueur caractéristique x*, u permet d'introduire une autre échelle de temps celui de l'advection x*/u. Et déjà sans avoir aborde la résolution ni numérique, ni littérale on peut avoir par l'approche mathématique des idées certes encore floues, sur les différentes échelles de temps et ainsi percevoir la possibilité de comportements asymptotiques.

Toujours sur cet exemple, nous insisterons sur un autre profit de mathématiques l'introduction d'un espace de représentation dual et virtuel par l'utilisation de la transformation de Laplace, exemple de ce qu'on appelait le calcul symbolique et que l'on retrouve bien sûr avec l'usage de la Transformée de Fourier et de bien d'autres transformées moins usitées comme la transformée de Legendre, la transformée d'Hermite, etc ... C'est sous une forme non dite comme un visage du calcul variationnel. On prend une image qui est la pondération intégrale de la fonction inconnue sur l'intervalle de variation d'une des variables du problème. Avec la transformation de Laplace, la pondération est généralement faite par rapport au temps

$$LC(x, t) = \hat{C}(x, p) = \int_0^{\infty} e^{-pt} C(x, t)dt$$

C'est en somme, une méthode de moments où le coefficient de pondération e-pt est évanescent pour des valeurs très grande de la variable temps. Avec des valeurs discrètes pi de la variable duale, pour une fonction C(x, t) **connue**, on aurait une description synthétique. Avec une infinité de valeurs p, on a autant d'information dans les fonctions images que dans les fonctions originales.

Avec un peu d'habitude, on lit les traits caractéristiques du phénomène dans les solutions images.

$$\hat{C}(x, p) = \hat{C}_o(p)\exp\left[-\left(p+h-\frac{h^2}{ap+h+h'}\right)\frac{x}{u}\right]$$

$$\hat{C}' = \hat{C}_o(p)\frac{h}{ap+h+h'}\exp\left[-\left(p+h-\frac{h^2}{ap+h+h'}\right)\frac{x}{u}\right]$$

Sans trop allonger, nous nous livrerons à cette exégèse à l'occasion du retour à l'original.

L'image s'écrit en séparant les exposants

$$\hat{C}(x,p) = \hat{C}_o(p)e^{\frac{-hx}{u}}e^{\frac{-px}{u}}e^{\frac{h^2x}{u(\alpha p+h+h')}}$$

Le premier terme exponentiel est un simple coefficient d'amortissement en fonction de la profondeur atteinte, le deuxième caractérise un retard, le troisième est le terme spécifique de l'image.

Faire le changement de variable p' = p + a, revient à multiplier l'original par $e^{-\alpha t}$. On peut

faire apparaître la translation sur p caractérisée par $a = \frac{h+h'}{a}$ et il reste à trouver l'original de

$\frac{1}{p}e^{\frac{b}{p}}$ avec $b = h^2 x / u\alpha$

Soit L-1 l'opérateur de retour à l'original, d'après ANGOT (1972), il vient :

$$L^{-1}\left(\frac{1}{p}e^{\frac{b}{p}}\right) = I_O(2\sqrt{bt'})$$

avec Io fonction de Bessel modifiée de première espèce.
Donc

$e^{\frac{b}{p}} = p\, L[I_O(2\sqrt{bt'}]$

La règle sur les dérivées permet d'écrire que :

$L^{-1}\left(e^{\frac{b}{p}}\right) = I_O(0)\ \delta(t') +$

la variable temporelle étant $t' = t - \frac{x}{u}$.

Toutes ces opérations en cascade ne sont pas bien difficiles mais elles demandent cependant un certain entraînement. Et c'est cet entraînement qui risque de s'étioler si on n'y prend garde devant la facilité d'emploi de l'ordinateur qui lui aussi implicitement sans l'afficher joue avec ces dualités de fonctions et d'espace à l'insu de l'utilisateur. Mais bien sûr, un moteur d'automobile fait bien de la thermodynamique appliquée à l'insu du conducteur!

La méditation devant les résultats analytiques amène aussi des lueurs sur le phénomène physique.

Expression de la concentration dans l'eau vive

Avec une condition d'impulsion à la surface à l'instant initial, l'image $\hat{C}_o(p)$ est l'intensité de la fonction de Dirac, soit M_o / ρuP, l'expression de la concentration est donc

$$C(x, t) = 0 \quad \text{si} \quad t < \frac{x}{u}$$

$$C(x,t) = \frac{M_0}{ruP} e^{\frac{-h}{u}x} e^{-\frac{h+h'}{a}\left(t-\frac{x}{u}\right)} \left[I_0(0)\delta\left(t-\frac{x}{u}\right) + \frac{d}{dt}\left(I_0\left(2\sqrt{\frac{h^2x}{au}\left(t-\frac{x}{u}\right)}\right)\right)\right]$$

L'utilisation de développements asymptotiques va nous permettre de concrétiser l'évolution de C. Mais d'ores et déjà, on voit apparaître (ou confirmer) certains comportements :

a) le phénomène de convection mis en évidence par le retard x/u

b) l'atténuation de la concentration en cours de percolation

c) la propagation atténuée de la fonction impulsion créée par l'épandage quasi instantané à la surface du sol, l'amortissement de la fonction de Dirac étant d'autant plus rapide que le coefficient d'échange h est élevé

d) l'effet de puits ou de disparition des pesticides marqué par l'amenuisement exponentiel $e^{-\frac{h'}{a}\left(t-\frac{x}{u}\right)}$

e) la possibilité de deux extremums de la concentration : le premier est lié au déplacement suivant l'advection de la fonction de Dirac, le deuxième provient du relargage des produits provisoirement adsorbés et correspond au maximum induit par la dérivée de la fonction de Bessel $I_o(x,t)$

Expression de la concentration fixée

Dans le cas d'un épandage brusque on peut encore écrire la solution image :

$$\hat{C}'(x,p) = \frac{M_o}{ruP} e^{-hx/u} e^{-px/u} \left[\frac{h}{ap+h+h'} e^{\frac{h^2}{(ap+h+h')u}} \right]$$

L'expression de C'(x,t) est donc finalement :

$$C'(x,t) = 0 \quad \text{si} \quad t < \frac{x}{u}$$

$$C'(x, t) = \frac{M_o h}{ruPa} e^{-\frac{hx}{u}} e^{-\frac{hx}{u}\left[t-\frac{x}{u}\right]} I_o\left[2\sqrt{\frac{h^2x}{au}}\, t - \frac{x}{u}\right] \quad \text{si } t \geq \frac{x}{u}$$

Là encore la conjugaison d'une exponentielle décroissante en fonction du temps et la fonction de Bessel croissante va amener l'apparition d'un maximum de la concentration adsorbée puis l'évanescence.

Bien sûr, pour que ces résultats soient plus "parlants", il faudra passer finalement au calcul numérique ... ou au développement en série de la fonction de Bessel. Alors dira t-on pourquoi ne pas entreprendre immédiatement le calcul numérique sur les équations aux dérivées partielles de base ? Plusieurs raisons militent pour ce détour analytique :

- la solution littérale obtenue est plus générale (une fois acceptées les hypothèses simplificatrices de base). Elle donne une vision panoramique de l'influence des divers paramètres

- elle est plus économique (si les tâtonnements et l'abolition des fautes de calcul analytique ne sont pas trop laborieux)
- elle élimine les artefacts parfois perfides introduits par la discrétisation du champ des variables en vue du calcul numérique
- elle met bien en évidence les situations asymptotiques et cela nous parait une réelle supériorité sur le calcul numérique.

3. L'évolution imminente

Les codes de calcul commerciaux prolifèrent. Il est déjà loin le temps où chaque laboratoire universitaire bricolait son logiciel pour résoudre un système linéaire, calculer une régression multiple. Comme le disait le Président de Matsushita : "Il n'existe que deux crimes majeurs dans le monde industriel : ce sont 1°) inventer quelque chose qui a déjà été inventée ailleurs et 2°) payer pour quelque chose qui peut être obtenue gratuitement".

Eh bien dans le domaine du calcul, ce réflexe est quasi unanimement acquis. Il y a bien encore quelques maniaques de la programmation qui s'obstinent à bichonner un algorithme personnel mais dans la plupart des cas, on fait confiance aux kits mathématiques qui existent à prix dérisoires dans le commerce. Pour 1 000 F. de 1999 vous trouvez des ensembles de calcul d'écoulement dont la mise au point par une société d'informatique aurait été facturée 400 000 F fin des années soixante, soit en monnaie constante un rapport de 1 à 1000 environ.

Bien que les performances actuelles des logiciels mathématiques soient déjà époustouflantes, et tout à fait suffisantes pour bon nombre d'utilisateurs scientifiques ou techniciens, les choses vont vraisemblablement aller encore vers un progrès accéléré. En particulier la technique des réseaux néuronaux risquent d'envahir des domaines débordant de l'automatique, l'identification et la régulation réactive. Déjà plusieurs essais ont été tentés pour représenter la fonction de transfert non linéaire intervenant dans la genèse et la propagation des crues.

4. L'avenir des mathématiques individuelles

L'ingénieur de l'eau qui dispose déjà de programmes de réseaux (construction, optimisation, fonctionnement, extension) de suivi biochimique de la qualité des rivières, de gestion de réservoirs en série et en parallèle, d'analyse régionale pluviométrique par la méthode des composantes principales, d'optimisation linéaire ou dynamique, pourquoi voudrait-il que ses futurs jeunes collègues fassent l'effort d'emmagasiner en mémoire tout l'échafaudage mathématique nécessaire avant de produire du neuf ? Pour plusieurs raisons dont la première essentielle est que les mathématiques sont une culture objective qui permet à l'ingénieur une lecture critique de la vie. Bien sûr, il ne faut pas en rester à la religion du symbole ou du chiffre mais Galilée ne disait-il pas il y a quelques quatre siècles : "La philosophie est écrite dans ce très grand livre qui se tient constamment devant les yeux (je veux dire l'univers) mais elle ne peut se saisir si tout d'abord on ne saisit point de la langue et si on ignore les caractères dans lesquels elle est écrite. Cette philosophie est écrite en langue mathématique sans laquelle on ne fait qu'errer dans un labyrinthe obscur".

Et il y a encore des efforts à faire pour rendre aux jeunes esprits scientifiques plus comestibles et plus efficaces des notions émergées au cours du dernier demi-siècle telles que les mathématiques du chaos, les fractals, les équations stochastiques, le calcul des possibilités et l'algèbre floue, linéarité et non linéarités et l'effet de celles-ci sur les valeurs propres, les méthodes asymptotiques (merci à Monsieur POINCARE, à KRYLOV et BOGOLUGOF), les équations aux différences servantes de l'analyse numérique, les chaînes de Markov généralisées dans le temps et l'espace.

Cette litanie à la Prévert de tous les aspects mathématiques qui peuvent intervenir dans la mise en forme et l'étude des problèmes d'eau n'est peut être pas très convaincante pour le praticien. Mais encore une fois nous prendrons un exemple particulier pour montrer l'intérêt de l'analyse littérale au service du traitement numérique. Il s'agit de la fameuse méthode dite de Muskingum. Ce grand hydraulicien est en fait une rivière des Etats-Unis qui servit de support à la mise au point d'un modèle de propagation de crue fondé sur la discrétisation (déjà en 1935 !) de l'équation désormais classique de l'onde cinématique

$$\frac{\partial Q}{st} + c\,\frac{\partial Q}{sx} = q$$

q débit d'apport latéral qui peut être nul si la genèse de la crue se passe haut en amont. Admettons cette hypothèse pour gagner du temps. La discrétisation dans le plan des variables espace-temps conduit au modèle de calcul et même de prévision
$Q(x + \Delta x, t + \Delta t) = aQ(x, t) + b\,Q(x + \Delta x, t) + cQ(x, t + \Delta t)$
L'analyse fine littérale permet de montrer la métamorphose d'un modèle mathématique continu de type hyperbolique de propagation pure en un modèle discret introduisant de la diffusion qui n'est pas choquante lorsqu'on observe la réalité. Cet avatar fameux ne pourrait pas être perçu sans l'humus culturel mathématique fondée à la fois sur la connaissance des systèmes hyperboliques et sur l'expertise des équations aux différences (où l'on retrouvera utilement les valeurs propres, signe sensible de la stabilité ou de l'émoussement des algorithmes confiés à l'ordinateur). Ainsi un artefact de discrétisation d'équations aux dérivées partielles a conduit à un nouveau modèle hydrologique de nature tout à fait différente de celle du modèle continu de base. Ce serait dommage que les futurs hydrologues perdent la possibilité de flairer et de dénoncer de tels artefacts (afin de mieux s'en servir éventuellement !)

5. Conclusion

Ce plaidoyer pour le maintien d'une culture mathématique chez l'ingénieur de l'eau est le fait d'un hydraulicien qui n'est pas mathématicien mais qui reconnaît avoir eu la chance au cours de son bivouac universitaire d'être initié aux ressources des mathématiques concrètes.

Les mathématiques sont d'abord une école de pensée. On en voit d'abord la rigueur mais on en découvre vite la richesse pour l'intuition quand la provision en mémoire est assez diversifiée et digérée.

Les nouvelles générations des hydrologues devront poursuivre le syncrétisme des mathématiques du déterminisme et des mathématiques de l'aléatoire. Ils devront aussi confronter les mathématiques de la complexité (dont il ne laisseront pas le monopole aux futurs ordinateurs) et les mathématiques de la simplicité, dilemme vu encore sous la comparaison de l'approche mathématique des systèmes et de la mathématique des phénomènes universels (propagation, diffusion et échanges) où l'on entend la résonance de la thermodynamique.

La solution idéale serait de faire progresser l'approfondissement en culture mathématique et technique de calcul littéral au fur et à mesure de l'augmentation des performances des codes de calcul littéral sur ordinateur. Mais cela semble une course poursuite très vite essoufflante. Plus réaliste est le souci d'une coopération entre l'ordinateur et le calculateur humain qui existe déjà depuis l'avènement de l'informatique. L'ordinateur soulage des tâches ancillaires de tabulation de nouvelles fonctions universelles. Il assure la vérification des résultats analytiques obtenus manuellement. Il permettra la consultation aisée des correspondance entre fonction original et fonction image. Il pourra être un pédagogue de plus en plus convivial tel un maître virtuel de plus en plus socratique aussi docte que livre mais plus vivant.

Mais sera-t-il réellement capable de créer comme le laisserait entendre et attendre la venue des réseaux néuronaux ?

Enfin dans ce monde où le loisir devient une entité économique, on ne pourra pas échapper à la question "mathématiques pratiques" et mathématiques ludiques" (analogue aux mots-croisés ou aux mathématiques de Pierre de Fermat).
Finalement l'ingénieur de l'eau, dans un monde à la Huxley (pas forcément meilleur), devra-t-il s'en remettre aux quelques survivants des mathématiques universitaires ou bien gardera-t-il l'autonomie (relative) dont certains de ses collègues ingénieurs du XX° siècle ont usé et parfois abusé ?

Environmental Education for Sustainable Development: UNESCO Chair for Water Problems

Tskhai A. A., Koshelev K. B., Mikhailidi I. M., Verevkine M. N., Yevtushenko A. T.
Altai State Technical University, Russia

Abstract
It is considered problems of interdisciplinary education in the university on the example of the decision support system "Hydro-manager" training with respect to water quality improvement in river basin.

Résumé
On a étudié les problèmes d'éducation interdisciplinaire à l'université sur un exemple d'enseignement au système du support des décisions "Hydro-manager" pour l'amélioration de la qualité de l'eau.

1. Introduction

During a transition period in Russia to the sustainable development model the problem of environmental education and training acquires a priority meaning.

Recently the Law "About environmental protection" and "Water code of RF" were adopted in Russia. Many of obsolete administrative procedures are replaced by new economic methods in environmental management like as developed market economies [5]. Now the enterprise's payments for environmental pollution form the regional ecological foundations (briefly, REF) resources used for support of water protective activity. The special state environmental committees had appeared in the regions. But serious changes in environmental protection are not observed. Why it is so?

The reason is not in absence of interested institutions and organizations. The waterquality protection in region is the object of activity for regional departments of Ministry of Natural Resources, State Environmental Committee, Federal Service of Hydrometeorology and Environmental Monitoring and many other water users. The reason is not in absence of Federal and Regional Water Resources Programmes. Many such documents appear in Russia nowadays. But the effect of these documents realization is small for the present, because the important aspect is not taken into account.

The reason is in absence of comprehensive and holistic approach to complex problem realization and in ineffective training of specialists of different profiles who are able to make decisions in interdisciplinary fields.

The top priority task for every water resources goal-related program is creation of effective management mechanism including as a minimum two main elements:

- the detailed order of interaction between water users themselves and with control organizations according to normative base in Russia (the procedure of economic regulators definition for water users behavior, the analysis and forecasting of consequences for different variants of "ecological bankruptcy" of insolvent enterprises, the special measures for reduction of prior contaminants concentrations in control river sections, the account of expenses for "long-term" projects investment , etc) ;
- training system of specialists with appropriate knowledge and skills for working in new conditions.

The UNESCO Chair in the Altai State Technical University, the first established in Russia out off Moscow and St.Petersburg, is working on the problem of water quality improvement in river basin taking into account both important factors.

2. Regional water quality management of river basin on the base of geoinformation system "hydro-manager" application

Scientific newness of elaborated approach [1-4] is
- *in the methodology field:* in the accounting interaction of the changes of economic, ecological and technological factors while investigating nature-technical complexes;
- *in the field of methods:* in application of the mathematical modeling and informatics with using as the input data the standard information of State services for assessment, forecasting and management of water quality in basin scale on the basis of Law of the transfer economic period;
- *in the mathematical modeling field:* in creating the complex of original simulation and optimization models which let, in particular, to assess ecological consequences of management decisions' implementations in the river basin.
- *in the technologies field:* in the elaboration of information technology of monitoring and regional water quality management for river basin.

Implementation of these results will become the real basis for the water-users bodies and water-protection institutions interaction. Now the results of such interaction are negligible, because its order isn't defined.

This integrated approach is oriented to resources conservation and sustainability and contains administrative and economic actions combination of exposure to water users. The scheme of regional water quality management in basin scale on the base of original information system "Hydro-manager" for decision making support is proposed (see Fig.1).

This information system is elaborated for optimization of water protective activity in basin scale, assessment and forecasting of ecological consequences under management decisions. "Hydro-manager" includes problem-oriented geoinformation system of water quality of river basin within boundaries of administrative region.

The information block of system consists of three parts: text data base; map-graphical data base; modeling data base. The text data base is intended for gathering, keeping and use of monitoring information of the river basin and data on the corresponding water-technical complex. There are the observed hydrological and hydrochemical data, the information on intensity and composition of the point and diffuse sources of the watershed pollution in this data base.

The map-graphical data base is realized on the example of the real watershed and is intended for holding the map-diagrams of administrative and river basin boundaries, the situations of towns and settlements, the posts of hydrological and water quality observations, the morphometry of rivers and so on.

Water quality is estimated and forecasted by means of mathematical modeling methods. The information of the text data base is used. Our water quality model uses standard data of the Russian State Service for Observations as input information. The dependencies of the model coefficients on hydrologic characteristics were calibrated by means of the monitoring data on the same river.

Water quality model simulates the river spatial distribution for the values of twenty contaminants: (1) BOD, (2) oxygen deficit, (3) suspended matter, (4) COD, (5) ammonia, (6) nitrite, (7) nitrate, (8) synthetical surface-active matter, (9) oil-pollution, (10) phenol, (11) hexachloran, (12) chlorine, (13) sulphate, (14) magnesium, (15) calcium,
(16) lindane, (17) iron, (18) copper, (19) lead, (20) phosphate for 18 periods of the year.

For construction of production functions the original optimization model of water user behavior is realized. This model is related to discrete and nonlinear type. The input data for this model is standard information about enterprise economic activity. The enterprise - water user pays for its water pollution in accordance with differentiated rates and receives the financial support for its water protective actions. The payments for permissible pollution refer to the manufactured product cost, for the beyond-permissible pollution - of the enterprise's profit. The optimality criterion is maximum enterprise net profit corresponding on level of subsidization of its water protective activity from all sources.

The economic norms of "Hydro-manager" - management parameters are coefficients of ecological situations (changing the interval of payment for pollution rates. The concrete values are determined by local authority); "return-parts" of payments for water pollution from REF on enterprises environment protection activity.

Further, the final best variant of water protection actions sets in basin scale is found with help of comparison criterion of the annual pollution minimum for few control river sections (for example, near sections where urban works intakes).

The scheme of regional water quality management in basin scale is formulated in accordance with modern Russian normative basis and taken into account real opportunities of control organizations in the field of water supply and protection.

In the research part of the project the simulation and optimization methods were used (including production functions methodology), apparatus of differential and algebra equations with applying traditional methods of their solving etc.

The software is made using Object Pascal and the developments of Inprise Delphi 3/4 for Windows 95/98/NT. DSS "Hydro-manager" is created as a project in ArcView GIS 3.0a of general version using Spatial Analyst and Dialog Designer. For input data standard information on water economy of investigated river basin is used.

3. Students educational programs on environment management on the base of information technologies

In the Altai State Technical University are developed to introduce given mechanism of natural resources usage into practice. Besides of traditional training programs on natural sciences, economics and management, students study such special programs on nature resources usage as: nature resources usage development retrospective in the world, establishment quality standards for environment; ecological control and expertise; economic assessment of natural resources; ecological damage assessment; assessment of cost effectiveness; planning and mechanism of nature resources usage and environment protection; ecological information and methods of its analysis under environment impact assessment.

In detail during lectures, practical seminars and laboratory works students study modern conception of geoinformation systems (GIS), presentation of spatial and attributive data in GIS, introduction in ARC/INFO GIS, introduction in ArcView GIS, combined use of GIS products with other software.
After such preparation students are ready for GIS "Hydro-manager" training.

4. Conclusion

The UNESCO Chair of Altai experience showed that introduction of modern methods of water resources usage in the regional needs uniform complex approach, including at the same time realizing one:

- development of scientific basis;
- goal-oriented training of the qualified staff;
- impact to the public opinion, integration of social powers;
- forming the necessary decision makers position.

References

Tskhai, A.A. Monitoring and Water Quality Management in River Basin: Models and Information Systems. - Barnaul: Altai Publ.-House, 1995. - 174 p. (in Russian)
Tskhai, A.A., O.P.Dorotshenkov. Model Approach and Application of Water Quality Management for Urban Areas. // Environmental Research Forum: Integrated Water Management in Urban Areas. Int. Symp., Sweden, Lund. 1996, Vol. 3-4, pp.197-207.
Tskhai, A.A., K.B.Koshelev, M.A.Leites. GIS "Hydro-manager" and its Application to Water Quality Management in the Ob River Basin. // Application of Geographic Information Systems in Hydrology and Water Resources Management - IAHS Publ., 1996, N 235, pp. 365-371.
Tskhai, A.A Model for Water Quality Management in River Basin // Water Resources, 1997, Vol.24, N5, p.699-706 (in Russian).
Tskhai, A.A. Information Systems for Environmental Management. Textbook. Part 1 - Barnaul: ASTU Publ.-House, 1997 - 140 p. (in Russian)

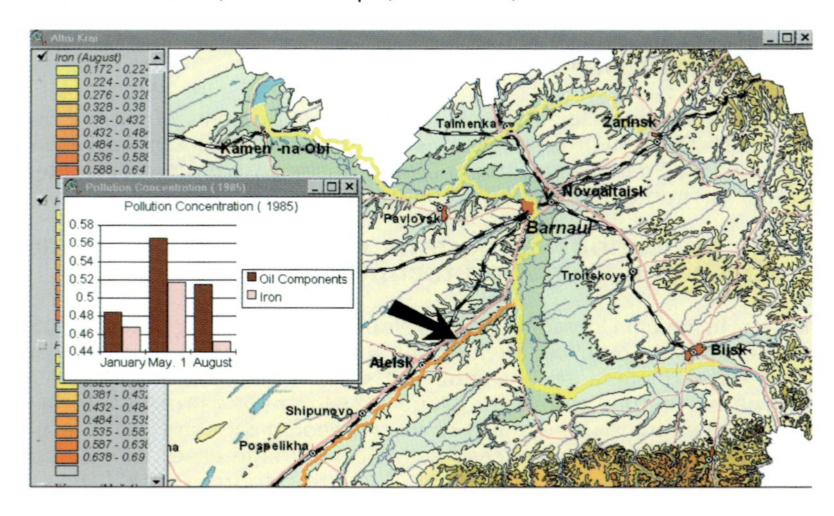

Figure 1. Interface of GIS "Hydro-manager" for water quality control in river basin

Capacity building in water sector in Africa. Proposal for a Nile river network

Développement des compétences dans le secteur hydraulique en Afrique

LUIJENDIJK J.[1], BOERIU P.[1], SAAD M. B.[2]
[1] IHE-Delft, The Netherlands
[2] Hydraulics Research Institute, Egypt

Abstract
Capacity building is now widely regarded as the key strategy in realising sustainable water sector development. Considerable attention has been given in the last few years to the importance of professional manpower development in the water sector as an essential tool for sustainable water resources development in the African Continent. Despite this continuing international attention, skilled manpower development in the African water sector is lagging considerably behind the rest of the world.

The need for capacity building at regional level is also strongly felt by TECCONILE. In their "Nile River Action Plan" it is stated that TECCONILE's actions will include a comprehensive training needs assessment and the design and implementation of a basin wide, long term human resources development programme as well as short term priority training programmes in specific fields and institutions.

The paper elaborates on an initiative to create the NILE RIVER NETWORK, a capacity building network for sharing ideas, best practices and lessons learned among the Nilotic states. The HRI regional training centre is seen as one of the focal points in such a network, specialised in Hydraulic Engineering.

Résumé
Le développement des ressources humaines et des compétences est maintenant largement considéré comme le facteur stratégique clé pour la mise en place d'un développement soutenable du secteur hydraulique. Une attention considérable a été accordée ces dernières années au développement des compétences humaines dans le secteur hydraulique en Afrique. Même avec cette attention internationale continue, le développement de ces compétences en Afrique est encore situé considérablement en deçà du niveau mondial.
La nécessité du développement des compétences au niveau régional est fortement reconnue par le TECCONILE. Dans le Plan d'Action récemment développé pour le Nil, il est précisé que les actions de TECCONILE devront inclure, parmi autres, l'évaluation détaillée des besoins en enseignement et formation, l'implémentation d'un programme de longue durée pour le développement régional des ressources humaines, et l'organisation des programmes d'instruction de courte durée dans des sphères d'intérêt spécifique.

Cet article présente l'initiative de créer un réseau international pour promouvoir la coopération interdisciplinaire et le développement des compétences au sein des états limitrophes du Nil – The Nile River Network. Le centre régional de formation HRI est considéré comme l'un des piliers principaux pour l'élaboration d'un tel réseau, spécialisé en ingénierie hydraulique.

1. Introduction

The concept of capacity building is now widely defined and accepted and considered as one of the main keys to improve water management in future. All conferences held on this topic recognized that people and institutions are the pillars of sustainable water sector development. In 1996 the 2nd UNDP Symposium on Water Sector Capacity Building was held at IHE in Delft, The Netherlands. The main objective was to exchange experiences and to identify constraints in the process of capacity building. One of the conclusions was that "networking" and the sharing of information and skills are key instruments in development and coordination of the knowledge and institutional resource base.

The UN's 1994 medium population projection forecasts that by the middle of the coming century, 4.4 out of the 10 billion people will live in 58 countries experiencing either water scarcity or water stress. Food consumption in developing countries will then be more than doubled, with an annual increase of food demand for North Africa of 3.8%. In addition municipal and industrial water demands will grow even faster, while at the same time urban and industrial pollution are limiting the use of available sources. This all will put an enormous pressure on the further development of land and water resources.

An evaluation of what happened in the last decades, revealed that despite of all the billions of investments, many projects have failed to deliver the expected benefits, particularly those that should have provided water services to individual users (irrigation, drinking water supply). Most failures can be attributed to systematic deficiences in the "institutions" responsible for the planning, design and management of water resources. Moreover the need was expressed for an integrated approach for the planning and management of water resources and its use. The answer to all this: "Capacity Building". A popular word nowadays, but how to turn it to real action?

"Capacity Building" is now widely regarded as the key strategy in ensuring sustainable water sector development. The concept of Capacity Building was defined in the Delft Declaration during the 1991 UNDP symposium by its three components:

- The creation of an enabling environment with appropriate policy and legal frame works;
- Institutional development, including community participation;
- Human resources development and the strengthening of managerial systems.

These three elements are equally important. If one of these elements is deficient, then capacity building activities in the other fields are due to be ineffective. All three mentioned elements of capacity building rely heavily on adequate staff. Staff to develop policy and legal frameworks. Staff to man and develop the necessary institutions. Hence, training and education is due to be an essential element of all capacity building efforts, irrespective of which of the three constituting elements has priority in a given situation. In all capacity building efforts it's the people that make the difference between success and failure.

Education is a long-term investment in the future and aims at transferring knowledge, insight, methodologies, skills and, importantly, new attitudes. Training implies shorter contact time and aims at solving more specific problems and development of more specialized and directly applicable skills. To create structural local capacity it is required to focus on typical capacity builders such as educational and training establishments, research institutions, NGO's and management consultants. Typically, such support will concentrate on staff development, curricular reform in order to deliver graduates with skills appropriate for the local requirements, and overall management assistance to run new regular and short courses and carry out problem-oriented research. In many developing countries academic institutes offer insufficiently specialized programmes and are as such a limiting factor in the performance of the sector. Therefore academic institutes need support to review their curricula in order to (i) focus better on local technical and multi-disciplinary problems, (ii) to introduce more interactive and stimulating teaching methodologies, (iii) involve practitioners in the teaching program, and (iv) shift from teaching factual knowledge to developing skills and attitudes. Similarly, the capacity of research establishments and universities to design and carry out research and development to solve the problems of the future, appropriate for the local conditions, needs to be strengthened.

Training is an important tool of management and should as such be an integral part of any management strategy. In such management strategy a balance is made between investments in infrastructure and procedures on the one hand and human resources development needs on the other hand. Changing political, physical and management environments, new technologies and growing awareness to use water effectively and efficiently will put a continuous demand on training and development of professional staff and consequently on the development of a training capacity.

Although there are numerous differences between countries, regions and economic zones, a number of *trends* can be discerned in higher education in recent years (see also UNESCO, 1995):

- quantitative expansion, which can hardly keep pace with the need for specialists in the water sector;
- diversification in response to the dynamics of society and speed of developments;
- individualization in response to industrialization and urbanization;
- globalization and internationalization of water resources and environmental problems and establishment of international networks for research;
- enhanced inter-disciplinarity of problems asking for integrated approaches.

A number of constraints have restrained universities in developing countries to meet these challenges. Again there are large regional and national difference in the extent to which the *constraints* mentioned in the following list apply to particular situations. But in general the constraints met by universities to educational reform include (see also UNESCO, 1995):

- Reduced public funding for higher education;
- Deficient quality of staff and facilities;
- Poor quality of students;
- Limited access to higher education;
- Inferior quality of education and conventionalism;
- Insufficient training and research capacity;
- Institutional factors;
- Cultural factors;
- Insufficient incorporation of appropriate indigenous approaches.

Interventions aiming at improving educational quality need to address the trends and constraints discussed in the previous sections. Where the improvement of educational quality needs to be sustained over a long period, it is best anchored in a local educational establishment. Strategies need to be developed to identify and equip this local institution for such task.

Staff development
The success in the development of training capability ultimately hinges on the trainers themselves. A well designed training programme, good materials, state of the art facilities and long-term co-operation agreements between networking institutions are of no value whatsoever without the educational and management capabilities, i.e. warm bodies in place to implement, evaluate and renew the training programmes.

The availability, development and particularly also the retention of highly motivated staff is the key to sustainable, high quality education and training. Where the situation, as is often the case, does not allow for arranging attractive primary packages (high salaries), a well-balanced secondary package must be designed, requiring a creative human resources management system, including a.o. for non-material incentive packages (sabattical leave, advanced training, staff exchange programmes, attractive working environment, interesting work, possibilities for research incl. assistance in publication of results, etc..), career development incl. outplacement perspectives, merit-based promotions.

Training of trainers in subject-specific areas, training methodology, curriculum development and training management is essential and a continuing requirement in view of both sector and educational development. Opportunities must be arranged for each trainer, in keeping with personal abilities and sectoral demands.

2. IHE-models for direct and indirect capacity building

Over the past ten years IHE has gained valuable experience in a substantial number of projects aimed at improving the quality of sectoral professionals in developing countries and countries in transition. These projects all had their specific objectives, and range from the implementation of a generic course offered at the IHE in Delft to an intensive multi-year co-operation with a University involving intensive staff development and modernization of graduate and postgraduate course programmes and related facilities (Savenije and Ramsundersingh, 1995). The analysis of this practical experience allows to distinguish a number of models for the improvement of educational quality, where each model answers a particular training requirement.

The models can be categorized first in two main groups, i.e. those aimed at resolving an immediate training need in the water sector, and those aimed at the development of indigenous water sector training capability. The first group is basically a short-term input by an external training institution aimed at the immediate removal of specific human resource constraints in the operational sectors, whilst the second one aims at developing the educational sector itself (e.g. the training of trainers to execute a certain training course), which in turn, and after some time, will remove the constraints in the operational sectors.

In considering the application of these two types of interventions, the first group provides an adequate solution in situations where the particular water sector training need is unique, non-repetitive and immediate and where local educational capability for the particular course(s) is absent and/or the course subject is not considered a priority area of development. The second group of interventions typically applies to situations where the sector demands a substantial number of trainees over a prolonged period of time, and where educational capacity is, at least potentially present, and where the subject area is considered a priority This provides a basis and rationale for the development of permanent, indigenous training capability (See figure 1).

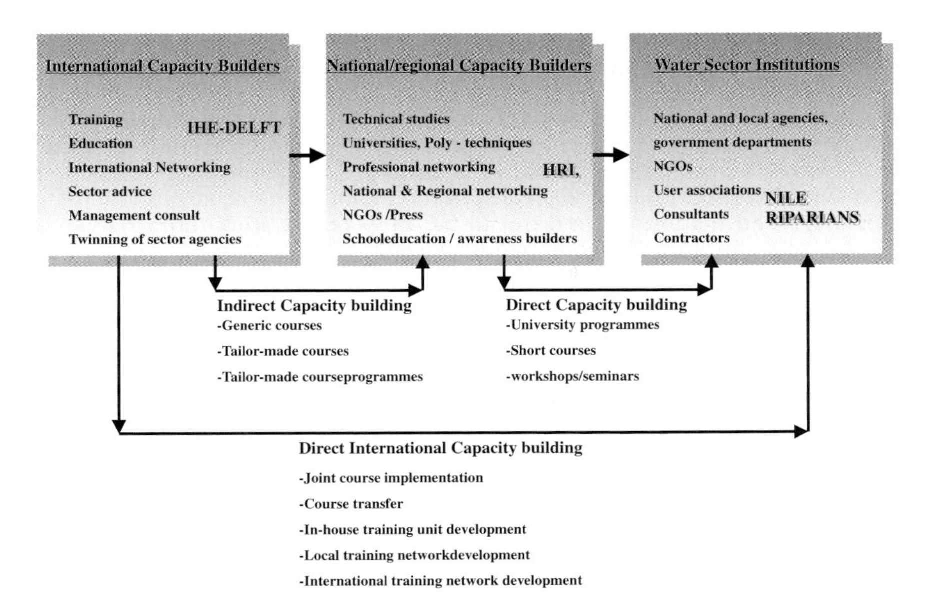

Figure 1. A model for operational direct and indirect capacity building activities

2. The HRI Regional training project (1995-1999)

As a special example of one of the capacity building models for developing water sector training capability, the establishment of a regional training node for the Nile region can be mentioned. The Hydraulics Research Institute (HRI) as one out of the 11 institutions under the National Water Research Centre (NWRC) of the Ministry of Public Works and Water Resources, Egypt. Since 1996 HRI has been organizing a three-month course and various short courses on Hydraulic Engineering in River Basins for especially professionals from the Nilotic States. These courses are supported professionally, by the International Institute for Infrastructural, Hydraulic and Environmental Engineering (IHE), Delft, The Netherlands and is co-financed by the Dutch Government.

The main objective of the HRI project was to _strengthen the capacity of the Nilotic states to develop their research infrastructure required for a sound and proper management of the Nile and other river basins_. The first phase of the project (1995 - 1999) focussed on the practical training of professionals active in the various (semi-) governmental authorities, institutes and projects dealing with water resources development in the region. Activities were tuned both to support HRI to become a regional training institution and to support the region to train professionals from the region and to provide a number of fellowships to enable participants to attend the regular courses at HRI.

On the very last day of this millennium the First Phase of the HRI-project in Egypt will officially expire. The training center became in a short time a *regional training center* well-known not only within African Nilotic countries but also further in the African Continent: it is a well-equipped regional training center with up-to-date educational facilities, a computer network, laboratory facilities and a library. A core of 15 well-trained senior lecturers are gradually taken over the implementation of the courses while a managerial team of the training center is ready to take over full responsibility for course management and marketing. During the first four years some 75 professionals from 8 Nilotic countries (including 9 female professionals) participated successfully in the 3-months regular training course. Another 120 participants joined the various short courses and workshops organized by HRI in cooperation with IHE-Delft.

The extension of the scope of the HRI Regional Training Center towards the whole African continent including the Arab Region was highly appreciated in both in Egypt and in the other Nilotic states. Training groups of young professionals coming from countries with different political background and culture fostered friendly relationships between them and actually turned out to be an excellent long-term investment. They learned not only how to design hydraulic works but also how to solve professional problems in a technical and environmentally sustainable manner. The above experiences seem worth elaborating further the concept and approach of the first phase through facilitating an even more intensive cooperation between professionals from the Nile region.

3. Main reasons for an extension of the HRI initiative

The justification for an extension of the HRI project is based on a mix of three main arguments. In the first place there is the general need for capacity building in the water sector in Africa. There is no doubt that there is no continent like Africa that suffers more from the lack of capacity to carry out research, studies and to establish policies for the region_s development in general and for the water sector in particular. This is evidenced by statements made by various regional and international organizations. Training needs are also felt strongly by TECCONILE: The Technical Co-operation Committee for the Promotion of the Development and Environmental Protection of the Nile Basin. Specific training needs were formulated also by the participants to the workshop "Hydraulic Engineering Training Initiative for Africa" held in Cairo in March 1996, within the framework of the HRI - Regional Training" capacity building project.

A second reason for an extension of the HRI project is the need for regional cooperation in the Nile river basin. There is now a strong call for coordination and integration of water resources development in international river basins. Co-operation in the region can save enormous costs both in manpower and in costly equipment. Co-operation at technical level has proved instrumental to improve the joint management of shared water resources and to mitigate and even prevent international conflicts. Informal and formal contacts at technical level often formed the basis for agreements and effective international cooperation. It is essential that sufficient capacity is available at national level to assess and evaluate possible impacts of interventions. Knowledge is power. It is in the interest of both powerful and less powerful riparian countries that they _level the playing field_ and so share the same data making it possible to analyse and interpret the consequences of certain measures on both national and river basin scale. Information and knowledge sharing is a critical issue in the development of transboundary rivers since it will build confidence and partnership among riparian states.

A third reason to continue with support to the HRI regional training is to secure the sustainability of the HRI-Phase-I project. More support is needed to adapt and develop new educational and training activities of regional interest; to support the course management in marketing the courses and sponsoring; and to offer fellowships to participants and trainers. A steady state condition has certainly not yet been reached at present. In order to keep the momentum and to create a more sustainable environment for further strengthening of the regional cooperation a longer input and more commitment is required. This should come from both the Government of Egypt and the other Nilotic countries with support from bilateral donors (like the Netherlands Government) as well as possibly from international donor organisations such as the World Bank and UNDP.

4. The new initiative: Nile River Network

The main concept on which the new project will be based is to create an environment in which professionals from the water sector sharing the same river basin would have the possibility to exchange ideas, their best practices and lessons learned. Such an environment can best be established by fostering a network through which education, training, research and exchange of information for and by professionals can take place.

To this end the NILE RIVER NETWORK (NRN) is to be established as a regional programme to build and strengthen capacity in the Nile riparian countries for an environmentally sound development and management of the Nile River Basin. The network is to be an open network of national and regional capacity building institutions and professional organisations active in education, training and research. The ownership of the Nile River Network is an essential issue to be seriously addressed. It is envisaged that the network will be owned by a group of information and knowledge suppliers, capacity builders and end-users, all under the umbrella of a regional organisation. Such an organisation could for instance be TECCONILE or any other regional organisation.

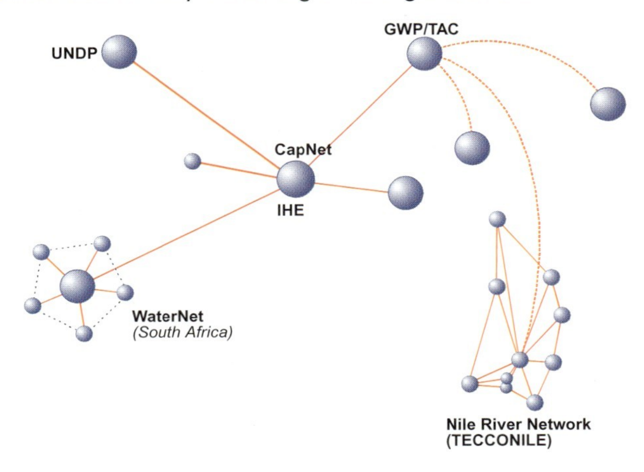

Figure 2. Global network for capacity building in the Water Sector

The initiative of establishing a network is fitting well within the mission of the Global Water Partnership (GWP) to create an International Network for Water Sector Capacity Building (CAPNET) as a multi-country, multi-donor undertaking. CAPNET, that will start its activities this year, will assist countries and regions in their efforts to build up their capacity for integrated water resources management (IWRM) through human resources development. CAPNET, that will have its secretariat at IHE in Delft, will provide technical support, stimulate networking and promote applied research. The first network that is to be linked to the CAPNET is □WaterNet□, a regional programme and network in Southern Africa. The Nile River Network could well become the next one (see figure 2). Through CAPNET the Nile River Network can become part of a global network and take advantage of the experiences, information and lessons learned of other linking networks.

The Nile River Network is a relatively ambitious project aiming at the establishment of regional centers connecting organizations active in education, training and research on different aspects of Water Resources Management in the Nile River Basin. The project aims at the establishment of one of the existing centers as a regional _focal point_ and as a node of the network. The project will build further on the success of the HRI-regional training center to become the center for Hydraulic Engineering for the Nile basin. The current scope of the project is limited to river processes, hydrographic surveys, monitoring, forecasting and simulation of river flows, engineering works such as river regulating works, hydropower plants, navigation works, etc. The HRI-phase II project will be based on the existing structure as developed during the first phase of the project, but will, apart from the training, extend its activities to research and knowledge dissemination. It could well serve as a pilot for similar activities in other fields of IWRM.

5. Capacity building components, instruments and activities

The Nile River Network has four distinct components through which it addresses its capacity building mission: (see figure 3): **Education & training,** including curricula development for educational programmes and organisation of professional training courses; **Research,** focussed on the execution of joint regional research; The set-up of a regional **Information Center** and data base; and a **Network secretariat**, for building the structure of the Nile River Network.

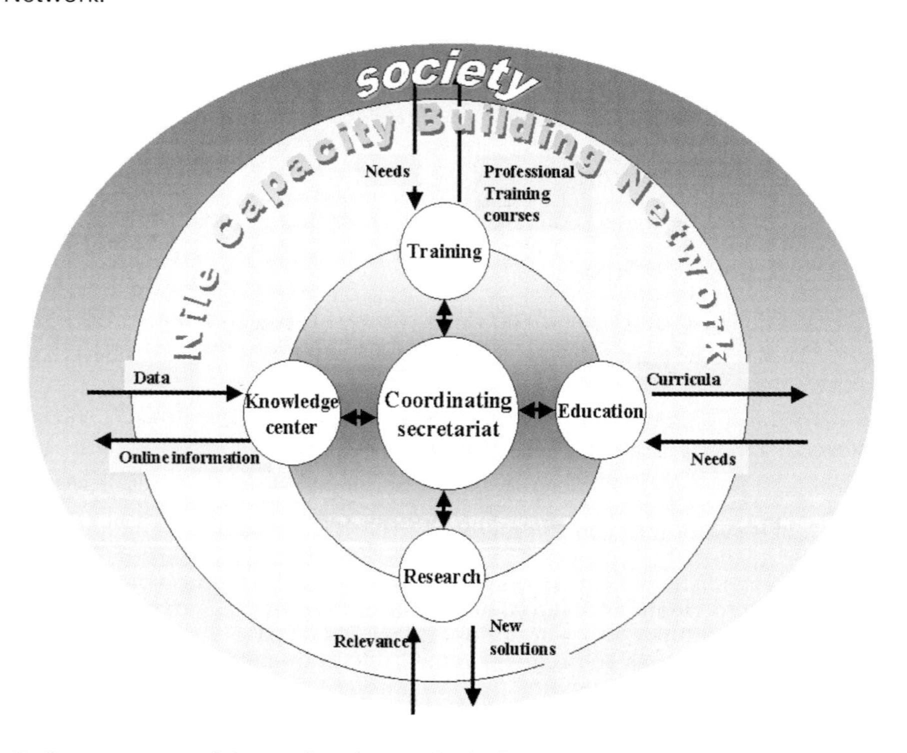

Figure 3. Components of the regional capacity building center

To support the implementation of these capacity building components, the following instruments are foreseen:
Fellowship fund: to support the participation in post-graduate programmes.
Node strengthening fund: to support institutions in the different Nilotic countries
Staff development fund: for "training the trainers"
Staff exchange fund: to stimulate the exchange of lecturers and professional staff
Research fund: to support researchers to perform joint regional research

Different donors will be approached to adopt one or more of the above components and to contribute to the various funds, which are a.o. World Bank, UNDP, UNESCO, CAPNET, FAO, ALECSO, CEDARE, NEDA, CIDA, etc.

6. Conclusions

Based on the HRI experience the following conclusions can be drawn:

"*Capacity Building*" is now widely regarded as the key strategy in ensuring sustainable water sector development.

Training is an important tool of management and should as such be an integral part of any management strategy.

The availability, development and particularly also the retention of *highly motivated staff* is the key to sustainable, high quality education and training.

The HRI became in a short time a regional training center well-known not only within African Nilotic countries but also further in the African Continent

Training groups of young professionals coming from countries with different political background and culture fostered friendly relationships between them and actually turned out to be an excellent long-term investment.

There is no doubt that there is no continent like Africa that suffers more from the lack of capacity to carry out research, studies and to establish policies for the region_s development in general and for the water sector in particular.

The initiative to establish the "Nile River Network" is aimed at creating a capacity building network in the Nile region through which education, training, research and exchange of information for and by professionals can take place.

The initiative can only be successful if it is fully supported by the initiators, all Nile riparian states and the international and bilateral donor community.

Bibliography

1. Alaerts, G.J., T.L. Blair and F.J.A. Hartvelt (eds.), *A Strategy for Water Sector Capacity Building*, IHE Report Series 24, IHE/UNDP, Delft/New York,1991.
2. Fathy El-Gamal, *"The Nile River Basin Action Plan"*, In: Proceedings of the Nile 2002 Conference, Kampala, Uganda, 1996.
3. IHE, Conference Proceedings *Water and Environment*: Key to Africa's Development, IHE Report Series 29, IHE, Delft, the Netherlands, 1993.
4. Patorni F.M. *Water Policy Reform Program*. Economic Development Institute of the World Bank. In: 2nd UNDP Symposium on Water Sector Capacity Building.
5. Saad M.B., Verwey A., *Institutional Capacity Building and Training Initiative for the Water Sector in Africa*, in Proceedings of the Nile 2002 Conference, Kampala, Uganda, 1996
6. Savenije, H.H.G. and A.S. Ransundersingh; "*International Linkages, a necessity for International Education*". in: Linkages Revisited. NUFFIC Conference, The Hague, The Netherlands, 1995.
7. UNCED (UN Conference on Environment and Development / Conference des Nations Unies sur l'Environment et le Développement), *Agenda 21*, UN, Geneva, 1992.

Rural community vs urban engineer

Bunker R.
Barefoot College, India

1. Present situation

In spite of billions of dollars being spent on meeting the basic needs of safe drinking water and sanitation in the 600,000 villages of India after 50 years the scenario is still grim and depressing. More than 40% of the villages still face an acute shortage of safe water for drinking.In spite of the bewildering variety of technologies available- ground water surveys through satellites,sophisticated drilling machines,hand pumps and piped water supply systems and water testing equipment the tragic fact is they have failed to solve the problem.Every year plans have been made,studies conducted and funds approved involving millions of dollars in the name of the poor and everyone seems to have benefitted-the geologists/geophysicists,the consultants,the private companies,the national governments(politician, bureaucrat,contractor)the international funding agencies but the reality is that for the time,money,energy and technology invested the poor have been the very last to benefit-that is,if they have tangibly benefitted at all:even that is an open question.

So for the last 50 years-if we honestly ask ourselves-its the Urban Engineer who has been responsible for the abysmal state of affairs.They -the so called"experts"-have always SOLD the problem of drinking water as a technical problem and because the administrator, the planner and the politician have not collectively applied their practical common sense to resist it the situation has come to this sorry pass.It is difficult to condone,excuse or forgive the colossal and incalculable damage this capitulation has resulted in by compromising the quality of life of the rural poor.By giving the technologist a free hand,by being made intellectually impotent and allowing technology to replace common sense and by virtually marginalising the knowledge,wisdom and awesome skills of the communities themselves,all the rash promises and commitments made in international conferences will always remain a remote distant dream.

Drinking water is not and has never been a technical problem alone.In the 21st Century it will be a social problem and unless we realise the need for a fundamental strategic shift in thinking,in attitude and collective action we will continue to make the same expensive mistakes.

The 20th Century has been the Year of the Urban Engineer.They have been given an open cheque.They have demanded and received a free hand for too long.And look where they have led us by the nose to an almost impossible situation.

Let the 21 Century be left to the Rural Community to apply their own skills and solve their own problems without interference and any more advice from the"experts".

2. New work

25 years ago the Barefoot College started with the belief that the Urban Geologist and Geophysicist was required to solve the problems of drinking water.But with the gradual exposure to rural communities and their practical wisdom,their incredible knowledge and skills in solving their own problems of finding and collecting and storing ground water as well as rain water the perception and priorities of the Barefoot College have changed.The

percentage of sucess and failure of the geologist/geophysicist to locate sweet water was almost the same as the water diviner whose powers the communities believe in.If we reduce it to calling it"mumbo jumbo"we are only displaying our own ntolerance,ignorance and narrow mindedness.

Most of the successful 1700 India Mark II hand pumps installed since l979 have been located with the help of the water diviners and counter-checked with the help of a terrameter-the best mixture of traditional and modern technology.20 years ago hand pumps was the latest technology available.Quite apart from providing safe water the hand pump solved other major problems-the social problem of access to the lower castes to take water from the same source not possible through open wells in India.It solved the problem of unemployed and unemployable rural youth men and women who were trained to become barefoot mechanics by the hundreds using village knowledge and skills to repair and maintain hand pumps thus rejecting the urban engineer's expensive solution.

But what the hand pump did was to cut off the communication and gossip channels of rural women who used the open village well as a vital source of information and acted as a rest and recreation centre from husbands and family chores.In addition what advanced drilling technology did through deep tube wells was to over exploit deep sources of ground water resulting in traditional open well sources drying up.This water was not eventually used for drinking water to meet a basic need in a rural area but for irrigation for commercial farming and industrial purposes located primarily in urban areas thus raising ethical issues.It was decided that hand pumps were NOT a long term solution.

In the 80s the Barefoot College after years of direct interaction with rural communities shifted their policy fundamentally away from installing hand pumps in hot desert regions to preserving,conserving, using and collecting rain water..By l998 12 million litres of rain water was collected through 207 fully constructed underground tanks in l54 remote rural schools and 53 community centres located in brackish water areas.Only local materials,local village masons and village labour was used giving employment to 6000 people(93,500 man days)at the height of the summer season(50C in summer) costing Re2_/litre (l998 prices)($0.04cents/litre)to fully construct them without any subsequent maintenance problems that plague the hand pumps and piped water supply systems.

In l998 using the method of diverting the water running off roofs to unused and disused open wells with a view to recharging ground water it is estimated a total of l5 million litres has been conserved which would otherwise have been lost through surface run off to the immediate region.

The idea of the Barefoot College was to demonstrate that illiterate villages using only common sense could plan and implement their own piped water supply schemes without blue prints and feasibility studies and totally without the help of the Urban Engineer.Between l985 to l998 a total ofl0 villages with an average population of 3,000 people,each village with 60 to 100 individual taps in each house are currently in operation.A Water Committee has selected an unemployed youth for training in the repairing and maintenance of their own piped water supply schemes with regular monthly contribution from the actual users.Today the total bank balance of all the Water Committees is about $ 20,000. In Rajasthan (India) the Public Heath Engineering Department full of hundreds of urban engineers has not managed to convince the community to pay for drinking water in any village in the whole State.

Its not difficult to locate the reasons why.

3. My vision

> Some men see things as they are and ask why
> I dream of things that never were and ask why not?
> -John F.Kennedy

What I dream about to some extent has been accomplished by the Barefoot College in my life time.That makes it easier to write about it with some authority.Its based entirely on practical experience.

I dream that all village communities have control over their water source.They should never be made to be dependent on people living and manipulating their lives from far away.This means that existing traditional water sources(open wells)should be reactivated and revitalised by whatever means so that all citizens of that village have equal access to water throughout the year.This is the real job of the engineer-NOT to replace, command,control,manipulate and increase dependency but to act as facilitator,motivator and confidence builder.Their role is to reactivate and revitalise.

I dream that village knowlege,skills,wisdom and materials is given much more importance than mere lip service because that alone is the most sustainable solution.Time and time again they have proved they need no help from engineers from outside.They need"confidence" building NOT capacity building.On the contrary its the engineers who need to build their capacity on how to communicate in a simple un-technical manner humble manner with communities.

I dream that rainwater harvesting is given the highest and only priority in desert regions where water is brackish and not potable for other reasons(iron,florides etc).Its the only long term solution for drought proofing villages.Water should not be allowed to flow away from where the rain drops.

I dream that thousands of disused and unused open wells in the villages of India can be used as vast recharge natural collection tanks.By channeling rainwater from roofs into these wells dry hand pumps can be revitalised and open wells can be recharged every year.In this way billions of litres of rain water can be collected underground as well used to recharge existing water sources instead of it being wasted.When compared to the incredibly expensive solutions the Urban Engineers have to offer this will cost nothing.

I dream that technology is not used as a tool to exploit or abuse or increase dependency but promote equality and enhance the self respect and dignity of the poor.Technology should never deprive the poor of work.

I dream that the issue of ethics,reducing corruption and waste of water,resources and manpower is given the importance it deserves and it features prominently in all future thinking and action on drinking water and sanitation.Thus every citizen in a village has the right to demand and receive information on how the funds have been spent for drinking water and sanitation in their village and who in the village has benefitted.

I dream that Transparency and Accountability of all partners-funding agency,national governments,private companies and local self government institutions-is an absolute MUST and this can be facilitated through the holding of Public Hearings when details of development expenditure is made public and made available for everyone to see.

4. Action required

I. Include the importance of traditional knowledge,skills and village solutions in the curriculum of all courses in Universities and Colleges in the North and the South for newly recruited water engineers.It will at least convince them that water is no longer a technical problem only but also a growing social problem.

2. Encourage in the courses a field trip to a grass root agency dealing with drinking water and sanitation issues so that they know they are not indispensable when it comes to looking for alternative solutions.This exposure will demonstrate the vast communication and credibility gap that exists(and growing wider by the year)between the Community and the Engineer.

3. Promote more active and visible interaction between practitioners who have proved that an alternative community based low cost approach is also possible at the national and international levels so that eventually it influences and changes policies towards funding and involving communities more directly in decision making for their own problems.

4. Make Transparency, Accountability and the Right to Information a major issue among all stakeholders in the Drinking Water and Sanitation sector.

Women for water sharing

Hasna J. Moudud
Bangladesh

1. Introduction

"Water is the source of all life."
The Holy Koran.

"The River of life, dark and deep, moves swiftly
The two sides are muddy, the middle is depthless."
Catillapda, 8th century Bengali Buddhist Siddha poet.

Life and river are seen as one. In all the religions of the world, particularly in South Asia, water is an important symbol. The journey of life here begins with a wash and ends with water cleansing the sufferings and impurities of life before the ultimate journey.
Science now has singled out one element, water, in search of life in the universe beyond. Life on Earth depends on water- water maintains and links the planets ecosystems. Water is the most precious life giving natural resources of the world, yet we continue to ignore it and abuse its use. Birth of a planet is marked by water, death of a planet will be marked by gradual withdrawal or disappearing of water.

In South Asia home of more than 400 million people with the highest concentration of population and poverty in the world, water plays an important role in the well-being and survival of its population and ecosystems. This part of the region is neglected and poorly developed but it is rich in its natural resources, particularly water.

Bangladesh drains the combined catchment areas of the Ganges, Brahmapura and Meghna rivers, an area of about 1.76 million square kilometers of which only 7 to 8 percent lies in Bangladesh. Bangladesh is an agrarian country, more than 84 percent of the people depend on agriculture and fishing for subsistence. Although Bangladesh is a water surplus country often suffering from widespread flooding, Bangladesh faces severe shortage of water due to India withdrawing waters from Ganges, the Teesta, Manu, Muhuri, Gumti, Knowali and many other rivers depriving Bangladesh of water during the seven months of dry season when there is no rain. India withdraws water through dams and barrages during the dry season but during the rainy season Bangladesh faces severe flooding due to sudden and unannounced opening of dams which flood the country as has been the case in 1998. In addition to barrages in upstream country, deforestation in the Himalayas are reducing total water flow.

Bangladesh known as land of water or better still water in land is now faced with acute water shortage during the dry season. A woman walks 2 to 3 kilometers to get drinking water from deep tubewell since the shallow tubewells have tried out.. desertification is taking place in the north western part of the country. Salinity is increasing at an alarming rate in the south west, threatening the Sundarbans, world's largest mangrove forest and UNESCO declared World Heritage Site. The country is also facing severe flooding as rivers are dying and are filled with silt they are unable to drain additional flood and rainy water during rainy season. In addition to the above arsenic poisoning is spreading in ground water in the border districts of Bangladesh in the north and south western parts. Experts say due to heavy metal pollution, industrial and agricultural pollution, arsenic contamination is spreading in South Asia at a rapid rate. In all this women and children are the worst victims.

The problem with water in Bangladesh started with India building the Farakka Barrage in India, 11 miles from Bangladesh border, with the stated purpose of flushing Calcutta port. The Farakka barrage was commissioned in 1972 on trial basis only, for diversion of a large portion of the downstream Ganges flow in India. It is causing untold human and ecological miseries.

2. How we see the problem

Water should be shared in such a way that none of the countries in the catchment as well as in delta disrupt the ecological balance or cause human misery. There should not be any adverse effect in the short term or long term of water withdrawal. All international convention and norms should be respected by regional powers who often exploit shared natural resources and create environmental problems which surpass local or regional scale.

What India may think as a short time arrangement to convert river to meet unlimited demand by constructing dam which in fact may in the long run harm India's agricultural and other economic development. In the short as well as long run the lower regime of the dam will be the loser, because of the scarcity of natural flow of water and secondly, the ecological disorder in the catchment area will seriously affect people in the lower reaches of the watercourse.

3. Background of Women for Water Sharing

In 1992 a meeting of UNEP Global 500 Laureates of South Asia (SAARC) was held in Dhaka, with the objective of proposing recommendations on an environmental action plan based on regional cooperation. Their following recommendations reflect the beginning of cooperation:

* Integrated, catchment-wide water resources development planning with the aim of ensuring sustainable use and controlled withdrawal of upstream flow for major riversystems in the common interests of co-basin communities.

* Consideration of cross border upstream constraints and downstream impacts of proposed development interventions should be carried out as part of integrated transnational Environmental Impact Assessments.

In 1993, a workshop on Women and Environment was held in Bangladesh attended by leading Indian environmentalist, Maneka Gandhi. A Task Force on Women and Environment was endorsed at the meeting which took up the issue of Ganges water sharing as a common concern which is causing great deal of hardship for women and children and ecological destruction. Together with women, leading water experts of the country joined Women for Water Sharing as the group called itself. Cross section of the civilian society and experts joined because it was an environmental initiative without any political and bi partisan basis.

Women for Water Sharing visited the affected area, carried out surveys and held workshops and seminars on the effects of Ganges water withdrawal. The findings have convinced the group that what we lose in 5 years we may not be able to regress in 50 years. Ganges water sharing is not a technical or political problem. It is a humanitarian and environmental problem. People all over the world are taking more interest in the environment and their representatives and government have to pay more attention to environmental degradation.

In May 1994 Women for Water Sharing addressed a seminar at the Gandhi peace Foundation in New Delhi where we presented Ganges water sharing problem and appealed to the Indian people to talk to their leaders towards an immediate solution of this immense problem for

us. We said, we are not asking India to halt its water development activities but it should be their responsibility not to kill this river holy to them and cause ecological and humanitarian disaster.
The quality of Ganges water is said to be dangerously unsafe for human for human use.
Ganges is said to die in Kanpur in India due to effluent discharge of leather industries. Ganges is the dumping ground of 17 major chemicals which are highly toxic. All of the 17 industries using these chemicals are located on the banks of Ganges. Eight permanent pesticides are used in India which are banned by most countries in the world including Bangladesh. A combination of 17 pesticides and fertilizers are poured into downstream Bangladesh. In Varanasi at Ganges, about 10,000 half burnt dead bodies are dumped into the river annually in addition to about 60,000 carcasses of domestic animals.
The Ganges flows through 700 cities in India, and about 120 million litters of waste waters from industries and municipalities including untreated sewage and garbage are dumped into the river.

At first it was not easy to get support within India because they were benefiting from Farakka barrage. We looked for anyone who would listen from housewives, writers citizens groups to human rights activists.

In 1996 we were invited by Ganga Mukti Andolon (Free Ganges Movement) based in Bihar, India who suffered due to Ganges water withdrawal. In other words water withdrawal was also creating ecological and livelihood disaster upstream. Their fishermen, boatmen and small farmers were hard hit because of adverse effect of Farakka. In Patna, Bihar we made a joint agreement for a joint movement to be led by Sundarlal Bahuguna, 80 year old leader of famous Chipko Movement and who protested against Indian Government constructing barrages on all our common rivers. Our meeting was inspired by the belief that our rivers should unite, not divide us. It made us realize that we must bring pressure on our governments for peacefully resolving conflicts relating to shared and scarce natural resources. The well being of the people of our countries are tied to each other. Environmental degradation will only increase tension in the region.

We cannot choose our neighbour. We as a small neighbour expect large gestures from a large country. It is the right of lower riparian country to get their equitable share of common river. There has been agreement on augmentation of flow of Ganges but there was no agreement on Ganges water until 1998. Even today the Ganges Water Treaty which has been signed between India and Bangladesh based on the guidelines set by People's Initiative for Water Sharing on 3rd June, 1994, it is not a perfect one. There is not enough water inspite of the treaty. There is no minimum water guarantee clause, it is an agreement on sharing during lean period not annual sharing, the treaty was not discussed in Bangladesh Parliament and there is no visible change in water availability following the treaty. But it is a historical treaty. CoChairperson of Women for Water Sharing and People's Initiative for Water Sharing (as the group expanded to include wider participation) Mr. Kuldip Nayar who not only led this movement in India but saw to it that his Government sign a treaty with Bangladesh on Ganges water. There is more awareness among ordinary people. Water is food, security and survival for all our people. It is not international laws and treaties which alone can solve the problem, people need to understand and agree to share scarce resources such as fresh water. In conclusion I have found that people to people cooperation is most important. Free Ganges Movement of hundreds of boatmen, fishermen, farmers and men and women from ordinary walks of life responded to human and ecological problems more readily and sympathetically because there is a stake in there for them. I quote a Bengali folk song:

> Cows are of different colours
> But milk is of the same colour
> All over the world I see sons of one mother.

Theme B

Training needs analysis. Curricula contents. Job competencies. Compendium of water related disciplines.

Analyse des besoin en formation. Programme des cours. Compétences professionnelles. Compendium des disciplines dans le domaine de l'eau

Towards improved learning in the water environment in developing countries

Tag E.M.Elseid
Sudan

Abstract
The radical changes associated with the new international economic system have their far-reaching effects on environmental reservation and the use of natural resources. They dictate new axioms for building knowledge and improving learning in the water environment in developing countries. The need to incorporate new, yet practical situation-based and society-oriented methodologies make the notion of "learning societies" to be useful. However, building the infrastructure for a learning society is a complex organizational and institutional process given the dramatic economic shifts. The creation of effective and dynamic water organizations, the support of creative and responsive research activities, the exploitation of technological capabilities and the formulation of effective linkage between water organizations, educational institutions and the society at large, are the principal critical success factors. This paper is based on the initiative on "leveraging water-related DSS in developing countries."

We have been observing the distinct features of the new international economic system including global and cost-serious competition, shifting consumer preferences, fragmented markets and the massive proliferation of technology. In such a burgeoning economic environment, the competitive advantage of many firms has diminished and managers across the organizational landscapes and managerial layers are struggling to adapt to the unfamiliar circumstances and absorb the dysfunctional consequences of the new strains of competition. The struggle for building lasting competitive edges in such an environment has been driven by cost-reduction and profit maximization motives that made natural resources (both renewable and non-renewable) proper candidates for exhaustion and degradation because of their low acquisition and opportunity costs and accessibility. During the last two decades environmental problems in general and water-related problems in particular are becoming global in scale and context. This has been accompanied with a considerably growing interest in learning and knowledge leveraging to ensure transformation and adaptation. However, the growth in awareness has raised a fundamental question: what's a learning society? Reaching a comprehensive and conclusive definition for such a term, is a complex task that encompasses a wide range of social and qualitative dimensions but one can view a learning society in the water environment to be:

> "A society that learns- in a continuous learning pattern- from formal instrument-based learning methods as well as from past trends, experiences and situational variability. It, also, has the ability and willingness to reshape, modify or transform its behavior and interaction in response to knowledge acquisition and/or situational changes."

The social, demographic, economic, political and spatial variables are the main change agents that shape the behavior and interaction patterns in the learning society. In this respect a holistic, integrated and society-oriented approach is needed. The formation and maintenance of such a society in water environments is a fundamental, yet complex long term change process particularly in developing countries facing spatial variability, escalating down-side risks and economic hardship. Without tremendous institutional, technological and societal arrangements the process will eventually go into disarray. Such complexity can be attributed to the following factors:

1. The lack of Effective Organizational Sets

Based on the General System Theory, the learning society in the water environment can be viewed in terms of interactive organizational sets. An organizational set is a cluster or group of organizations that are connected to each other through some common activity or through a need to interact (Evan, 1967). The interconnection among and inside the organizational sets fosters specific learning styles, patterns and orientations. Thus, the learning capabilities and responses in the society shape the slope of the learning curve. All organizations have some sort of learning patterns that reflect their internal processes and behavioral adjustments throughout the development life cycle. Within the context of water environment, water-related organizations, educational institutions and societal organs can be considered as the main organizational sets that form a learning society. However, for organizational sets to be effective entities in the learning society, they should be themselves learning organizations that are interactive enough to maintain constructive linkages to ensure information partnerships and highly dynamic to foster societal orientations and shared visions of conservation.

Water-related organizations in many developing countries are highly bureaucratic, line-structured and inflexible to form an interactive responsive learning organizational set. The formulation, communication and implementation of strategy within and across organizations is a complex learning process. It constitutes structural and behavioral changes and requires collective involvement and commitment of the different organizational units to meet the consequences of the organizational defensive patterns (Argyris, 1990). However, the need for high level learning is of paramount importance to install a mechanism for adaptation and transformation. Many of these organizations in developing countries have a "merchant's" image rather than a strategic, collaborative, cross-organizational and societal orientation. Their failure to think in strategic forms, initiate and maintain communication across their internal platforms and among themselves has limited their learning capacities. They have been characterized by fragmented efforts, sub-optimization and diverse orientations that did not support the enforcement of a shared vision and responsibility for knowledge acquisition, communication and utilization for collaborative conservation.

Educational institutions, on the other hand, form an important organizational set in the learning society in the water environment based on their distinct features. Their role in knowledge diffusion has been threatened by the dysfunctional ramifications of the economic transformations that resulted in considerable financial difficulties and unprecedented brains drain particularly during the last two decades. However, they are not only unable to create a conducive-atmosphere for their internal learning processes but are also unable to engage in organizational sets and networks that provide scientific and technological accessibility and society-related research. The assets are inventories of business-oriented, rather than society-oriented, research that hardly result into the necessary adaptive and transformational change of behavior in the learning society.

Despite the regional and international efforts, the massive economic hardships has endangered the learning capacities of educational institutions in developing countries where many of them are unable to build the appropriate infrastructures, integrate environmental sciences into their educational programs and generate a societal rather than a technical and engineering domain for water-specialization. Under the new economic system, environmental studies are low-ranked on the "professionalization" scale - compared to medicine, business administration and computer sciences- and are not yet viewed as mature candidates for building competitive advantages in the intensive job hunting arena under the high unemployment rates.

Despite their heterogeneous structure, societal organs can be considered as the important organizational set in the learning society in the water environment. It includes a diversity of formal and informal organizations, groups and gathering. The importance of such organs, stem from their dual role in conservation. They are the users of water resources and the bearers of the dysfunctional consequences of shortage and misuse on the one hand and the target of the overall development process. Their role in monitoring and developing water resources is of paramount importance.

Due to their contextual variability, the components of such organizational set (if can be considered like that even for methodological considerations) often differ significantly in their learning styles, patterns and adjustment capabilities. Improving learning in this organizational set will be eventually reflected on the learning capabilities of the whole society in the water environment. It is essential, therefore, to focus on knowledge diffusion particularly in developing countries where water-related problems are not yet that critical and obvious and the general attitude is that water conservation is a governmental responsibility. Without improved learning in such environment there is no way for any transformation and adaptation.

2. The Lack of The Appropriate Technological Platforms

The international economic transformations have been accompanied with technological developments in the areas of knowledge acquisition, sharing, and communication that, if effectively employed, can improve the learning in the water environment. However, despite the promising news and future potentials, technological projects are not delivering operational and learning promise in developing countries. The economic shifts made the market for technology to be highly imperfect and oligobolistic that many countries are not able either to acquire the technology or meet its escalating operating costs. It's known that water-related technologies fall in the category of "interdependent technology" and the lack of sustained inputs jeopardize the operational and technical interdependent feasibility. Hydrological and hydraulic monitoring technologies, for example, depend on input data from GIS-based instrumentation that can't be fully maintained in many developing countries. This not only defeated the learning processes of organizational sets, but also defeated the knowledge base of the whole society.

On the other hand, many developing countries lack the organizations capable of managing technology-based integrated investments that support effective organizational learning and cross-organizational interactions. Under man situations, the decision regarding technological investments is individualistic and oriented towards acquisitional rather than the utilizational concerns. Each organization is free to acquire, configure and use the technology it sees appropriate. The lack of a strategic technological intent that guarantee the situational fitness, the appropriate degrees of operational precision and the contextual feasibility, made many technological acquisitions in many organizational sets to be outright failures. The communication, knowledge sharing and organizational orchestrations have been challenged by the diversity of the learning styles and patterns dictated by the different technological platforms. The prevailing-high and rising costs of technology, the shortage of skilled manpower and the high rates of unemployment have further complicated the problem.

3. Towards an Improved Learning in the Water Environment

In the face of the tremendous spatial variability and the fundamental economic shifts, creating a learning society in the water environment in developing countries is contingent upon the formation of dynamic and flexible organizational sets that are mature enough to make transformational or adaptive behavioral adjustments. The necessary change in the societal behavior pumps out of involvement, interaction and participation. To make this a reality, there is a need for "responsible", dynamic and effective water organizations that don't have the "merchant's" image but are concerned, instead, with enforcing shared responsibility for generating and communicating knowledge for conservation, continuous learning and behavioral adjustment. Within the context of the information age, rational technological investments can facilitate the learning patterns and processes of organizational sets and the society at large. The intent should be on "what to communicate, to whom and in what format" rather than on the acquisition of technology. This will facilitate the formulation of the effective training and development programs that produce knowledge workers. However, it is also essential to create a harmonized network of communication between the different organizational sets in the water environment. This will facilitate information partnership and exchange; relate the research domain to the societal pressures and/or the behavioral adjustments necessary for the adaptive and transformational shifts in the water environment. However, supporting educational institutions will enhance their functionality and the overall learning curve and considerably move the environmental specializations many steps ahead the professionalization ladder.

References

Argyris C. (1990): *Overcoming Organizational Defenses: Facilitating Organizational learning*, Boston Allyn and Bacon.

Elseid, T. (1998): "*Leveraging water-related DSS in developing countries*", The Methodological Report, Phase One, Limited Distribution, The Netherlands.

Evan, W.M. (1967) "The Organization Set: Toward a Theory of Inter-Organizational Relations." In J.D. Thompson (Ed.) *Approaches to Organizational Design*, University of Pittsburgh Press.

Towards a Compendium for water related education and training
Recent information about needs and expectations

Vers une Compendium pour éducation et formation professionelle sur l'eau. Des informations récentes sur exigences et expectations

Wessel J.
Faculty of Civil Engineering of Delft University of Technology,
The Netherlands

Abstract
This contribution gives an overview of the objectives of an ETNET.ENVIRONMENT-WATER project directed at composing a Compendium of water related disciplines and of the results of its feasibility study. The project intended to fill a major gap in easily accessible structured information on the main topics for future education and training in the water sector. Its goal was the composition of a computer based Compendium of topics and terms used in water resources sciences (water related disciplines) against the background of educating and training for integrated water management. The project produced two distinguishable results:

a) an overview of (recent) needs and expectations for education and training in the water sector;
b) an indication for an outline for the composition of a water-related computer stored data base.

Keywords: education and training, compendium, water-related disciplines, computer stored database.

Résumé
Cette contribution donne une panorama des objectives d'un projet ETNET. ENVIRON-MENT-WATER dirigé vers la construction d'une Compendium des disciplines de l'eau et des résultats d'une étude de preuve. Le projet avait l'intention de remplir l'absys majeur dans des informations structurés qui sont facilement accessible sur les sujets principales pour l'éducation et la formation professionelle futurs dans le secteur de l'eau. Il était dirigé vers la composition d'une Compendium en ordinateur sur des sujets et expressions appliqués par des sciences sur l'eau (disciplines de l'eau) sur le fond des exigences d'éducation et de la formation professionelle sur l'aménagement integrée de l'eau. Ce projet a produit deux résultats assez divers:

a) une panorama des exigences et expectations (récentes) pour éducation et formation professionelle dans le secteur de l'eau;
b) une indication d'une schéma pour la composition d'une data base dans une ordinateur.

Mots clef: *éducation et formation professionelle, compendium, disciplines de l'eau, data base dans une ordinateur.*

1. Introduction: purpose and scope of a compendium

This contribution reports on the results of a feasibility study of ETNET (= European Thematic Network of Education and Training).ENVIRONMENT-WATER (A guide to the needs of Education and Training in the Water Sector. Towards a Compendium for Water-related Competencies) for a project directed at preparing a Compendium of water resources sciences (water related disciplines). In preparation of this report many colleagues, especially those composing the ETNET.ENVIRONMENT-WATER Scientific Committee (Anonymous, 1998) were consulted. The goal of this project was to establish a computer stored data base of topics and terms used in the water related professions and disciplines in fields of practice, research and education. It would contain - in a rather structured way - definitions of disciplines, their topics and related information (possibly) to be included in course units of students aiming at a profession in the field of hydrology, water resources management and environmental sciences. It also could especially become a tool to describe the education and training requirements for specific competencies, including research competencies, in a case to case way.

The reasons why a project like this feasibility study after a Compendium was launched and included in the activities of ETNET.ENVIRONMENT-WATER were:

(a) As water education is becoming one of the nuclei in future directed sustainable development environmental education and training hydrology and related disciplines should consider themselves more critically and delineate more properly the disciplines presently and desirably involved in educational and training programs.
(b) When dealing with education and training in the field of hydrology, some common ground should be covered when discussing the value of courses offered and the professional skills which a trainee may have obtained by following them.
(c) Thus far the absence of a universal program for hydrology and related sciences has made it difficult to compare individual teaching programs and the resulting diplomas, both at the national and the international level.

During the last decades, due to actual shortages, technological progress and environmental concern a new awareness of water as a specific field of activities, research and policies has been raised. Starting from this awareness, which has been reinforced by the new emerging paradigm of integrated water management, something as the global water sector has emerged. The water sector includes all phenomena in the hydrological cycle and its context, all related living phenomena, their processes and habitats, all relevant aspects of water uses and water users, all problems encountered by and related to activities of water managers and other water professionals and all phenomena connected to water related disasters.

The main concern of the author was with a basic approach to the present and future education and training (ET) info-needs in this newly developing sector. In this area already many studies describing curricula needs exist [Van der Beken (1993), Gilbrich (1994), Kobus, et al. (1996), Maniak (1993), Nash (1990)]. Also many decision support systems and computer stored databases [WMO, (1977)], even on the Internet, exist. But these only cover parts of the scene.

The aim of this project was to compose a thesaurus of all the topics of concern in this broad field. This thesaurus was named Compendium. Ideally it should contain:

(a) all (relevant) objects, processes, problems and activities in the water sector;
(b) all disciplines related to objects and processes in the water sector;
(c) all water-related topics covered by these disciplines;
(d) all principles, paradigms, theories, methodologies, methods, concepts and terms;
(e) all professions related to problems, objects and processes in the water sector;
(f) all activities and competencies of these professions.

2. Contents of the feasibility study

The ETNET-project had five target-areas:
(a) Delineating the water sector (areas and problems)
(b) Delineating the field of old and new water related disciplines and professions
(c) Structuring this field into a number of identifiable (groups) of professions and disciplines including new specialisations
(d) Collect the relevant information for each profession/discipline and problem area
(e) Store this information in a user friendly way in a computer database.

The said feasibility study showed that the project as it was originally envisaged would be too ambitious. The study however contained much information which is of value for education and training in the water sector. For this reason it has been published. Especially the description of the actual and future information needs in the perspective of an integrated approach to the water related area seem of importance. Moreover some interesting thoughts on the structure of a computer based database are given.

The report has eight chapters and 5 Appendices. After an introduction (Chapter I) the reader is first invited to explore the water sector and its jobs (Chapter II). Then the focus shifts to professions, sciences and disciplines (Chapter III). Thereafter education and training are dwelled upon (Chapter IV). In Chapter V the trends and developments in the water sector are described, resulting in a search for new professions and disciplines (Chapter VI). Chapter VII only gives an outline of what a computer stored database would look like and gives some suggestions for its completion, further applications and related studies. A final Chapter VIII presents some perspectives on further applications and related efforts. Appendix I gives an overview of traditional water related professions and disciplines. Appendix II contains definitions of older and newer disciplines, Appendix III a glossary of concepts and terms. In Appendix IV titles of water related thesaurus are given. Finally Appendix V gives a list of all experts which were consulted and contributed to the project. In preparing the report recently a workshop was held at Delft University on New paradigms in water management. One of its conclusions was that we should speak about an integrating approach to water issues instead of speaking about a complete integration of all their relevant aspects.

3. How to build the water-related compendium

Building this Compendium asked for a number of different activities. A first goal was a basic reflection on the scope of the water sector and its info-needs for the purpose of education and training for science and professions. Approaching the water-sector can be done in many ways. Preparing the feasibility study choices were made. The following six approaches seemed feasible:
(a) starting from elements of the hydrological cycle (hydrology-oriented approach)
(b) starting from protection of environment/sustainable development (natural resources approach)
(c) starting from activities of the water users and water consumers (socio-economic approach)
(d) starting from activities of water professionals-practitioners (profession-oriented approach)
(e) starting from activities of water-related researchers (science-oriented approach)
(f) starting from calamities and looking for insufficient training of those involved (disaster-approach).
After consultations and considerations primarily the fourth and fifth approach have been followed. The former is closely related to the engineering professions, the latter to the scientific academic tradition.

Engineering education has shifted from the application of technology to the employment of science to analyse and solve engineering problems. The study distinguishes between disciplines and professions and approaches the education and training requirements from both sides. Disciplines are windows, dealing with specific aspects of the real world. Professions deal with problems.

Modern water-related education and training has to be given in a broad field on a large number of topics. In education the principles to be applied in training have to be taught. These principles can no longer be taught exclusively in traditional disciplines, but also new disciplines should be given attention to. And specific attention should be paid to the need to instruct the principles of integrated water management and to multidisciplinary, interdisciplinary and transdisciplinary training. Also the field of ecology and environmental problems should be covered.

4. Developments in the water sector

There have been some very important developments in the water sector, such as a change of paradigm, which ask for new approaches. We see the creation of new emerging disciplines like integrated water resources management, which could be included into university curricula. Trends and developments of importance are: a technological drive, a shifting from pure technology towards management, a societal drive, asking for more participation in decision making, an ecological drive and attention for the water problems on a global scale. The technological drive has led to an (global) information explosion.

Actual water management is approaching a stage in which for every major decision there must be discussion with the general public. Managers are no longer to decide on the basis of financial profits only, but they are also expected to yield judgments on ethical, social and environmental values. Water resources management has become a profession in its own right.

In the water sector new professionals have emerged. We also find new water related disciplines, e.g. eco-system water resources management, the use of decision-oriented systems approach, the use of policy analysis and risk analysis, and hydro-informatics. Specially the introduction of environmental ethics and more specifically of water ethics should be mentioned. Without an opinion on ethics and values visions and scenarios on future water management seem void of contents. Also the field of social sciences, economics and (environmental and water) law is gaining importance on the water scene. Attention should also be given to anthropology and cultural awareness. In future the author dreams of a new macro-discipline of integrated water resources management (cf. also Bogardi, 1990).

5. A computer stored database

The final goal of the project was to build a computer database accessible on Internet on an ACCESS-basis. Because of the enormous amount of data to be collected, analysed and stored in a structured manner only an indication could be given for an outline for such a database. At first a clear picture should be obtained of what kind of information (and approximately how much and how accessible) should be stored in the system.

One of the characteristics of this database is its more dimensional character. In analysing complex systems a reduction of the complexity must be achieved. This calls for orthogonality, i.e. that one should look for sub-systems whose components are independent of the components of other subsystems. Each independent subsystem should have a separate description. For sake of completeness reference should be made to other related systems.

6. Contents of the database

For this Compendium quite independent sub-systems seem necessary on:
(a) philosophical and ethical concepts. All general basic concepts, like time, space, life should be entered here as well as general scientific principles like objectivity, reproduction of findings, and some specific paradigms. Among ethical values relevant for practice and science we count: respect for others and life, and responsibility for the environment.
(b) natural science concepts. Natural science concepts which have not yet been listed under (a) are entered in the data base. All the specific theories, methods, etc. of the natural science disciplines should be stored. A list of all relevant natural science disciplines should be entered, looking at the formal objects instead of at the material ones. For each discipline specific research after water related info-needs has to be performed. Here we may list: general principles/paradigms, approaches/windows, theories, methodologies, methods, tools, basic concepts, concepts and terms. And also the elements of the scientific cycle, like: exploration, description, analysis, explanation, simulation, evaluation, prediction, prescription/design.
(c) natural abiotic phenomena. In this third section a catalogue of natural abiotic phenomena should be entered. Here we list the elements which should be covered to satisfy the needs of the professionals and scientists in that area. These technical (mostly material) aspects again can be subdivided into objects, conditions, processes. We also could enter more categories, for instance: events and phenomena like the greenhouse effect.
(d) natural biotic phenomena. In another, fourth, section the same should happen for the biotic phenomena.
(e) social science concepts. In a fifth section those socio-economic concepts which have not yet been listed under (a) should enter the database. Here also a list of all relevant social science disciplines should be entered.
(f) human water related activities. In the sixth section all human water-related activities, among them those of professionals in the water sector, should enter the database. With each activity also a list of the tools necessary for its performance will be given. Many of these activities can be found in practical engineering or systems analysis guides. They include the elements of the engineering cycle: assessment, research/planning, screening, design, decision making, financing, implementation/construction, use, operation and exploitation, maintenance, evaluation (of performance), removal of object/decomposition and eventually: crisis management.
(g) skills and competencies. In a seventh section all skills and competencies related to activities and jobs are entered. Here we may list general skills, like oral and written expression, managing uncertainty and risk and ability to perform in project teams and taskforces. Examples of competencies are self-direction, problem solving, team-building and innovation, leadership.
(h) problem areas. Finally a list is entered of problem areas in the water sector, which should be investigated by multidisciplinary or interdisciplinary efforts/teams. Here specific phenomena are stored in connection to their causes and possible solutions. For instance eutrophication is entered along with agriculture (cause) and change of crops (solution).

In each of the 8 separate sub-data bases specific indications and links should be provided towards the other sub-databases.

7. Desirability of continuity

The efforts made thus far are only a preliminary theoretically based approach. They are also meant to convince the scientific community of the fact that recent trends and developments ask for a reorientation of ET in the water sector with due attention to new tools and technical possibilities like an online computer stored database of terms applicable to the science and the profession. An advanced pilot project is needed to test the above given scheme. And one should keep in mind that this feasibility study for a Compendium throughout is a first attempt. The applicability of the information contained in this study should not be rigidly used to build European-wide or even country-wide curricula. There might be a different blend of scientists/engineers according to the national physiography, the economy and culture and educational and research traditions of the countries involved.

References

Anonymous, *General Report of the Second Year of Activity* 1997-1998. ETNET.ENVIRONMENT-WATER, Brussels 1998.
J.J. Bogardi, *Op weg naar integraal waterbeheer/towards integrated water resources management*, Wageningen University of Agriculture, 1990.
W.H. Gilbrich, Hydrological education during the fourth IHP-phase (1990-1995), *Technical Documents in Hydrology*, UNESCO, Paris 1994.
H. Kobus, E. Plate, H.W. Shen and A. Szöllösi-Nagy for UNESCO/IAHR, Education for hydraulic engineers, *SC.96/WS/4*, UNESCO, Paris 1996.
U. Maniak, Curricula and syllabi for hydrology in university education, *W-IV Project E-2-1*, UNESCO, Paris 1993.
J.E. Nash, P.S. Eagleson, J.R. Philip and W.H. van der Molen for IAHS and AISH (1990), The education of hydrologists, *Hydrological Sciences Journal*, Volume 35, nr 6, pp. 597-607.
A. Van der Beken, Continuing education in hydrology, *Technical Documents in Hydrology*, nr. 47, UNESCO Paris 1993.
J. Wessel, H. G. Wind and E. Mostert, Paradigms in Water Management, RBA Research report nr. 11, Delft 1999.
WMO, *The international glossary of hydrology*, Geneva 1997.

Some key issues of the development of continue education and training (CET) for hydrology in Romania

Viorel Alexandru Stanescu
National Institute of Meteorology and Hydrology, Sos. Bucuresti-Ploiesti 97, Bucharest 71552, Romania

Abstract.
Considering the present and foreseeable organizational structures, the development of the water management and the policy of human resources in this domain in Romania, the reasons to stress the needs for continuing education and training are presented. The newly emerged hydrology topics have led to the identification of CET needs. The main CET forms on-the-job training, post graduate courses classroom education, workshops, seminars etc., are presented in the paper. For each one of them one considers: the expectations, the planning of the educational programmes, the activity and the strategy of learning, the formative and global evaluation of the trainees and the feedback for the CET improvement.
Problèmes du développement de l'éducation et de la formation continue en hydrologie en Roumanie

Résumé.
En considérant l'actuel et prévisible développement de l'utilisation de ressources en eau et la politique des ressources humaines dans cet domaine en Roumanie, on présente les raisons pour une éducation et formation continue. Les actuels et nouveaux problèmes de l'hydrologie ont déterminent l'identification des besoins en formation et développement des ressources humaines. Les principales formes de l'éducation continue instruction au lieu du travail, l'enseignement post - diplôme, les ateliers, les séminaires etc., sont présentées dans le papier. Pour chaque activité et forme on considère: les espérances, la planification du programme de l'éducation, l'activité et la stratégie de l'enseignement, les tests formatives et globales et la validation des connaissances et des compétences ainsi que le feed-back pour améliorer le processus de l'éducation et de la formation continue.

Introduction

Romania is a country with relatively weak water resources (1700 m_/inhab. year) and the water demand has continuously increased. Yet, even though a noticeable progress in improving water quality has been achieved since 1989, this has remained under the standard limits. After the political changes (the restructuring of the economy, market-oriented production and steps toward an European integration) the structure and the number of the hydrological personnel involved in the operational activity of River Basin Authorities and in the research has been subject to noticeable changes. New trends in the development of hydrological sciences and water resources monitoring as well as in the policy of human resources are reasons to stress the need for a continuing education program. Therefore, an informal continuing education and training (CET) is to be pursued for all those who are newly recruited or are already employed in the water resources management system.

HYDROLOGICAL NEEDS IN THE PRESENT AND FUTURE SOCIO-ECONOMIC DEVELOPMENTS AND THE POLICY OF HUMAN RESOURCES

The acquisition, transmittal and processing of data have been increasingly carried out by automatic equipment. New abilities and skills to know the functioning principles of these high-level devices, to ensure an adequate maintenance and use of them as well as to make use of computer software for primary data processing and data base constitution should be learned by technicians. On the other hand, nowadays, there are newly emerged needs of the modern hydrology which are not subject to the curricula in the universities where only a general hydrological education is given. Among them one can mention:

(a) The assessment of the synthetic hydrological characteristics under man - influenced regime and the study of physical chemical and biological aspects of water.
(b) A deeper understanding of the water cycle at several time and space scales.
(c) The development of the real time monitoring and the hydrological forecasting models of the water quantity and quality as well as of the hazards.
(d) The assessment of climate change impact on the water cycle components.
(e) The interdisciplinary projects to improve integrated water management.

To keep abreast to these hydrological needs, advanced newly employed methods and models have been successfully used in hydrology. Among them, one can mention the fractal analysis, the entropy methods, the artificial neural networks, the multivariate stochastic analysis of temporal series, soil-vegetation-atmosphere models and so on.
Hydrology has come closer and closer to decision making, water planning and water works operation. Therefore, technicians and especially senior hydrologists should become familiar with water management procedures and techniques. Yet, in the field of applied research, some hydrologists should become acquainted with procedures such multicriteria modeling techniques for decision making, multi-objective water resources conflicts, and the incorporated risk and hydrological uncertainties reflected in the decision making. In order to participate at the water resources monitoring, technicians should know the structure of the water management network and the hydraulic and operational characteristics of the water use works.

The policy of human resources has followed the replacement of a centralized economy by a descentralization of all activities and the self-management of the economic units. Under these new circumstances of increased challenge the personnel working in the hydrological field has had to cope with broaden problems such as an extension of knowledge of the water consumption of the users and a better and a more correct information on the water resources and the hazards at basin level.

An important attention is paid to the integration of the hydrological products in the environmental problems, especially to those concerning evaluation and forecast of the surface and groundwater flows and their vulnerability. Consequently, the hydrological staff of the River Basin Authorities has shifted from a "producers of hydrological data" to a "contributing personnel" at the activity of their enterprise. So, a "shift" of the knowledge and skills from the center to the basin level has to be deemed. On the other hand, an accretion of the role of the central unit (Hydrology Division of the National Institute of Meteorology and Hydrology-NIMH) in producing and transferring new methods and technologies, to replace the obsolete ones, has to be emphasized.
As far as the policy of human resources is concerned, there is the need to harmonize the age of the hydrological staff which has become older and older as a result of a scarce employment of the young people in the last years. A policy to rejuvenate the personnel will simultaneously request reinforced programs for training of newly recruited personnel either technicians or university graduates.

THE CONTINUE EDUCATION AND TRAINING (CET) NEEDS ANALYSIS

Taking into account the above mentioned conditions of the hydrology development in Romania an identification of CET needs refers to three phases of application:

(a) **Phase I** implies the hydrological education and training focusing on the newly recruited young people which left the formal education in secondary schools or universities. This CET activity, according to the case, is carried out as follows:

(i) The young university graduates have to learn in a reasonable time interval (4 - 6 month) general knowledge and skills concerning the overall domains of research and operational hydrological activity carried out at the National Institute of Meteorology and Hydrology (NIMH) or at any unit of the River Basin Authority, namely: hydrometry, operational hydrological services (hydrological forecast, data processing, data validation and constitution of the data base) and basic and applied research in the field of surface and groundwater hydrology. In general, the training of the university graduates is of the on-the-job training and self-learning type. Also, they are trained to use computer procedures and programs. A tutor is commissioned by the leading staff of the NIMH- Hydrology Division or of the River Basin Authority to survey and to conduct the training. At the end the of the training period, the tutor makes an evaluation report concerning the activity of the trainees.

(ii) During a period of 2-4 months the young graduates of the secondary schools who are recruited in view to become technicians, have to get familiar with the basic hydrometry (how to use and maintain the instruments, how to select a suitable site and measurement method, how to apply all the safety requirements), with the methods and procedures for primary data processing (manually and by use of the computer facilities) as well as with the functioning of the information system for hydrological warning and forecasting. The training is conducted by a tutor (a hydrologist or an experienced technician) and in general is performed on the job.

The first phase of CET is needed in view to:

- Allow the employer to make a correct assessment of the intellectual capacities of the graduates in order to direct them in the right direction of activity according to their preliminary evaluation.
- Get closer to the precise domain where the newly recruited graduate has the strongest vocation in order to make a good choice for his/her future activity.

(b) **Phase II** is the continuing professional development which follows up the first phase having in view the transfer of the knowledge skills and attitudes (Van der Beken, 1993; Nash et al., 1990).Further on the CET needs of this phase are presented.

(i) **Needs for broadening and developing a specific research activity** in which the trainees are already involved, in order to fill the gaps, to refresh instruction and to accumulate further knowledge according to newly emerged methodology and computer facilities. The most appropriate forms of this type of CET are : on-the-job training, post graduate high level courses and short courses.

On - the - job training is prepared at the beginning of every year and includes: objectives of the sessions, list of the individual (trainees), nomination of the supervisors, consideration of the background of the envisaged trainees, time schedule of the training session and the content of the selected topic. In the training program (syllabus) some initial background knowledge is included too (e.g.: statistics, hydraulics, numerical calculus etc.) depending on the initial evaluation of the trainees. For such "initial instruction" self learning and part-time education are often deemed. The final result of on the job training is materialized in a "diploma project." This often belongs to the research or operational annual plan of the NIMH-Hydrology Division.

An alternative of on-the-job training is the training work abroad on a specific topic (especially in the EU Countries) during 3-6 months. Over 40 young hydrologists fellowship granted from the NIMH - Hydrology Division and the "Romanian Waters" Company were trained in France Germany, Italy, Spain, Switzerland, Belgium and the USA. This type of training has demonstrated a very good effectiveness.

Post graduate high level courses in water resources engineering are of a combined type of classroom instruction and open seminars for practical works. During 1994-1999 a two-year MSc.-Course devoted to the use of GIS in environmental protection was organized by the Bucharest-Technical Civil Engineering University (UTCB). This course has had the financial support of Sörös Foundation and supervision of UniGIS Consortium - U.K.). A number of about 50 trainees graduated the course.

During the academic year 1998-1999 a Post graduate high level course has been organized by UTCB with the support of the World Bank and with the scientific and host laboratory assistance of NIMH-Hydrology Division, and "Romanian Rivers" Company. This course has two main topics: Geographical Information System and Integrated Physical, Chemical and Biological Models. A number of 18 trainees have attended the course. The biggest merit of the post graduate courses consists in the fact that due to their well selected curricula, they provided trainees with a large fan of subjects, on a multidisciplinary basis. So, the graduates have had the possibility to respond to a broader choice of opportunities for employment. Moreover, a part of them, who are employed in the same hydrology-water planning and management unit have had the opportunity to form together multi-disciplinary teams for integrated projects, fact which enable to heighten the effectiveness of CET (Van der Beken, 1995).

Since 1990 a number of 15 fellowship granted students followed the international Post University Training Courses in Hydrology in Laussane, Budapest, Padova, Prague, Madrid, Moscow and Tel Aviv (Gilbrich, 1991). In general, this form of courses does not integrally meet the individual needs of the trainees but they have been useful for developing their basic education.

Short courses in hydrology covering a month (2-4 hours weekly) were organized from 1992 to 1996 by the NIMH-Hydrology Division and "Romanian Rivers" Company. During this period they have had various topic as: deterministic hydrological models (VIDRA Model) and models for diffusion and transport of pollutants.

(ii) **Needs concerning the knowledge on the use of new technologies** for automatic data acquisition, transmission and primary processing. The education programmes to accomplish these needs are especially destined to the hydrology technicians. The expectations are: the knowledge of the functional principles, manipulation and maintenance of the automatic devices and data loggers as well as the knowledge of the new facilities for computer primary processing of data. The most appropriate forms of this type of CET are: on-the-job training and short courses. In order to follow national standard procedures these forms are assisted by manuals and guides carried out by the NIMH - Hydrology Division.

(c) **Phase III** is oriented towards modern research approaches in order to keep abreast with new emerged major problems of hydrology and water management. This activity regards the research staff (senior hydrologists and even young hydrologists deeply and very successfully involved in the research activity). The most appropriate forms of this type of CET are: workshops, seminars and tailor-made courses.

The workshops offer opportunities for senior researchers to undertake works concerning special hydrological needs of "avant-garde" type, as: implications of the climate change and its variability on the water balance components, water management in urban zones, new

stochastic methods applied in hydrology and biological aspects of the hydrological cycle (BAHC). More than 20 seminars were organized by the Romanian Academy-Commission of Hydrology. They were focused on developing special domains as: spatio-temporal analysis of hydrological series, multivariate statistics in hydrology, fractals in hydrology; and transfer of the hydrological data from small to large scales.
For all forms of CET activity the self-learning is seldom associated, particularly to the on-the-job training and the classroom courses. This method has proven to be very efficient provided that the tutor and/or the professor has had special care to select carefully the text books and the printed material and to ensure the necessary number of copies for the individuals grouped in a certain CET form.

THE EVALUATION OF CET ACTIVITIES

This evaluation is achieved at each step of a CET program: project evaluation, evaluation of implementation and evaluation of final impact of the program.

In the first step (project evaluation) the following curriculum components are analysed: objectives, target team, trainees background, contents, execution supporting techniques and strategies of teaching.

In the second step (evaluation of the implementation) the following aspects are considered: (Allaburton, 1991; Stanescu, 1995) hierarchical selection of the priorities and the methods of teaching (to explain why they need to know and understand, how will they apply in their job, to teach at the learners level and to prepare intermediate knowledge the trainees need to have before the lecture, to relate the new content of the lesson to previous lectures or prior knowledge, to apply multi-sense learning, quality of continuing evaluation. and operative feedback).

In the third step some approaches to evaluation have been used, as follows:

(a) For the postgraduate courses the evaluation of the global theoretical knowledge of the students for each discipline is organized during the classroom course period, short after the teaching of the discipline has been completed. The evaluation generally consists in written tests covering the main issues of the whole material. The written test comprises the general and specific objectives of a certain discipline and their corresponding items, aiming to evaluate abilities of the creativity development of fluidity, flexibility and interdisciplinary thinking. The items are derived from the general and specific objectives and they are questions with short answer (true/false or yes/no) or with "structured answer" (description of the steps to achieve the objective).

(b) Just before the completion of the classroom courses the trainees should present a project description of the "diploma project": The presentation of the "diploma project" and its evaluation by a quality-judgement commission is the last phase of the global evaluation.

(c) For some types of CET an "external" evaluation is annually performed. For example, a visiting committee of hydrology specialists from NIMH-Hydrology Division makes annually an evaluation of the on-the-job training performed at the hydrological services of the River Basin Authorities. This external evaluation has become regular practice and is included in the official targets of the NIMH-Hydrology Division.

GAPS AND WEAKNESSES OF CET

The gaps and weaknesses of CET may be grouped according to the following issues:

(a) **Organizing issues**. CET does not have a sufficient financial support.. The accreditation at the end of a CET activity has not yet been generalized.
The policy of the decentralization in the domain of the water resources management has not yet been followed by a special but very necessary CET activity, namely "the training of the trainers". A first training of trainers in the field of hydrology for professional schools has been achieved in 1997 in the framework of a PHARE-Vocational Education and Training Program (VET-RO-9405 PHARE project).

The assessment of CET quality is thus made to meet the expectations of the customer and its satisfaction concerning the products of hydrological synthetic parameters and forecasts. According to the current view for the perception of CET quality the starting point in the appreciation of quality is the customer needs. The practice to advise the customers concerning the increasing complexity of the hydrological parameters and forecasts should be a preoccupation of the NIMH–Hydrology Division

During the process of CET the self-checking evaluation and the feedback for the improvement (trail-evaluation-redesign) of curricula and syllabus as well as the execution-supporting techniques for a given form of training is newly but partially implemented.

(b) **Execution-supporting techniques**. All the forms included in the generic class of open learning: Distance Learning Systems, Computer-Aided-Learning etc. are not yet possible to be implemented due to the low ratio between the available computer supporting techniques and the requested advanced learning technology. There are some "electronic classrooms" but they are far from being sufficient.

INSTEAD OF CONCLUSIONS

(a) The present methods for continuing education and training used in Romania are adapted to the needs of the hydrological activity and to the available technical supports, taking into account the social and economic development of the country and the actual context of institutional organization of hydrology.

(b) The CET activity has a good tradition in the field of hydrology and it has been performed since the systematic hydrology activity has occurred (45 years ago) in spite of the financial difficulties.

(c) Although there exists an experienced personnel which can ensure a good level of training either as supervisors or instructors, the level of educational technologies does not allow to develop high level methods for CET execution.

(d) The main present preoccupation should be focused on establishing the criteria to assess, in an as much as possible objective manner, the quality of CET activity in the field of hydrology. Also, the development of standard criteria for the assessment of efficiency and especially of effectiveness under the Romanian present social and economic conditions should be considered.

References

Allaburton, R. (1991) Effective on the job-training in hydrology. *Technical Documents in Hydrology*, IHP-UNESCO

Gilbrich, W.H. (1991) 25 years of UNESCO's programme in hydrological education under IHD/IHP. *Technical Documents in Hydrology*, IHP-UNESCO

Nash, J. E., Eagleson, P. S., Philip, J. R. & Van Der Molen, W. H. (1990) The education of hydrologists. *Hydrol. Sci. J.*, 35 (6)

Stanescu, V. Al. (1995) Workshop Report on some items of Continuing Education and Training for Hydrology in Romania. In: *Evaluation and Assessment of present and future activities of Continuing Education and Training (CET) in Water Resources Engineering, TECHWARE Workshop Document*, 151-166

Van Der Beken, A. (1993) Continuing education in Hydrology. *Technical Documents in Hydrology*. SC-93/WS.27, UNESCO

Van Der Beken, A. (1995) Efficiency and effectiveness in course evaluation. In: *Evaluation and Assessment of present and future activities of Continuing Education and Training in Water Res. Engineering, TECHWARE Workshop Document*, 167-173.

A new paradigm for education in water and environmental management

Abbott M. B., Price R. K.
International Institute for Infrastructural, Hydraulic and Environmental Engineering, Delft, The Netherlands

Abstract
There are far-reaching changes taking place in present society that indicate a new paradigm for education in water and environmental management. Particular issues reveal the emergence of the paradigm. For example, there are the importance of socio-technical solutions to management problems and the wide-spread distribution of computer- and Internet-based tools. Then there is the development of new solutions first in developing countries and then transferred to the developed countries. Further, there is the growth in education and training of engineers from third-world countries. A philanthropic strategy for implementing the paradigm more effectively is introduced. It is argued that this offers a significant way forward for the human community.

1. Introduction

On a world level we are becoming seriously short of fresh water resources. This presents radical challenges to human society to remedy the situation and to ensure a sustainable resource. Remedies are being sought using a variety of means, including technological, institutional, social, economic and political. It is, however, apparent that within areas of water and environmental management there are rarely nowadays any purely technical or social problems: nearly all problems are of a socio-technical nature. It follows, for example, that any introduction of technology cannot normally be effective without more general social and institutional changes. Correspondingly most social changes cannot be realised without the introduction of new technical means to catalyse these changes. As a consequence the hydraulic engineer and environmental scientist can no longer be regarded as 'homogeneous', but 'heterogeneous'.

The growing world-wide crisis in environmental and water resources management has lead to the 'developed' ('first-world') countries seeking to help the 'developing' (or 'third-world') countries through activities such as aid programmes promoted by the World Bank and other similar institutions. In the past, an engineer or scientist from the 'developed' country would simply travel to a 'developing' country as an 'expert', to 'establish the facts'. This person was not called on to make judgements of the social desirability of the project to which the facts applied, even less to make arrangements for such judgements to be expressed, and even less again to design the socio-technical arrangements for this purpose. Today the engineer and environmental scientist is in the first place one who persuades the rest of society to follow one line of action rather than another, whether individually or (which is much more common) as part of a team with this same purpose. Increasingly, this line of action concerns the provision of the technical and social means for other persons to choose one line of action rather than another.

It follows that it is essential that the individual or at least some members of the project team speak the same language and, more importantly still, share the culture of those who are being persuaded. This is a matter of education, and highlights the need for academic institutions involved in education to provide these sorts of persons. To the extent that these persons have acquired knowledge of the facts, so they are empowered, or empower themselves, in relation to those whom they seek to persuade. In effect, the engineers or scientists who originate from a third-world society enter into a relation with their own society that places them in a first-world situation. These persons enter into a 'first-world' that is less and less demarcated by political-geographic divisions and is instead much more demarcated by the realities of knowledge-power relations. These relations transcend national boundaries, regions, and all other such geographical constructs. What is more, through their adoption of electronic networking such as Internet, these engineers and scientists are now less than a heart beat away from their colleagues around the world. We are seeing the emergence of knowledge 'boundaries' that are beginning to supersede geographical boundaries.

This process is now being greatly accelerated and strengthened through the ongoing process of providing tools, working environments and languages for water management. In effect the engineer and environmental scientist have become 'tool-using animals'. An increasing flow of equipment, primarily computer-based but increasingly Internet-based, is changing the way of working of the professional in these fields as never before. In the case of hydraulic modelling tools alone, the few major European suppliers have now equipped some 4000 organisations in some 80 countries with the most advanced knowledge-encapsulating facilities available. At the same time there is growth in new agent-consulting companies to support this market. Complete 'knowledge centres' have been formed to support hydraulic and environmental modelling tools and more general hydroinformatics activities in all regions except Central Africa. In effect, a whole new structure of knowledge/power relations has been set up, and one in which there are less and less clear or discernible differences in these relations between persons of different geographical origins and current locations.

This situation is reflected by the growth of interest world-wide in 'knowledge management'. A proper management of knowledge is recognised as being important, if not critical, for many commercial and business activities in the first-world. In 1998 the World Bank acknowledged in its Annual Report (http://www.worldbank.org/wdr/) that knowledge transfer and knowledge management is at the heart of development in the third world (see also the reactions to this report by PANOS (1998) at http://www.oneworld.org/panos/home/knowlpap.htm).

2. A new paradigm

This paper claims that there is a new paradigm of knowledge acquisition, generation, dissemination, application and production that has radical implications for the relations between first-world and third-world countries, at both the educational and implementation levels. A new trans-national community of engineers and environmental scientists is being formed within this new paradigm, constituted by a common class of tools and a common mode of their application, and which largely meets in a virtual manner through modern communication facilities. The community consists of both geographically first- and third-world engineers and environmental scientists. Those coming from the geographical first-world regard the community as essential, simply because a major part of all development in the field is first applied in the geographical third-world. This is because there appears to be a vested interest of consultants and contractors in the first-world to keep technological advances as slow and incremental as possible, thereby minimising the rate of investment in knowledge acquisition for a given rate of circulation of knowledge. Projects in what are geographically third-world countries have proven to be decisive in 'unlocking and releasing' technological and scientific developments in first-world societies. So the sustainability of technical and scientific developments, at least within a socio-technical context, is every bit as much a requirement for the first-world as it is for the third-world.

Ostensibly, the interests of a first-world region such as Europe have to be reconciled with those of the third-world. These can be said to be reconciled through the principle of philanthropy, which is understood as the sum of all those acts of generosity, as expressions of human solidarity, that serve to hold the most disparate parts of society together. In this sense philanthropy is essential to the stability of civil society, and can be said to be of the essence of such society. Traditionally all persons and groups of any wealth and social responsibility have practised philanthropy. Today, for example, the 'charity business' of the United States has a turnover bigger than the gross national product of China, while half of all Americans choose to devote some of their time, as well as their money, to charitable work. This is partly because, in order to maintain the stability of a civil society within the USA, philanthropy has been increasingly directed to the poor within its own frontiers. This is particularly true for education (see Financial Times 9/10 May (1998)) where "the value of [charitable] foundations to a democratic society is precisely that they are undemocratic". In Europe however, money-flows towards education are primarily directed through a state apparatus that is influenced by political (democratic) considerations. There is therefore a difference even within the first-world as to the practice of philanthropy, both within a particular first-world society and its relations with third-world societies. This brings tensions in educational institutions within Europe that are seeking to facilitate knowledge transfer between the first- and third-world societies within the new paradigm outlined above. The purpose of this paper is to explore how such tensions can be resolved.

3. Towards a strategy

A tacit assumption made within foreign aid educational programmes is that there needs to be a sort of 'technological apartheid', with 'high-tech' technology best suited for the first-world and 'low-tech' technology for the third-world. The notion that there could be a high-tech socio-technical community transcending geographical divisions and that can become self-sustaining through business enterprise has been little entertained by the aid-giving institutions in Europe. Consequently, existing funded projects are couched in terms that are largely seen as ineffectual and even self-defeating when seen from within the new paradigm, while the new paradigm itself is all too often regarded with suspicion by persons outside it as being too commercial, technically one-sided and inappropriate.

There are also problems encountered with the traditional European approach to education with its Cartesian mode of thinking as expressed in terms of 'objectivity-methodology-theory-experiments-results-conclusions'. This approach is often inappropriate and counter-productive when applied outside the rigid structure of a post-1816 Ecole-Polytechnique-like structure. Such a narrow, authoritarian, 'this-is-the-way-to-do-it' approach comes into direct conflict with basic Buddhist and Hindu philosophies, and other philosophies also. There needs to be a restructuring of knowledge traditionally accepted within Europe so that it will coincide more with the cultural diversity of those to whom it is transmitted.

What is at issue is the building of an educational community that is self-sustainable through its dependence on business viability. It needs to be open to areas that are subject to rapid socio-technical change, where openings to business opportunities present themselves at fairly frequent intervals. This leads to a circumspective observation of the socio-technical market for future alumni of the community and the development of appropriate teaching and research plans. In turn, this would lead to the involvement of the major water-industrial management and communication companies and the encouragement of alumni seeking to establish their own business ventures. Business administration should be as much a part of the learning environment as the engineering and the science. This implies the need for building business and professional networks that facilitate knowledge management in support of the emerging socio-technical community. Such networks are primarily networks for encapsulation, transport, distribution and processing of knowledge. They are constructed incrementally and for the most part opportunistically. What they do is provide a means for mobilising the community (including alumni) to constrain, redirect and otherwise modify the political and institutional objectives of state- and community-financing and other interested agencies, which are subject to philanthropic forces.

In turn, there is a need to 'tool-up' education with the most appropriate simulation and other modelling systems, necessarily supported by Internet. This carries with it the need to adopt the new ways of thinking that these tools justify, and without which they will be of an unnecessary limited application. In particular, whereas the use of tools can be said to be constraining in the sense of how things are done, they open up a far greater range of what things can be done. It appears therefore that the emerging subject of hydroinformatics (Abbott, 1991) can be claimed as having a central role in implementing a suitable strategy, though clearly its success depends on the active participation of those in more traditional subject areas of water and environmental science and engineering. It also depends on active collaboration with various industrial organisations who are the specialised, commercial tool producers and users. They are involved in solving real world problems using tools and tools sets, and therefore themselves part of the socio-technical community.

4. Conclusions

Education in water management should be seen increasingly in terms of a new paradigm. Most problems are seen within this paradigm as socio-technical problems. These require more-or-less heterogeneous engineers and scientists for their solution and necessitate the appropriate use of tools, tool sets, networks and all other such equipment. Generally hydroinformatics provides the source of such tools and methodologies which need to be diffused to other groups internally and externally through Internet. The community that begins to form in this way must be seen as an integral part of the much wider community that is forming itself in similar ways. It encompasses tool producers, other expert tool users, and commercial organisations that are committed to a similar path, and especially those in what are, geographically speaking, third-world societies.

The motivation of the educational institution concerned with water and environmental management comes from regarding technology as creation within the world and science as understanding the workings of the world. This is admittedly a Christian ethic that is largely dominant in Europe. The interventions of creative actions and understanding generated by the teaching in the educational institution modify and catalyse the knowledge/power relations. These interventions have the capacity to lead to a reduction of misery and suffering in the world of both humans and nature, and are the means for protecting and even restoring the natural world.

The question 'why?' therefore becomes of growing importance in the new paradigm compared with the traditional 'how?' Studies in ecology require some understanding of its relevance to design and other decision-making processes; geographic information studies should include some awareness of the knowledge/power relations that are at the heart of geography; water resource teaching should be supported by the effects and consequences of bio-diversity and human-social arrangements.

In addition, the Christian ethic, in company with others, encourages the sustainable growth of community, no matter where it is focused. This is now more feasible in, say, third-world countries through setting up and maintaining 'knowledge centres'. These are socio-technical enterprises that depend on electronic knowledge transmission, distribution and processing networks. Setting up such enterprises is non-trivial in that resources allocated to such projects most likely have to be minimised while the efficiency of the socio-technical complex is maximised. In turn the proclaimed and contracted objectives have to be achieved. Such enterprises cannot be set up by an educational institution alone; it must have the support and co-operation of industrial and other partners.

A socio-technical community focused on education in water and environment management is held together both by its social and technical relations. It succeeds through the promotion of philanthropic values. It is at this level that the Christian ethic becomes associated with philanthropic elements in other philosophies, whether in Islam, Buddhism or Hinduism, and indeed with such elements in all religions, theistic or atheistic.

References

Abbott, M B (1991) Hydroinformatics: Information technology and the aquatic environment. Avebury Technical, Aldershot, UK, p145

Continuing education in water sciences and the environment. Tradition and the new social and economic framework in Romania

L'éducation permanente dans les sciences de l'eau. La tradition et le nouveau contexte social et économique en Roumanie

Damian R. M.
Technical University of Civil Engineering Bucharest, Romania

Abstract
Continuing education (CE) has always been a traditional activity in Romanian universities. Under the communist regime the state-owned enterprises were obliged by law to train their specialists. For university graduates the role of universities as providers of continuing education was formally recognized, but several adverse factors prevented both parties, learner and provider, to make full benefit of such activities. Following the dramatic changes in 1989, Romania is slowly adapting to the requirements of a society based on the free market economy. Higher education has become a very dynamic sector, with new curricula, new post-graduate study directions such as advanced studies and doctoral programs and an explosive offer for CE. Several leading universities have included water sciences and environment among their offer for CE, many in close relationship with professional organizations. The current implementation of CE in both domains faces many challenges in the present economic and social context.

Résumé
La formation permanente a toujours été une activité traditionnelle dans les universités Roumaines. Sous le régime communiste les entreprises d'état étaient obligées par la loi d'assurer une formation a leur spécialistes. Pour les diplômés de l'enseignement supérieur le rôle des universités comme fournisseurs de services de formation permanente était formellement reconnu. Toutefois, plusieurs facteurs empêchaient les deux parties de bénéficier pleinement des résultats de leurs activités. Par la suite aux changement dramatique de 1989 la Roumanie est en train de s'adapter lentement aux sollicitations d'une société basée sur l'économie du marché. L'enseignement universitaire est devenu très dynamique, avec l'élargissement de domaines d'études, les nouveaux études approfondis et programmes de thèse ainsi que l'ouverture quasiment explosive pour la formation permanente. Les sciences de l'eau et de l'environnement sont offertes comme direction de formation permanente par plusieurs universités d'élite, parfois en étroite collaboration avec autres associations professionnelles. L'implémentation de la formation permanente pour ces domaines est confronté avec plusieurs provocations dans le contexte social et économique actuel.

1. Introduction

In the immediate after-war period the educational policies in Romania and the public interest have been mostly focused on undergraduate studies. In the late forties and early fifties, higher education experienced new developments, as well as other traditional components of the "formal education". There are several explanatory reasons for such policies, such as the need of the "Party-State" to control politically the educational institutions and subordinate education to ideology, according to the analysis in a recent PHARE publication (1998). Although the law imposed on the state-owned companies to organize training activities for their specialists, the whole process was rather bureaucratic and often a mutual lack of interest for it was manifest for both parties, learner and training/education-provider. One should also remember and mention the existence of a universal form of a substitute "Continuing Education(CE)", namely the "Political Education", addressing all individuals irrespective of their position in the society, which induced a general reluctant attitude towards this altered idea of CE. In spite of these unfavorable conditions, the need of real continuing education was recognized by the enterprises which, especially in the early sixties, took some measures to organize their own internal training process or to send specialists, mostly university graduates, to the few state-supported training courses offered by the higher education institutions. However, it is only fair to say that in many technical universities the "postgraduate courses" were also partially supported by some enterprises which understood their real importance and significance to their staff policy. Damian (1998) underlines that in the water sector this cooperative action was a common case with many higher education institutions, such as the Technical University of Civil Engineering Bucharest, the University "Politehnica" Bucharest, the technical universities of Iasi, Timisoara and others. The links with the Romanian Waters Authority and with the local authorities in charge with water distribution and sewerage systems were quite close in training and research. Also, collaborative work with design and research institutes in the water and environment sector has lead to several joint training activities involving the participation as associate professors in training courses organized by universities of many specialists from institutions such as the Institute of Environmental Protection Research, the Institute of Hydrotechnics Research, the National Institute of Meteorology and Hydrology etc. Unfortunately, if the need for adequate water management with priority given to power generating and water supply was considered to be important, the environmental component was approached mostly in a declarative way. This attitude reflected on the contents of the CE processes in which the environment-related component was rarely a central point.

2. Educational policies in a new economic and social context

In a recent contribution, Damian (1999) points out that under the communist regime the main characteristics of the Romanian economy were, in the name of "building the socialism", a policy of extensive industrialization and the almost total destruction of private land property in agriculture. This is probably the key of the differences between the starting point of Romania in 1989 with respect to those of other countries in the area, such as Hungary, Poland, the Czech Republic etc., which are now better placed in the effort to join the EU and NATO. Whereas in these countries the seeds of the free-market economy were still kept in place and even stimulated to develop, in Romania even the concept of a free-market economy, not to speak of "capitalism", was banned not only from the books but from everyday practice. Thus, Romania is one of the Central and Eastern European Countries (CEEC) in which the transition of the economy towards the free-market is slow and very difficult, due to the inherited structure of the economy, with many inefficient enterprises and lack of a "free-market" culture, but also to the lack of consistent support from the international finance and investors. However, in the last weeks of 1998, when it became obvious that the economic situation of the country was to become very difficult, the present government was forced to move on in what is thought to be the right direction: closing down a number of inefficient enterprises, privatization and other economic measures aiming to keep the economy alive.

In the author's opinion, in spite of the present difficult economic situation, one of the first sector which shall recover is the infrastructure. In fact, the rehabilitation of existing water supply and sewerage networks of 16 cities has already started, financed by the Romanian government and a World Bank loan. If we think that only 10% of Romanian villages have running water and less than 5% have centralized sewerage systems, with more than 40% of the country's 22 millions population living in villages, we have a first dimension of the problem.

One major action which is in progress is the harmonization with the EU in water policies and environmental legislation. In Romania the water management, based on the river basin concept, is now becoming an "integrated water management", along with the restructuring of the Romanian Water Authority. Then, if environmental protection is to become a priority, as it should, the objectives would be to allocate resources to new technologies and for protective measures, to re-train the specialists and to educate the general public.

Thus, the transition must be supported by education at all levels. The challenges education faces are manifold, one important reason being the fact we are living in a changing world, in which even better-off economies and societies experience difficulties. In the CEECs, such as Romania, one of the most difficult processes is to adapt adults to a different mentality, which has little to do with the egalitarian approach very much praised under the former communist regime.

Romanian universities must consider at least the following basic realities related to continuing education programs they intend to offer:

a) internationalization of the market, leading to the constant emergence on the market of new technologies, products and materials, at a rate which does not compare to any previous situation;
b) diversification of the offer for education, with the emergence of newly created private universities and other training organizations, public or private, which eliminate the monopoly-status the public universities were used with;
c) displacement of the interest of the younger generation from technical studies to other fields; for instance, it is significant that in the private universities economics and law studies share almost equally 80% of the students whereas less than 1% study engineering;
d) difficulty to adjust the educational offer to an uncertain labor market.

Here we come to a basic question: is Romania a "learning society", in the perspective of the next millenium? As it is known, this concept was introduced by Husen (1974) who pointed out four main characteristics of it:

- the social status of the individuals will depend more on his/her "educated ability" rather than on inherited social and/or material conditions;
- increased role of experts and technocrats;
- generalized access to knowledge due to the spectacular evolution of information technologies;
- the principle of "equal opportunities" becomes effective at all levels and for all individuals and communities.

At this moment, I have doubts if any country really fits these idealistic criteria. More recently, at a joint meeting of presidents/rectors of European universities with owners and managers of big companies, the concept of "learning society" was somewhat redefined, on the basis of the following five items:

- long-life learning;
- responsibility towards own social and professional evolution in life;
- evaluation focused on confirmed progress rather than on failure;
- team-working and participation to civic activities;
- social partnership.

To try to answer our question we may look at three criteria which simplify the task: legislation; basic education for all; offer for continuing education.

a) *Legislation.* In the lack of a special law for continuing education, a number of provisions of other laws, including the Law of Education 84/1995 refer to this activity. One important missing point is the obligation of companies, public or private, to spend a percentage of their income on training, which leads to a permanent shortage of financial support for CE;
b) *Basic education for all.* The transition in Romania induced a number of favorable evolutions, such as diminishing of the ratio teachers/pupils from 20.3 in 1990 to 16.8 in 1996, which contributes to a better quality in the teaching process, diversification of the educational offer etc. In the same time, an involution process is taking place, with an alarming drop-out rate at the level of general compulsory education, increased disparity between regions, higher functional illiteracy etc.;
c) *Offer for continuing education.* In the lack of a coherent legislative framework and of sound financial support the offer for CE could be considered as spontaneous and adapted to some immediate needs. It is expected that the implication of the universities will offer more credibility and consistency of CE programs.

Hence, the real "learning society" in Romania is still in its early stages, as it is in many other countries, remaining still a challenge for this generation and for the generations to come.

3. Recent evolutions in continuing education in Romania

Dinca & Damian (1997) show that after 1989 Higher Education was the most dynamic sector of education in Romania. Under the framework of the Higher Education Reform program, initiated by the Ministry of Education and supported by the Romanian Government and the World Bank, the component for Continuing Education raised very much interest, the universities having already initiated many new study directions and curricula. More information can be found in the paper *Development of Continuing Education in Romania* (1998), produced by the National Council for Higher Education Financing of the Ministry of Education. The council has developed a database with the majority of programs offered by public universities. It is interesting to note that, from 347 reported on-going CE activities in 27 public higher education institutions (HE) (total public HE in Romania: 52; not all universities contributed to the database), 180 are in medical areas (data from five medical schools; very detailed topics taught), 30 in Economics, 80 in Engineering etc. but only four have as the main topic Water sciences or Environmental protection.

One very important support to develop CE programs was the access to TEMPUS PHARE funds from the EU. Many of the new curricula and educational techniques, for instance distance learning, in Romanian universities were implemented with participation of representatives of higher education institutions from the EU countries. In the water sector, one important program is the Postgraduate school in Water resources engineering developed by the Technical University of Civil Engineering Bucharest, addressing 25 students per academic year for a two-year curriculum. To be noted that, with the contribution of a TEMPUS project all the learning material has already been produced, partly in Romanian and partly in French.

Networking activities, with an European dimension, have started to develop either as follow-ups of TEMPUS programs or as a consequence of the eligibility of Romania for SOCRATES and LEONARDO programs. One such project is, of course, ETNET ENVIRONMENT WATER, which networks a large number of professionals and trainers. It is also very important to mention TEchnology for WAter REsources (TECHWARE) as a successful University Enterprise Training Partnership (UETP) with more than 200 partner institutions, in which we find five Romanian technical universities as well as seven enterprises and professional associations in the water and environment sectors. This is another proof that, in this important direction of development of the infrastructure, Romanian higher education institutions are fully involved and ready to meet the quality requirements for continuing education and training.

Participation of other non-academic professional organization offering training in the Water sector becomes more and more significant. One good example is the foundation created under the auspices of the National committee of water producers and distributors in Romania, namely the *Center for Professional Formation and Documentation in the Water Sector*. Of course, collaborative work with higher education institutions is already effective, their competence and role in the training process being well understood and considered to be beneficial.

4. Conclusion

In Romania, we are witnessing in education a transition from state-controlled policies and formalism to realistic approaches. The offer of Continuing Education and Training is becoming diversified, being provided by universities, professional organizations, local private organizations, international organizations related to companies trying to promote their own technology, products and know-how etc. As it should be expected, it is not an easy task. Many difficulties are encountered in the attempts to change mentalities and in reshaping curricula to meet the expectations of the learners. It is very encouraging that more and more people understand they are going to live in a competition-based society and in a global world. The challenge for the educational system is twofold since it must equally address both young and mature people in a very difficult transition period for the society.

References

*** *Învățământul superior într-o societate a învățării (Higher Education in a Learning Society)*, PHARE-Universitas 2000, București, 1998.

Damian, R.M., *Continuing Education and Training. Specific Needs in Central and Eastern Europe, TECHWARE- Euro-Workshop: Towards the learning Society in the Water Industry,* Antwerpen, 12 June, 1998.

Damian, R.M., *New developments in economic and technical education in the transition to the free-market economy in Romania*, Relazioni Internazionali, 48, ISPRI, Milano, 1999.

Husen, T., *The Learning Society*, Methuen, London, 1974.

Dincă, Gh., Damian, R.M., *Financing Higher Education in Romania*, Editura Alternative, București, 1997.
*** *Dezvoltarea educatiei permanente in Romania (Development of Continuing Education in Romania)*, Editura Alternative, Bucuresti, 1998

Education and Training in Environmental Protection - the Polish experience as Exemplified by the Activities of the Gdansk Water Foundation

Z. Sobocinski

"Education and research policies form the cornerstone of European integration..."

Edelgard Bulmahn,
Minister for Education and Research, Federal Republic of Germany

"Éducation et Formation pratiques dans le domaine de protection de l'environnement - expériences polonaises - exemple de la Fondation de l'Eau de Gdansk"

"Éducation et politique des recherches créent les fondements de l'integration européenne..."

Edelgard Bulmahn,
Ministre allemand de l'éducation et des recherches

Abstract
The political and economic changes in Poland have brought in their wake an educational system whose object is environmental protection. The introduction of modern technologies and new legal requirements have made it necessary to improve the professional knowledge and skills of all employees, especially those directly involved in environmental protection. One of the institutions providing training in environmental protection in Poland is the Gdansk Water Foundation. This body originated in co-operation of Denmark, France and Poland. Its activity is based on the conducting of training in the form of lecture series, complemented by workshops which often end with a presentation of the achievements of western countries. Lately, training through the Internet has been introduced. In this paper we will try to present the experience of the Gdansk Water Foundation in organising and conducting training courses in environmental protection broadly understood.

Résumé
Les processus des transformations politiques en Pologne ont entraîné la création d'un système d'éducation dans le domaine de la protection de l'environnement. L'apparition des technologies et dispositifs modernes a imposé le perfectionnement des qualifications et le besoin de la mis à jour du savoir-faire par tous les employés, surtout ceux qui s'occupent direstement de la protection de l'environnement. Une des institutions qui s'occupe de l'organisation des cours de formation dans ce domaine est la Fondation de l'Eau de Gdansk. Cette organisation a été créée en résultat de la collboration de Danemark, France et

Pologne. Son activité est basée sur la réalisation des cours de formation comprenant les conférences et les exercicies pratiques complétés par la présentation des expériences des pays occidentaux. Les formes non-conventionnelles de formation par l'Internet ont également été introduites. La conférence présente les expériences de la Fondation de l'Eau de Gdansk en matière de l'organisation et gestion des cours de formation liés à la protection de l'environnement dans le sens large de ce mot.

Introduction

The great political and economic changes of the late eighties clearly indicated the necessity of devising an adequate educational system specifically devoted to environmental protection. A crucial factor was the new way of perceiving environmental protection that was slowly becoming an integral part of the economic development of the country and every investment undertaken. Environmental protection slowly but surely started to attain its rightful position. An increasing concern with the worsening condition of the environment resulted in the introduction of more demanding legal regulations. There was a growing realisation that the world where we live and work is important not only for us but also for future generations.

On the other hand changes in Poland's political system resulted in new equipment and technologies becoming available on the domestic market, which offered lower exploitation costs, were more environmentally friendly, and made it possible to meet the more stringent legal requirements. They required new knowledge and a higher degree of professional skills, while at the same time changing the philosophy of designing and planning of new investments. All this in turn necessitated the introduction of new legal regulations. The constant adaptation of law in conditions of rapid economic development and the use of new technologies made it imperative for those engaged in a variety of fields, not only in administration, to supplement their professional knowledge and skills to the required level.

It is indicative that at that time there was an immense growth of interest among Poles in the legal regulations applied in West-European countries - the result primarily of a desire, and later a conscious wish, to join the European Union. In the absence of Polish legal norms the regulations of the European Union were quoted and the argumentation here was obvious: in the prospective integration process Polish law must be in conformity with the European solutions. The EU directives and recommendations should by now be regarded as the basis for solving our problems - and not only legal ones - as they arise.

At the same time it has become apparent that a constant increase in knowledge is necessary if one wants to play a role in society, an awareness that is constantly growing.

The importance of training institutions

This need, widely felt among employers, for staff with higher qualifications and greater knowledge created a market for training bodies. Many new firms were established in Poland whose activity aimed at the organising of professional training. Universities, research and science institutes, social and professional organisations and other firms also started to provide this type of service. For them training activities held out the possibility of learning to prosper in a new reality.

The remarkable growth of institutions organising training in the field of environmental protection was also caused by the activities of numerous educational firms and institutions from Western Europe and North America that established themselves on the Polish market in the early nineties. It was the time when Poland received extensive help in many forms to assist the transformation of the country's political and economic system. A great role was played by the training and educational programmes which were adjusted to the specific problems of the different regions. Thus in Upper Silesia the emphasis was put on protection of the atmosphere and the disposal of waste products, in the Mazurian Lakes Region priority was given to protection of water and the natural environment, while in the Gdansk Region the focus was especially on the protection of the shore waters of the southern Baltic. That period was characterised by gratuitous training, its cost covered by the financial resources granted to Poland. Many courses were conducted in foreign languages, and hence considerable attention was paid to learning these languages. Despite some problems in co-ordination of training action undertaken by foreign bodies (courses were not always organised where they were really needed), it should be stated that those efforts generally brought very positive results. It is clear that on the one hand these activities constituted an initiation and enhancement of public awareness in the field of environmental protection as well as a developing of educational requirements in that field; on the other hand many of those trained went on to become speakers or trainers in subsequent courses, even choosing that activity as their new profession.

As mentioned at the beginning of my paper, changes in Poland's political and economic structure caused a significant growth of interest in professional education among employees and employers. This interest gave rise to many new firms which have organised and conducted training in environmental protection. The slogan "environmental protection" became very popular and fashionable. Firms dealing with water and sewage treatment, solid wastes or air pollution found themselves receiving hundreds of invitations to take part in training courses. Efforts were made to convince management in the public and private sector of the benefits of such courses. Efforts were also made to make such courses as attractive as possible, especially by inviting persons to lecture who were widely regarded as leading authorities in given subjects, and who might be expected to draw large numbers of participants. This was to ensure commercial success, while the effectiveness of training was less important.

Receiving lots of invitations for courses but lacking the necessary discernment as to the educational value of the services offered, firms were deciding upon participation on the basis of names of known experts, scientists, heads of state administration units, representatives of foreign research institutions, international organisations or firms active in given fields. Generalising, it can be said that the nineties were characterised by a large number of different educational enterprises mainly of conference character, with the participation of large numbers of trainees. In those courses the "conference method" of knowledge-transfer tended to dominate. It was based on numerous, concise presentations related to the problems discussed, followed by short discussions. This method of transferring knowledge did not always satisfy the participants.

In Poland there are no legal regulations as to continuous training of personnel in a given post, nor training requirements when the post is being changed. Participation in training depends mainly on the awareness of the management and a conviction that that training will be advantageous for the employee and profitable for the company. Similarly, there are no regulations stating that a part of a company budget ought to be specially set aside for staff training purposes. As a result of this situation it is mainly people at managerial level and less the technical staff of a firm that are participants of training courses.

Activities of the Gdansk Water Foundation

In 1994 the representatives of the Region Limousin in France commenced activities aiming at the establishing in Poland of a training centre for people engaged in the water sector. This centre would be in close touch with a similar training centre run by the International Office for Water in Limoges, France. Negotiations were fruitful and resulted in the establishing of the Gdansk Water Foundation by partners from Denmark, France and Poland in the framework of the LIFE BALTIQUE Programme. The European Commission financed the participation of the Danish and French partners.

The aims of the GWF are:

1. to promote knowledge of water economics, in particular the principles of water management in catchments
2. to provide additional professional training in fields connected with water and sewage economics
3. to facilitate the exchange of information and contacts between water users, central and local-level administration
4. to promote water saving and ecological awareness in general
5. to facilitate the exchange of information between scientific research centers as well as to promote the results of their work
6. to co-operate with other foreign countries in order to propagate new technologies, to unify standards as well as exchange experience and information in the sphere of water economics.

The training activities of the Foundation are addressed mainly to all those interested in issues of environmental protection, particularly to:

- central and local-level administration
- water and sewage enterprises, communal enterprises
- design offices active in the field of environmental protection
- institutions which monitor the state of the environment and compliance with legal regulations relevant to environmental protection
- operators and designers of specific projects
- producers and distributors of equipment and installations.

Training in the Gdansk Water Foundation

From the beginning of its activity, the Foundation adopted the principle that the training should consist of lectures and workshops, in which theoretical knowledge is linked to the skills necessary for applying that knowledge. The decision to accept such a form of training was the result of an analysis of the opinions of participants of other kinds of training courses, for these stressed the need for broad discussion of the presented problems, greater contact with the lecturer, or presentation of their own professional problems in order to obtain help in solving them. The adoption of such principles entailed specific forms of organisation. It should be mentioned that in determining those forms, Danish and French experiences were taken into account: methods used in these countries were adapted to Polish conditions.

Firstly, it was decided that the groups of participants should not consist of more than 15 to 25 persons, as this was the number that made possible the greater degree of contact with the lecturer and ensured that the participants did not remain anonymous; at the same time the participants had more opportunities for exchanging opinions, which in professional life often plays an important role. It was also accepted as a rule - and stressed at the beginning of each course - that the lecturers should be treated as colleagues who merely possessed a greater degree of knowledge and experience in the same profession. This makes it possible to create better relations between lecturer and trainees. In smaller groups participants are less inhibited and more open and do not feel that their knowledge disqualifies them from being equal partners in professional discussions. Sometimes it happens that the prestige of a known authority in a given field intimidates the participants, the atmosphere becomes tense, and much organisational effort is needed to change the situation.

Introducing a moderator from the staff of the Foundation was another way to activate course participants. The Foundation employs three high-class specialists in water and sewage management and analytical chemistry. In creating a proper atmosphere during the training such an expedient creates the possibility of exploiting their expertise.

Securing the above conditions has decidedly enhanced the effectiveness of the training. Much more readily than before the trainees become involved in the exchange of news, they propose lecture-modifications in order to discuss problems which they are interested in, they actively take part in the discussion, they learn to formulate their ideas and arguments when presenting them. At the beginning of the Foundation's activity it was no more than an abstract notion that what was really needed was a form of training that offered the possibility of discussing and solving individual problems; soon, however, it became a certainty that such an approach was simply indispensable for structuring the knowledge and improving the qualifications of trainees. For it results in learning to solve problems with the help of other people and in learning to share one's knowledge and experience with them. These points need to be stressed, as people were slow to recognise the merits of such an approach and to make it standard practice.

Our Foundation organises on average 30 to 34 different courses a year. As a rule the courses last 2 to 3 days, although some last four days. The duration of the courses continues to be a source of friction given the conflict of interest that exists between the firms sending their employee to a course and its organiser. Obviously the absence of an employee can cause difficulties for the company. Hence the pressure to shorten the course duration, while a course that truly does justice to its subject requires longer duration. A compromise satisfying to both parties is not easy. Such an inference may be drawn from the evaluations which trainees submit at the end of each course.

In our conditions the recruiting of trainees is a complicated business. As mentioned above, firms in Poland receive a deluge of invitations to take part in training. Which is to be chosen and what is their value? In Poland organisations dealing with training in environmental protection do not require the accreditation of any official bodies such as the Ministry of Environmental Protection, Natural Resources and Forestry, or the Ministry of Education. The bodies organising training issue a certificate of attendance at a course – but this certificate is not a qualification as such. Such certificates cannot be used by an employer as a basis for the promotion of an employee. Quite apart from being a formal qualification, a certificate issued by the Foundation is professionally valuable because the knowledge and skills acquired during the course were used in practice: companies have benefited from participation in our courses. The growing interest in the activities of the Gdansk Water Foundation can be illustrated by the growing numbers of trainees. In 1995 the figure was 111, while in 1998 it rose to 723. In 1999 it will increase further, as our courses will also be conducted on the premises of our Foundation's Technical Training Centre. At present the Centre consists of the following:

- the waste-water treatment plant (WWTP), a pilot scheme based on activated sludge,
- the water and sewage test laboratory.

When the second stage of development is completed, the Centre will additionally consist of:

- a model of a potable water pressure distribution network
- a pumping station
- a station containing additional network equipment.

The Technical and Training Centre is located on the premises of the "Gdansk-East" WWTP, which treats 180,000 cubic meters per day. The pilot plant, apart from having a broadly understood educational function, allows for the conducting of different types of pre-design research for the construction of a WWTP, as well as researches in improving the work of existing technical equipment. The exploitation of the pilot station will make it possible to organise training courses of a very detailed and precise nature.
The combination of theoretical training and workshops is considered to be a principal advantage of our Foundation as the centre for training for all people involved in the water sector. Obviously, the extensive form of training presented above still remains the

Foundation's key advantage. It is a recognised truism that examples, especially practical ones, constitute the best method for assimilating the studied material. That is why the Foundation supplements its training programme in Poland by on-site training at its partners' premises in Denmark, France and Germany. Practical experience has proven that foreign examples are a perfect supplement for the knowledge transferred during courses held in Poland.

Courses in foreign countries are organised with the aim of presenting the achievements and experiences of the countries visited. Thus we organised courses in France and Germany dedicated to communal and industrial sewage treatment technology and sludge management based on the experiences of those countries, as well as a course in Denmark entitled "Water, sewage and waste management in the Danish experience". Polish specialists had an opportunity not only to see state-of-the-art technology and equipment, but also to get acquainted with modern corporate management, its approach to the problems of environmental protection and personnel training. They saw what challenges lie ahead for them and are now more aware of the immense work that needs to be done during the integration process with the European Union.

The above remarks were intended to emphasise that the recruiting process of trainees in Polish conditions will be successful and effective only if the prospective participant is convinced that:
- his time will not be wasted
- the organiser has prepared an interesting programme
- the organiser has secured suitably qualified lecturers
- the participant will have an opportunity to address his problems
- he will have an opportunity to discuss issues that interest him
- he will have an opportunity to become familiar with new technologies and modern equipment
- he will have an opportunity to meet specialists from other countries.

When the above conditions are fulfilled prospective participants will avail themselves of an opportunity to enlarge their knowledge, to enhance their professional qualifications and thereby their chances of promotion. Providing as it does such opportunities, our Foundation strengthens its position on the environmental protection training market.

In general, Polish training bodies exploit the traditional methods of transferring knowledge. Most of them offer courses organised within the working day. However, it is observed that people interested in problems of environmental protection are spending more and more time on self-education, not only by availing themselves of an ever-richer professional literature but also by making use of distance learning centres and the Internet, which give them opportunities for self-education in their free time. The future of educational activities lies in the exploitation of new techniques. While the traditional methods of learning and imparting knowledge have many adherents, open and distance learning are becoming more and more popular in Poland, and there is no doubt that they will play a considerable role. We too have begun to introduce this form of training. The Gdansk Water Foundation is a member of a consortium composed of 7 partners from 5 countries, namely France, Poland, the Czech Republic, Romania and Lithuania, whose purpose is to prepare a series of courses in water and sewage management with a view to European integration. The final aim is to enhance the general awareness of the legal regulations in the European Union in respect of environmental protection and professional knowledge of water and sewage management. The conducting of these courses is financed by the European Training Foundation in Turin in the framework of the PHARE programme. In co-operation with the Distance Learning Centre at Gdansk Technical University our Foundation has also introduced that form of training into its programme. We are well aware that the first training courses are likely to give rise to various problems – after all, in the Polish context self-education is strictly speaking a very innovative concept, yet we have no doubts that this scheme will constitute an ideal supplement to our present activities.

Summary

The following conclusions may be drawn from the activities of the Gda_sk Water Foundation – a centre of training for people involved in environmental protection, especially in water and sewage management:

- an ongoing analysis of the wishes of prospective participants of training courses is a basis for effective activity on the part of firms engaged in environmental protection education
- it is becoming increasingly obvious that the trainee considers the lecturer as a partner who helps him to expand his knowledge and develop his needs for self-education rather than as a typical teacher who merely imparts his knowledge in the form of a lecture
- the most effective method of transferring knowledge is a combination of the traditional lecture and workshops allowing for independent solving of problems aided by a familiarisation with modern technical solutions
- while the improving of employees' professional qualifications has a serious and noticeable effect on the economies of water and sewage enterprises, this cannot be noticed to the same degree in the field of administration. What this means is that efforts should be made to make administrative staff aware of the benefits that will ensue from an increase in their knowledge and an improvement in their qualifications.
- measures should be undertaken to change Poland's system of professional training, which also covers issues of environmental protection, by introducing a system of certification and diplomas of training completion as evidence of real improvement in the professional qualifications of the employee
- legal regulations should be introduced obliging firms to earmark a certain percentage of their budget for professional training of employees. This will enforce the recognition of qualification development as an element indispensable for the enhancing of corporate efficiency.

Experience and problems of vocational training of water sector's specialists in Central Asia (Aral Sea basin)

Experience et problems de l'ensignement professionnel pour les specialistes en l'eau de l'Asie Centrale

Shapiro A. M
Scientific-Information Center of Interstate Water Coordination, Commission of Central Asia, Uzbekistan

Abstract
Agriculture is responsible for main share of water consumption and represents a source of imployment and income for bigger half of mankind. Along with the rural population growth tendencies of development become descending. As agriculture development is competative mean to eliminate the poverty and a solid base for economic development, water and land resorces, which constitute natural basis for this development, should be saved, conserved and protected for sustainable development of agriculture.In the past, planners of water and land resources management proposed structures which should response to demand of new lands and, respectfully, water. Now we understand, that this approach has changed because new methodologies of water and land resources planning become more and more complex.Training could be one of important instruments for introduction of new approaches to solve the conflicts in demand for land and water, giving attention to complex systems with due regard for natural and managed by man environment, physical and non-physical aspects, tracking water streams and related land resources starting from the upper reaches according to their different uses and users.

Résumé
Le secteur agricole est responsable pour la plus grand proportion de la consommation de l'eau et constitue une source d'emploi et de revenu a' plus de la moitié' de la main d'œuvre. Avec l'augmentation de la population rural , le standard devie suivra une trajectoire descendante.
Puisque le développement de l'agriculture est reconnue d'être un moyen compétent pour l'élimination de la pauvreté', et une bonne base solide pour le développement de l'économie, les ressources en eau et en terre qui constituent le système de soutien primaire de ce développement, doivent être gérées, conservées et protégées pour une développement durable de l'agriculture.
Dans le passe' les planificateurs de l'aménagement de l'eau et de la terre proposaient surtout des structures pour répondre aux besoins de nouvelle terre et relatifs a' l'eau. Nous somme maintenant plus conscients que cette approche est a' changer, puisqu'il y a une meilleure perception sur le ressources en eau et en terre et les méthodologies en planification, conception et exécution sont devenus de plus en plus complexe. Enseignement

pourrait devenir une instrument importante pour pénétration de les nouvelles voies d'approche de conflits de demande en eau et en terre relative, donnant une attention sur les systèmes complexes et en corrélation avec l'environnement naturel et manie' par l'homme, des aspects physique et non-physique, en commençant par la ligne de partage des eaux en amont et en traçant le courant avec la terre correspondante a' travers leur différentes utilisations et utilisateurs.

The Aral sea basin is located within the arid zone and encompasses 5 states of Central Asia: Kazakhstan, Kyrgyzstan, Tadjikistan, Turkmenistan and Uzbekistan with about 50 millions of inhabitants. Since time immemorial water has been one of the most important factors determining life and development in Central Asia, one of the world's six most ancient civilizations. Two big rivers AmuDarya and SyrDarya are main source of water in the region. Irrigation has long been one of the principal means of water use in the region dating back to the 6-7th millennium B.C. Through the improvement and expansion of irrigation 3.5 mln.ha of irrigated land were available at the beginning of the 20th century. This land, which area is now 7.9 mln.ha, irrigated by network of varying degree of efficiency constitutes the base of economic development in the region. The construction of a large number of dams, reservoirs, canals, pumping stations and drainage systems gave Central Asia a highly-engineered water management system.

At the same time the large-scale and rapid expansion of irrigated area, technological potential and consumption requirements led to the deterioration of environment bringing on ecological crisis: irrigated lands salinization and waterlogging, growing desertification, reduced water level and aggravation of water quality. By the early 90-es Central Asia faced serious socioeconomic and environmental problems due to the exhaustion of water resources as well as the reduction of irrigated farmland productivity-the base of the region's prosperity, because about 60% of population is busy in agriculture.

After the Former Soviet Union collapse, aware of the destruction of centralized management for water resources, regional water specialists realized the need for a joint effort to solve water management problems. The Interstate Water Coordination Commission of Central Asia was established in February 1992. In the past 7 years it has prevented conflict over water distribution between states and governments. Conflict-generating factors still exist today and their solution needs systematic and cautious negotiations.

During last 5 years the world community has been intensively involved in assistance for solving these problems. Under WB, EU, ADB and countries-donors funding, the Aral sea basin Program has been established and started to be implemented. Basic Provisions for the Development of the Regional Water Strategy in the Aral sea basin has been developed, which states: "To develop a common strategy of water distribution, rational water use and protection of the water resources in the Aral sea basin and prepare interstate legislative documents, regulating issues of common water use and water resources protection against pollution with due regard for the region's socioeconomic development". Along with the regional strategy development institutional potential strengthening is under progress.

Starting since 1998 the project supported by the Global Environmental Facility, the Netherlands and European Union has been initiated including 5 components: A-regional water strategy development, B-public awareness, C-transboundary waters monitoring, D-dam safety, E-wetlands restoration in the AmuDarya delta. Within this project it is expected to develop crop water requirements, methods of water saving, transition to water pricing, transfer of water rights and management functions to new established associations of water users. Total cost of the project is 22 mln.US$, project duration 4 years.

Many critical changes took place during recent several years in existing technologies and methodologies for water and land resources management. Among them creation of GIS managed regional water and land database, development of the water management mathematical models, participation in the world-wide information exchange through the IPTRID network, FAO CROPWAT methodology introduction, development of the legal base for the regional collaboration, etc. With assistance of the Canadian Agency for International Development automatic water management and control system SCADA prototype has been installed on the head structure Dustlik in the SyrDarya river basin.

Different pilot projects are being initiated all over the Central Asia relating to on-farm water management, water management automation, water saving, agriculture productivity improvement, drip and sprinkler irrigation introduction, land salinization and waterlogging prevention, etc. It is worth to note, that all these technologies and methodologies are developed by the specialists of the leading water institutes of the region but the practical specialists of the middle and lower level are not aware about changes occurred and they mostly need to be trained, because they are responsible for proper performance of irrigation and drainage systems. It is well known, that successful performance of any system is impossible without permanent staff training, which provides knowledge about new technologies and techniques, methodologies and world-wide experience in irrigation practice.

Human potential of water-related sector of Central Asia remains rather high thanks to perfect education and vocational training system in the FSU. But due to destruction of this system after the FSU collapse and some local and external factors this potential is significantly undermined.

For example, in Uzbekistan potential decrease within the period 1990-1995 is on average about 30%, for scientific-research institutes it is 40%. Number of young specialists graduated in 1997 compared with 1992 (engineers:1300 against 1652, technicians: 420 against 1500) show significant fall.

Local factors lowering potential are as follow:
a) absence of permanent vocational training system; b) low labor cost; c) "brain drain" (specialists' emigration abroad); d) personnel aging; e) insufficient education level of young specialists and absence of willingness to work in water-related sector due to reason, mentioned in item b.
External factors lowering potential are the following:
a) leaving centralized irrigation management (basin management, transfer of management rights by irrigation systems to the farmers and their associations);
b) increasing role of economic mechanism in water saving and environment protection.
c) appearance of new technologies and irrigation technique and irrigation systems' management.

All above mentioned factors not being compensated by timely vocational training adversely influence water-related sector's performance causing such negative phenomena as:
a) low efficiency of agriculture production and respectively population living standards;
b) low water supply (sustainability) of irrigated farming;
c) river and irrigation water quality aggravation and, as a consequence, environment and population health deterioration;
d) aggravation of the region's energy supply due to imperfect management by water reservoirs and hydropower stations' cascades.

At present time joint management by water resources in the region is kept with assistance of such interstate structures as Interstate Council (at the level of states'

Presidents), International Fund for Aral sea saving (IFAS), Interstate Coordination Water Commission (ICWC) (at the level of Ministers of Agriculture and Water Resources) and its Scientific -Information Center(SIC ICWC), Basin Water Associations BVO "SyrDarya" and BVO "AmuDarya".

Management is executed over the transboundary rivers and inter-farm systems. Because irrigation consumes almost 90% of water, special attention should be paid to irrigation systems operation. Most vulnerable is water allocation within on-farm irrigation systems, their operation and maintenance, where the role of water users themselves is very important. Their vocational training is necessary because systems operation efficiency depends on their activity.
At the middle and high managerial level there are many shortcomings too because of certain reasons: absence of legal base for transboundary resources management, imperfect information system and tools for its use for purposes of forecast and planning, weak public awareness and involvement in decision-making, planning and water saving.

All mentioned shortcomings in water resources management could be eliminated by specialists' training. Joint efforts of the regional organizations allow to undertake certain measures for specialists of middle and high level thanks to financial support of such international organizations as the World Bank, EU, UNDP, UNESCO, USAID, CIDA by means of workshops, study tours, short-term courses, etc.

For example, within the framework of TACIS Program WARMAP Project workshops on water right, regional information system are carried out on a regular base. Within the Program 7, funded by CIDA, with support of Israeli Government the workshop "Experience of water resources management in Central Asia " held in May 1997. In November-December 1997 the Roving Seminar for CROPWAT methodology adaptation was carried out together with FAO and WMO. For personnel of high level (ministers, deputy ministers) study tours are organized regularly in Israel, Egypt, Spain and other countries. With support of ILRI (the Netherlands) courses on computerized library creation were carried out for two persons from Scientific Information Center. Within the framework of Program 7 special course for training management were organized in Calgary (Canada) for specialists from SIC ICWC and BVO "SyrDarya" in November 1997 and study tour in summer 1997. Every year 2-5 persons attend CINADCO courses in Israel on different aspects of agriculture and water resources management.

But radical solution of training problem is possible only by means of permanent Training Center establishing. Experience of similar centers in Egypt (together with ILRI), in France, etc. shows their high effectiveness. Understanding of necessity of such a center is spreading among the Interstate Water coordination Commission members, but its equipment and successful functioning is impossible without assistance of the international financing organizations.

At present time consideration of such Center creation is under progress on the basis of SIC ICWC Center at expense of the IFAS members' contributions. After its establishing attempts to get international support, mainly for equipment purchase and foreign lecturers' reimbursement, will be continued.

The region possesses sufficient number of highly qualified specialists which can be used as a lecturers in technical disciplines but updated technologies and technique in irrigation and drainage, water system management, legal and juridical questions, market economic approaches and incentives in water management, water market development, management functions transfer to water users and their associations could be in the best way highlighted and explained by foreign experts having free access to the world information centers and expertise in world wide practice in irrigated farming.

SIC ICWC is an ideal base for the Center establishing because it has the following advantages:
- broad information exchange through the IPTRID network;
- highly qualified staff in water management, modeling and environment working in close collaboration with foreign experts within the framework of WARMAP, GEF, USAID, CIDA Projects;
- high level of computerization and communications;
- regional information center with GIS system ;
- teachers trained abroad (in the Netherlands, Israel, Canada, USA);
- direct personal contact with all levels officials in water-related sector of Central Asia;
- participation in preparation of ICWC decisions on the regional water resources management;
- experience in the international courses and workshops arrangements.

The main objectives of the Center are as follow:
a) Regular vocational training by means of short-term courses of middle-level staff at provincial and district level.
b) Low-level staff training by the specialists trained in the courses (knowledge dissemination).
c) Demonstration of the Training Center's possibilities with purpose to form sustainable demand in Central Asia and NIS.

For that purpose the following tasks should be resolved:
- to purchase necessary equipment;
- to prepare necessary working space;
- to define possible constraints in system operation which could be eliminated by staff training;
- to select teachers and categories of trainees;
- to prepare program of study process with regard for study objectives and specific features of trainees;
- to develop approaches and methodology of study process;
- to work out criteria of study process effectiveness assessment and methods of its results dissemination;
- to carry out initial phase of study process and introduce its results;
- to perfect center's financial activity by means of study pricing with subsequent transition to self-financing.

Following main topics constitute a study process:
- water system management under new conditions;
- reclamation systems' operation and maintenance;
- water resources management's legal aspects;
- water resources management's economic aspects;
- economic mechanism of water resources management;
- computer proficiency improvement;
- English proficiency improvement.

The following groups of specialists should be encompassed by training:
- Ministries, basin water organizations, provincial and district level staff responsible for efficient operation and maintenance of systems and structures.
- ICWC and its branch's personnel directly connected with ICWC and BVO's activity.

Study process is supposed to be continuous (course duration 2-3 weeks, group size 25 persons, interval between courses 1 week for results assessment and next course program preparation). Thus, annually about 300 persons can be thought. Twice a year courses will be carried out outside of Tashkent: one for Karakalpakstan-Khorezm- Tashauz

(Turkmenistan) (under BVO "AmuDarya" supervision) and another for Fergana valley-Osh province (Kyrgyzstan) - Leninabad province (Tajikistan) (under BVO "SyrDarya" supervision).
Administration, lecturers and experts will be hired on contract basis. Expenses for foreign experts could be reduced due to their participation in ongoing projects (GEF, WARMAP, CIDA, FAO, etc.)

2 years duration initial stage of the Center activity is supposed to be divided into 3 phases:
1. Preparatory phase including organizational and financial arrangements, selection of trainers and categories of trainees, preparation of working space, equipment purchase and mounting, development of training strategy and program. Duration 12 months.
2. Initial phase of study process. Duration 12 months.
Study process efficiency assessment and its results dissemination. Duration 6 months.

The implementation of the project will allow to achieve the following results and benefits:
As a result of a well established education and specialists' qualification increase through the water resources management improvement certain benefits would be obtained in the most important fields:
1. Rise of water supply (sustainability) of irrigated agriculture.
2. Improvement of irrigation and river water quality and, as a consequence, population health (in first turn, women and children).
3. Rise of living standard of population through the agriculture efficiency increase, population health and environment improvement.
4. Improvement of ecological situation in the region as well as at the global scale because 25% global pollution originates from the Aral Sea basin.
5. Improvement of energy supply to the region through the proper development of the reservoirs' cascade operation regime.

To the end of the first year of educational process the appropriate measures will be undertaken for gradual transition to the self-financing through study charges establishment and paid management and computer training for the non-water organizations.

Goals and Methods for teaching integrated water management to engineering students

Martine Ruijgh-van der Ploeg
School of Technology, Policy and Management,
Delft University of Technology
Jaffalaan 5, P0 Box 5015, 2600 GA Delft, The Netherlands
E-mail: tinekep@sepa.tudelftnl

Annemiek J.M. Verhallen
Sub-department of Water Resources, Wageningen University
Nieuwe Kanaal 11, 6709 PA Wageningen, The Netherlands
E-mail: Annemiek Verhallen@users whh. wau. ni

Introduction

Education, training, and technology transfer are key issues in raising awareness and seeking solutions for current and future water management problems. A lot of attention is being given to the subjects that should be taught in formal education or training of professionals in water resources management (Dyck, 1990). The array of subjects is wide and covers many disciplines: hydrology, water-related aspects of civil and environmental engineering, agronomy, ecology, spatial planning, economics, law, and policy sciences. The importance of designing a curriculum which emphasizes integration of these disciplines has been suggested (Wessel, 1999). How can this integration be achieved? At some universities, new curricula have been designed for that purpose, e.g. the program in Systems Engineering, Policy Analysis and Management at Delfi University of Technology (Wessel et al. 1999). Other universities, however, offer specific integrated water management classes. These classes are designed to challenge students to integrate knowledge and skills they have developed in specific disciplines, most often in engineering and the sciences.

This paper addresses goals and methods for teaching integrated water management. We base this paper on our experience in teaching senior level classes in the educational programs of Wageningen University (the former Wageningen Agricultural University) and Delfi University of Technology, both located in the Netherlands. At both locations, the students have different backgrounds and ambitions with regard to the type of engineer they would like to be. It's exactly in those differences that we find our rationale for teaching obligatory classes in integrated water management in our respective educational programs.

We focus on (1) knowledge-building required for the analysis of the water system and the human activities it supports, and (2) development of skills for problem formulation from a multi-stakeholder perspective. But how to challenge students to integrate knowledge from different disciplines, especially when they come to our classes with a monodisciplinary education and way of thinking? We have found solutions for this by incorporating the systems-based, policy analysis approach and other principles in our teaching program. In this paper we give some examples of this.

Rationale for teaching integrated water management

We all look at the world differently. And so, we do not see the same problems when we look at water resources either. Three pictures we found in literature on sustainable management of river basins (Newson, 1997) illustrate our point. These figures show the different perspectives of the hydraulic engineer, the hydrologist, and the river manager, respectively (Figures 1-3).

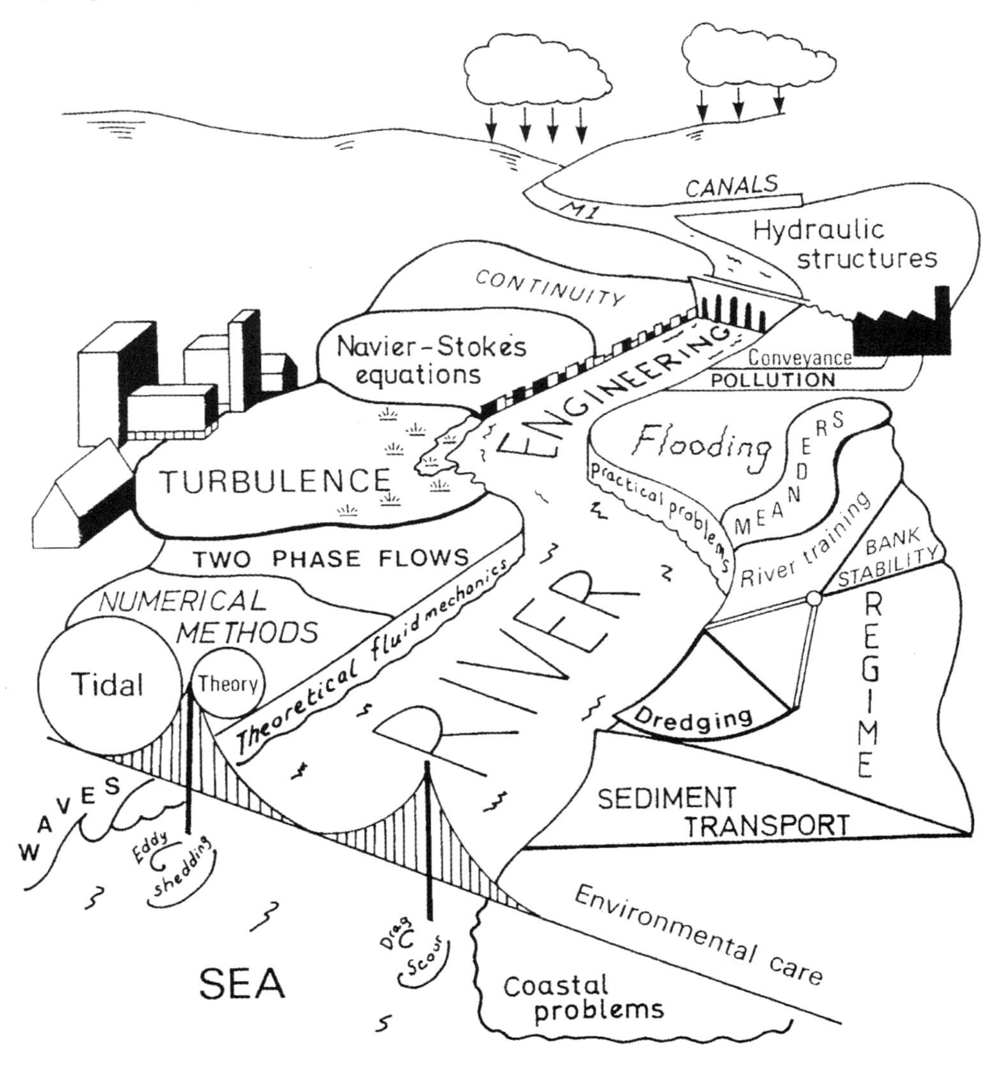

Figure 1 An engineer's view of a river basin: a series of hydraulic problems (Knight,1937).

Figure 2 A hydrologist's perspective on a river basin, emphasizing transformations and controls in the drainage basin (Schumm, 1977).

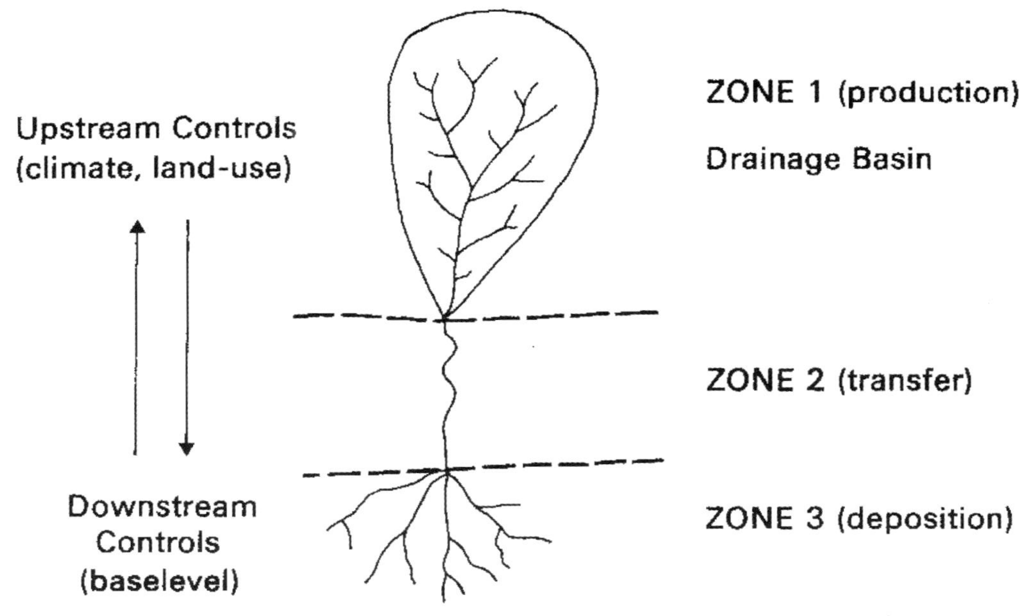

Figure 3 A water manager's view of a river basin focusing on land- and catchment use without consideration of natural boundaries (Newson, 1997).

For hydraulic engineers, river management is concerned with planning, design and operation of water works and thus with a set of hydraulic problems (Fig. 1, Knight 1987). In their view, a river can be transformed to meet the needs of people and their economies: navigation, recreation, irrigation, and safety from flooding. However, without knowledge of hydrology and the natural system, the hydraulic engineer will not succeed. Hydrologists (Fig. 2, Schumm, 1977) consider groundwater as well as surface water, looks at sediment transport and deposition as natural processes. The hydrologist sees human activity (e.g. water works) and hydraulic engineering as a threat to the natural functioning of the river. Engineers working in a government agency perceive a river differently once more (Fig. 3, Newson 1997). They emphasize the importance of the social and economic activities in the river basin. They are aware of the multiple, and often-contradictory requirements for water availability and water quality. They see the river as a scarce resource. In their view, water works are important for the fluictioning of the river, but so are the market economy, rules and regulations.

The differences in these perspectives translate into different water system descriptions with regard to
(1) The physical aspects of the water system to be considered.
(2) The human aspects of water system use and management to be considered. (3) The choice for system boundaries, temporal and spatial scales of water system definition. These differences in perspective have direct consequences for the goals different engineers may have in water management, for the criteria by which they evaluate the impact of human activity, and for the choice of alternative solutions to water management problems. Students must be able to analyze these perspectives with regard to all of these aspects. It will assist them in developing their own perspective on water management, in the influence of (mono-disciplinary) training on this perspective, knowledge and skills in solving water management problems.

The engineers we educate today have to be able to do their work with a long-term view, not only for current but also for future generations (AS CE et al, 1998). These engineers will need knowledge of natural processes taking place in the river basin. Knowledge of and skills in planning, design and operation of water works are an important part of their academic luggage (Wessel, 1999). They need to know about the different technical, judicial, economical and adminstrative options available for water management. In this respect, we find it especially important that engineers learn to communicate with the different users of a water system and to be able to analyze the different points of view. Solving water management problems requires more than knowledge and analytical skills. It requires a cooperative attitud&to be able to explore, find, or create win-win solutions in decision-making. Perhaps water management is not so much about water as it is about people.

Challenges in teaching integrated water management

Engineers serve human kind in various ways: each approach has its own value. However, to be able to work in the field of integrated management, it is necessary to reframe disciplinary analytical frameworks so that knowledge and skills of other fields may be included. Therefore, it is important to first show students the limitations of their current perspective. Then, students must be challenged to expand their existing knowledge base and to develop new skills so that they will be able to consider different perspectives simultaneously. Students with a background in natural sciences or engineering are stimulated to develop knowledge of the human sciences: law, economics, policy sciences, administrative sciences, anthropology and sociology. Students of the human sciences should learn to understand hydrology and ecology, the functioning of water works etc.

The most important challenge is to teach students to recognize and understand the context in which water management problems occur. Regardless of the future positions they may hold, students must understand how the different perspectives of water professionals and water users relate to each other and why. Therefore, their understanding of natural processes and works must be expanded with knowledge of societal developments. They must understand decision-making processes. Communication with other professionals and politicians is important and it should be part of their training to be able to talk and listen.

How can education contribute to this contextual knowledge? What is the minimal knowledge-base that students must build? What should they know of human and of natural sciences? How much do we want an economist to know about hydrology and water quality? What should the engineer master with respect to sociology? The field of water management is very broad, not only with regard to the different types of water systems (river basins, coastal areas, urban water systems) but also with regard to issues of scale (local, regional, transboundary and even global water issues). What aspects of water management should be covered so that students understand the dependencies within and between water Systems? How much attention should be given to the national and European situation in comparison with water management in different climates or countries of different economical development?

We do not have the answers yet but are confronted with these questions on a daily basis. However, we have found methods to deal with these challenges.

Paradigm, goals, and methods for teaching IWM

Paradigm

The following teaching philosophy proved useful to answer the challenges mentioned earlier:

Our teaching focuses on knowledge and skill-building necessary for the identification of stakes and stakeholders in water management. To accomplish that, we teach students to adopt a systems perspective on water management. First, we teach them the elements that make up a systems perspective as explained above. Then we challenge them to become flexible in adapting the different disciplinary perspectives. We add one perspective, namely the perspective that the natural system supports all human activity. In this perspective, the natural system delivers the so-called goods and services that make social and economic activity possible. Those goods and services are scarce, and scarcity is felt as a problem in the human system. It is in the human system that problems occur, not within the natural system itself. Problems are experienced, understood and solved by people, the stakeholders. In that regard, we also share our view on sustainability: every member of the current generation owns the future and can be held responsible for it. This means that it is very important to consider the long-term impact of solutions to water management problems.

Teaching goals and methods

How does this philosophy translate in teaching goals? The first goal is that students can apply the systems analysis approach in problem solving. This means that students must be able to define problems in water management, to indicate general characteristics of solutions and to develop a frarnework for the evaluation of alternative solutions. We teach them to define problems from a multi-stakeholder perspective. Also, we teach that the design and evaluation of solutions to problems must be presented in terms of trade-offs between the different objectives aimed at in problem-solving. Students must learn to identi~ the criteria for judging the impact of alternative solutions, and to identify the temporal and spatial dimensions of the trade-offs (Brooks et al. 1997). They learn to ask and answer questions such as "Who will benefit from the implementation of a new water treatment plant? How will they benefit and when? Who is going to pay for construction and operation: the sewage producers of the current or the next generation? How are costs and benefits distributed in the river basin? Do the solutions affect the efficiency of human activity in the upstrearns or downstreams areas?"

We ask students to describe what stakeholders experience as problems when working on a problem definition. Who has identified the problem? Why is the issue at stake a problem? How do other stakeholders look at the situation? Answering these questions requires leniency or flexibility from the student in being able to change perspectives. We would like for them to be acrobats in that regard so that they can write problem-definitions from a multiple-stakeholder perspective. We find that skills in conceptual modeling are essential for developing skills in problem definition. It is a very good way of bringing together knowledge of the processes in the natural and human systems, to identify the interdependencies between these systems (Brooks et al. 1997). We also use computer aids developed for decision-support and simulation games to demonstrate these interdependencies. We focus on teaching ways to think through problems before the actual calculations are started or complicated models are run.

We feel very strong about bringing the real world into our classroom. In fact, we consider the world outside the university to be our classroom. All assignments are related to cases, to actual problems or tasks in water management. The variety in the cases we use is wide. We do this for several reasons. First, students need to develop a framework of reference and they can do this by studying different cases. Second, students must be given the opportunity to explore in what area of water management they would like to work. Another advantage of working with case-material is that when students are confronted with the complexity of the real-world, it helps them discover their discrepancies in knowledge and motivates them to deal with them.

Our students do MSc thesis research "on the job." They work for organizations such as local or regional water authorities, engineering companies or water services companies.

Often they do the acquisition for the internship themselves. We coach them in identifying the type of problem they want to work on and in finding an organization to work for. Students always work on a specific problem and write a thesis after 4-6 months of research. Of course, we assist them in specifying a research proposal and supervise them during the period of research.

The student's perspective

From the perspective of a student, the learning goals can be summarized as follows:
(1) Students can recognize, define, analyze and solve water management problems from a multi-stakeholder perspective.
(2) Students are able to translate a problem description in conceptual models.
(3) Students are able to analyze complex, real-world problems by applying the concepts of integrated water management concepts.
(4) Students are able to communicate with all stakeholders during all stages of planning and decision-making

They can add personal learning goals to this list, such as exploring specific topics of interest (e.g., water scarcity, water law, river basin commiffees) or broadening their knowledge-base in either the human or natural sciences. We can support them in this effort of learning integrated water management by asking them "What kind of water manager or scientist would you like to be?" or "What contribution do you want to make to sustainable water management?"

References

American Society of Civil Engineers, Water Resources Planning and Management Division and UNBSCOIIHP IV Project M-4.3 Working Group. 1998. *Sustainability criteria for water resources Systems.* American Society of Civil Engineers, Reston, Virginia

Brooks, K.N., Ffolliott, P.F. , Gregersen, H.M. 1997. *Hydrology and the management of watersheds,* 2nd ed. Ames: Iowa State University Press, 502 p.

Dyck, 5. 1990. *Inte grated planning and management of water resources: guidance material for courses for engineers, planners and decision-makers.* Paris : UNESCO, 76 p.

Knight,D.W. 1987. Dissemination of information to practicing engineers and researchers in the water industry. *Journal of the Institution of Water and Environmental Management, 1 (3)315-24.*

Newson, M. 1997. *Land, water and development. sustainable management of river basin systems,* 2nd ed. Routledge, London. 423 p.

Schumm, S.A. 1977. *Thefiuvial system.* Wiley, New York.

Wessel, J. (1999) A guide to the needs of education and training in the water sector. *European Thematic Network ofEducation and Training, Environment-Water.* Brussel.

Wessel, J., H.O. Wind, andE. Mostert. 1999. Paradigms in water management. Series on River Basin Administration, Research Report 11. RBA Centre, Delfi University of Technology.

Theme C

Testing and validation, Quality assessment and accreditation

Test et validation. Evaluation et accréditation

Theme C: Testing and validation. Quality assessment and accreditation.

Rapporteur: A. Van der Beken (Belgium)

The keynote lecturer W. Van Den Berghe has presented his view on self-assessment techniques for quality management in education and training. His long experience in numerous European education and training projects allows for a theoretical approach as well as for a well-balanced expert analysis on the advantages, disadvantages and practicalities of different quality assessment techniques.

He states "*...the real source of improvement lies in the internal processes. These can only be measured adequately through self-assessment*". And it applies to both the training providers and the training users. In this way we really move towards the **"learning society"**.

The contribution of G. Neveu and C. Toutant on competencies/qualifications in the field of water and their evaluation and recognition (certification) has two clear messages: 1. lets merge competency and qualification, 2. lets break down competency/qualification into a facet "knowledge" and a facet "behaviour" (related to the competency). The authors recognise that the second facet is much more difficult to evaluate than the first one. They propose the model ESC'EAU.

The contribution of J. Feyen et al. presents a case study of the research-orientated long-term (2 years) MSc-programme IUPWARE. The programme evolved from two separate programmes since 1981 towards a merging in 1994. A continuous self-assessment has allowed for the improvement of the structure and contents of the programme with the human and financial resources available.

Ziglio et al. present an example of quality assessment of a series of five training courses of a 5-day duration each. Evaluation was organised systematically in the same way for all five courses among participants and among the lecturers/tutors. This procedure allowed for clear conclusions of benefit towards further analysis of training needs in the given field.

A final remark of this report: it must be recognised that only three submitted papers were selected for this THEME. Of course, many other contributions have touched the quality issue of education and training. Nevertheless, it can be stated that the concept of quality assessment and evaluation is not yet always fully integrated in education and training projects.

Keynote lecture

Self-assessment as the cornerstone of quality management in training

Wouter Van den Berghe
Department of Organization & Quality , Deloitte & Touche, Belgium

Introduction/Abstract

It is increasingly recognized that self-assessment has to play an important role when implementing quality and quality assurance arrangements in a continuing training environment. In this contribution we first define a number of quality concepts and discuss the implications for continuing education and training. Subsequently, advantages of self-assessment are discussed. The final section of the text includes a number of reflections and areas for debate.

Quallity terminology

Quality

There is widespread confusion over the real significance of terms like "quality" and "quality control", in particular in service environments such as education and training.

This situation is linked to the fact that "quality" is a multi-dimensional and relative concept. As such, it is not possible to give a unique definition that fits all circumstances. It is possible, however, to distinguish between some important viewpoints and perceptions to quality which people use when they assess:
- quality as excellence, as something special
- product-oriented quality (quality can be measured)
- quality as the fulfilment of customer expectations
- process-oriented quality (quality is conformance to specifications)
- price / benefit-oriented quality (the value approach).

Such quality perceptions may apply, alone or in combination, to any type of product or servic0e, including education and training.

When we use the term "*quality*" on its own, we refer to the "output" quality of products and services. In other words: the characteristics of the end result of the production and/or service development process. In "modern" quality thinking, these characteristics are perceived from the perspective of the "customer": the user of the products or services. In that logic, a product or service delivered is perceived as a quality product or service, if the customers are satisfied with it.

Quality is not a new subject in education. Institutions, teachers, administrators and policy makers have always been concerned with quality. Even without adopting a formal 'quality' approach, education and training providers have needed to develop methods, norms, procedures and standards that allowed them to ensure the quality of their provision. However, the notion of quality has often been illdefined, defined in a narrow sense, or not defined at all. Moreover, it is only fairly recently that quality concepts and approaches adopted successfully in business envi-ronments have started to penetrate the education and training world.

This recent trend reflects a more general phenome-non, which is that the dominant quality concepts in education and training tend to change over time. The different viewpoints from which quality in education and training has been considered can be summarized as follows:

- Quality from a ***didactic and/or pedagogical*** point of view, e.g. issues like teaching and train-ing effectiveness, appropriateness of flexible learning, and compensation -programmes - education quality seen as the optimization of the teaching and learning process.

- Quality from a ***(macro)-economic*** point of view, with considerations on the return on investment of education and training (also by com-panies), and topics like the effects and costs of class size - education quality seen as the opti-mization of the education and training costs.

- Quality from a ***social or sociological*** point of view, including issues like providing equal opportunities for disadvantaged groups - education quality seen as the optimization of the response to social demand for education.

- Quality from a ***customer*** point of view, e.g. the capacity of schools and training providers to respond to particular demands from clients (stu-dents, pupils, parents, employers,..) to deliver the education and training required - education quality seen as the optimization of the demand.

- Quality from a ***management*** point of view, with the focus on effective schools and ***Total Quality Management (TQM)*** methods in education institutions - education quality seen as the optimisation of the organisation and processes of education.

The order in which these different viewpoints are listed reflects to some extent the shifts in emphasis over the last decades in many European countries. However, it is not so much a question of replacing 'old' paradigms, but rather one of adding 'new' dimensions - which reflects the growing complexity of the education system and the objectives it has to meet.

It should be noted that the "modern" way of looking at quality – customer-oriented and based on a TQM-approach – presents some difficulties when applied to a training environment, since there are often three types of customers, sometimes with conflicting interests:

- the direct user (trainee, student)
- the indirect user paying for the training (company, parents, ...)
- the society at large (in particular for publicly supported training measures).

Quality assurance and quality control

The definition of 'quality assurance' according to ISO 8402 is: "All the planned and systematic activities implemented within the quality system, and demonstrated as needed, to provide adequate confidence that an entity will fulfil requirements for quality". In practice, in order to be able to speak about quality assurance, the following needs to be in place:

- quality standards are defined
- suitable procedures are available
- these procedures are monitored for conformance
- causes of -non-conformance are analysed
- causes of problems are eradicated through ap-propriate corrective action.

Quality assurance' mechanisms can be considered as a subset of a '***quality control system***' (see below): they con-cern the pro-cesses that ensure conformity of out-put with specifications. The application of quality assurance principles requires, of course, some consensus about what are considered as the main quality attributes - which is, as we recall, not always straightforward in the provision of training programmes.

Quality assurance grows in importance when moving from a product-oriented focus on quality - which is still predominant in European education and training - to a process orientation of quality. It assumes an understanding of the various processes which lead to a quality result, of the input and output factors which should be controlled, and of the process factors on which control can be exerted. It can easily be understood that, given its complexity and set of interrelated processes, there is no easy or generally applicable method for quality assurance in education and training.

Although 'quality control system' and 'quality assurance system' are sometimes used interchangeably, in the context of this report, quality assurance is seen as part of a wider quality control system.

The term '***quality control system***' refers to a large number of aspects and their interconnections, which together ensure the quality control of a number of related processes. The key characteristics of a quality control system can be considered to be:
- design and planning are based on identified needs and requirements
- operation and implementations are in conformity with what was planned and designed
- effective mechanisms exist for assessment of outcomes
- there is continuous improvement of design and operation.

Such a quality control system might apply to a product, a service, a process, an organisation or even an industrial sector.

Quality control systems require a range of things to be in place and operational:
- clear ownership and identified responsibilities of the system, its various components, stages and processes
- an understanding of needs, in particular when these needs are changing rapidly
- the availability of an efficient design and planning methodology for responding to identified needs
- the resource and competence capacity for effective implementation of what is planned, and production of the outcomes specified
- an understanding of the internal and external factors which may influence the process
- continuous assessment of what is being achieved
- identification and utilisation of process control factors
- availability of resources for the implementation of changes in the design and operation phases.

This list clearly shows that the concept of a 'quality control system' is applicable to education and training. In fact, the concept can be applied in different ways and at different levels. One may consider of quality control system for education and training provision at the level of countries or regions, exemplified by the state-controlled initial education systems. Another use-ful level for considering quality control is the institutional level.

Although quality control systems require a clear identification of who is in charge of what aspect of the system, it is not necessary that all processes are under the responsibility of the same body. For instance, when considering quality control at the institutional level, several of the assessment processes (and sometimes also the needs analyses and programme design) may be undertaken by bodies which are external to the one which is at the heart of the quality control system under investigation.

A modern way of implementing quality control systems at the level of an education or training provider is to apply the principles and methods of '***Total Quality Management***' (TQM). TQM refers to an organisational approach for continuously delivering high quality. Its main basic principles are:
- customer orientation
- continuous improvement
- carefulness
- prevention instead of detection of quality problems
- process orientation.

Making the link between quality, training and the learning society

What does all of this mean for training? And how does it relate to the learning society, which is the topic of the conference?

First of all, one should understand that quality in training – like in some other service environments – depends on two parties:
- the training provider: the organisation and people who produce and deliver training programmes and courses
- the training users: the people (and often organisations) who use and consume the training produced.

Both parties contribute to the quality of the training: training is only successful if:
- providers understand training needs of the users and have the capability and mechanisms to address those needs
- users are clear about what they need and are willing to invest in the learning process.

Secondly, one cannot sustain and improve quality if it is not measured or assessed on a regular basis. However, it is difficult to evaluate and to assess the quality in training. In modern quality thinking, the bottom line for quality in training is its effect or impact after a certain period of time. Simplified one could state:

"Quality in training is measured by its impact six months later"

In reality, however, such measurements are very difficult and costly to undertake and therefore seldom happen – neither with the provider or the user. The most frequently used "quality measurement" is the measurement of the degree of satisfaction by the trainees at the end of the training course.

An alternative to such measurements is to initiate – both with the training provider and the training user – a process of regular "self-assessment" about the type and nature being produced or consumed. In a self-assessment process an organisation will first define a number of quality standards and desirable input, process and output characteristics for the training. It well then use a number of appropriate methods in order to assess to what extent these quality standards and characteristics have been achieved. These methods may involve the trainees; however, the organisation remains in charge and sets the standards.

Self-assessment may serve a double purpose:
- first of all, through regular self-assessment an organisation will gradually come to understand better the effects and impact of training delivered – and thus come closer to the real measurement yardstick of quality in training
- secondly, a good self-assessment process will yield many ideas and suggestions for quality improvement, which contributes to the development of a dynamic training and learning environment.

That, in itself will support the development of a "learning organisation": an organisation that has the intrinsic capacity to learn and develop as a whole – rather than as a set of individuals.

Thus, whilst self-assessment has a number of direct advantages (see also below), it is also a tool which supports organisation in becoming a learning organisation. And the existence of many organisations that are really learning organisations is a prerequisite for the emergence of a learning society.

Why self-assessment can provide advantages

All types of quality assessment have advantages and disadvantages.

Let us first consider "external assessment", i.e. when the quality of training is assessed by an independent "third party", for instance a public body or a certification organisation. Advantages of such an approach are:
- it has high credibility
- it ensures a neutral view and original perspective
- it allows comparability and benchmarking

The most important disadvantages are:
- in general, it is very expensive
- assessors may not always fully be qualified
- it may interact with other activities

Another type of assessment is when the customers or users are asked to assess the training received. The advantages of such customer assessment are:
- it meets the real interest of the customers
- it is cheap when done immediately after or during the training
- the real impact may be measured (if undertaken at the right moment)

Disadvantages are:
- customer assessment often only covers output characteristics
- the possibility of high variability of satisfaction of individual customers
- customers may not understand there own needs.

Self-assessment can have the following advantages:
- it is relatively cheap
- it may cover the totality of the organisation
- it may involve everyone

The main disadvantages are:
- it may not be credible (inside and outside the organisation)
- it may lack rigour and reliability
- it may interact with other activities

From a quality management perspective, self assessment is the preferred way. Indeed, assessment by externals and customers tend to focus on input and output-characteristics of training. Although output and input characteristics are relatively "easy" to measure, the real source of improvement lies in the internal processes. These can only be measured adequately through self-assessment.

Moreover, any form of external or customer-based assessment may lead to a defensive, rather a constructive reaction of the people assessed. Quality improvement requires a positive motivation towards improvement. This is more easily supported by self-reflection than by external evaluation.

Last but not least: it may not be obvious in the beginning, but eventually self-assessment will be the most economical way of assessing quality as an effective mechanism for quality improvement.

In summary, self-assessment applies to both partners in training: the training providers and the training users. Self-assessment with both parties is important for:
- achieving quality improvement
- becoming a learning organisation
- moving towards the learning society

Some final points for reflection

Quality assessment is not a goal in itself. Whenever the quality issue comes up, it remains important to respond to two series of questions.

The first set of questions concerns the definition of quality and the scope of the underlying quality management system: which processes and outputs are concerned? What type of quality is pursued? What ought to be the depth of any assessment?

A second set of questions concerns the responsibility for quality in of the training. What is the role of the provider? How does the customer organisation affect quality? Are all quality expectations of users/beneficiaries legitimate? What is the responsibility of the individual?

A clear response on such questions will facilitate the process of carrying out an effective and successful self-assessment.

But even then, a number of areas for discussion and policy debate remain open, such as:
- Who determines quality standards?
- What is an acceptable cost for self-assessment and quality management?
- What are the quality requirements for assessors?

References (publications of the author on related topics)

1. ***Achieving Quality in Training***. European guide for collaborative training projects. Tilkon, Wetteren, 1995 (ISBN 90-75427-01-8). Also published in Italian as ***La qualità della formazione***, Diade, Padova (ISBN 88-87157-01-4)
2. ***Quality issues and trends in European vocational education and training***. Cedefop Document. Office for Official Publications of the European Communities, Luxembourg, 1997 (ISBN 92-827-8194-1). (also available in French "La qualité dans la formation et l'enseignement professionnels en Europe: aspects et tendances" ISBN 92-828-1189-1, in German "Qualitätsfragen und -entwicklungen in der beruflichen Bildung und Ausbildung in Europa" ISBN 92-828-1188-3, and in Spanish "La calidad de la enseñanza y formación profesional en Europa: cuestiones y tendencias" ISBN 92-828-0976-5)
3. ***PROZA*** (a comprehensive self-assessment instrument for higher education) (with P. Gar--ré, J. Leniere, J. Schoofs, K. De Boevere and others), PROZA-groep p/a Ehsal, Brussel, 1998 (ISBN 90-804031-1-3).

Vers un modèle inédit d'évaluation et de reconnaissance (certification) des compétences/qualifications des travailleurs de l'eau

Towards a new model for evaluation and recognition (certification) of workers competencies/qualifications in the field of water

Neveu G., Toutant C.
OIE, France

Résumé
S'inscrivant dans les perspectives du programme d'initiative ADAPT 1997 du Fonds Social Européen, le projet ESC'EAU veut développer un modèle inédit d'évaluation et de reconnaissance (certification) des compétences/qualifications des travailleurs de l'eau. A la croisée des expériences françaises, anglaises et allemandes dans le domaine, le modèle ESC'EAU s'intègre parfaitement au cheminement mondial actuel de reconnaissance des acquis expérientiels et des formations professionnelles, préoccupation majeure des grands acteurs nationaux tels que le MEDEF (consultation nationale sur les compétences), l'ANPE (mise en place du ROME), l'Education Nationale (intégration des compétences dans le cursus de formation traditionnelle) et la branche professionnelle proprement dite (création éventuelle de certificats de qualification professionnelle).

Côté contenu, le modèle ESC'EAU vise les métiers transversaux d'exploitation/gestion d'entités plus ou moins importantes (stations, secteurs, etc.) du domaine de l'eau qui nécessitent d'associer plusieurs types de compétences/qualifications. Le modèle ESC'EAU évalue ces compétences/qualifications suivant 2 axes principaux. Chaque compétence/qualification est décomposée en un volet « connaissances de base » et en un volet « comportement » (afférent à la compétence), le poids relatif (pondération) du comportement étant supérieur à celui des connaissances de base.

Avec cette approche, on répond aux préoccupations des praticiens (qui insistent sur la mise en situation) et des formateurs (qui misent plutôt sur les savoirs). Cette fusion des compétences et des qualifications (au sens strict) dans l'évaluation, qui se rapproche de la certification américaine, présente l'avantage de cerner l'ensemble du profil multi-dimensionnel des métiers de l'eau. De plus, le modèle ESC'EAU définit des niveaux/classes inhérents(es) à la verticalité des métiers de l'eau puisque la technicité évolue d'un(e) système/station à l'autre demandant ainsi (pour une entité plus complexe) des compétences plus vastes.

D'un point de vue technique, on peut facilement mesurer une connaissance: la réponse est souvent vraie ou fausse. Un QCM est alors satisfaisant pour évaluer les savoirs. La mesure de l'adéquation comportementale est plus difficile à réaliser, mais les outils informatiques offrent maintenant des possibilités de simulation tout à fait satisfaisantes, en particulier pour décrire des procédures d'intervention. Chaque « mise en situation » peut alors être appréciée par une approche multi-critères (compétence proprement dite, compétences associées, sécurité, temps de réaction, etc.). Le modèle ESC'EAU, outil d'autoévaluation accessible à distance (via Internet), intègre donc ces 2 approches techniques, combinant un QCM (pour les connaissances de base) et un logiciel d'évaluation comportementale, le premier étant prérequis au second.

Reste la reconnaissance proprement dite de ces compétences/qualifications. Le modèle ESC'EAU propose de visualiser le niveau atteint par le travailleur à l'aide d'un graphique où chaque compétence/qualification est bien identifiée (par une courbe) autant en faciès « comportemental » qu'en faciès « connaissances de base ». On peut facilement imaginer que le graphique afférent à un travailleur donné lui serve de certificat et/ou évalue ses lacunes. Dès lors, il sera facile de déterminer la formation (sur mesure) nécessaire à l'atteinte de la(les) compétence(s)/qualification(s) manquante(s). Le modèle ESC'EAU est un outil convivial d'évaluation et de reconnaissance des compétences/qualifications adaptable à tous les métiers multi-dimensionnels, quel que soit le secteur. Il permettra de répondre aux attentes des professionnels intéressés à bien connaître leur main d'oeuvre, qualitativement et quantitativement.

Abstract
Included in the perspectives of the ADAPT 1997 Programme (European Social Funds), the ESC'EAU project wants to develop a new model of evaluation and recognition (certification) of workers competencies/qualifications in the field of water. At the confluence of French, English and German experiences in this matter, the ESC'EAU model becomes integrated perfectly to the actual world-wide progression on practical experiences and vocational training recognition. In the same time, this model meets the principal concerns of major national actors in France like the MEDEF (national inquiry on competencies), the ANPE (implementation of the ROME), the Education Nationale (integration of competencies in traditional education) and the specific professional branch (possible creation of vocational qualification certificates).

On the content aspect, the ESC'EAU model aims at transversal jobs for exploitation/management of more or less important entities (plants, sectors, etc.) in the field of water. The mentioned entities need to associate many types of competencies/qualifications. The ESC'EAU model makes the evaluation of these competencies/qualifications following 2 principal main lines. Each competency/qualification is broken down into a facet « basic knowledge » and a facet « behavior » (relating to the competency), the relative weight of the behaviour being superior to the basic knowledge one.

With this approach, we answer the practicians needs (who are more interested by situations on site) and teachers needs (who are more interested by knowledge). That fusion of competencies and qualifications (from a strict point of view) in the evaluation, similar to the American certification, presents the advantage of defining the whole of the multi-dimensional profile of water workers. In addition to that, the ESC'EAU model defines levels/classes inherent in the vertical side of water jobs because the technicality changes from one system/plant to another asking (for a more complex entity) larger competencies.

On the technical aspect, it is easy to measure a knowledge: the answer is right or false. A MCQ is enough to evaluate the knowledge. Measuring the behavior facet is more difficult to realize but the computer tools give us real possibilities of satisfying simulation, particularly to describe intervention procedures. In this way, each on site situation can be estimated by a multi-criterions approach (related competency, associated competencies, safety, reaction time, etc.). The ESC'EAU model, self-evaluation tool accessible to distance (with

Internet), integrates in this manner 2 technical approaches, combining a MCQ (for basic knowledge) and a software for behavior evaluation, the first one preceding the second one.

Remains the recognition of the competencies/qualifications. The ESC'EAU model proposes to visualize the reached level by the worker with the help of a graph where each competency/qualification is well identified (by a curve). This is valid for the 2 mentioned facies. We can easily imagine that the worker graph could be used like a certificate or could sirve to assess the worker deficiencies. After that, it will be easy to determine the requested training (tailored training) to fill the gap between existing and wanted competencies/qualifications. The ESC'EAU model is an easy tool for evaluation and recognition of competencies/qualifications adaptable to all multi-dimensions jobs, whatever field, sector or branch. This model will help the concerned professional to know better their manpower, from a point of view quality and quantity.

1. Contexte

Le projet ESC'EAU, issu de l'initiative communautaire ADAPT qui est consacrée à l'aspect "ressources humaines" des mutations industrielles et en parfaite synergie avec le programme communautaire LEONARDO, s'inscrit évidemment dans la logique du Fonds social européen (FSE) qui vise à développer les ressources humaines et à améliorer le fonctionnement du marché de l'emploi dans l'Union Européenne (1). Comme toute action de ce type, ESC'EAU se veut transnational, puisant à plusieurs sources nationales, et tourné vers une harmonisation/intégration de toutes les tendances en la matière.

Le projet ESC'EAU veut donc développer un modèle inédit **d'évaluation et de reconnaissance (certification) des compétences/qualifications des travailleurs de l'eau**. Exceptionnel en soi puisque la dualité compétences/qualifications a toujours été vue en opposition et non en complémentarité, le modèle ESC'EAU se situe à la croisée des expériences françaises, anglaises et allemandes dans le domaine et s'intègre au questionnement mondial actuel sur la reconnaissance des acquis expérientiels et des formations professionnelles.

ESC'EAU veut être un outil essentiel non seulement pour le DRH de l'entreprise (ou pour la promotion des compétences de l'entreprise) mais aussi pour les travailleurs qui y trouveront leur compte puisqu'ils pourront donner une valeur marchande à leurs propres compétences.

Cela se fera dans la plus grande convivialité via des outils performants tels que nous les offrent maintenant les NTIC.

Au niveau national (France), la reconnaissance des acquis expérientiels et des formations professionnelles est une préoccupation majeure des grands acteurs nationaux. C'est le cas du Mouvement des Entreprises de France (MEDEF) qui a organisé en octobre 1998 à Deauville un forum sur le thème "Entreprendre et compétence" auquel ont participé plus de 1500 responsables en formation.

Pour le MEDEF, il est clair que "le premier capital de l'entreprise sera de moins en moins ses machines ou ses procédures mais ses hommes et leurs capacités, individuelle et collective, à faire évoluer l'organisation et à mieux satisfaire les clients. La gestion des compétences est au cœur de ces nouvelles conditions de productivité"(2).

Dans sa réflexion, le MEDEF est conforté dans son approche par l'Agence Nationale pour l'Emploi (ANPE) qui, avec son répertoire opérationnel des métiers et des emplois (ROME) axé sur les compétences, vient ériger en système établi l'approche par compétences (3) (4).

L'évolution de l'Education Nationale sur cette thématique sera probablement plus lente mais, à moyen terme, on peut penser à l'intégration des compétences dans le cursus de formation traditionnelle comme cela se fait actuellement au Québec, via le Ministère de l'Education de la Belle Province, région de tradition similaire à celle de la France en terme d'éducation.

Quant à la branche professionnelle proprement dite, elle sera fort intéressée par les expériences menées dans d'autres secteurs (5) qui ont conduit à la création de certificats de qualification professionnelle (CQP), là où un "vide" existait.

Le milieu de l'eau se rendra vite compte de l'importance d'une telle certification puisqu'il est naturellement porté vers l'exportation et qu'il aura de plus en plus besoin de référentiels transnationaux, reconnus en France et ailleurs.

2. Contenu

Le secteur de l'eau se caractérise par des métiers parfaitement identifiés (électromécanicien, chimiste, plombier, etc.) et par des métiers transversaux d'exploitation-gestion d'entités plus ou moins importantes (stations, secteurs, etc.) qui nécessitent d'associer plusieurs types de compétences/qualifications.

Avec les métiers traditionnels, on se retrouve en terrain connu et la qualification va de soi.

Les métiers pluridisciplinaires en eau, par essence plus complexes, sont difficilement assimilables à un diplôme "eau" bien établi puisque l'expérience professionnelle et la capacité individuelle sont des atouts majeurs dans la responsabilité qui incombe aux travailleurs visés.

Figure 1

DIPLÔME	Expérience professionnelle -------------------------------> Comportement individuel	Conducteur/responsable d'installation

L'enjeu est de taille et les approches diffèrent. On pourrait s'installer dans un statu quo pédagogique et reconduire le système par qualifications, particulièrement prisé dans le monde latin. On peut aussi choisir de se positionner à l'opposé du spectre en oubliant à toute fin pratique la formation initiale classique lorsqu'il s'agit d'établir ou de reconnaître une spécialisation des travailleurs.

Le cas anglais est fort instructif à cet égard (6) (7). En créant ses NVQs (National Vocational Qualifications), l'Angleterre a voulu remédier aux faiblesses de son système d'éducation dans le secteur technique. Mais cette apparition des NVQs a du être soutenue par un ensemble organisationnel extrêmement important tel que le BETWI (Board for Education and Training in the Water Industry) et le CABWI (Certification and Assessment Board for the Water Industry) pour l'industrie de l'eau. De plus, la méthodologie d'évaluation pour l'obtention des NVQs comporte aussi une certaine dose de subjectivité puisque l'évaluateur, qui doit lui-même être certifié, s'appuie essentiellement sur des éléments fort difficiles à quantifier.

La certification américaine en eau apparaît dès lors plus simple puisqu'elle passe par un examen d'évaluation pour chacun des niveaux de certification existant (8). Par contre, l'examen afférent à une qualification donnée présente des lacunes évidentes au niveau du spectre des connaissances (4 savoir-faire modulés en fonction des classes) et l'exigence de diplômes (formation initiale) combinés à un certain nombre d'années d'expérience ajoutent une dimension qui ne tient plus compte des capacités intrinsèques des travailleurs qui n'ont pas nécessairement appris leurs métiers dans des conditions classiques.

S'appuyant sur ces constats et poussé par les besoins d'évaluation et de reconnaissance des compétences exprimés par les entreprises et les travailleurs de l'eau en France (9) (10), le modèle ESC'EAU a été défini en respectant une sorte d'équilibre entre les compétences et les qualifications afin que ce distinctif ne soit plus en opposition mais qu'il agisse en complémentarité au sein d'un progiciel performant d'évaluation (et de reconnaissance des compétences).

Le modèle d'évaluation, et l'algorithme qui en découle, que nous proposons considère que le savoir-faire des travailleurs de l'eau, concept qui pour nous inclut savoir et savoir-être, peut être décomposé en une somme pondérée de compétences élémentaires, chaque compétence étant elle-même la somme de connaissances de base (approche théorique) et de comportements (approche pratique), dans un ratio à définir, mais qui est plutôt favorable aux comportements (40%-60% ?).

Les connaissances de base sont mesurables par un questionnement de type QCM, et les comportements le sont par l'observation des réactions face à une situation donnée ; ces deux modes de mesure sont automatisables, le premier de manière triviale par des quizz, le second en s'appuyant sur les capacités multimédia de l'informatique aujourd'hui, grâce à la vidéo en particulier.

Chaque « mise en situation » peut alors être appréciée par une approche multi-critères (compétence proprement dite, compétences associées, sécurité, temps de réaction, etc.). Le modèle ESC'EAU intègre donc 2 cheminements, combinant un QCM (pour les connaissances de base) et un logiciel d'évaluation comportementale, le premier étant pré-requis au second.

Avec cette approche, on répond aux préoccupations des praticiens (qui insistent sur la mise en situation) et des formateurs (qui misent plutôt sur les savoir). Cette fusion des compétences et des qualifications (au sens strict) dans l'évaluation, qui se rapproche de la certification américaine, présente l'avantage de cerner l'ensemble du profil multi-dimensionnel des métiers de l'eau. De plus, le modèle ESC'EAU définit des niveaux/classes inhérents(es) à la verticalité des métiers de l'eau puisque la technicité évolue d'un(e) système/station à l'autre demandant ainsi (pour une entité plus complexe) des compétences plus vastes (voir figure 2).

Figure 2

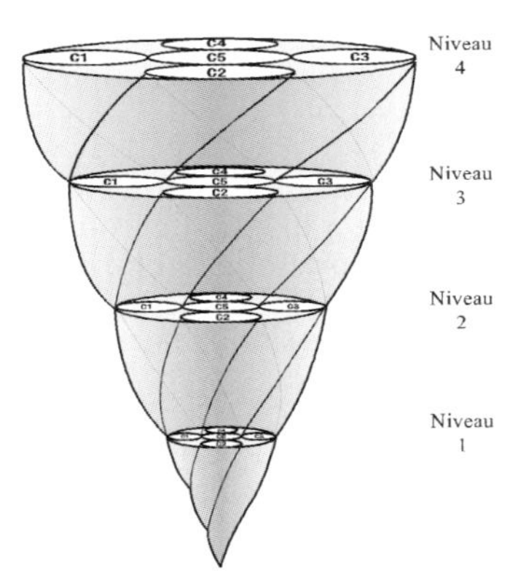

La figure 2 montre que :

1°) Les compétences se prolongent d'un niveau hiérarchique à l'autre en s'élargissant et/ou en s'approfondissant. Parallèlement, de nouvelles compétences apparaissent et se rajoutent aux autres pour un niveau hiérarchique supérieur.

2°) Le travailleur devient "responsable" (occupe le poste) d'un niveau donné si et seulement si il a la maîtrise de toutes les compétences ; si non, il ne peut être qu'assistant ; on peut penser également qu'il ait besoin d'une probation sous la supervision d'un tuteur qui participe à la décision finale de lui reconnaître le niveau (la classe) atteint après un certain temps de pratique (junior/senior).

3°) Chaque composante (connaissances de base et comportements) d'une compétence donnée doit être évaluée. L'évaluation des connaissances de base précède, en toute logique, celle des comportements.

Le modèle ESC'EAU, tel que défini, ne se préoccupe pas du distinguo compétence/qualification puisqu'il l'intègre. Par ailleurs, les travailleurs et leur hiérarchie pourront identifier leurs lacunes (compétences non maîtrisées) de manière assez pointue, de telle sorte que le vide constaté pourra être facilement comblé par une formation individualisée.

3. Aspects techniques

Techniquement, le modèle ESC'EAU peut adopter différentes configurations. Dans sa structure la plus simple, on vérifie chaque compétence, une à une, d'abord en faciès "connaissances de base", puis en faciès "comportements". On imagine donc un certain nombre de questions (10 par exemple) portant sur les « connaissances de base » associées à une compétence donnée. Si on répond correctement à un nombre minimum de questions (8 par exemple), l'ensemble est réussi et on peut alors passer aux questions (10) sur les « comportements » afférents à la compétence. On a réussi l'ensemble « comportements » quand on atteint le niveau requis.

Le niveau requis pour les « comportements » peut être de 2 ordres. On peut, à l'instar des « connaissances de base », fixer une barre à atteindre et à dépasser (8/10). On peut aussi, considérant que le travailleur n'a pas droit à l'erreur dans son fonctionnement, définir un système tout ou rien où la moindre erreur rejette le candidat. Quel que soit le choix arrêté, la réussite de l'ensemble « comportements » est essentielle à l'atteinte d'une compétence donnée.

La réussite des 2 ensembles détermine normalement la maîtrise de la compétence.

Pour un niveau de métier donné, chaque compétence peut être ainsi évaluée et le métier est dominé si on a réussi tous les ensembles « connaissances de base » et « comportements » des différentes compétences.

Côté pratique, pour s'éviter les lourdeurs inhérentes à une approche trop segmentée, La mesure des « connaissances de base » et des « comportements » peut se présenter dans un même questionnaire (un pour chaque ensemble). Si c'est le cas, la différenciation des différentes composantes (et des niveaux) du métier se retrouve dans un ordonnancement du questionnaire (comme cela se pratique dans la certification US) et on pourrait même se rendre à une classification de type TOEFL.

On peut même raffiner le processus et introduire un 3e test, appelé test synthèse, qui à l'instar d'un projet de fin d'études ou d'une thèse, synthétise les acquis du candidat via une mise en situation "complexe" qu'on évalue par des tests (globaux) de comportement. Le tout est illustré à la figure 3.

Figure 3

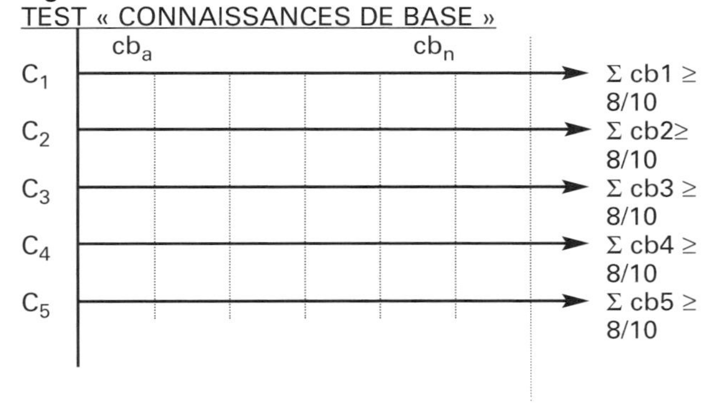

Test « cb » centré sur C_1

Test « cb » centré sur C_2

Test « cb » centré sur C_3

Test « cb » centré sur C_4

Test « cb » centré sur C_5

Test global « cb » centré sur C_1 à C_5

CONNAISSANCES à maîtriser pour la classe

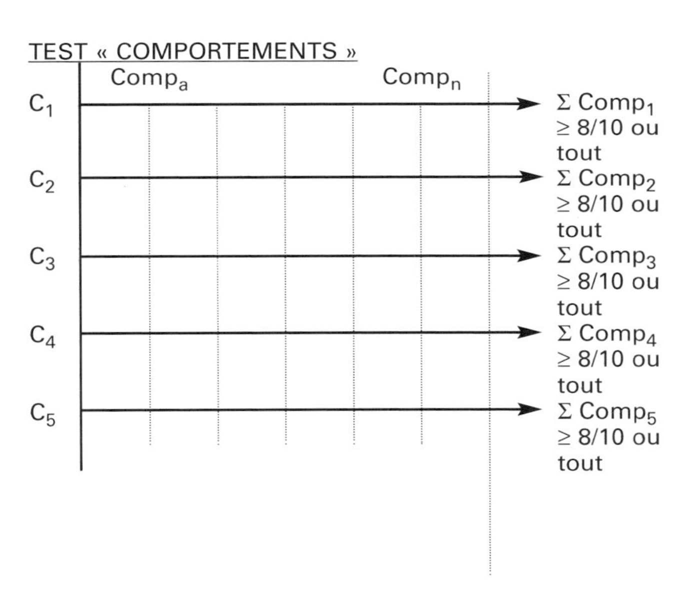

Test de comportements centré sur C_1

Test de comportements centré sur C_2

Test de comportements centré sur C_3

Test de comportements centré sur C_4

Test de comportements centré sur C_5

Test global de comportements centré sur C_1 à C_5

COMPORTEMENTS à maîtriser pour la classe

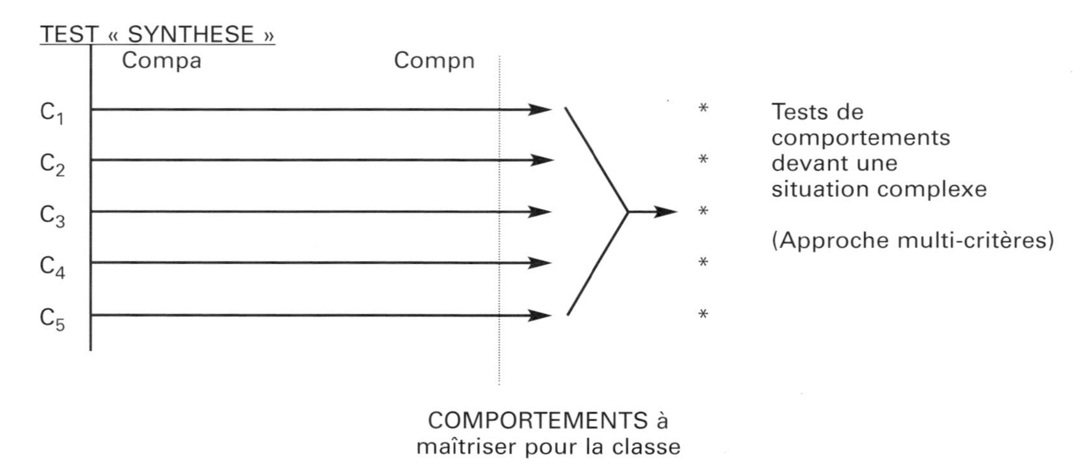

Légende
$C_1 \ldots C_n$= COMPETENCES
$cb_a \ldots cb_n$ = CONNAISSANCES DE BASE allant de a à n pour chaque compétence (C_1 à C_5)
$Comp_a \ldots Comp_n$ = COMPORTEMENTS allant de a à n pour chaque compétence (C_1 à C_5)

4. Conclusions

On aura compris que le modèle ESC'EAU est avant tout un système convivial d'évaluation, et d'autoévaluation, accessible via Internet ou sur CD-ROM, et qu'il intègre les Nouvelles Technologies de l'Information et de la Communication (NTIC) avec tout ce que les média peuvent offrir en terme d'outils: logiciels de simulation, images vidéo, dessins, etc. Dans un proche avenir, ESC'EAU pourra, en plus d'évaluer, reconnaître les compétences des travailleurs visés et émettre un relevé officiel de celles-ci que l'ensemble des intervenants de l'eau reconnaîtront.

Références

(1) http://europa.eu.int/comm/dg05/esf/en/public/brochure/brochfr.html
(2) http://www.cnpf.fr/francais/actua/htm1/fta_fram3quad.htm
(3) Antoine, A.-M., Le nouveau répertoire opérationnel des métiers et des emplois (ROME) de l'ANPE, Actualité de la formation permanente no 143, Centre Inffo, juillet-août 1996, pages 34 à 40
(4) http://www.anpe.fr/
(5) Charraud, A.-M., Personnaz, E. et Veneau, P., Les certifications de qualification professionnelle (CQP), Construction des référentiels et mise en œuvre dans la métallurgie, la plasturgie et l'agro-alimentaire, CEREQ, Document no 132, Série synthèse, mars 1998, 100 pages
(6) Perker, H., Leplâtre , F. et Ward, C., Identification et validation des savoir-faire et des connaissances acquises dans la vie et les expériences de travail, Rapport comparatif France/Royaume-Uni, CEDEFOP panorama, 1994 (1ère réimpression en 1996), 63 pages
(7) BETWI et CABWI, NVQs around the World, 49 pages
(8) WEF/ABC, Certification Study Guide for Wastewater treatment Personnel, 1992 (réimpression en 1995), 45 pages
(9) Réunion du 27 octobre 1998 sur le projet ESC'EAU à Marseille
(10) http://www.eaufrance.tm.fr/aeap/infos_120697.htm

IUPWARE : a case study of interdisciplinary education and training

IUPWARE : un cas d'étude d'une formation interdisciplinaire

Feyen, J.[1], De Smedt F.[2], Raes D.[1], Batelaan O.[2]

[1] Institute for Land and Water Management, Katholieke Universiteit Leuven, Belgium

[2] Laboratory of Hydrology, Vrije Universiteit Brussel, Belgium

Abstract

The Interuniversity Programme in Water Resources Engineering (IUPWARE) evolved from the integration of two existing 2-year postgraduate programmes in Irrigation Engineering and Hydrology. The paper gives a brief description of the evolution of the two separate postgraduate programmes into a single interdisciplinary 2-year postgraduate programme. Emphasis is given on the changes introduced in the programme to make it truly interdisciplinary. The changes required to make postgraduate training interdisciplinary are highlighted by describing the evolution of the IUPWARE programme since 1981, the starting date of the two postgraduate programmes, which led to the establishment of IUPWARE. It is concluded that achieving an interdisciplinary programme requires more than the simple amalgamation of existing programmes and the introduction of new courses. The complete programme environment and structure need to be adjusted. The foregoing needs to go hand in hand with the introduction of measures to assure that the training is given in an interdisciplinary way, requiring a drastic change in the mentality of the lecturers, the way courses are taught and the programme operation.

Résumé

Le programme interuniversitaire en Ingénierie des Ressources en Eau (IUPWARE) est le résultat d'une intégration de deux programmes de troisième cycle respectivement en Ingénierie en Irrigation et en Hydrologie. L'intégration de ces deux programmes résulte d'un besoin de concevoir de nouvelles approches multidisciplinaires dans le domaine de la gestion des ressources en eau. Cette approche doit permettre d'apporter des solutions durables. Ce manuscript donne un aperçu de l'évolution des deux programmes de troisième cycle à partir de 1981 vers le programme actuel IUPWARE. Le document accentue les évolutions introduites dans le programme visant une interdisciplinarité. Il est clair que le développement d'un programme interdisciplinaire exige plus qu'un simple regroupement de programmes existants et l'introduction de nouveaux cours. C'est surtout le cadre de travail et la structure du programme qui doivent être ajustés. Ceci exige un changement de la mentalité des enseignants et du déroulement du programme. Cette approche a résulté dans un programme multidisciplinaire d'enseignement de deux ans en ingénierie des ressources en eau.

1. Introduction

The Belgian Administration for Development Co-operation (BADC) took in 1981 far-reaching measures as to assure that from 1982 the budget for university cooperation, as well as for other sectors of development cooperation, between the Flemish and the Walloon

Communities was adjusted in proportion to the population of both regions. In the past most of the funding for development cooperation went to the French speaking region of the country. The foregoing was the logical consequence of Belgium's colonial past whereby most aid for development cooperation went to the former colony, the Belgian Congo, and other French speaking countries in Western and Northern Africa. The new measures gave enormous opportunities to the Flemish Community, in particular the Flemish universities. The Flemish (Vlaamse Interuniversitaire Raad, VL.I.R.) and Walloon (Conseil Interuniversitair des Universités de la Communauté Française, C.I.U.F.) Interuniversity Councils co-ordinated from the early beginning the activities of development cooperation at university level. Whereas C.I.U.F. continued focusing on the training of overseas students at the undergraduate level, as a way to maintain government funding to the universities by keeping up the number of students in the 1st and 2nd cycle programmes of the universities, the VL.I.R. pursued a completely different approach.

The university development cooperation activities selected by VL.I.R. consisted of :
- the establishment of research oriented projects between a research unit of a Belgian University and a department or school of a university in a developing country, with the overall objective of strengthening overseas local training and research capacity;
- the establishment in Belgium of short and long-term postgraduate training courses for young graduates of the developing countries; and
- the establishment of a grant programme to enable overseas graduates to enrol at a Flemish University for the PhD-programme.

Further, VL.I.R. proposed and implemented the restriction that 65% of the budget available for interuniversity cooperation should be allocated to projects in the developing countries, and that development-oriented research and training projects in the developing countries and Flanders are not automatically extended, and that with time training programmes in Flanders are transferred to the developing countries.

Over the years many overseas projects between a department/laboratory of a Flemish University and a university in Asia, Africa and Latin America were established, whereby the majority of projects were located in Sub-Saharan Africa. Those projects were mainly research oriented, with a small extension component and in most cases co-ordinated by one academic staff of a Flemish university. Projects had traditionally a narrow scope, were commodity oriented, had a maximum duration of 4 years and a total budget ranging between 8 and 16 million Belgian Francs. In general, the budget was just large enough to accommodate the salary cost of one full-time expatriate, responsible for the project coordination and implementation. With the exception of the travel expenses of the responsible Belgian professor/scientist and some administrative expenses varying in cost between 10 and 15% of the total project budget, the remaining budget was spent in the overseas university. Over 100 projects have so far been implemented, spread over an equal number of universities and in as many developing countries.

VL.I.R. soon recognised the difficulties in getting those projects properly administered and evaluated, and the limited long-term impact of these types of cooperation projects in the receiving countries. Instead of having the budget for overseas university cooperation spread over too many locations, the idea arose to concentrate the limited means to a small number of priority university campuses in the developing countries. Therefore, as from 1996, VL.I.R. changed drastically its policy and terminated most projects of the type "Own Initiatives of the universities". These are individual projects set up by different Flemish universities at their own initiative. Instead, VL.I.R. selected 10 universities in the developing countries (5 in Africa, 3 in Asia and 2 in Latin America), where future overseas university cooperation will be concentrated.

In contrast to the 3-4 years duration of the previous projects ("own initiatives of the universities"), the duration of the new cooperation projects with the priority universities was extended 3 times to 12 years. The annual budget per university was fixed to an average of 30 million Belgian Francs per year. The individual professor/scientist can still play his role in the institutional university cooperation by participating in one of the many projects funded at the priority partner institutions. The average number of sub-projects per institutional partner ranges between 8 and 12. By having more projects in a university, different professors/scientists of Flemish Universities are encouraged to cooperate with several professors/scientists of the overseas partner institutions. This approach has certainly contributed in getting interdisciplinary training and research going among the Flemish universities.

So far, the Institutional University Co-operation has been implemented at 10 priority university campuses in Africa (University of Nairobi, Kenya; Sokoine University of Agriculture, Tanzania; University of Dar es Salaam, Tanzania; University of Zambia, Zambia; and the University of Harare, Zimbabwe), Asia (Can Tho University, Vietnam; Hanoi University of Technology, Vietnam; Network of the Saint Louis University and Benguet State University, the Philippines) and Latin America (University of Cochabamba, Bolivia; Escuela Superior Polytécnica del Litoral, Ecuador). Although the main part of the annual budget for overseas project development cooperation is consumed by the Institutional University Co-operation, there is still a budget of 100 million Belgian Francs for the funding of individual projects ("Own Initiatives of the universities"). However, this budget item, as all other budget items for the State Secretariat of Development Co-operation, is subject to the restriction that 50% of all means should be allocated to projects in Sub-Saharan Africa.

In addition to the funding of a limited number of carefully selected partner universities in the developing world and a limited number of "Own Initiative of the universities" type of projects, VL.I.R. controls also the development of postgraduate level training initiatives in Flanders. Over the years, VL.I.R. gave the green light to the establishment of 4 International Training programmes (short-courses) and 13 International Courses with duration of 1 to 2 years. The International Course Programmes (ICP) follow the same pattern as the regular academic programmes. Five of the 13 ICPs organised by VL.I.R. are truly interuniversity courses, whereas the remaining 8 ICPs are courses organised at one university. It is VL.I.R.'s policy that with time all ICPs become interuniversity courses, i.e., organised by different universities in Flanders, so as to guarantee the interdisciplinary approach and to make sure that postgraduate training is based on a wider group of prominent researchers. Furthermore in the long term, VL.I.R. aims to transfer fully developed and successful ICPs to the developing countries, so that the available funding for postgraduate training programmes can be used to set up new ICPs and to allow existing ICPs to evolve to full-fledged doctoral programmes.

2. Brief history of the ICP in irrigation engineering and hydrology

One of the interuniversity ICPs is in Water Resources Engineering, being organised by the Katholieke Universiteit Leuven (K.U.Leuven) and the Vrije Universiteit Brussel (V.U.B.). In the following sections, the history, and the changes in the programme's philosophy, structure, study programmes and organisation are highlighted, illustrating the evolution of the Interuniversity Programme in Water Resources Engineering (IUPWARE) since its inception more than 18 years ago.

IUPWARE arose in 1994-95 from the integration of two postgraduate programmes, the ICP in Irrigation Engineering being organised by the Katholieke Universiteit Leuven (K.U.Leuven) and the ICP in Hydrology organised by the Vrije Universiteit Brussel (V.U.B.). Both programmes operated successfully since their formation in the early 80's. Although both programmes were organised independently and run by two different universities, they had a lot in common. They were set up for the same purpose; had similar objectives and programme structure; and the training thrust of both programmes was related to water resources. The reasons why both ICPs were established almost simultaneously could be best described as follows:

- The awareness that training capacity and facilities at academic level in the field of irrigation and hydrology in many developing countries are largely insufficient, often non-existing;
- The increasing demand for institutions/organisations/universities to organise sandwich courses in the field of irrigation engineering and management, surface water and groundwater hydrology for developing countries. Over time, this placed time and financial burden on the staff of both institutions;
- The desire of both institutions to internationalise some of its programmes, and the wish to improve its international image;
- The pressure from the society to integrate in a better and more effective way the competence and experience in the field of irrigation at Katholieke Universiteit Leuven and in hydrology at Vrije Universiteit Brussel ;
- The readiness of the Belgian Administration for Development Co-operation (BADC) to finance international postgraduate programmes which are directed to priority problems of developing countries; and
- The increasing pressure from inside the institutions to increase the share of soft (project) money with respect to the hard (government funding) money so as to assist in modernising infrastructure for teaching, research and the generation of new employment for junior faculty staff.

In addition both ICPs had the following characteristics and needs in common:
Both study programmes were primarily organised for young academicians and engineers from the developing countries;

The programmes, lasting 2-years, were taught in English by university staff and international visiting staff, and led to the degree of Master of Science in Irrigation Engineering or Hydrology;

The trainees to the ICP in Irrigation Engineering were a mixture of civil and agricultural engineers, whereas the trainees to the ICP in Hydrology had a first degree in engineering, geology or geography;

The main aim of the first year was to bring the trainees coming from different continents and having different backgrounds to the same level of understanding in a number of basic courses;

Courses in the 2nd year were mainly applied courses in the field of irrigation or hydrology;

Both study curricula addressed primarily in a quantitative way the engineering aspects of irrigation and hydrology, and paid little attention to the socio-economic, cultural, legal and institutional-related aspects. In the developing countries, these non-engineering factors are the primary bottle-necks why irrigation schemes and hydrology projects do not perform as anticipated at design stage; and

In addition, both programmes did not pay attention to the interaction between irrigation and hydrology on one hand, and the impact of engineering designs in the field of irrigation and hydrology on the environment on the other.

In summary, the programmes of both ICPs was straightforward, very classic in concept and structure, and formal in the way courses were taught. Most lectures were given ex cathedra, with few practical sessions and home tasks. Instead of offering the trainees hands-on techniques for solving common problems, typical for the developing countries in the field of hydrology and irrigation, the trainees received an excellent academic education in the theory and practice of hydrology and irrigation. Some minor improvement in the way of teaching was achieved by the introduction of computers and software tools in the middle of the 80's. By lack of vision and incentives and the fact that both programmes recruited relatively well internationally, both ICPs remained mainly purely disciplinary and academic-oriented postgraduate training programmes. In the ICP in Hydrology the main emphasis was placed on understanding the hydrology of complex systems and the interaction between surface water and groundwater bodies. Little emphasis was placed on studying the impact of human interference on the quality of the earth's water resources. Similarly, in the ICP in Irrigation engineering the main thrust of the programme was on the engineering aspects of irrigation at field and project scale. The study programme poorly focused on teaching the interaction between the irrigation processes and other disciplines such as the socio-cultural and socio-economic aspect of the indigenous communities operating and farming the irrigation schemes, nor paid much attention to the interaction between the irrigation scheme and the surrounding environment. Further both programmes paid little attention to incorporating in the different courses tools for analysing in a quantitative way the impact-effect relationship of human interference in engineering designs of irrigation and water resources projects.

3. The educational and training concept of IUPWARE

As indicated earlier, the ICP in Irrigation Engineering (K.U.Leuven) and the ICP in Hydrology (V.U.B.) were merged as from the academic year 1994-1995 for the following reasons:
- to generate the basis for an interdisciplinary training programme covering the field of hydrology and irrigation, the interaction between both disciplines and aspects of water quality management;
- to optimise the restricted volume of training resources aiming at the same time at an improvement of the study programmes, i.e., making the programmes more appropriate to conditions in the developing and industrial countries and providing the trainees more hands-on training;
- to anticipate the eventual decline in government funding, which due to the reduced number of grants from BADC, resulted in a direct reduction of number of attendants;
- to anticipate to the general decline of interest in irrigation matters by international donors and overseas graduates, and the growing interest of the society and trainees in water quality environmental problems.

The integration of the two postgraduate programmes resulted in a new interuniversity programme for postgraduate training in Water Resources Engineering (WRE). The main objective of this interuniversity programme was the organisation of a 2-year advanced academic study programme in WRE for young academicians and engineers of developing countries leading to the degree of Master of Science. At the same time it was hoped that by bringing together the two ICPs would result in a pooling of expertise in water resources related issues within K.U.Leuven and V.U.B forming a solid foundation for post-doctoral training and continuing education, overseas project development and consulting activities. The merger of the two ICPs resulted in a more attractive, effective and interdisciplinary study programme, an increasing number of students, and a reduction in the organisational costs.

The 1st year of the ICP in WRE integrated the basic courses of the previous ICPs in Irrigation Engineering and Hydrology, with the objective of bringing trainees with different educational backgrounds to the same level in mathematics, statistics, operational research, hydraulics, surface water and groundwater hydrology, irrigation, and water quality and treatment. Whereas the 1st year curriculum of the newly ICP is common for all participants, offering a prerequisite course in basic calculus, 10 subjects and 6 workshops, the 2nd year offers three curricula in Hydrology, Irrigation and Water Quality Management. The structure of the three curricula is similar, consisting of 7 courses, seminars, an integrated project design and a thesis research project. The interaction between the three curricula is limited to one course, the seminars and the integrated project design. All other courses in the 2nd year are topic specific, i.e., related to hydrology, irrigation or water quality. In contrast to the previous replaced ICPs, in the new ICP in WRE, extensive use is made of microcomputer applications as to achieve a more problem-oriented and interactive training of the participants.

The new programme is still focusing on the engineering aspects of water and land resources (resources that are often difficult to define, control, manage, negotiate and adjudicate) but it is also addressing the negative side effects of water and land resources development. The new programme makes intensive use of mathematical and operational techniques and new technologies. Notwithstanding these positive changes, the programme still remains too academic, discipline-oriented, i.e., the interaction between the options and the individual courses within the options are too limited to have a full interdisciplinary programme. Further it was noted that insufficient use in different courses was made of numerical modelling, being nowadays considered as the most rational and systematic method of quantifying many of the environmental impacts caused by human interference on the land and water resources.

While numerical modelling does not give all the answers, it helps to find answers to the short-, medium- and long-term impact of alternative designs and solutions. Numerical models can best address 'what if .... questions', for example:

- What happens if the sewage to be discharged through a marine outfall is primary, secondary or tertiary treated?
- What happens if the tailwater is extended by 100m, 200m, or 300m?
- What happens if groundwater levels are lowered by 1m, 5m or 10m?

The point is that more than ever before engineers in the field of water resources should thoroughly be prepared to cope with present and future problems, which mainly will emerge from the conflicting demands of the fast growing society for the scarce natural resources. Today it is well recognised that future productivity increase will have to be achieved while at the same time, conserving and enhancing the natural resources base on which we all depend. The truth is, we have to feed nearly six billion people worldwide, whereas as time passes the water is becoming scarcer and demand for it is increasing in every sector of life. This result in a great deal of competition for water from agriculture, industry, domestic uses, forestry, etc. Therefore, on one side there is big demand of water in every sector, while on the other side its overuse, the lack of maintenance of water delivery systems and the contamination of the surface water and groundwater resources have caused a host of socio-environmental and economic problems.

In recognition of this, the steering committee of IUPWARE being in charge of the short, medium and long term programme objectives and the organisation and daily operation of the MSc in WRE, recently decided to introduce as from next academic year (1999-2000) the following modifications in programme structure, particularly in the programme structure of the 2nd year. Changes in the 1st year programme have been limited to a reduction in study load for the trainees, an upgrading of the workshops, and integrating in the different basic courses as many links to water resources problems as possible, as to make those courses more meaningful. Measures are taken as to assure that the lecturers of the basic courses make a sound balance between theory and practice, and that practice is related to a variety of water resources problems, typical of the developing countries. The foregoing undoubtedly will help to convince the trainees about the purpose and the relevance of the basic courses. Further, making in the 1st year a link to the courses in the 2nd year and the field of practice and letting the students work in teams will contribute in developing in the trainees' mind a holistic and multidisciplinary way of thinking and functioning.

In the 2nd year programme far more reaching changes have been proposed and will be introduced, as indicated above, from next academic year in order to prepare the graduates better for the challenges they will face in their professional career. Instead of splitting the students from the beginning of the academic year in different modules, all students of the 2nd year, with their different backgrounds and expertise, will be kept together and obliged to take 5 common courses. In those courses strong emphasis will be given to the use of mathematical modelling and new technologies with application to the hydrologic cycle, hydrologic transport, and the management of water use and re-use. To keep balance between theory and practice, the trainees will have to work in team on an integrated project design, in which all the different aspects of water resources engineering are presented in a real-life context. In addition to the compulsory common courses and the integrated project design, the trainees will have to choose between one of the following four modules: hydrology, irrigation, water quality management and aquatic ecology. Each module is similarly structured, consisting of two working courses on the engineering and technological aspects of the module and a thesis research project. Also in the thesis, students will be requested to use intensively mathematical models. Furthermore, lecturing staff will be urged to emphasise in courses and workshops the holistic and multidisciplinary dimension of problems and solutions. The proposed measures will not only result in making the WRE MSc programme more academic and professional, but also lead to considerable optimisation of local and visiting lecturing staff, organisational and related costs.

4. Conclusions

IUPWARE today is the result of a continuous evolutionary process that started since the inception of the former ICPs in Irrigation Engineering and Hydrology, two international postgraduate programmes organised by the K.U.Leuven and the V.U.B., primarily for students of developing countries. A number of factors led to the merger of the two ICPs and the continuous adjustment of objectives, goals, content and structure of study programme. These factors include regular internal and external evaluations; the fact that the capacity and competence of many overseas undergraduate programmes are improving; the increase in the magnitude and complexity of water related problems; the decline in government funding and the taking over of the organisation of ICPs by the Flemish Interuniversity Council (VL.I.R.) from the Belgian Agency for Development Co-operation (BADC).

After 18 years of operation and regular programme adjustments the authors are convinced that IUPWARE has reached a balanced and sustainable status. The 2-year Master of Science postgraduate training programme prepares in an academic and professional way its trainees for the many and complex water challenges the society will face in the 21st Century with increasing frequency. The present interdisciplinary programme status is not only the result of changes in number of courses and course structure, but also the consequence of a growing interaction between staff, students and the professional world, and change in the way knowledge and expertise are transferred. Realising that the conditions for IUPWARE, and the society as a whole, are far from static, the steering committee must remain alert and dynamic, in order that the MSc programme keeps its present status of performance and competitiveness.

Acknowledgements

First the authors like to thank the universities of K.U.Leuven and V.U.B. for providing the environment and infrastructure for the establishment of the former ICPs in Irrigation Engineering and Hydrology, and the present Interuniversity Programme in Water Resources Engineering (IUPWARE). Also the financial support from the Belgian Administration for Development Co-operation (BADC), today provided through the Flemish Inrteruniversity Council (VL.I.R.), is very much appreciated, because without this financial support and the several external evaluations, IUPWARE would never have reached its present level of functionality and professionalism. The authors are also very much indebted to the national and international lecturing staff and the administrative support from the secretarial staff, and this since 1981-1982. Last but not least, the authors are very grateful to Kenneth Wiyo for the constructive comments and corrections made in reviewing the manuscript.

References

Feyen, J., 1986. Irrigation Engineering: 1981-1986. Institute for Land and Water Management, K.U.Leuven, 16 pp.
Feyen, J., 1990. Evaluation report of the ICP in Irrigation Engineering for the period 1981-1990. Institute for Land and Water Management, K.U.Leuven, 122 pp.
Feyen, J., 1991. Brochure at the occasion of the 10th anniversary of the Center for Irrigation Engineering. Institute for Land and Water Management, K.U.Leuven, 46 pp.
Feyen, J., 1992. The Center for Irrigation Engineering: mission, history, students, staff, financial situation, achievements, issues and concerns. Institute for Land and Water Management, K.U.Leuven, 24 pp.
IUPWARE, 1994. Programme brochure (69 pp.) and Course syllabi (71 pp.). Institute for Land and Water Management, K.U.Leuven, and the Laboratory of Hydrology, V.U.B.
Van der Beken, A., 1991, Programme-Brochure IUPHY-Document No 1, Interuniversity Postgraduate Programme in Hydrology, Brussels, 40 pp.
Van der Beken, A., 1994, Hydrology at the V.U.B. 1979-1994, a register including IUPHY Alumni. V.U.B.-Hydrologie No. 29, Brussels, 334 pp.

Quality assessment and evaluation of a series of five transnational training courses on water quality measurements

Évaluation de la qualité d'une série de cinq cours de formation transnationaux en matière de mesures de la qualité de l'eau

Ziglio G.[1], Heinonen P.[2], Karayannis M.[3], Pilidis G.[3], Quevauviller P.[4], Van Den Berghe W.[5], Van Der Beken A.[6]

[1] University of Trento,
Dipartimento di Ingegneria Civile ed Ambientale, Italy
[2] Finnish Environment Institute, Finland
[3] University of Ioannina-Department of Chemistry, Greece
[4] EC-DG XII (Standards, Measurements and Testing), Belgium
[5] Tilkon bvba, Belgium
[6] Co-ordinating TECHWARE Bureau, Belgium

Abstract
In compliance with the Standards, Measurements, Testing (SM&T) Workprogramme (1994) and its "Preparatory, Accompanying and Support Measures", a series of five short courses prepared by a Specialist Group "Measurements" of TECHWARE (TECHnology for WAter REsources) was sponsored by EC DGXII (Contract SMT 4-CT 96-6501). It was anticipated that the target groups had to be managers, professionals and technicians of water quality laboratories with monitoring responsibilities, from the regulator organizations, the water sector and the sanitary control institutes. The outcomes of the five courses were very positive, following both the assessment made by the participants and the assessment made by lecturers. The industrial water sector was weakly involved and not reached. This might be a consequence of the prevailing theoretical, scientific approach in the course delivery. However, staff working in the water sector could have been potentially interested to most of the courses both from a theoretical and from a practical point of view. Some explanations are proposed for such a lack of interest. A follow-up activity is planned.

Résumé
En conformité avec le programme de travail du programme "Normes, Mesures et Essais (1994-1998) et ses "Mesures préparatoires, d'accompagnement et de soutien", la DGXII de la Commission Européenne a sponsorisé une série de cinq cours preparés par un Groupe de Spécialistes faisant partie de l'association TECHWARE (TECHnology for WAter REsources). Cette activité a été financée dans le cadre du contrat SMT4-CT96-6501. Les groupes de participants ciblés pour ces activités de formation étaient composés de dirigeants, de professionnels et techniciens de laboratoires pour la qualité de l'eau, d'or-

ganisations de réglementations du secteur de l'eau et d'institutions pour le contrôle sanitaire. L'évaluation des cinq cours a globalement été jugée très positive par les participants, comme par les professeurs. Le secteur industriel de l'eau n'a pas participé beaucoup à ces cours, tant sur le plan de l'enseignement que sur celui de l'apprentissage. Ce manque de participation pourrait être en partie lié à l'approche théorique dominante du contenu des cours. Pourtant, le personnel travaillant dans ce secteur aurait pu être potentiellement intéressé par les aspects théoriques et pratiques des cours. Différentes explications pour ce manque d'intérêt sont examinées dans cette contribution. On est en train de programmer une activité de follow-up.

1. Introduction

Quality monitoring and control of water is a wide-range responsibility guided by EC directives and regulations, such as the EC Directive on the quality of water for human consumption and the recent new regulation, the EC Directive on waste water discharge or the EC Directive on dangerous substances in surface waters.

Such an important activity needs scientific research to improve efficiency through new instrumentation and better assessment methods. A corollary to research is training for improving efficiency, for promoting "best practices" and, in general, for transfer of the latest proven scientific knowledge to the field.

2. The five courses

The EU SM&T Workprogramme selected and founded several training courses: Quality asurance for chemical analysis (SMT4-CT97-6514; Prichard 1995), Preparation of reference materials (SMT4-CT96-6504), Sampling and sample handling (QUASH project, Wells et al., 1997) and Water quality measurements (SMT-CT96-6501; Techware 1995).
The last initiative consisted of a series of short courses (5 days' duration) prepared by leading organizations in the EU, under the coordination of a Specialist Group "Measurements" of TECHWARE (TECHnology for WAter REsources).
The topic, place and dates of the five courses (from A to E) on Water Quality Measurement are indicated in Table 1. The five courses are shortly descripted and evaluated in the following chapters.

Table 1. Description of the five courses

Course	Subject	Place	Period
A	Monitoring and measurements of lake recipients	Helsinki, Finland	25-29 August 1997
B	Measurements of heavy metals and anions in drinking and surface waters	Ioannina, Greece	5-9 May 1997
C	Measurements of organic compounds i n drinking and surface water	Ioannina, Greece	6-10 July 1998
D	Use of Biotic Indexes to evaluate the quality of freshwater streams:) a comparison among four different European methods (IBE, BBI, BMWP, RIVPACS)	S. Michele a/ Adige (Trento), Italy	22-27 June 1998
E	Analytical Methods for Algae, Protozoa, Helminths in fresh water	Genoa, Italy	10-14 Nov., 1997

Training needs and target groups

The training needs were expected to cover the following categories:
a) the use of advanced analytical instrumentation and related new tools for data acquisition and handling;
b) "best practices" and quality assurance of the sampling and the measurements procedures;
strategy and methodologies for water quality control;
c) training with respect to the legal aspects of directives and regulations.
It was anticipated that the target groups had to be managers, professionals and technicians of water quality laboratories with monitoring responsibilities.

Objectives and structure of the courses

Two general aspects are important when deciding what will be the elements of the training and what will be the general structure of these courses:
Exchange of knowledge, not only between the teacher and the participants of the course, but also between the different European countries mutually.
Potential solutions for problems related to the field of water quality taking into account the European dimension and harmonization of the activities.
Each local course organizer provided for a good mix of the above training aspects through state-of-the-art plenary sessions (seminar type), workshops on specific topics and real case discussions and practical training in the laboratory facilities. Field activity was also included.

Evaluation

Using standard procedures (Van den Berghe, 1995), evaluation questionnaires were distributed among both participants and teachers and analysed as reported below.

2. Quantitative evaluation of the five courses by participants

This evaluation includes the opinion of almost 120 course participants for about 90 lectures and/or laboratory training sessions.
Breakdown of attendants
Overall, 2/3 of participants came from the host country. Only in one course the number of participants from outside the host country exceeded that from the home country. However, all courses had a very international lecturer profile.
The predominant profile of the course participants, despite common elements, was different for each course.
Overall assessment of the courses
An overall assessment has been provided using the response distribution to the different evaluation criteria of the standard questionnaire.
How relevant was the course programme for your current and future professional needs? Overall, the course programme was relevant for the target groups of the different courses. The best match between needs of the course participants and the course programme was achieved in course D; the least (but still at an acceptable level) in course B. In assessing the different topics of the courses, the relevance of the lectures and training was much more "homogeneous" for courses C and D than for the other courses.
To what extent did the course improve your knowledge? The overall appreciation for this criterion was good. The median response was: "I feel my knowledge about the field has considerably improved".
To what extent did the course enhance your practical skills? The overall assessment regarding skills enhancement was less positive than for learning new knowledge. In fact, the typical statement was "The course enhanced a range of practical skills". This result

should be seen within the context of the courses: skills enhancement was likely not the prime purpose for some courses.

In regard to your initial expectations when you registered for the course The overall result is very good: only 1 course participant stated: "my expectations were only partially met", while all the others said at least that "the course corresponded more or less with my expectations"; more than 10 % even stated that "the course exceeded my expectations".

3. Quantitative evaluation of the five courses by lecturers and tutors

Breakdown of the lecturers/tutors

The number of lecturers and tutors ranged in each course from 11 to 27. The courses lasted for a similar period but were differently structured. On the average, about 2/3 of lecturers and tutors came from the host country. The absolute numbers of "other country" lecturers and tutors were quite similar (4-7) for each course. Budgetary constraints can justify these figures. With regards to the predominant profile, seventy-five percent of the lecturers and tutors were from Public Research Institutions (Academia, Research Institutes) and only 25% were from Water Industries and Local Water Associations. The predominance of the scientific profile was common to all courses.

Overall assessment of the courses

An overall assessment has been provided using the response distributions to the following evaluation criteria:

understanding of the overall objectives of the course

adjustments of the training approach to the course objectives and contents in preparing the lecture

after having delivered the lesson, realization that the lecture should have been modified.

Lecturers and tutors were contacted and informed about their tasks and the overall objects of the course. Nonetheless, an acceptable proportion of them (25%) proposed after their presentation/tutorial activity some modifications that were for the major part (64%) related to the content and the training approach.

Combining together all the outcomes coming from the responses, the overall evaluation was highly positive. To try to compare the overall assessment of the courses, these following criteria were applied: the sum of the lowest scores and the sum of the highest scores (normalised to the number of the respondents). Accordingly, Course A and Course D seemed to reach the minimum and respectively the maximum score, where the remaining courses occupied a much more similar rank of positions (see table 2).

Table 2. Comparison of the overall assessment of the five courses (see text)

Lowest absolute scores						**Highest Normalised Scores**				
	A	B	C	D	E	A	B	C	D	E
	2	0	0	0	1	0.4	0.3	0.6	0.9	0.4
	2	0	0	0	0	0.9	1.0	1.0	1.0	1.0
	6	2	3	1	2	0.6	0.8	0.7	0.8	0.8
	0	0	1	0	0	0.4	0.6	0.5	0.5	0.5
Total	10	2	4	1	3	2.3	2.7	2.8	3.2	2.7

A substantial fraction (74%) agreed on the repetition of the course series. The courses have been adequately targeted and delivered: only 12% and, respectively, 25% of the lecturers/tutors recognised the need to modify the course objectives and the course training approach.

4. Conclusions

The outcomes of the five courses were very positive, following both the assessment made by the participants (as regards relevance, expectations, knowledge and skill improvements) and the assessment made by lecturers (objectives, training approaches and course repetition).
The courses addressed different aspects of water quality measurements. Therefore the participants had differentiated background and expectations when registered. In 4 out of 5 courses a national dimension, as regards participants (and, to a lesser extent, lecturers) was reached. This can be explained as a consequence of:
a local specific interest in the content/objectives of the organized courses, and
the cost for travel and subsistence.

The industrial water sector was weakly involved (only two lecturers) not reached (no participants from this sector). This might be a consequence of the prevailing theoretical, scientific approach in the course delivery. The practical (lab/field) activity included by 4 out of 5 courses in most of the cases can be considered as a demonstration, rather than as a direct involvement of the single participants. Constraints and limitations due to the number of participants, the needed facilities, the duration of the short courses and the broad content, all can preclude a person-to-person training approach. However, staff working in the water sector could have been potentially interested to most of the courses both from a theoretical point of view (knowledge enhancement) and a practical point of view (skill enhancement).
Some explanations could be proposed for such a lack of interest:
The backgrounds of the technical personnel in the water sector are not sufficiently adequate for the level of the course.

There are requirements of more focused specific practical objectives, to be better fulfilled with a individual "on the job" training than with a general group course.
Costs, language (only English as official one), duration and continuity cannot be adequately managed by the water sector organization.
For all these reasons, a follow-up activity on the analysis of training needs for water quality measurements with industrial applications is presently under way.

References

E. Prichard, Quality in the Analytical Chemistry Laboratory, E.J. Newman (Ed.); John Wiley & Sons Ltd., Chichester, ISBN 0-471-95470-5 (1995).
Techware (1996). Training for Water Quality Measurements, A project proposal under EC-DGXII, SM&T, Coordinating Techware Bureau, Brussels, Belgium.
Van den Berghe, Wouter (1995). Achieving quality in training. European Guide for collaborative training projects, Tilkon, Wetteren, Belgium.
Wells D.E., Cofino W.P., Mar. Pollut. Bull., 35 (1997) 146.

Theme D

Continuing Professional Development (CPD) and tailor-made training

Formation continue et à la carte

Thème D : Formation continue et à la carte

Rapporteur : Gilles Neveu (France)

Trois idées force semblent se distinguer des papiers présentés et des discussions qu'ils ont engendrées :

- gestion intégrée de la ressource, globalisation des approches, multiplication des « petits mais nombreux » problèmes (en opposition avec les grandes infrastructures d'antan), prise en compte des besoins et de la demande (en complément de l'approche traditionnelle par l'offre), autant de concepts qui induisent d'autres connaissances et d'autres savoir-faire de la part des techniciens et ingénieurs.

Il leur faut élargir leur vision, en prenant en compte non seulement les aspects scientifiques et techniques, qui déjà se multiplient, mais aussi les aspects sociaux, sociétaux, institutionnels, etc, du monde qui les entoure.
Comment répondre à cette explosion des besoins de connaissances et de compétences ?

Les structures de formation doivent faire évoluer les méthodes de formation, en facilitant la réflexion des étudiants, en leur faisant partager et confronter leurs expériences, en leur facilitant l'accès à la documentation, etc, pour les préparer à l'évolution permanente de leurs compétences. De nombreuses initiatives ont été présentées lors du Symposium.

La formation continue apportera aussi une réponse, en sachant que la formation initiale ne pourra pas tout apporter sinon les fondations de l'apprentissage tout au long de la vie, et que les besoins de progression existent aussi à 30 – 40 – 50 ans et + .

Il faudra pour cela que les Universités et les Ecoles acceptent de ne pas tout couvrir pendant le cursus de base, mais qu'elles se mobilisent pour proposer, en osmose avec les professionnels, des cursus à la carte qui répondent aux besoins « quand il le faut » avec toute la qualité nécessaire (obligation de résultats).

Il faudra aussi favoriser l'échange d'expériences à plus grande échelle en soutenant des réseaux efficaces entre les institutions.

- plusieurs orateurs ont insisté sur le fait que les ingénieurs et techniciens ne choisissent pas (plus ?), ils proposent des solutions aux utilisateurs ou à leurs représentants, qui eux font les choix politiques qui leur semblent les meilleurs pour la collectivité.

Deux conséquences :

d'une part les techniciens doivent accepter que le point de vue des décideurs puisse différer de leurs certitudes scientifiques et qu'il leur faudra expliquer et défendre leurs options ;

d'autre part les décideurs doivent comprendre le langage des ingénieurs et techniciens, ce qui implique de les informer, de les former (c'est l'exemple des villageois indiens ou des agriculteurs tunisiens, ou encore des maires ruraux français) pour qu'ils puissent choisir en analysant eux-mêmes les avantages et les inconvénients des solutions proposées.

La formation continue doit alors être étendue à l'environnement des professionnels de l'eau, pour faciliter la discussion et l'acceptabilité des projets.
Il s'agit bien de réintroduire l'approche participative de la collectivité à son propre futur, comme cela se passait autrefois avec succès (des solutions locales pour des solutions durables).

- l'articulation entre recherche et formation universitaire est - elle satisfaisante, qui plus est lorsque la recherche devient pluridisciplinaire ? le transfert de connaissances se fait - il suffisamment et comment cela remet - il en cause l'organisation traditionnelle des cours ?

Les enseignants et les chercheurs ont le devoir de répondre à ces questions, à l'instar des scientifiques indiens qui vont vers les villageois pour les aider à comprendre et à améliorer leurs propres réalisations.

La question a été abordée rapidement lors des discussions, mais elle semble pertinente dans un contexte de valorisation et de légitimisation de la recherche par la résolution des problèmes.

La formation continue est sans conteste l'une des principales clefs d'une meilleure gestion de la ressource en eau, tant par la diffusion des bonnes pratiques qu'elle permet, que par l'ouverture des esprits qu'elle favorise, en rapprochant les décideurs - utilisateurs et les professionnels.

Réinventer le couplage formation initiale – formation continue est certainement l'un des enjeux majeurs des années à venir, impliquant de re - penser l'apprentissage sur une autre échelle de temps.

Keynote lecture

Continuous Professional Development (CPD) from the demand side, a prospective

Cabrera E., Izquierdo J., Espert V., García- Serra J., Pérez R.
Grupo Mecánica de Fluidos.
Universidad Politécnica de Valencia. Spain

Summary
The final aim of Continuing Professional Development (CPD) is to update the knowledge required by professionals to improve their performance, allowing them to find valid solutions to the problems they have to face every day. CPD will be more effective as long as it is addressed to those knowledge areas that the labour market demands and especially those shielding the new employment sources. Nevertheless, high demand orientation, even though it shows necessary to achieve a good response within the labour market, does not guarantee success. It is of paramount importance to perfectly balance fundamentals and applications, so that professionals be able to assess the different alternatives. This task is far from being easy and demands a continuous dialogue between education sender and receiver.

It is a fact that the labour market within the water industry requires especially experts in management techniques form the demand side. At the same time, the traditional background of the Hydraulic Engineer of the 20th century, driven by the universal and generalised water policy exerted from the supply side, has been oriented to Civil Engineering, neglecting and even ignoring technical aspects that have proved to be strongly demanded by the labour market. These circumstances, under the authors point of view, give outstanding importance to subjects closely related with efficient use of water, form the perspective of both R+D and the shear transmission of knowledge, that is, CPD.

Sommaire
L'objectif final de la Formation Professionnelle Continuelle (FPC) est l'actualisation des connaissances dont le professionnel a besoin pour un meilleur développement de son travail quotidien, de telle sorte qu'il serait capable de résoudre les problèmes qu'il puisse rencontrer chaque jour. Elle est d'autant plus utile qu'elle se dirige vers les matières demandées par le marché du travail à un moment donné et, tout spécialement, celles sur lesquelles est basé l'emploi. Cependant, bien qu' une orientation avec une demande élévée de formation soit nécessaire pour trouver une bonne réponse dans le marché du travail, elle ne garantit pas le succès. Il faut trouver dans l'activité l'équilibre nécessaire entre les bases de la matière enseignée et l'application des contenus, si bien que le professionnel soit capable d'analyser les différentes alternatives possibles. Cette tâche n'est pas facile, elle exige un dialogue permanent entre l'enseignant et l'apprenant.

Le fait que le marché du travail de l'industrie de l'eau demande surtout des experts en techniques de gestion du point de vue de la demande, lié au fait que la formation traditionnelle de l'ingénieur hidraulique du XXe. siècle, en réponse obligée à la politique de l'eau qui a été en vigueur d'une manière généralisée dans tout le monde, ait été ori-

entée en grande mesure vers le génie civil, oubliant, parfois même ignorant, les aspects techniques demandés actuellement par le marché du travail, attribue, selon l'avis des auteurs, un avenir heureux aux matières en rapport avec un usage plus efficace de l'eau, tant depuis le point de vue de la R+D, que depuis la transmission de ces connaissances, c'est à dire, la FCP.

1. Introduction

Through this vanishing 20th century, Water Engineering has devoted all the attention to management from the supply side. Surface and ground hydrology, system engineering applied to the optimisation of the available water resources, inter-basins transfers, regulation and analysis of natural water courses, maritime and coastal engineering, sediment transport and, finally, the big hydroelectric constructions and the transformations of dry land into irrigated land have attracted the Society attention regarding water. As a consequence, also the interest of the Hydraulic Engineer has been attracted in the same direction. These big engineering constructions and the consequent important water transfers explain clearly why Hydraulic Engineers have been assimilated into the Civil Engineering field.

Nevertheless, the end of the century has witnessed very important changes. During the last decades, engineers, using computer technology, have been able to face once-unthinkable calculations. On the other hand, instrumentation and electronic development has offered the possibility of measuring with great accuracy and velocity certain physical magnitudes to extremes. Certainly, this has dramatically changed the classical ways to face certain problems, but has not contributed at all to change the framework in which the Hydraulic Engineer has developed his activity.
The change in this direction has a different origin not linked to technological progress. In fact, it appears as a counterweight for the practically infinite potentiality of the technological development. This origin has an economical and social character reflecting priority changes in our modern society. The inflexion point must be located in 1987, when the Bruntland commission issues its conclusion (Bruntland, 1987). Emphasis is placed on keeping on exploiting and taking advantage of the natural resources of the Planet, but under the command that development do not compromise the future, that is, it be sustainable. It is important to underline the decisive influence that this report has exerted so far and will keep on exerting in the Water Engineering field. The concept of sustainability of a natural resource has triggered the updating of a number of aspects closely related to the water field, completely ignored a few decades ago. It is unanimously accepted that water management must be undertaken in an integral and global way, without the possibility of forgetting environmental, ecological and biological aspects. The water world has turned an interdisciplinary subject demanding co-operation (Lopardo, 1995, Kobus, 1997). The ASCE Task Committee (ASCE 1996) underscores clearly this point: "Hydraulic Engineers have to think bigger and broader".

This new framework, together with the pressure that available water resources must withstand due to the rampant competence between traditional uses (mainly agricultural) and new uses (consequence of demographic evolution, surge of traditional industry and leisure activities) provoke that water is changing its exclusive social character giving room to an important economic facet. This fact has important consequences regarding water policy.

To achieve resource sustainability the only alternative is to balance supply and demand. Supply management has enjoyed, especially in developed countries, an era of outstanding splendour during the 20th century. This era has repeatedly qualified as Golden Age (Rouse, 1987). As a consequence, those constructions of higher technical viability and higher economic productivity have been undertaken. The current pressure exerted by environmental conditions, formerly ignored, turn these constructions more expensive if not impede their viability. This limits strongly the possibilities of stretching the supply. This limitation to continue exploiting natural resources forces the adoption of rational use perspectives, turning demand into a highly important current issue. This new philosophy on saving water will bring possibilities that not only need to be exploited but even explored. Some objective opinions, -unrelated in our view-, point in this same direction. For example, these changes have affected substantially the research line of the Bureau of Reclamation of USA. Burgi (Burgi, 1.998) synthesises it in this way: "As public values have shifted from emphasis on water resource development to management of western waters, the bureau's contemporary hydraulic research program has also changed from water development to water management". This shift in trend begins to prevail in developed countries, causing an upsurge in the demand of practitioners in this field, thus offering wide future possibilities in water management related questions, both in its R+D branch and in CPD within the area.

A better understanding of the turning point represented by the Bruntland report in 1987 can be gained by revising the relationship man-water through the History. Excellent and passionate travel guides can be found. As Levi (Levi, 1995) affirms perhaps there are no other branches in Engineering enjoying so a prolonged and rich history as Hydraulics. Without doubt, the reason is that water is integrated in the very life of humans. Human never existed and will never exist without water. This fact forces such an unbreakable link.

2. Management and use of water resources through the history

Men, and also animals and plants always needed water to live on. The most ancient irrigation practices we have evidence date back to 30,000 years ago (Bonnin, 1984), while actions to facilitate human consumption (for example, family cisterns to store rainwater) can be traced back to several millenniums before our era. These individual cisterns fell into disuse when, through the time, transfer works were performed to transport water from resurgence to consumption points. It is worth mentioning here that two thousand years ago Rome had a system of aqueducts with an impressive transport capacity: 600,000m^3/day. It amounts to an equivalent of 500 litres per day per person, a level of consumption unthinkable nowadays for a distribution system performing a reasonable management. The passionate world of the use of water in the antiquity can be followed in detail in (Bonnin, 1984) and summarised in (Garbrecht, 1987).

We have inherited scarce samples of pressurised water transport from the antiquity. The main reason comes from the difficulties on getting materials and techniques that would be watertight to moderate pressures. Also the problems derived from changes between regimes (water hammer) contributed to a poor Hydraulic of pressure in ancient times. The siphon is perhaps the most ancient realisation. Such a device was used to supply water to Jerusalem and its construction (Bonnin, 1984) is attributed to Salomon, the son of King David. There is evidence that latter, in the 2nd century, the acropolis of Pergamo was supplied by a siphon made out of lead, with inner diameter of 22cm, thickness of 8cm and withstanding a pressure of 20bar (200mca). In Rome, the downstream ends of the water distribution systems were pressurised, although only to approximately 10mca. As far as pumps concern, those of positive displacement are the first to appear in History (Raabe, 1987) and it was not until the 17th century that turbopumps made their appearance. All the known pressurised pipes are realisations of the siphon o gravity mains. Thus the ancient culture of the Hydraulic of pressure was really poor, came later into oblivion and it was necessary to wait for centuries for it to get some interest.

Due to the difficulties to convey water through pressurised systems, open channel flow is much better known in the antiquity. Human tends to mimic the way Nature performs water transport. Technologically is less complex, since only slopes must be taken into account. On the other hand, regulation is really simple under these circumstances. And, finally, the materials required by open channel flow are better known. Romans constructed several aqueducts that can be qualified as works of art along the countries they occupied. They were used to transport high flow rates. This was impossible for them by using pressurised pipes. Thus, open channel flow was familiar for Romans. Also, taking into account that river floods were relatively frequent, -the prophet Jeremiah reports about them in the Bible-, it is not surprising that the first examples of river channelling date back to some millenniums before our era.

Regarding hydraulic machinery the Archimedean screw is one of the best known. But, with no doubt, it is the waterwheel or chain pump that predominates during the antiquity. The first reference can be set in the 3rd century BC (Raabe, 1987). It was very popular during the Middle Ages. It was used not only as a pump to raise water, but also as motor taking advantage of the river flow energy. Arabs used the waterwheel widely and their word al-na'ura, meaning wonder, is the etymological origin of the widely used word 'noria'. The waterwheel rises water to a point where it can flow by gravity. Thus, most ancient transport and water distribution systems where based on open channel flow. Pumping water through a pressurised conduit was infeasible even during the Middle Ages.

Another proof that shows that open channel flow Hydraulics was widely known during the antiquity is the fact that river regulation by using dams is a technique already long ago. According to Schnitter, the first reference to this aspect dates back to 2600BC and was located in Memphis (Schintter, 1994). In this reference a history of dams is presented. Most dams were built to allow irrigation, although there are examples of dams devoted to supply water to urban nuclei and, also, for flood control or energetic purposes (by rising water irrigation areas were better dominated).

After Roman splendour, regarding water distribution, supply to urban nuclei is limited to a few cubic meters to satisfy the very vital needs. On the other hand, is worth pointing here that for many centuries population is concentrated in small urban nuclei. Less needs and lower level of comfort than Romans explain the strong limitations of water supply systems in the Middle Ages. A graphic example is posed by one of the most outstanding medieval cities, Paris. In 1553 only 399m3 per day were supplied to Paris, a 260000 strong city (Thirriot, 1987). It means a ridiculous amount of over 1.5 litres per person per day. In 1669, only 1800m^3 were supplied for 500000 inhabitants, it amounts to an individual consumption of less than 4 litres per day. It is even worse in other European cities. In fact, the 80000 inhabitants in Lisbon, in 1740, almost a century later, were supplied with 560m^3 per day, that is 7 litres per day per person (Thirriot, 1987). By then, Madrid exhibits a similar situation, since only 3600m^3 per day were distributed to supply its whole population (Paz Maroto and Paz Casañé, 1969). With no doubt the process was heavily burdened due to the scarce technological knowledge and limited resources available to convey pressurised water.

It was not until 1738, when Daniel Bernouilli published in Strasbourg his work Hydrodinamica (Rouse, 1963), that the basic principles on the Hydraulic of pressure began developing suitably. At the same time, the fundamentals of turbopumps were established, since, even though the original idea must be attributed to Leonardo, only two centuries later, in the 17th century, the French Papin proposed the first impeller with radial blades. In any case, the first industrial realisation of this kind of machine, including a nail-shaped diffuser, was carried out by the Massachusetts Pumps in 1818 (Mataix, 1975).

This would have been worthless without suitable conduits to transport water under pressure. In 1672, Franzini manufactures the first pipes cast in iron (Paz Maroto and Paz Casañé, 1969). This enabled water supply systems in cities, which began to resemble ours. It represented a great breakthrough, since water shortage made that the death rate due mainly

to cholera and typhus surpassed the birth rate in the big cities of the 17th century. Nevertheless, urban nuclei did not stop growing due to the people flow coming from the countryside (Steel, 1.972). On the other hand, open channel transport of water for human consumption was inadequate, unless a suitable treatment is performed, what was not the case by that time. The vulnerability, in case of conflict, of a conduit open to everybody was also evident.
Urban supply, an absolute necessity for cities of the 19th century, became a reality when metallic pipes able to transport pressurised water were available at reasonable prices and when the hydraulic power of turbopumps (product of flow and pressure) became significant. (Kenn, 1.987). The first urban water supply in USA performed through pressurised pipes dates back to 1754 and was laid in Bethlehem, urban nucleus in Pennsylvania (Griegg, 1986). The first filter to produce potable water was installed in London in 1829, while the first use of chlorine is made in 1908 (Steel, 1972).

Anyway, satisfying the Society's need of water involves the mobilisation of discrete amounts of water, sensibly lower than volumes demanded by irrigation, a use completely dominated by open channel Hydraulics. On the other hand, it is important to underline that the sanitary problems linked to urban supply, do not exist practically in irrigation. Thus, open channel Hydraulics and Hydrology, necessary to assess availability of resources and to suitably plan water shortage and floods, were clearly preponderant. The big hydraulic works, devoted essentially to satisfy irrigation and hydroelectric needs, dominated Water Engineering in a world configured by isolated countries. "Irrigation allows nourishment and, as a consequence, it holds power", says at the end of the 19th century Spanish honest politician Joaquín Costa (Costa, 1975). His words were completely appropriate for the time they were pronounced.

Water Management preponderance during this 20th century has been overwhelming. One of the works of higher prestige of this century, the Handbook of Applied Hydraulics edited by Davis and Sorensen, devotes 14 out of 42 chapters, a third of its 3rd edition (Davis and Sorensen, 1965) to the most representative water management works: dams. On the other hand, the number of chapters somehow related to Hydraulics of pressure is just one half (7). Thus, it is not surprising that even Rouse had commented "Hydraulicians are human too" (Rouse, 1987) and that this century had been considered, according to the general feeling, the last golden age of Hydraulics (Rouse, 1987 and Plate, 1987). This general feeling is deeper in countries like Spain, in which two facts have coexisted: an important temporary isolation forcing to run a self-sufficient agriculture and an irregular climate favouring critical scenarios in Hydrology, like floods and droughts. Hence, that Spain is one of the countries with more dams in the world and, depending on the indicator considered, leads the list of countries, is not at all surprising. To be fair, we must say that right policies were applied.

It is curious to observe that the historical revision carried out by the IAHR in 1985 (IAHR, 1987), which in our view is an exhaustive sample of what we have synthesised in this point, contrasts with the Bruntland Commission inform published the same year, which, as said, is a real turning point for Water Engineering. The best way to bring into evidence this contrast is by comparing what Kennedy (Kennedy, 1987) thought of the trend of Hydraulics towards the turn of the century in the closing session of the mentioned historical revision in 1985 (the paper was published in 1987), with the opinion of the Task Force of ASCE regarding this same trend (ASCE 1996). To begin with the adjective environmental is added in the latter. This is only one example of the many changes produced in so a short lapse of time. These changes must be considered a consequence of the work developed by the Brundtland Commission and also of its later development, for example the Agenda 21 of the Rio Meeting in 1992 (UN, 1992). Evident as they are, these facts have gone deeply into our Society.

Let us present that synthesis.

a) Hydraulics, according to Kennedy (1985, published in 1987), on its way to the year 2000 will have to face the:

- Development of economical means of reducing water loss due to seepage and evaporation from canals and reservoirs.
- Invention of strategies for management of reservoir sedimentation to prolong reservoir life.
- Design and production of very large pumps and turbines that are free of the vibration, cavitation and bearing-wear problems that have plagued many large installations; but still achieve the economies of scale resulting from use of fewer, larger machines.
- Implementation of extensive, if not complete, computer controlled automation of complex, multi purpose water projects.
- Development of improved design methods for hydraulic structures, to reduce engineering and model study time and costs, as well as the cost of structures.
- Formulation of improved predictors for the effects on rivers- especially on their channels and water quality- of large scale intervention in their hydrologic and sedimentary regimes; and development of improved means for dealing with these effects.
- Development of improved strategies for operation and maintenance of existing water systems, to prolong their useful lives.

b) Environmental Hydraulic towards de 21st century, according to the ASCE Task Force Committee on Hydraulic Engineering Research Advocacy (ASCE, 1996).

It is, under our view, a really self-critical and honest analysis. Four basic causes are pinpointed that, according to the Commission, make the civil Engineer to lose his leading role within the nowadays Society. The clarity and significance of the assertions do not need further development:

- We have not clearly articulated the relationship between research and education.
- Researchers are not connecting to social needs as determined by policymakers.
- Traditional hydraulics programs have fallen out of step with the needs of the profession.
- Hydraulic Engineers have to think bigger and broader.

Certainly, these two approaches are completely open to debate and discussion, since they are just approaches. But their significance is clear and evident. In any case, they openly lay bare the gap between two different ways of thinking supported by prestigious Civil Engineers only a few years apart.

3. CPD from the demand point of view

From what has been said it is clear that an important shift in trend regarding Water Engineering is well under way. History, tradition, decision centres and even the more developed bodies of teaching back the open channel hydraulics. But awareness of the need for a different way of water management (Beard, 1994) is pervading our Society and it is evident that 21st century policies will necessary meet a balance between supply and demand. Both points of view far from enter into competition should be complementary. Nevertheless, the real fact is that one of them has been clearly pampered, especially through this 20st century. On the other hand, the complementary alternative mode hardly is looming on the horizon. Some countries are showing themselves very diligent to adopt the changes. Others are more prone to the inertia of history.

It is an unquestionable fact that rational use of water is based on the Hydraulics of pressure.
In minority uses (supply and industry) there are no alternatives. But the main use, irrigation, has been clearly linked till recently to open channel Hydraulics and currently is immersed in a process of deep revision. The new technologies of drip irrigation and controlled scarce irrigation force this process.

Traditional water policies have been insensitive to expenditure. Consumption was not measured and the necessary maintenance demanded by pressurised systems, with the complexity derived form the fact that they are buried, was systematically consigned to oblivion. Even recently, the solution to overcome leaks has consisted on providing more resources ignoring completely the loss. This accounts for the low efficiency in most urban water supply systems. The fact that the price of water has been based on political reasons and that distribution Companies, supported by public money, have shown no reluctance to blindly increase the supply has contributed to consolidate such a culture. As pointed out by Lambert and colleagues (Lambert et al., 1.998), *"leakage management has - until the recent world wide trend towards sustainability- been the Cinderella activity of water companies"* .

From what has been said it becomes apparent that the level of knowledge on water management from the demand side (water management) is by far lower than that of classical management (water development). Literature provides clear evidence of this. For one thing, water management seems to be a monopoly of private companies having in their know-how one of their main assets, thus being reluctant to disseminate results. That has been clearly pinpointed at the recent AWWA Conference on water conservation, CONSERV99, by Vickers (Vickers, 1.999): *"One of the results of increasing privatisation is that the free flow and sharing of information between water utilities and the public may become limited compared to the past".* For the other, University (academia) has shown almost no interest in the subject, perhaps because of its novelty or because of lack of tradition or because of its inertia to accommodate rapidly to the Society demand. All in all, it is frequently difficult to find educational publications of quality.

Finally, taking into account that, regarding the labour market, management is the field of water industry demanding most jobs and that Universities do not respond rapidly to the change, the need for education and for availability of information sources of practical utility on the subject becomes evident.

It is important to pinpoint here that in the absence of University Education and of practical publications regarding the subject, professional Associations have played a fundamental role, especially in countries with strong technological tradition and with water management mainly in public hands. The reason is that they have shown some willingness to disseminate results and interchange experience and knowledge. Among them the DVGW in Germany and the AWWA in USA stand out. Perhaps this last is the Association, related to water, having the bigger number of associates in the world. This confirms that job demand comes, and evolves upwards, mainly from water industry. To deepen into these questions resource must frequently be made of publications unrelated with the University world. Even in France, a reference country as far as private water management concerns, the AGHTM, Municipal Engineers Association, which has published excellent applied works, would have enjoyed a more brilliant activity if it would have been backed explicitly by the powerful industries in water management.

Summing up, high employment demand within the sector, unusual specific training and education from the University and scarce or very expensive available information shape a panorama endowed with great future possibilities open to those who take a chance in R+D in this field. And certainly, these possibilities will be empowered if time sustainable policies, instead of the prevailing expansive management habits, are firmly adopted by Governments and Countries.

4. Future trends in CPD within the water industry

According to a recent study by the European Union (European Commission DGXII-Science, 1996), the biggest challenges that the European Society must face regarding research and development within the water field are:

- Pollution fight
- Rational Use of Water
- Water shortage fight
- Prevention and management of crises: floods and droughts

This list, according to the purposes of this paper, can be split into two categories. The first two items are clearly connected with water industry and, as a consequence, demand a higher number of experts. They are more closely related to management from the demand point of view and raise subjects whose need has appeared as a counterweight against an uncontrolled, thus not sustainable, development. Also, they are more interdisciplinary subjects. Associations gathering practitioners concerned with these aspects (AWWA, DVGW, IAWQ, IWSA, etc.) appear to position themselves distant to University and close to water industry and professionals related with this industry. Some of them (AWWA and DVGW) undergo noteworthy activity in CPD and promote the generation and dissemination of technological information with marked applied imprint.

The other two subjects are more in line with classical Hydraulic Engineering. Also, Academia has been closer to them. Hence, publications and information about them are abundant and their quality more widely checked than for the other lines. Besides, they deal with more ancient problems. It is well known that floods and droughts have always existed. Also, humans must have faced water shortage. There is no other explanation for the millennial 'Tribunal de las Aguas' of Valencia.

Also, we cannot forget that the problems posed within this context are more complex and are still absolutely open. Hence the interest that there exists from an academic outlook. The most representative Associations of this wing are ASCE and IAHR. Their associates are many Professors, big manufacturers of hydraulic machinery, very important consultancy companies and relevant hydraulic laboratories. Education promoted from these Associations has a marked University character. Even so, it is noteworthy that the IAHR's president in his 1998 new year greeting message (Kobus, 1998), clearly expresses the need to include professional aspects and engineering practices within the objectives of the Association, *"by bringing its activities near the real demands of the Society".*

Since the technological world moves so fast and, as said, the priorities of the Society follow this rhythm, CPD needs in the first two thematic areas are very important. Several reasons, already poured into this paper, support this assertion. In short they are:

- They demand more professionals.
- They are less developed since they are new. There is important shortage of R+D materials and consequent lack of dissemination.
- There are not too many University curricula developing these subjects.
- Available information is really scarce.

Taking for granted, as Edward Gibbon writes in the Fall of the Roman Empire, that 'any human activity that does not improve becomes decadent', it is clear that the more pressing needs within the water field will come from these areas.

Since the first two lines are closely related to management, water industry predominates on them. In contrast, the University locked in its ivory tower, reluctant to multi-disciplinary activities can not easily incorporate into its structure the wide fan of possibilities offered from within these lines. Finally, the fact that water administration, as explained above, leans toward water management from the supply side (England and its OFWAT is perhaps a noteworthy exception), explains why CPD within this water field is so poorly promoted. Big Water Companies, jealous to safeguard their knowledge, have developed their own CPD structures (Wattier, 1988) and (Booth, 1988). Countries with powerful professional organisations in this field have also solved the problem. In contrast, a number of countries, among them Spain, exist in which there is a lack of offer and, at the same time, present strong CPD demand. The greatest the interest of the administration to implant a more efficient demand management the bigger the possibilities. The pace of implementation will greatly depend on the intensity and frequency of new crises in the form of droughts.

In any case, an important number of subjects have recently gained importance within this context. They are problems concerning researchers and professionals as well. Once solved, their solutions will be disseminated across both the scientific and the technical communities. Among other relevant questions we mention the following: methodologies to develop efficient water audits in water supply systems, rehabilitation and renewal of networks, methods to evaluate and locate leaks, sustainable water economy, design of devices allowing more an efficient use of water, environmental impact of different spills, co-generation of energy from the energy generated by waste-water purification, recycling and reuse of water, use of management indicators to improve the efficiency of supply and irrigation systems, application of benchmarking techniques, integration of GIS systems and improvement of systems of remote control in real time and a wide spectrum of subjects, -the list is far from being complete-, more and more interdisciplinary that can be sheltered under the umbrella of water management.

Certainly, the future of the applied research and subsequent transmission of results to the end users through CPD within the different fields covered by the water industry is increasingly promising as our consolidating modern Society makes its way trough the turning point marked by the Brundtland Commission. Under the authors' point of view, that will be the Golden Age of Water Engineering in the 21st century.

5. Conclusion

The growing pressure over water resources, the need of using them in a way that benefit a bigger number of citizens and the need of deeper respect to environment, that is to say, the mandate of our Society decided to establish time sustainable water policies has introduced a shift in the current an future trends of Water Engineering, breaking a policy having its roots in the very Antiquity. There is no doubt that our predecessors developed the right hydraulic policy for their moment. Also, this policy has contributed to the industrial and economic growth -frequently uncontrolled- that has helped to implant the modern consumption Society. In turn, this has triggered countermeasures to balance the uncontrolled evolution. Under the authors' point of view, it is only the very start of these changes.

Taking into account the lack of tradition in the University regarding this aspect, the fact that technology within the water industry is still to be developed, the lack of training and education in this field despite the great job demand and the future consolidation of this trend with new crises due to lack of water resources and/or loss of water quality, it becomes clear that wide possibilities are open to applied research tending to develop more efficient techniques of management from the demand side. Logically, research would be worthless without result dissemination through CPD activities reaching the professionals in charge of implementing them in practice. Academy, in substantial debt within this subject, sooner or later will have to undergo the process of courageously assuming the guidelines given by the ASCE Task Force in 1996.

REFERENCES

ASCE Task Committee on Hydraulic Engineering Research Advocacy (1.996)
"Environmental hydraulics: new research directions for the 21st century".
Journal of Hydraulic Engineering. ASCE. April 1.996. pp 180 - 183.

Beard, .P. (1994)
"Remarks of Daniel P. Beard, Commissioner of the U.S. Bureau of Reclamation before the International Commission on Large Dam"
Durban. South Africa. November . 1.994.

Bonnnin J., (1.984)
"L'eau dans l'antiqueté. L'hydraulique avant notre ère".
Collection de la Direction des Etudes et Recherches d'Electricite de France. Editiones Eyrolles. Paris. 1.984

Booth T., (1.988)
"Training of personnel in water distribution".
European Regional Conference. IWSA. Proceedings of the Conference pp S3-9, - S3-15. Lisbon. 1.988.

Brundtland, Gro H. et al. (1.987)
"Our common future"
Report of the World Commission on Environment and Development. Oxford University Press, 1.987

Burgi P.H. (1.998)
" Change in Emphasis for Hydraulic Research at Bureau of Reclamation".
Journal of Hydraulic Engineering, July 1.998, pp 658 - 661

Costa J. (1.975)
"Política hidráulica. Misión social de los riegos".
Published by the Colegio de Ingenieros de Caminos, Canales y Puertos. Madrid. 1.975

Davis C.V., Sorensen K.E. (1.965)
"Handbook of applied hydraulics 3rd edition".
Mac Graw Hill, New York, 1.965.

EU, European Commission DGXII -"Science", (1.996)
"Task Force Environment Water".
E.U. 200, rue de la Loi. Brussels. Belgium.

EU, European Commission, (1.997)
"Propuesta de la Directiva del Consejo por la que se establece un mafco comunitario de actuación en al ámbito de la política de aguas"
Framework Directive Draft 97/0067 (SYN).E.U. 200, rue de la Loi. Brussels. Belgium.

Garbrecht G., (1.987)
"Hydrologic and hydraulic concepts in antiquity".
Hydraulics and Hydraulic Research. An Historical Review, pp 1 - 22, published by the IAHR and edited by Günter Garbrecht. Ed. Balkema. Rotterdam. The Netherlands

Griegg N.S. (1.986)
"Urban Water Infraestructure. Planning, Management and Operations".
John Wiley & Sons. New York, 1.986
IAHR, International Association of Hydraulic Research (1.987)
"Hydraulics and Hydraulic Research. An Historical Review"
Edited by G. Garbrecht and published by Balkema. Rotterdam. The Netherlands. 1987

Kenn M.J. (1.987)
"Advances in hydraulics and fluid mechanics in the 19th century"
Hydraulics and Hydraulic Research. An Historical Review, pp 159 - 172, published by the IAHR and edited by Günter Garbrecht. Ed. Balkema. Rotterdam. The Netherlands

Kennedy J.F. (1.987)
"Hydraulic trends towards the year 2.000".
Hydraulics and Hydraulic Research. An Historical Review, pp 357 - 362, published by the IAHR and edited by Günter Garbrecht. Ed. Balkema. Rotterdam. The Netherlands

Kobus H. (1.997)
"Desafíos en hidráulica en el Siglo XXI".
Revista Ingeniería del Agua. Volumen 4, n° 1, pp 27 - 38.

Kobus H. (1.998)
"New Year's Message from the President".
International Association of Hydraulic Research, IAHR, Newsletter. Volume 15/1.998, pp 1-3.

Lambert A., Myers S., Trow S., (1.998)
"Managing water leakage. Economic and technical issues"
Published by Financial Times Energy. London. UK.

Levi E. (1.995)
"The science of Water. The Foundation of Modern Hydraulics".
American Society of Civil Engineers. 345 East 47th Street. New York, New York 10017 - 2398.

Lopardo R. (1.995)
"La formación del Ingeniero Hidráulico para el Siglo XXI".
Revista Ingeniería del Agua. Volumen 2, n° 4, pp 67 - 76.

Mataix, C. (1.975)
"Turbomáquinas hidrúlicas".
Editorial ICAI. Madrid. 1.975.

Paz Maroto J.M., Paz Casañé J.M. (1.969)
"Abastecimientos urbanos de agua".
Published by the Colegio de Ingenieros de Caminos, Canales y Puertos. Madrid. 1.969

Plate, E. (1.987)
"Opening address"
Hydraulics and Hydraulic Research. An Historical Review, page IX, published by the IAHR and edited by Günter Garbrecht Ed. Balkema. Rotterdam. The Netherlands

Raabe J. (1.987)
"Great names and the development of hydraulic machinery"

Hydraulics and Hydraulic Research. An Historical Review, pp 251 - 266, published by the IAHR and edited by Günter Garbrecht Ed. Balkema. Rotterdam. The Netherlands.

Rouse, H. (1.963)
"History of Hydraulics".
Ed. Dover. New York. 1.963
Rouse, H. (1.987)
"Hydraulics' latest golden age"
Hydraulics and Hydraulic Research. An Historical Review, pp 307- 314, published by the IAHR and edited by Günter Garbrecht. Ed. Balkema. Rotterdam. The Netherlands.

Schnitter J. (1.994)
"A History of Dams. The useful pyramids"
Ed. Balkema. Rotterdam. 1.994

Steel E.W., (1.972)
"Abastecimiento y Saneamiento urbano".
Editorial Gustavo Gili, Barcelona 1.972.

Thirriot C. (1.987)
"L'hidraulique au fil de l'eau et des ans à travers les XVIIe et XVIIIe siècles"
Hydraulics and Hydraulic Research. An Historical Review, pp 117 - 143, published by the IAHR and edited by Günter Garbrecht Ed. Balkema. Rotterdam. The Netherlands.

UN, United Nations, (1.992)
"Agenda 21: Chapter 18: Protection of the quality and supply of freshwater resources".
Conference on Environmental and Development, Rio de Janeiro June, 1.992

Vickers A.M., (1.999)
"The future of water conservation"
Proceedings of the AWWA Conference CONSERV99. AWWA. Denver. CO. USA

Watier P., (1.988)
"Typologie des systemes de formation dans la distribution d'eau"
European Regional Conference. IWSA. Proceedings of the Conference pp S3-1, - S3-8. Lisbone. 1.988.

Particularities of post-graduate education of oceanography in aspects of unique Caspian Sea

Kulizade L.[1]**, Mamedov R.**[2]
[1] Baku City Committee on Ecology
[2] Institute of Geography of Academy of Sciences of Azerbaijan Republic

Abstract
On its geographical location the Caspian sea is very important climatic and ecological indicator of the World. The first characteristics, differing the Caspian sea from other large reservoirs, are the changing of its level in a wide range periodically. It is discovered the impact climatic anomalies on changing Caspian sea level. The other individual peculiarity of the Caspian sea is biological diversity. The main factors making influence biological diversity are waste waters from municipals, industrial waste waters, flow from agriculture areas (waste water with pesticide), drilling oil offshore sea, level changing.

All these particularity were taking into account for completing 2-years and 3-years programs of postgraduate education and research training. The complex model of the level changes is complied defines the research methods for working out the prognosis. In these programs to geological and anthropogenic processes, water balance and solar activity the prognosis of the level changes was worked out. The additional mathematical methods are developed and also may be used in process of postgraduate education of research specialists.

1. Introduction

Caspian sea is deep water reservoir having a large offshore area. It extends on 1200 km in sub-meridional bearing with a medium width 350 km. The sea sets on the border between two parts of indivisible Eurasian continent. Five countries surround Caspian sea - Azerbaijan, Russia, Kazakhstan, Turkmenistan and Iran.

Uniqueness of the Caspian sea is caused by next main aspects:

SETTING This natural objects is placed in large and deep depression within largest internal water drain area in Europe and Asia and has no connection to World Ocean. Its surface level is at ~27 m lower that the Ocean level. According to all geographical definitions Caspian is a lake but because of its extents and reason of dynamic hydrological, hydro-chemical and biological processes it is called 'SEA'.

MORPHOLOGICAL STRUCTURE There is very specific Partition of areas taken by different bathymetric lines in Caspian sea.

BIODIVERSITY The majority of the world's sturgeon fish are concentrated in the Caspian basin. It is here more than 90% of the total world sturgeons are caught. There are also significant range of another endemic species of fauna and flora.

INDUSTRIAL IMPORTANCE Besides biological value the Caspian sea has great importance for economy. There are substantial reserves of hydrocarbons in the basin and new large oil and gas fields are discovered and began to develop during recent years.

At the same time Caspian is one of most polluted water reservoirs in the world. The main factors making negative influence biological resources are:
- waste waters from municipals;
- industrial waste waters;
- flow from agriculture areas (waste water with pesticide);
- drilling oil offshore sea;
- level changing.

SEA LEVEL FLUCTUATIONS. The most representative characteristics differing the Caspian sea from other large water bodies is the changing of the level in a wide range periodically. There have significant fluctuations in level in the short and long term.

All efforts to solving environmental problems in the term of continuing change of the sea level (rapid rise since 1978) assumes the role of modelling of optimal ways reaction to estimate aftereffects to climate warning.

2. Objectives and methodology

As it was mentioned above, existing educational programmes of oceanography for postgraduate researchers do not allow use them adequately in a case of the Caspian sea.

First lectures explaining causes making processes in the Caspian sea different from other large water reservoirs, and dynamical processes having place in the basin, were organizing and conducted in Azerbaijan State University in Department of geography.

Special attention in educational sea level programme was paid to sea level fluctuations. The last period of rise that has been begun since 1978 to 1994 (and slightly continuing till present time) has brought great damage to economy, society and ecology of Azerbaijan, estimated approximately in US $ 1 billion.

Study concerning this problem begins from retrospective analysis of sea level fluctuations and introduce different view points for these processes. The diagram of the Caspian sea level fluctuation depending of climate changes by Dr.Ramiz Mamedov was made.

Taking into account that the Caspian sea is uniform ecosystem, where interact all factors shown in the diagram, the programme of long-term forecast for three year study was developed.

FORECASTING THE CHANGE OF THE CASPIAN SEA LEVEL AND ITS SOCIAL AND ECONOMIC, ECOLOGICAL AFTEREFFECTS (Research programme)

1. Retrospective analysis and research methods
   - 1.1 Paleographic, historical, archeological data about the changeability of the Caspian sea level and their restoration.
   - 1.2 Critical analysis of the works devoted to the prognosis of the level change.
   - 1.3 Determination the factors exerting influence on the level change.

Research methods and materials would be used.

2. Time and space changes of the level
   - 2.1 Short range change of the level.
   - 2.2 Seasonal, annual and long-term changes of the level.
   - 2.3 Space changes of the level.
   - 2.4 Statistic and spectral characteristics of the level change.

3. Investigation of the components formed the water and possible prognosis of their changes.
   3.1 Surface currents to the Caspian Sea.
   3.2 Evaporation from the surface of the Caspian sea.
   3.3 Precipitations falling on the surface of the Caspian sea.
   3.4 Flow to the Kara-Bogaz-Gol bay.
   3.5 Water balance of the Caspian sea.
4. Influence of geological factors on present changes of the level
   4.1 Horizontal tectonic movements of the Caspian sea region.
   4.2 Vertical tectonic movements of the Caspian sea region.
   4.3 Present seismotectonic conditions.
5. Influence of the antropogenic factors on the level change of the Caspian sea
   5.1 Influence of pollution with oil products on evaporation from the surface of the Caspian
   5.2 Change of the amount of the river waters at the result of anthropogenic influence.
   5.3 Possible changes being taken place in the climate of the Earth planet because of the influence of anthropogenic factors in the level changeability of the Caspian sea.
6. Forecasting the level changeability of the Caspian sea by different methods
   6.1 Forecast according to water balance.
   6.2 Forecast according to atmospheric processes.
   6.3 Forecast according to solar activity.
   6.4 Comparison of the result obtained by different methods and final forecast.
7. Forecasting of social, economic, ecological changes being taken place at the coastal zone in different levels and preparation of re-commendations
   7.1 Forecasting of the social and economic changes.
   7.2 Forecasting of the ecological changes.
   7.3 Results and recommendations, utilization strategy of the coastal zone.

The second main aspects of Caspian sea is its ecological stage. Extensive development of industry, active exploitation of mineral biology resources in basin of Caspian sea in period 1970-90 came to strong degradation of ecological condition.

Due to reports of state water protection authorities, only in 1992 in the Caspian sea were discharged 6799 million m^3 sewage, including fron Russia – 3422 million m^3, Azerbaijan – 1708, Kazakhstan – 1655 and Turkmenistan – nearly 13 million m^3.
Over 60 million m^3 of sediment composed of 2 to 40 per cent oil compounds have accumulated in Baku bay. The levels of phenols and mercury are also very high (0,2 – 1,0 and 5,0-140 g/kg of sediment respectively). From a biological point of view, those bottom area are considered virtually dead, affecting sturgeon and other fish that feed mostly on the benthic fauna in shallow areas.

Moreover, nearly 75 million ton chemical compounds were carried out with rivers flow: Volga (60 million ton), Kura (5million ton), Terek (3 million ton), Ural (2,5 million ton), Samur (1,5 million ton).

Among discharge components for sea predominate oil compounds –95%, phenols – 2-3 %.

As a result of rapid sea level rising over 800 km^2 of coastal areas were flooded. Many industrial plants and factories were submersed as well as old oil fields in coastal zones. New substantial source of pollution appeared. The use of pesticides and fertilizers in agricultural regions also leads to pollution of the sea water.

Later relative lecture course for students and post-graduated researchers were organized and conducted in accordance with the structure of pollution of the Caspian sea.

On result of the researches was completed the condition of ecological system Caspian sea (fig.1).

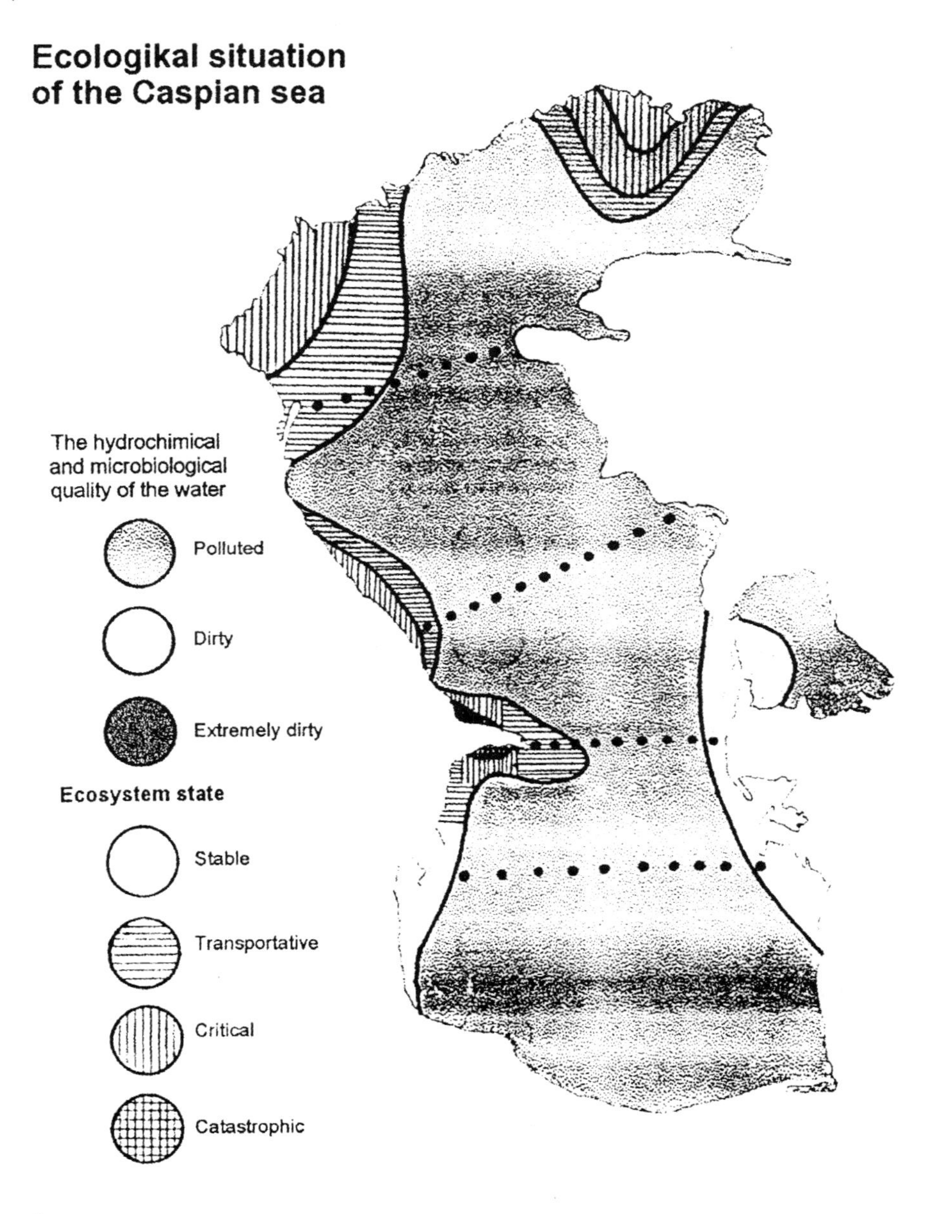

Figure 1.

There is needed 1 year for study above mentioned programme.

Group of scientist under leader professor Ramiz Mamedov developed computer programme for calculation and forecast of carrying out containment and cleaning-up of oil spills. This special model allows forecast spread discharge in different area of Caspian sea for different hydro-meteorology conditions taking into account composition and capacity of source. On fig. 2 is presented block-scheme of complex model for calculation this programme.

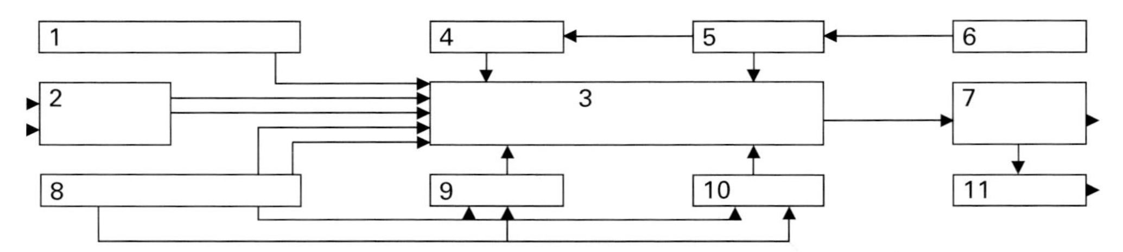

Figure 2.

The legend:
Information input (coordinates and composition of spillage).
Velocity and density of water.
Velocity of contaminating particles on water system.
Contaminating particles own velocity.
Definition of casual increment.
Generation of casual numbers.
Mass of particles' coordinates.
Input stream velocity and water density for definition of turbulence exchange coefficient.
Definition of horizontal coefficient of turbulent exchange.
Definition of vertical coefficient of turbulent exchange.
Definition of concentration in define point.

There is half year needed for studying of this programme. The programme widely used by oil companies developing new oil structure offshore the Caspian sea.

3. Conclusion

Scientific work mode allowed increase and supplement the system of postgraduate education in sphere of Oceanography.
- Influence change climate on fluctuation Caspian sea level
- The ecology of Caspian basin
- Fundamental Oceanography
- Hydrology
- Forecasting the change of the Caspian sea level
- Mathematical modeling.

References

R.Mamedov, L.Kulizade, Y.Gadiyev "Impact climate anomalies on the change Caspian sea level" 2nd International Conference on Climate and Water, august 1998.
"Long-term prognosis of the Caspian sea level "Regional Workshop of coastal zone management, Chalahar, Iran, Feb.1999, p.79-83.
R.Mamedov "Investigation Fields and their Influence on Distribution of pollution in the Caspian sea". Abstract of dissertation for degree of Doctor of Technical Sciencces.Tbilisi.1996.p 44.

The Realization of a permanent structure for continuing vocational training in water field: the Sardinia case

Stefania Nascimben, Giuseppe Cane', M. Cristina Melis
Hydrocontrol, Research and Training Centre
for Water Systems Control,
Strada 52, Poggio dei Pini, 09012 Capoterra, Cagliari, Italy

Abstract

The Adapt project "Permanent structure for continuing vocational training in water field " aims to develop a system of assistance and consultation for continuing vocational training . It is for those who work in the water field in Sardinia. This initiative is designed to meet new and ever-changing company needs in the world of water. Project Adapt with its emphasis on research, the training for trainers, awareness and the spread of results, opens up new opportunities within Hydrocontrol's training structure. The various project activities in progress also express themselves through the setting up of a Learning and Support Centre, based on Hydrocontrol's *know-how,* in itself a further growth factor. Consequently, Hydrocontrol can, therefore, become a reference point and focus for training in water system management.

1. Introduction

The creation of a permanent structure for continuing vocational training in water field requires a strategic programme in an area such as Sardinia (Italy) considering its historical lack of natural resources. As with all regions in the country, these requirements can only be met through a significant change in investment within the entire water industry.

The innovative nature of this project is closely related to the water management situation in Sardinia. Several institutions work within and are responsible for the water industry in Sardinia : three large Local Authorities, twenty-two Land Improvement Agencies, one Water Works Authority, fourteen different Authorities and Industrial Corporations, ENEL (Electricity Board), local town councils which distribute resources and SME (Small Medium Enterprise) which deals with the management of processing disposal and maintenance services.

The Sardinian Council has primary competence in the water field, initiating the restructuring of the water sector in accordance with the national law 36/94, based on the following criteria;

1. All water is to be declared public.
2. The use of water for human consumption takes priority over all other uses.
3. The reorganisation of all water services must be in line with efficiency, cost effectiveness and economy.

According to the law draft presented to the Regional Council Committee, the main management bodies will undergo a natural transformation.

During 1997 the European Community (Adapt Initiative) funded a project, submitted by Hydrocontrol, to develop an assistance and consultative structure for continuing vocational training, for boards operating in the water sector. By the end of 1999, the following will have been established :

1. The study and close examination of continuing vocational training in Sardinia and the analysis of the state of water industry reorganisation.
2. *Acqua on Line*, a water documentation centre, to provide information and data for people operating in the water industry.
3. Training strategies for trainers to improve the competence of Hydrocontrol staff already operating in the sector.

The project will provide a reference point for how to achieve the necessary *know-how* in restructuring; this concerns not only the organisational aspect but a "new culture" in the management of water resources as well. Vocational training, seen as professional support to the work force together with the spread of information, will allow the growth of this "culture".

In the early stages, it became clear that there was a need to undertake a careful analysis of existing systemised and normative models within the water industry and in the realms of professional training. As a consequence of this work, there is a wealth of useful literature now available to whoever works in these two sectors, highlighting the latest trends in each.

Secondly, we have occupied ourselves with improving present skills which have already been acquired through the ad hoc introduction of top class trainers, providing in-house training to Hydrocontrol personnel. This planned and successful scheme has been of fundamental importance in that it has helped develop other related items within the entire project.

As with all projects financed by the U.E (European Union), it is of prime importance that the results reach as many of the interested parties as possible. In order to realise such ends, it is not only predicted that project results will be published but also that a centre for water documentation will be set up ; Acqua on Line, via Internet (http//: www.hydrocontrol.com/cda). This centre will provide information and data of great interest to whoever works within the water industry. Through the project findings, there will be particular emphasis on collaboration with our European partners in the creation of a web page holding data on national and international projects.

2. Study and research

The report is divided into 4 sections:

1. The analysis of certain European experiences.
2. The organisation of the water industry and professional training in Italy.
3. Aspects and problems in Sardinia at present.
4. Training initiatives.

The main findings of the study have demonstrated how, as with other European cases, there's a tendency to go beyond catalogue training in offering services more designed to client needs. Here, advice services also have an important role to play, apart from pure training itself. Recent trends in training methodology tend to take the form of traditional lessons in the classroom coupled with other learning techniques such as on-job Training, distance learning and self-training.

On the national level, and from the normative point of view, 1999 may well prove to be a crucial and critical year considering recent innovations established by the Italian Government, namely The Social Pact (signed in December 1998 by the Government, employment organisations and interested parties), present regulations arising from law no. 196/98 ; and above all due to allocated funds for 1999. As a consequence, it is expected that vocational training will play an important part in the national education system. While still awaiting the full impact of these reforms, a slow but constant increase in demand and supply for training can, even so, be noted.

The report has identified main local and national training organisations, both scholastic and beyond. If we simply consider water resources, and examine the relevant data provided by the Department of Employment on vocational training courses, carried out in 1997 throughout the country, the following emerges;

- the most common themes concern water treatment and waste water treatment plants. The motivating principle lies in the will to create qualified personnel able to solve the various maintenance and upkeep problems connected to the water plant, and in the will to finally establish a regulation, (still in a transitional phase), having been issued in 1991 by the U.E (directive no. 271 21/5/91).
- The primary importance of organising professional courses in the South, compared to those in the North (which have far greater funds at their disposal). To this, one must add the attempt to create job opportunities, and a greater awareness of water usage in regions where it is often lacking.

The picture is completed with an analysis of new trends of training through experimentation on methods best-suited to company needs.

By analysing the water industry's structure, and seen in context, Italy appears to be going through a transitional phase. The sector, characterised by a great fragmentation of management organisations, is moving towards a management which will oversee the complete water cycle. This management will be one body operating within Optimal Management Areas, according to economy, efficiency and cost-effective principles.

3. Training for trainers

For years Hydrocontrol has been operating in the training and research field in the water industry. Training for trainers strategies have been studied to improve Hydrocontrol trainer competence so that they may possess the relevant knowledge to perform both studies and training. The Centre is the main force behind this "new culture" because it combines both training and research structures in this field.

The training programme which has been planned is developing along the following lines and with the following aims;

1. The acquisition of methods and techniques which are typical of training programmes cycle.
2. The qualitative and quantitative development of the trainer's role, integrated and synchronised with that of the researcher.
3. The creation of certain structural means as training supports and the setting up of a *Learning and Support Centre* (LSC) in particular.
4. Experimentation within the conceptual field and within training methods.

To encourage such objectives an ongoing work programme in the classroom (of 20 days) has been devised in order to assist and verify work in this field. The very same availability of training resources in terms of materials, means, contacts and human resources have been sought within Hydrocontrol. Didactic training methodology, inspired by the 'empowerment model' predicts four principle courses of action (orientation, skills, elaboration, actions). It has identified the following areas of intervention, subdivided into the subsequent approaches:

Visioning area

Objectives in this area are to verify and plan activities within training schemes. Moreover, they identify the best training methods, making possible visits to centres of excellence to compare and contrast training theory and practice.

Empowerment area

The aim is the qualitative and quantitative development of identifying Hydrocontrol workers' duties and roles at the conclusion of their training programme.

Methods area

The aim here is to have an overall vision of both training cycles (with a more profound understanding of methodology) and the roles of those involved.

Course of action area

In this area a series of activities are predicted on-site:

- The setting up of a Learning and Support Centre.
- The setting up of a pilot scheme for existing clients.
- The development of relationships with existing clients.

Consolidation area

The objective of this area is to consolidate the positive elements achieved during the training process, in terms of skills, new experience, new methods and means of support, in order to successfully carry out the training programme.

Two aspects emerge from a study progrfamme on the training of trainers:

- The first represents the qualitative and quantitative development of Hydrocontrol trainers in the field of more traditional phases of training (analysis of training needs, planning, setting up, checking and relations with existing clients).
- The second is linked to the development of skills inheriting new training methods and using the resources available within the LSC, resources which also offer other services (research, assistance and advice).

When one talks of a particular organisation's training needs, those responsible within such an organisation desire a training course with certain prerequisites to satisfy their new requirements and objectives. Above all, they hope for contact with informed experts with *know-how* who, having worked in the same field as them, have the same experience and/or the same problems. Moreover, in order to learn, new didactic means and methods are called for (self-learning, training on the job, distance learning). Project Adapt aims to develop such training methods, using efficient and cost effective methods. In the case of Hydrocontrol, this network of *know-how* may represent an important resource for training while, at the same time, be an efficient and cost effective way of exploiting systems and structures, already in existence, used for other purposes.

The new figure of the trainer within Hydrocontrol must demonstrate an ability to continually observe his/her surroundings, identifying both within the company and without it, that which can be taken advantage of and used again for training purposes (in the field of *know-how*, and technology).

In this scenario, the LSC, being Hydrocontrol's adopted centre of knowledge, becomes the place (both physical and virtual) where various resources are concentrated and made available to the learner through various learning avenues. He or she is able, therefore, to make the most of the training course, eventually turning to the support of a tutor as well. Once Hydrocontrol's know-how has been systemised, and has taken its place in the whole structure, its voice may be heard on various topics, each one headed by a recognised expert, trainer or tutor capable of directing the user towards the desired resources available. These resources may be physically present (e.g. books, handouts and leaflets, audiovideo materials, software teaching aids and instruments) or virtual (e.g. reference to outside experts, links to similar structures, documents held in other places, information referring to further opportunities). What takes on greater significance as project Adapt develops, and which concerns Hydrocontrol's growth, is the concept of the LSC and the chances of setting it up for potential clients and users, offering them alternative forms of training, consultation, assistance and access to Hydrocontrol's considerable know-how.
By dividing potential clients into two categories, *internal and external*, the services offered by the LSC become diversified.
Internally, referring to those working within Hydrocontrol in different capacities (e.g employees, temporary workers, product/service trainers, consultants) the services offered may be;

1. The spread of information and diffusion of findings based on Hydrocontrol's competence.
2. The use of resources within company procedure.
3. Keeping employees up-to-date with latest trends etc.
4. The successful introduction of new staff into the system.

Additionally, the LSC may be able to guarantee an external client/user the following;

1. The spread of information and diffusion of findings based on Hydrocontrol's activities in the world of water.
2. The use of known resources, skills and competence possessed by Hydrocontrol through pre-set avenues of learning (self-learning as well).
3. Advice and assistance.

Despite all that's been said, the effort needed to establish this know-how within Hydrocontrol in order to immediately identify new resources and possibilities at any given moment is evident. Company adaptability and change will be imperative in meeting clients' present and ever-evolving needs. From this standpoint, project Adapt offers the chance to open and augment new possibilities, to empower the training set up at Hydrocontrol:

1. The ability to interpret with greater skill the needs of clients, reaching ad hoc solutions at reasonable costs, and guaranteeing clearer and more incisive findings over time.
2. The ability to identify new internal resources within the company for learning/training (The LSC).
3. The ability to maximise the use of identified resources for training, also for the benefit of other internal company procedures.

It's also evident, from what's been said, that only by broadening its horizons will Hydrocontrol be able to find the most efficient solutions in its relations with clients.

4. Awareness and its spread

One of Project Adapt's aims is the spread and diffusion of its findings.

To create awareness about continuing vocational training in the water field, and in order to attach greater importance to it, this continuing vocational training will be viewed as an instrument for managing human resources and for the updating of companies.
The project also aims to promote a campaign of information awareness for all those operating within the water field considering legislation changes.

This awareness campaign will consist of ;

1. Two workshops to guide the work force towards new assets for managing water resources, to be held at the project's conclusion, one locally and one nationally, to which the key actors will be invited.
2. Realisation of a Water Documentation Centre, *Acqua on Line*, to provide information and data on research subjects, technologies, legislation and training for those operating in the water industry.

To be brief, the Water Documentation Centre (WDC) that Hydrocontrol is working on and which will be set up by the end of 1999, hopes to be a point of reference not only for Sardinia but also for dealing with general problems connected to the management and correct use of water resources. From this point of view, the WDC should prove to be an additional source of knowledge for whoever may be interested in consulting it : Public administrators, managers, researchers, trainers, companies and industries. It has, therefore, never been more important to make the centre's resources available through the opportunities offered by Internet (www.hydrocontrol.com/cda) in addition to the traditional methods of information available on site. Other attempts at information awareness have been made through the publication of projects by way of the media.

5. International partnerships

Co-operation with our international partners, the Office International de l'Eau (FRANCE) and Bidungszentrum fur die Entsorgungs and Wasserwirtschaft GmbH (GERMANY), (who have already developed advanced training methods and technologies), has represented a useful learning experience.

Another success has been the setting up of a WEB page, the fruits of this co-operation.

Contacts developed with our partners constitute an important premise for future collaboration.

Formation et Recherche en vue d'une participation plus active des agriculteurs dans la gestion des périmètres irrigués

Training and research for a more active farmers participation in the management of irrigation schemes in Tunisia

Hedi Daghari
Institut National Agronomique de Tunisie

Abstract
Tunisia extends over a surface area of 16.4 million hectares. Agricultural areas are approximately 5 million hectares, Whereas the potential irrigable land Tunisia, considering also the water resources already exploited, is tody of 0.4 million hectares The irrigated area is about 0.35 million hectares in 1998 and their annual amount to the total agriculture production exceeds 30%. In Tunisia, more than 80% of water resources are used in the irrigated areas. Building capacity must be a priority. At least three distinct training levels; Training for farmers, Training for engineers in practice and extensionists and Academic and research training; have to be envisaged for the success of any irrigation policy. Each training form has its specificities. They are complementary to each other. To each training type a well defined working methodology does correspond. The economic interest has to be well kept in mind although to the detriment of the technical interest since profit is the primary factor in the farmer's choice. A best knowledge of the native expertise is necessary in order to enable a rational choice. A praiseworthy effort must be made for extending AIC Association d'Intérêt collectif (public concern associations) to all the irrigated schemes in order to make users responsible and aware of the every day management of irrigation.

A creation of an observatory is necessary and its funding can be paid from extra taxes on water.

Résumé
En Tunisie, la surface agricole utile est de l'ordre de 5 millions d'ha alors que les superficies irriguées ne sont que de 0.35 millions d'ha, mais leur contribution dans la production nationale agricole dépasse les 30%. D'autre part plus que 80% des ressources en eau sont consommées par l'irrigation et on assiste à un accroissement continu de la demande alors que l'évolution des ressources en eau s'est fortement atteigne.
Une gestion rationnelle du secteur irrigué ne peut se faire sans le consentement des agriculteurs d'une part et un effort important de recherche et de formation d'autre part. Cette dernière doit toucher nécessairement les agriculteurs et les ingénieurs en exercice.

La généralisation des associations d'intérêt collectif (AIC) contribue à la responsabilisation des agriculteurs mais aussi elle facilite toute forme de contact avec ces derniers y compris

la formation et la vulgarisation.
La création d'un observatoire de l'irrigation peut contribuer efficacement à la coordination des différents efforts déployés dans le secteur. Son financement peut être prévue sous forme d'une redevance sur la vente de l'eau.

1. La gestion du secteur irrigué en Tunisie

La Tunisie s'étend sur une superficie de 16.4 millions d'hectare. Les surfaces agricoles utiles sont de l'ordre de 5 millions d'hectares alors que le potentiel en sol irrigable pour la Tunisie et compte tenu des disponibilités en eau mobilisée, est aujourd'hui de 0.4 millions d'hectare. Les taux de mobilisation des eaux dépasse les 75% et pour plusieurs nappes souterraines, on assiste déjà à une surexploitation. L'évolution des ressources en eau commence à s'atténuer et on tend vers l'asymptote 4500 millions de m3 malgré que le volume d'eau annuel provenant uniquement de la pluviométrie dépasse les 33 milliards de mètre cube. Par contre, pour la demande, on assiste à un accroissement continu et on estime qu'elle passera à 3165 millions de mètre cube en l'an 2010 contre 1920 millions de mètre cube en 1990 et 2592 millions de mètre cube en 1994 (Ministère de l'Agriculture, 1998). La recherche d'un équilibre entre l'offre et la demande ne peut être imaginée sans la prise en compte des plusieurs mesures:

- investigation des nouvelles ressources et mobilisation des ressources difficilement accessibles
- recours aux eaux non conventionnelles (saumâtres, usées et de drainage)
- utilisation des techniques de récolte des eaux
- vulgarisation des techniques économes en eau et meilleure valorisation de la ressource eau dans les périmètres irrigués.
- renforcement des capacités (formation, recherche)

Les superficies équipées pour l'irrigation sont passées de 0.05 million d'ha en 1960 à 0.34 million d'hectare en 1998 et leur contribution dans la valeur de la production totale agricole dépasse à l'heure actuelle les 30%. Deux grands types de périmètres irrigués, occupant chacun approximativement la moitié des superficies irrigués sont rencontrés en Tunisie. Le premier concerne les périmètres publics irrigués (PPI) où tous les grands travaux hydrauliques ont été réalisés par l'état mais les terres appartiennent à des privés qui assurent l'exploitation. Le deuxième groupe concerne principalement les parcelles réalisées par les paysans eux mêmes avec en général une aide de l'état, sous forme des subventions ou de crédits. Leurs taux d'intensification moyen avoisine les 110%; les situant en première position, bien avant les périmètres publics irrigués.

En Tunisie, l'irrigation se taille la part de lion dans la consommation d'eau (82% pour l'agriculture et uniquement 18% pour l'industrie, le tourisme et la distribution d'eau potable). Dans plus de 80% des périmètres irrigués, on pratique l'irrigation de surface. Ces dernières années, tout le réseau de transport a fait l'objet d'une vaste opération de réhabilitation et de modernisation afin de réduire les pertes qui sont encore énormes au niveau de la parcelle. Les encouragements de l'état sont multiples pour inciter les paysans à recourir aux différentes formes d'économie d'eau et à améliorer l'efficience qui reste encore très faible. Les subventions accordées par l'état peuvent atteindre 60% du coût des équipements d'économie d'eau utilisés .

Au niveau central, c'est la Direction Générale de Génie Rural (DGGR) qui a la charge de coordonner les principales activités d'irrigation. A l'échelle régionale, la gestion était assurée par des Offices de Mise en Valeur couvrant tout le pays. Ceux-ci jouissent d'une certaine souplesse. Toutefois, ils menaient leurs activités en parallèle avec les

Commissariats Régionaux au développement Agricole (CRDA). Pour éviter cette dualité, il a été procédé à la fusion de ces deux structures au sein des CRDA qui ont été renforcés par la création des arrondissements des périmètres irrigués, en plus des arrondissements de Génie Rural qui existaient déjà. En fait, on visait une décentralisation mais surtout un transfert progressif de certaines responsabilités vers les agriculteurs

Problématique de la formation et de la recherche dans le domaine de l'irrigation

Pour réussir une politique de gestion rationnelle de l'eau dans le secteur irrigué, il faut au moins que ces trois conditions se réunissent, à savoir la responsabilisation et la participation de la population d'irriguants, la vulgarisation des acquis de la recherche et la formation et/ou le renforcement des capacités. En parallèle, la tarification de l'eau est une mesure incitatrice à l'économie d'eau mais difficilement applicable dans certaines situations tels que puits de surface et oueds par exemple.

Certains modes de tarification sont à exclure tels que celui utilisant la tarification forfaitaire, car même s'il permet une certaine sécurité dans les recettes; il est à l'origine des pertes énormes d'eau. Par contre, la tarification par binôme et par tranches peut au contraire inciter à une économie en eau. La consommation de base peut être fixée à une limite garantissant un revenu convenable pour l'agriculteur. Les autres tranches peuvent être fournies à l'agriculteur à des prix même supérieurs au coût réel de l'eau sans pour autant nuire à l'efficience économique.

2 - Problématique de la formation et de la recherche dans le domaine de l'irrigation

Le renforcement des capacités nécessite de prendre en compte trois acteurs distincts, à savoir l'agriculteur, le technicien ou le vulgarisateur en exercice ainsi que la formation académique. Chacune de ses formations a ses spécificités. L'une ne peut pas se substituer à l'autre mais elles sont complémentaires. A chaque type de formation correspond une approche de travail bien déterminée:

2.1 Premier type: La formation destinée aux agriculteurs

Il s'agit d'une population importante, très hétérogène avec des pôles d'intérêt très variés. Seuls les acquis (produits finis) sont à mettre à la disposition de cette population car un échec à ce niveau a des retombés très négatives. L'intérêt économique doit être bien pris en considération même si c'est au dépens de l'intérêt technique car c'est le gain qui prime dans le choix de l'agriculteur. La logique de l'irriguant repose souvent sur un savoir-faire et des contraintes spécifiques.

On prendra comme exemple l'adoption des techniques de jessours (petits ouvrages de stockage d'eau) dans la région de Matmata Nouvelle et de l'épandage des eaux de crues à Sidi Bouzid, pourtant deux régions semi-arides se trouvant à moins de 150 km l'une de l'autre. On se demandait toujours pourquoi les jessours ont été adoptés dans la région de Matmata Nouvelle alors que l'épandage des eaux crues a été pratiqué dans la région de Sidi Bouzid. Effectivement, à l'examen des histogrammes d'apports d'eau correspondants (fig. 1 et 2), il se dégage clairement qu'avec des petites quantités d'eau , il est difficile de réussir l'épandage dans la région de Matmata et vice versa. En effet, les débits annuels moyens enregistrés aux stations de Matmata Nouvelle et Sidi Bouzid sont respectivement de 0.057 m3/s et de 1.30 m3/s.

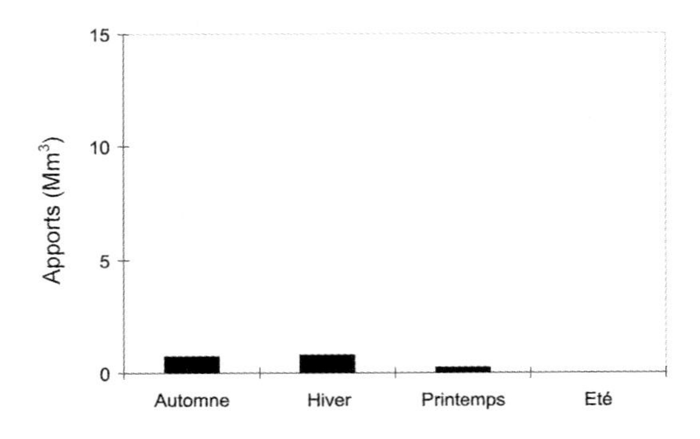

Figure 1. Apports enregistrés à la station du barrage Matmata Nouvelle (1980/1992)

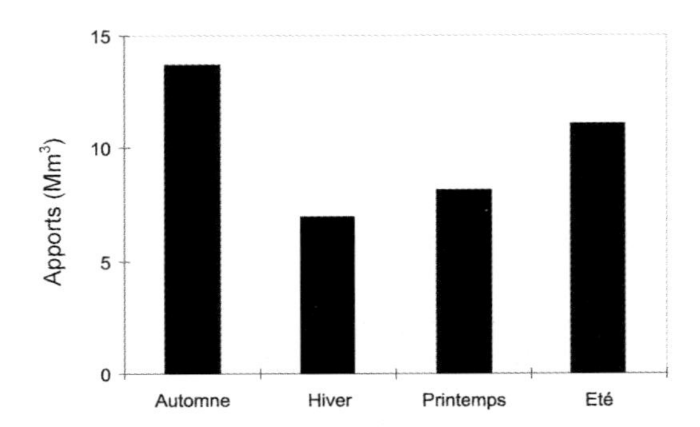

Figure 2. Apports enregistrés à la station Khanguet Zazia (1956/94), région de Sidi Bouzid

Il est aussi nécessaire de prévoir des mesures d'accompagnement dans un but d'améliorer les conditions de vie des paysans (électrification, scolarisation des enfants, pistes, sociétés de service).

2.2 Deuxième type: la formation destinée aux ingénieurs en exercice et aux vulgarisateurs

Cette deuxième population constituée par des salariés est assez homogène et a l'avantage de concerner un nombre limité d'intervenants. Aussi, devra-t-elle participer effectivement à l'adaptation des résultats de la recherche à la réalité terrain, même si la disponibilité en temps risque de constituer une contrainte.

Aujourd'hui avec les disponibilités de moyens de communications et d'outils pédagogiques, et pour inciter les techniciens en exercice à une mise à jour régulière, cette formation doit être menée sous forme d'unités ou de modules et une attention particulière doit être donnée aux aspects pratiques. Cette formation doit être aussi à la base de toute promotion de ces agents.

2.3 Troisième type: la formation académique et de recherche destinée aux élèves ingénieurs

Il s'agit d'une population très réduite, homogène mais avec une expérience du terrain "très limitée". La formation académique se heurte à plusieurs contraintes qui sont en voie d'être surmontées par l'adaptation des programmes, la souplesse introduite par l'enseignement modulaire et l'ouverture de plus en plus grande sur le milieu extérieur (stages, visites du terrain, etc.....)

Dans le domaine de la recherche, plusieurs insuffisances scientifiques sont encore à signaler et si on prend uniquement comme exemple les différentes composantes du bilan hydrique aucun de ces termes ne peut être évalués avec une précision suffisante. Les comparaisons menées entre les mesures neutroniques de la variation du stock d'eau et celles fournies par un lysimètre pesable de précision montre de divergences importantes (Fig. 3).

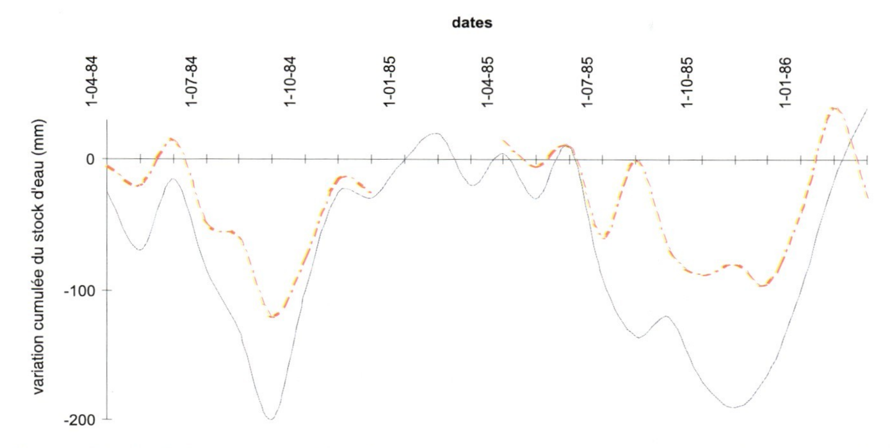

Figure 3. Variations cumulées du stock d'eau obtenues en utilisant le lysimètre pesable (____) et la sonde gamma- neutronique (- - - -), (Daghari et De Backer., 1988)

Aussi le flux profond de drainage ou de remontée capillaire est souvent négligé. Des très rares travaux ont été menés sur le sujet mettant en évidence son importance dans le calcul du bilan hydrique (Daudet t VALANGOGNE, 1976.). L'évapotranspiration réelle d'un sol en jachère a été de 140 mm et de 245 mm suivant que ce flux profond a été pris égal à la conductivité hydraulique ou totalement négligé (Fig. 4).

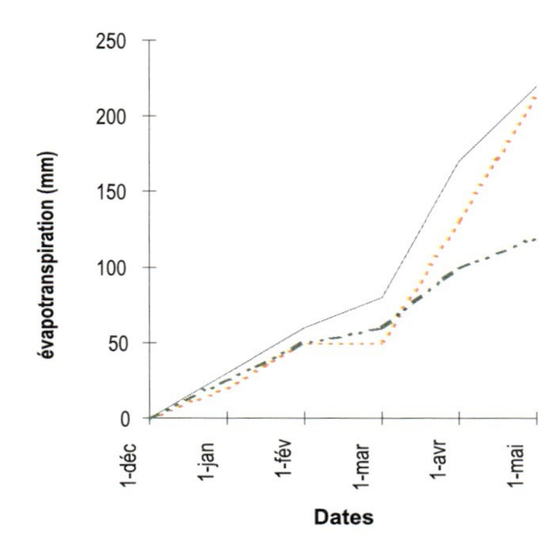

Figure 4. Evapotranspiration réelle du sol en jachère. Le flux profond étant pris: nul (___),égal à la conductivité hydraulique (_. ._-. _) et égal au flux du Darcy (- - -), (Daghari et al., 1989)

La formation doit nécessairement contribuer à une gestion durable de la ressource eau dans les périmètres irrigués moyennant:

- la recherche d'une efficience optimale (technique, agronomique et économique) Dans tous les cas, il est impératif que l'efficience technique de l'irrigation soit respectée. Les mesures (formation, vulgarisation, encouragement etc...) permettant d'atteindre un tel objectif sont à prendre. Par contre, pour les deux autres efficiences (agronomique et économique) il y a lieu de bien analyser le contexte socio-économique. A titre d'exemple, Pour qu'un pays puisse subvenir aux besoins de la population en cultures stratégiques, il est généralement difficile de se conformer à ces deux dernières efficiences simultanément.
- la préservation des acquis et des droits de ceux dont l'irrigation constitue une pratique très ancienne à condition de se conformer à une efficience technique optimale de l'irrigation.

- l'équilibre budgétaire du gestionnaire et l'équité entre usagers sans perdre de vue qu'il serait plus juste de ne pas accabler les régions où la seule possibilité du développement repose sur l'irrigation.
- la survie du monde rural (objectif indirect) où il faut reconnaître un rôle social incontestable de l'irrigation.

3. Experience tunisienne en matière de formation dans le domaine de l'irrigation

La Tunisie a oeuvré dans ce domaine a différents niveaux:

3.1 Formation des agriculteurs et des usagers:

Plusieurs Centres de Formation Professionnelle Agricole assurent la formation des jeunes agriculteurs et de la main d'oeuvre agricole spécialisée. Des journées de formation et d'information sont régulièrement organisées par ces centres, par les Commissariats Régionaux au développement Agricole, par l'Union Tunisienne de l'Agriculture et de la Pêche et par les Groupements Professionnels au profit des agriculteurs. Au cours de ces différentes journées, il est nécessaire d'utiliser des exemples concrets pour plus d'efficacité. Des agriculteurs étaient habitués à utiliser une forte dose d'eau en irriguant un sol lourd, voulaient maintenir la même dose dans un nouveau périmètre sablonneux. Pour les convaincre d'utiliser une nouvelle dose, il a fallu prendre deux échantillons correspondant aux deux sols, y ajouter la même quantité d'eau et observer avec ces irriguants les pertes énormes enregistrés dans le cas du sol léger.

Chaque année, des lots de terres domaniales sont mis à la disposition d'ingénieurs et des jeunes sortants de Centre de formation agricole sous forme de location de longue durée. Ces exploitations devaient servir de modèle pour les agriculteurs environnants. Une expérience pilote consistant en une prise en charge de la vulgarisation sur terrain par la profession (Union Tunisienne d'Agriculture et de la Pèche UTAP, structure représentant la profession) est en train d'être testée.

Les agriculteurs sont appelés à s'organiser sous forme de groupements ou d'Associations d'Intérêt Collectifs (AIC). L'exploitation en commun des ressources en eau étant une pratique très ancienne en Tunisie. "Tous les ravins et tous les cours d'eau permanents sont barrés par des ouvrages en maçonnerie ou par des fascines et des pieux qui arrêtent le courant, en élevant le niveau et obligent l'eau à s'écouler dans le canal ouvert à côté du barrage. Celui-ci est presque toujours la propriété de plusieurs arrosants qui se partagent l'eau et en usent aux heures convenues par un règlement" (Jaubert de Passa, 1847). Aussi,

un système rationnel de partage des eaux dans les oasis du sud tunisien remonte au XIIIème siècle. Le nombre important des anciens ouvrages, de mobilisation, de transport et de gestion de l'eau, parsemés à travers tout le territoire, reflète bien l'importance accordée par le tunisien à travers l'histoire au problème de l'eau.
Récemment et au cours de ce siècle, plusieurs lois et décrets visant une utilisation efficiente de l'eau, ont été élaborés; dont:

- décret du 24/5/1920 portant création, à la direction des travaux publics, d'un service spécial des eaux, constitution d'un fonds de l'hydraulique agricole et industriel et d'un comité de l'eau.
- décret du 5/8/1933 portant règlement sur la conservation et l'utilisation des eaux du domaine public
- code des eaux (Journal Officiel de la République Tunisienne (JORT n° 22 du 1/4/1975)) dont un chapitre a été réservé aux associations d'usagers de l'eau.

Un statut type de ces AIC a été adopté (JORT du 12/1/1988) fixant leur constitution, leur fonctionnement et leur gestion. Ces associations peuvent avoir en charge l'ensemble ou une partie des activités suivantes:
"1) L'exploitation des eaux du domaine public hydraulique dans leur périmètre d'action,
2) L'exécution, l'entretien ou l'utilisation des travaux intéressant les eaux du domaine public hydraulique dont elles ont le droit de disposer,
3) L'irrigation ou l'assainissement des terres par le drainage ou tout autre mode d'assèchement,
4) L'exploitation d'un système d'eau potable, (code des eaux de la république tunisienne, 1975)"

Les associations d'Intérêt collectif sont dotées de la personnalité civile, (et elles sont placées sous le contrôle d'un Groupement d'intérêt Hydraulique "GIH", organe présidé par le gouverneur et regroupant les représentants de l'administration et des usagers. Des cellules AIC et des comités d'économie d'eau sont crées au sein de chaque CRDA en plus d'un service central.

Une stratégie nationale de création et de suivi des AIC a été adoptée à partir de 1992. Leur nombre n'a cessé d'augmenter au fil des années et il dépasse actuellement les 700 AIC.

Vu la complexité et la diversité des situations, un bon fonctionnement de ces associations nécessite un effort de longue haleine et des approches appropriées. A titre d'exemple, le mode de gestion d'un puits de surface propriété d'une seule famille diffère de celui à prévoir dans le cas d'un réseau collectif.

Plusieurs contraintes entravent le bon fonctionnement de ces AIC dont principalement:

- Morcellement des terres

Bien que 91% des exploitants sont propriétaires de leurs exploitations, le morcellement des terres constitue un handicap majeur. De plus, au sein du même périmètre, les superficies des exploitations agricoles varient énormément. On compte 471000 exploitations dont 193000 ont une capacité de production faible et où l'agriculture n'est pas l'activité principale du propriétaire (Tableau 1). Par ailleurs, l'irrigation est pratiquée dans 124000 exploitations

Tableau 1. Répartition des exploitations agricoles en Tunisie

Gouver norats	Sans terre	0 -1 ha	1 -2 ha	2 -3 ha	3-4 ha	4-5 ha	5-10 ha	10-20 ha	20-50 ha	50-100 ha	100 et +	Total	Total 1000 ha
Tunis		19.4	11.3	12.9	6.5	6.5	21	11.3	6.5	1.6	1.6	98.6	9
Ariana	23.1	6.7	10.6	5.8	4.8	5.8	19.2	13.5	6.7	1.9	1.0	99.1	120
Ben-Arous		31.1	13.3	11.1	4.4	4.4	15.6	11.1	4.4	2.2	0.9	98.5	39
Nabeul		14.9	18	15.5	14.2	8.9	16.8	8.2	2.5	0.3	0.3	99.6	198
Bizerte	0.5	16.4	16.8	14	8.9	6.5	16.4	12.6	5.1	0.9	1.4	99.5	238
Béja	2.0	26.3	15.1	8.3	7.8	5.4	14.1	10.7	6.3	2.0	1.5	99.5	283
Jendou-ba		19.4	15.3	10.9	12.1	7.3	20.2	10.9	3.6	0.4	0.4	100.5	169
Kef		6.5	6.5	7.7	6	7.1	22	19.6	15.5	5.4	3.6	99.9	381
Siliana		12.2	9	7.4	8.5	6.9	19	19	10.6	4.8	2.1	99.5	340
Zagh-ouan		24.3	12.2	7.8	8.7	5.2	20	12.2	6.1	1.7	2.6	100.8	211
S/T Nord	1.7	16.7	13.7	10.7	9.4	6.8	18.2	12.7	6.4	1.9	1.4	99.6	1989
Sousse		11.3	12.8	9.9	8.9	7.4	25.1	16.7	5.4	1.5	1	100.0	195
Monas-tir		21.8	19	14.1	9.2	7.0	16.2	7	4.2			98.5	90
Mah-dia		16.5	15.7	10.7	9.1	6.6	22.5	12.6	5.2	0.5	0.3	99.7	252
Sfax		16.0	10.2	9.8	6.4	6.4	17.4	19.3	11.9	1.9	0.7	100.0	602
Kair-ouan	0.7	8.1	10.5	11.2	9.3	6.9	22.9	15.8	12.4	1.7	0.5	100.0	484
Kasse-rine		4.3	8.3	8.3	8.6	9.0	23.7	19.8	14.4	2.5	0.7	99.6	388
Sidi Bouzid		5.6	6.7	8.3	6.7	7.8	25	24.7	12.8	2.2	0.8	100.6	519
S/T Centre	0.1	11.3	11.2	10.1	8.1	7.2	22.0	17.4	10.2	1.6	0.6	99.8	2530
Gafsa		4.5	8.4	4.5	3.9	5.2	24.7	22.7	17.5	6.5	1.9	99.8	267
Gabès Mede-		22.4	6.8	5.7	5.7	7.3	19.3	15.6	13.5	2.6	0.5	99.4	231
nine		18.9	12.2	9.4	6.7	6.3	20.9	15.0	7.5	3.1	0.4	100.4	261
Tozeur		75.3	18.2	2.6	1.3		1.3					98.7	14
Kebili		78.3	10.9	6.2	0.8	0.8	1.6	0.8				99.4	32
Tata-ouine						1.9	11.5	28.8	44.2	7.7	1.9	96.0	133
S/T Sud		30.0	9.9	6.1	4.2	4.7	16	13.9	11.1	3.1	0.7	99.7	937
Total	0.7	16.6	11.8	9.6	7.9	6.6	19.6	15.1	9.0	2.0	0.9	99.8	5456

- faible taux d'encadrement:

Le nombre total d'ingénieurs et de techniciens affectés au secteur de l'eau à l' échelle régionale, donc en contact direct avec les paysans , est de 693 (dont 144 ingénieurs),(Tableau 2); soit un taux d'encadrement de 0.002 ingénieur ou technicien /ha irrigué; ajouté à cela le problème de morcellement

Tableau 2. Effectifs du personnel affectés au secteur de l'eau dans les Commissariats régionaux au développement agricole

Gouvernorats	Ingénieurs	Techniciens
Tunis		2
Ariana	11	56
Ben Arous	9	13
Nabeul	4	53
Bizerte	10	31
Béja	8	34
Jendouba	12	36
Le Kef	5	10
Siliana	9	23
Zaghouan	3	14
Sousse	8	25
Monastir	1	25
Mahdia	6	19
Sfax	6	29
Kairouan	7	30
Kasserine	4	26
Sidi Bouzid	9	17
Gafsa	6	27
Gabés	6	21
Medenine	5	20
Tozeur	5	15
Kebili	4	12
Tataouine	6	11
Total	144	549

- contraintes inhérentes à l'exploitant lui même:
Le pourcentage des agriculteurs ayant une formation secondaire ou supérieure n'est que de presque 11% environ. Pour ce qui est de l'âge, 37% des agriculteurs ont plus de 60 ans. Il faut noter aussi les limitations des ressources financières et l'impossibilité de bénéficier des crédits bancaires en raison de l'absence de titres fonciers.

3.2 Formation des ingénieurs en exercice et des vulgarisateurs

Plusieurs actions de formation ont été menées. Des cycles de formation continue sont organisés en Tunisie depuis plusieurs années déjà et des stages dans des centres spécialisés permettaient l'échange d'expérience et d'expertise entre les bénéficiaires de la formation continue.

3.3 Formation académique (élèves ingénieurs, MSc, PhD)

La création depuis une vingtaine d'années de centres de recherche et plusieurs écoles d'ingénieurs spécialisées en équipement rural, en horticulture, en grandes cultures et en gestion agricole et leur localisation autour des pôles de recherche implantés dans les principales régions naturelles du pays permettent aujourd'hui une approche régionalisée de la formation et de la recherche agricole (Fig. 5).

3.4 Formation professionnelle

Plusieurs lycées agricoles, centres de perfectionnement et de recyclage spécialisés pour former les jeunes agriculteurs et la main d'oeuvre spécialisée dans les différentes activités agricoles sont implantés dans toutes les régions. La reforme introduite dans le système de formation professionnelle témoigne la volonté de l'état pour donner une nouvelle dynamique au secteur agricole afin d'en améliorer la compétitivité (Fig. 5).

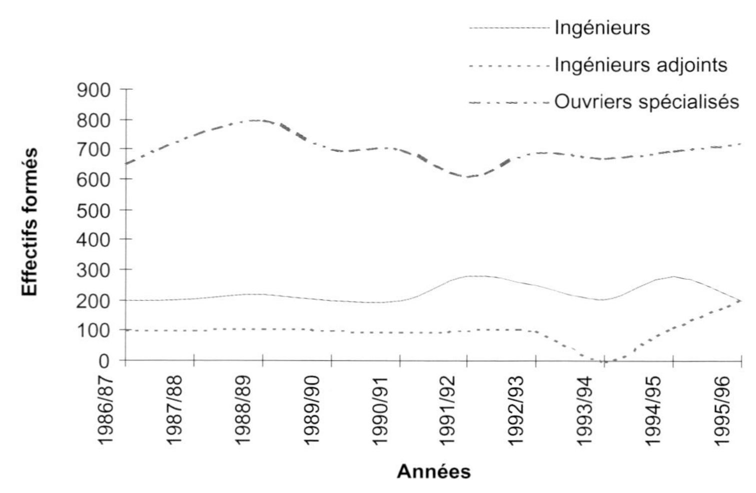

Figure 5. Formation agricole en Tunisie

4. Financement de la formation des agriculteurs et des ingénieurs en exercice

La formation académique est prise en charge par l'état dans les structures de l'enseignement officiel étant donné qu'il est difficile de savoir à l'avance les techniciens qui seront effectivement affectés à l'irrigation.

Quant à la formation des agriculteurs et des ingénieurs en exercice, il faut une structure, genre d'observatoire d'irrigation (Daghari, 1991) dont le financement peut être prévu sous forme d'une redevance sur la vente d'eau avec un maximum de 1 millime /m3 d'eau vendu. Ce que générerait annuellement au mois l'équivalent d'un million de dollars USA. Pour le lancement de cet observatoire, il est possible de recourir à une aide internationale. Cet observatoire aura pour rôle

- collecte, organisation des données dans le domaine de l'eau
- coordination entre les différentes formations et coordination de la politique de recherche dans ce domaine.
- financement et/ou participation à la formation des agriculteurs et des ingénieurs en exercice
- définition des indicateurs, vérification des points forts et détection des faiblesses de la politique de gestion de l'eau
- encouragement des nouveaux emplois et des petits métiers permettant une meilleure utilisation de l'eau (petit tâcheron de nivellement pour pallier aux pertes d'eau en irrigation de surface par exemple)

Références

Daghari, H. et L.W. De Backer, 1988. Utilisation de la méthode des profils hydriques pour la détermination du bilan d'eau (Comparaison avec des mesures lysimétriques). Revue de l'INAT 3(2), 153-166

Daghari, H., M. Maalej, Ch. Laroussi et L. W. De Backer 1989. Bilan hydrique sur jachère et sol nu dans les régions semi-arides. Journal of Hydrology 105, 197-204
Daghari, H. 1991. On farm water management policies, research and practices, prospects for improvements, note présentée à la FAO, Tunis
Daudet, F. A. et Valancogne, CH. 1976. Mesure des flux profonds de drainage ou de remontées capillaires. Leur importance dans le bilan hydrique. Ann. Agron., 27(2): 165-182
De Passa J., 1847. Les arrosages chez les peuples anciens, Bibliothèque Nationale de la Tunisie
Ministère de l'agriculture, 1975. Code des eaux 38 pages
Ministère de l'agriculture, 1991. Stratégie Nationale des ressources en eau , 95 pages
Ministère de l'agriculture, 1997. Résultats des Enquêtes sur les périmètres irrigués, DGPA
Ministère de l'agriculture. Etude du secteur de l'eau, 39 pages, DGRE

The experience of intensive training courses for the enhancement of professional competencies in natural disaster mitigation

Parodi U., Ferraris L., Colla L., Siccardi F.
CIMA, Centro di ricerca In Monitoraggio Ambientale, Italy

Abstract
Development in the field of natural disaster mitigation involves both the improvement of scientific knowledge and the enhancement of the community perception of danger. It is necessary to develop specific learning structures meant to spread knowledge regarding rare natural events, give guidelines to impact reduction and keep alive memory of rare events.
The interaction of the scientific community with the action makers helps to focus research attention on those problems that most matter in forecast and prediction, and is vital to update professional competencies. Due to the interdisciplinary character of the field it is necessary to develop courses that merge different kind of expertise.
The paper will present the structure and achievements of two intensive courses organized by the CIMA (Centro di ricerca in monitoraggio ambientale) for the local authorities and private professionals. The first of these experimental courses regarded the civil defense plans and the second the mapping of flood areas. The first course overviews those aspects of hydrology, meteorology, geology, law and logistics that were of interest for the civil defense problems. The second course gave the necessary knowledge of hydraulics, hydrology and mapping software to produce a first draft of historical mapping for flood prone areas of a set of basins in the Liguria Region of Italy. Both courses were successful; not only to update the skills of the participants, but also due to the feedback of information regarding the territory and problems encountered in practice.

Résumé
Le développement dans le domaine de la modération des désastres naturels à besoin de la synergie entre le progrès scientifique et la perception de danger de toute la communauté. Il est nécessaire de développer des structures spécifiques pour la diffusion des connaissances relatives aux les événements naturels rares, qui donnent des lignes de conduite pour réduire leur impact et garder vivant leur souvenir.
L'interaction entre la communauté scientifique et les agences de management du territoire aide à mettre au point l'attention des chercheurs sur les problèmes de majeur intérêt pour la prévision et la prédiction de les événements rares. El est vitale pour la mise à jour des compétences professionnelles du personnel technique des le agences de management du territoire. En raison du caractère typiquement interdisciplinaire du problème il est nécessaire de développer des cours qui mélangent les différents domaines d'expertises.
L'article présente la structuration et l'achèvement des deux cours intensifs organisés par le CIMA (Centro di ricerca in monitoraggio ambientale) pour les agences locales de management du territoire et les professionnels projeteurs. Le premier de ces cours expérimentaux regarde les plans de protection civile, le deuxième regarde les cartes des zones a risque d'inondations. Le premier cours repasse les divers aspects de l'hydrologie, la météorologie, la géologie, le droit et la logistique qui sont importants pour les problèmes de protection civile. Le deuxième cours regarde les aspects de l'hydraulique, de l'hydrologie et les

software pour la création de cartes thématiques dont le but est la détermination de zones inondables et la production d'un exemple de carte des zones historiquement inondées sur des bassins situés dans la région Liguria en Italie.
Les deux cours ont eu du succès, non seulement en raison de la mise à jour des participants mais aussi parce que les informations qui concernent le territoire et les problématiques du management ont donnés une précieuse contribution aux chercheurs.

1. Introduction

The Center for environmental monitoring research (CIMA), has site in Savona and was born due to the initiative of several departments and institutes of the University of Genova to better co-ordinate research and interventions for the development of science and technology in the field of environmental monitoring.

The Center promotes activities to transfer the research achievements to practitioners. The research activities of CIMA are in the fields of air, soil and water monitoring, territorial planning and water resources.

The Center has several co-operations with other universities and national and international agencies both private and public that operate in the environmental field. Moreover it promotes services for consultant activity in the field of water technologies, regulations and environmental logistics to help industries compel with regulations.

The Center carries out also curricular activities for the experimental bachelor courses activated in Savona with successful student tutoring and organization of the main part of the curricula in environmental monitoring and infrastructures.

2. Research and university formation activities

Hereafter are listed in briefs the research activities of CIMA. The three main research fields regard hydro-meteorological processes, water quality assessment and environmental risk analysis.
In the area of hydrometeorology the most important studies regard extreme rainfall events and space and time properties of rainfall fields.
Modeling of space and time rainfall fields using statistical methods and fractal analysis are part of a study for optimal localization of a meteorological radar in the Liguria Region and the design of a regional real-time rain gauge network. These activities are keystones of the INTERREG II and INTERREG IIC EC programs in which CIMA is actively involved.

Expertise of real time monitoring developed with the design and experimental management of the civil defense monitoring network ARGO enhanced expertise in calibration techniques and maintenance strategies for rain gauges. Moreover the center acquires and pre-processes satellite digital images. It has developed cloud tracking techniques to integrate use of remote sensors (Meteosat, SSM/I, radar and rain gauges) and other meteorological information coming both from ground sensors and atmospheric circulation models that are now in use with the Hydro-meteorological Center of the Liguria Region. The National Group for defense from Natural Disasters (GNDCI) and the Liguria Region finance the activities devoted to civil defense.

In the field of water quality monitoring the main projects regard chemical, physical and microbiological parameter analysis for the definition of water quality, rainwater analysis for acidification assessment and basin scale transport of organic matter. The activities are in the framework of studies for the rehabilitation and re-industrialization of chemical plants.

Studies for vulnerability and risk assessment regard the evaluation of the consequences of human transformation of the environment and the evaluation of the effect of extreme precipitation events on urban settlements. Of particular interest are studies for the progress of flood mapping techniques to use in the development of urban planning and civil defense action guidelines.

Among others have been carried out within the research group analysis of climate change effects on local hydrology and is being completed an atlas of Mediterranean climatology.

In the research activities of the center are involved the Ph.D. students of the doctorate in Technologies and Methodologies for Environmental Monitoring and the Ph.D. students of the doctorate in Hydraulic Engineering. The first doctorate course is held in agreement with the universities of Pisa, Firenze and Basilicata and the second course is held in agreement with the Universities of Padova, Firenze and Trento.

Besides the traditional teaching activities within the civil and chemical engineering curricula the faculty of Engineering of the University of Genova activated two experimental three-year bachelor courses in Environmental and Resources Engineering and in Infrastructure Engineering with site in Savona. The Liguria region university structure gained, in such a way, an intermediate formation level between the college diploma and the full university traditional curricula, aimed at the preparation of specialized technicians in the field of the environment, infrastructures, and resources management. The bachelor engineer will have a number of working opportunities within public administrations, public and private enterprises, environmental protection services, water related enterprises and enterprises in the field of production and management of monitoring and waste treatment systems.

3. The services and technology transfer activities of CIMA

The most important technology transfer is carried out within three groups of CIMA.
The first is the Hydro-Meteorological Center of the Liguria Region (CIMIRL) that was born in 1995 thanks to a regional law and is financed by the Liguria region. The second technology transfer project is called ECOSERVER and the third is the GIS Excellence project

3.1. The Hydro-Meteorological Center of the Liguria Region (CIMIRL)

The CIMIRL is meant to satisfy the regional need to forecast intense precipitation and dangerous discharges with sufficient advance to allow early warning. The CIMIRL is characterized by a close co-operation among meteorologists and hydrologists and gives strong emphasis to the quantitative aspects of warning in respect to qualitative ones. The group is subdivided in three branches one within the Liguria region structure itself that has the task of alerting local civil defense authorities in case of intense meteo-hydrologic events. The second branch lives within the CIMA and the department of physics of the University of Genova. It is the nucleus of the CIMIRL and has several tasks: acquiring useful meteorological data and remote sensor data; modeling local atmosphere circulation using a Limited Area Model; forecasting the meteorological situation for a time window of 72 hours; identifying potentially dangerous meteorological conditions and with the help of rainfall runoff models estimate land effect of intense precipitation; issue of hydrological alerts.

Finally the third branch that lives within the structural and geotechnical engineering department (DISEG) of the University of Genova acquires meteorological data used for statistical and climatological analysis, with special regards to wind data. Moreover it controls time series to build an homogeneous database of climatological information.

3.2. The ECOSERVER project

The project is developed within CIMA and financed by SPES S.c.p.A. (Società degli Enti Savonesi per l'Università) with funds coming from the regional environmental rehabilitation and re-industrialization plans. It foresees the development of a support system for small and medium sized industries. The system heart is a database divided in the two main branches of legislation and technologies. The first part of the database will give information regarding the Italian and EC regulations on environmental matters with special care for: air pollution, water resources management, water treatment, rehabilitation, waste reuse and disposal, water pollution, acoustic pollution, environmental impact evaluation, monitoring, energy, working security, Ecoaudit and Ecolabel. The database law texts are enriched with links to related laws and specialized comments. The second part of the database collects descriptions of the most diffused technologies available on the Italian market regarding the environmental sector that are more important to small and medium size industries. It will inform about the characteristics of the processes, positive aspects and drawbacks, limits of application, initial and management costs and main producers present in Italy. This database will be accessible both for direct search and through guided theme paths within a web-site. Moreover the ECOSERVER system will allow on-line retrieval of guidelines and forms for the compliance with regulations in the environmental field and will give news of interest and an e-mail legislative and technical support for environmental problems related to small and medium size industries.

3.3. The GIS excellence project

In the spirit of incrementing the connections among university and enterprises, the CIMA started a co-operation with Datasiel S.p.A. of the Telecom group to create an integrated information system for efficient management of the civil defense plans at local level. The creation of joint ventures among university and enterprises is believed to help technological transfer from research activities to the application field. The intent of this third group is to create a first embryo of an enterprise that merging the software know-how of Datasiel and the environmental engineering and civil defense expertise of CIMA can become leader in the field of information software for civil defense purposes. This structure has not only a productive character, but also the one of a service to aid with reference standards local authorities and consultants working at the realization of civil defense plans. The purpose is that of minimizing the effort of updating the plans and of merging plans together and with wider, but less detailed province and regional authority plans. The other stronghold of this kind of joint ventures is the direct feedback to bachelor and master students that focuses their curricula on market needs and boosts their introduction in the productive system. The experimental application, is aimed at flood risk management, that is the most important natural risk in the region, and will be extended later on at other targets like seismic risk mapping. The choice of user friendly environments and GIS interfaces will allow a very fast integration of higher (ex: regional level) and lower levels of information (ex: security plans for industrial sites or fire defense plans for airports, ports and gas deposits). The system will therefore aid both in the monitoring and warning phase and in emergency management. It will also be a useful instrument for storage and publication of revealed damages during and after emergency.

The above mentioned activities are backed by the academic structure of CIMA and develop higher technological standards in environmental planning and management. These activities help to form new competencies to be introduced in both public administrations and the private sector and are supported by the permanent formation program.

4. Permanent formation programme

The permanent formation program was meant to represent the tool for updating expertise of technicians of public administrations and private consultants in the framework of important changes of the Italian environmental and civil defense laws. The program at present counts two courses; the first oriented to civil defense and the second to territorial planning and risk assessment.

4.1. Civil defense school

This formation tool was meant to create a synergy among the natural risk assessment research community and the action makers in the civil defense field.
The school had a restricted number of selected participants chosen among the employees of the local authorities and consultants in the hydraulic field. The purpose was to give the basic knowledge for the layout of local level civil defense plans. This thematic school was meant to be given on annual basis and organized in three weekly workshops of thirty hours each. The course was financed by SPES and organized with the co-operation of GNDCI.

Contents
The final products of the course were drafts of civil defense plans oriented to flood risk management. The study cases were selected according to the participants' interests and the entire work supported by theoretical lessons and practical follow-up. The theoretical support included review of laws and regulations and lessons regarding meteorology, hydrological monitoring, hydrology, hydraulics and mapping techniques. The follow-up included the preparation of support material like a collection of laws and regulations, the assisted use of mapping software and the collection of commented documentation regarding damages produced by past extreme floods.

The work was carried out mostly within the participant's administrations since all of them were at the time actively involved in the layout of new civil defense plans as recommended by the latest Italian law and following regional laws and regulations.

By law the plans have to be made up of three parts: prevention and prevision programs, emergency plans and rescue plans. The last point was not treated by the school that focused its attention on a correct structuring of the prevention plan rather than on the logistics of rescue actions.

The importance of a prevention plan lies in its efficiency in transmitting information regarding both vulnerability and flood risk. This means that flood maps coming from historical records and hydraulic modeling, vulnerability databases and flood event scenarios must be merged in the same information system. The prevention plans moreover have to propose structural and non-structural actions to reduce vulnerability.
The first part of the course reviewed laws and regulations. This part of the course involved local and state coordinators for civil defense and gave origin to discussions and comments regarding the expected efficiency of the prevention tools proposed by regulations and problems encountered in their application.

The second part of the course overviewed the basic principles of meteorology and hydrometeorological monitoring including both remote sensing and gauge network systems.

The third part of the course gave elements of hydrology and hydraulics.

This last two parts of the course focused the attention on the difficulties encountered in forecasting extreme events for the region of interest. The region due to a rough topography near to the sea shore is characterized by small drainage basins prone to flash floods and long dry spells making land based monitoring difficult and insufficient for effective flood warning. The region forecasting system for flood warning must therefore be meteorology based.

The fourth part of the course reviewed statistical analysis of rainfall and discharge maximums and gave elements of risk assessment including tools for flood risk mapping.

Organization
The course started with four days of workshop in part devoted to theoretical lessons in part to discussion. To the four days workshop followed a week of introductory work to the civil defense plan during which eight participant groups were formed each followed by an expert of the teaching group. Afterwards the draft plan was completed by the participants within their own offices and the results reviewed in a conclusive day of debate and discussion.

Outcomes
The school helped the administrations to follow a coherent program in the layout of the civil defense plans and gave the basic knowledge needed to produce the required hydrologic and hydraulic analysis. The participants pinpointed the need for the development of reliable and simple tools for flood prone areas mapping. The feedback of the participants induced the CIMA to experiment and compare different flood area prediction tools and was one of the reasons for the promotions of a project for the creation of database system for risk management and mapping.

4.2. Preparatory course for basin scale planning

The course introduces to the legal context of the basin scale planning and reviews some basic themes of interests for hydraulic and geological risk assessment, soil defense and territorial and environmental planning.

The provincial public administration to give an advanced update to their personnel in view of the new laws regarding territorial planning and application of risk mitigation tools have financed the course. Attendants were technical employees of the public administrations with college level preparation and experience in territorial planning or master level employees of the architectural and geologic branches.

Organization and contents
The formation action was distributed over five months and occupied full time the three central days of the week with both theoretical teaching, discussions regarding practical problems, application exercises and analysis of some study cases. Due to restrictions imposed by working timetables of the participants the course was doubled; i.e. each teaching and exercise unit was repeated to allow greater attendance flexibility. At the end of the course one day was dedicated to plenary discussion of practical problems and course outcomes.

A full course lasted twelve working days. Since the greater gaps regarded hydrology and hydraulics, the course treated the following themes.
a) Basin scale planning in Italy and Liguria region: laws and regulations.
b) The study cases of the Centa and Leira basins.
c) Quantitative Mediterranean meteorology.
d) Rainfall duration-intensity curves and flood discharge evaluation.
e) Distributed and concentrated hydrological models.

f) Channel hydraulics: roughness parameters, permanent flow profiles, abrupt section changes, and sediment transport.
g) Field measurements for the Molinero stream.
h) Flood analysis and risk assessment for the stream Molinero.
i) Experimental basin scale plan for the stream Chiaravagna.
j) Near shore marine sediment transport.

The teaching team was composed by University professors and research staff in the hydrologic and hydraulic field and experts in the field of geological and hydraulic risk assessment that completed the first experimental basin scale study case plans. The course was coordinated by the university teaching staff of CIMA. The attendants received a rich package of course material organized as follows:

a) Laws and regulations.
b) Publications of interest to the risk assessment.
c) Case study basin scale plans material.
d) Notes regarding climate, atmospheric circulation, hydro-meteorological monitoring, hydrology, hydraulics, sediment transport and near shore sediment dynamics.
e) Extracts from a hydrodynamic book and bibliography.

All the exercise sessions were complete of the necessary material, oriented to solution of problems in part solicited by the participants with whom the results were discussed.

Outcomes

The course gave the participants an efficient update in the hydraulic and hydrologic field and focusing on practical problems posed by the participants allowed to touch most of the basic aspects concerning basin scale planning outlining a proper planning oriented hydrologic and hydraulic analysis.

The participants highlighted the importance for a simple but correctly tailored scheme for flood analysis and moved the CIMA staff attention to the need for a standard statistical analysis tool to evaluate extreme rainfall and discharge events over the Liguria Region.

5. Conclusions

The two courses have been successful both with regards to the update of the technical staff of public offices and private consultants and due to the feedback coming to CIMA from the participants. The courses focused the attention on the need for better tools in the field of flood risk evaluation and risk mapping. CIMA therefore is developing within its technology transfer groups a system for risk mapping and with the backup of research is evaluating the reliability of different flood zone identification methods.

The courses organization proved to be sufficiently flexible to allow good attendance and the wide discussion produced by lessons and exercises indicated that the level of the courses and the subjects were of interest. The work produced during the courses was a useful guideline for the new basin and civil defense plans and helped to give a more homogeneous cut to the work produced by the different public administrations. The authors believe that these kinds of courses are an efficient mean to maintain a strong link among research and application in the field of water resources management and will be a good example in order to improve the continuing education programs in the field of hydrological risk assessment and planning.

Interactive education of villagers in water conservation and quality improvement

Athavale R. N.
National Geophysical Research Institute, India

Abstract
Although India is well endowed with water resources, the spatial and temporal distribution of rainfall over the subcontinent is highly uneven. Education of the villagers in the principles and simple techniques of water management can, therefore, bring about a major change in terms of material well being and improvement in the quality of their life. The scientist/technician has to help in this process, not only by suggesting appropriate water management technology but also by adapting himself to an appropriate role of an initiator or a catalyst or a helper, depending on the specific situation. Some case studies of this type of non-formal education of farmers and villagers in attaining sustained supply of water and improving its quality are described. It is felt that there is a need for starting a one year's interdisciplinary post graduate diploma course on "Water Harvesting and Conservation" to train geologists, hydrologists and engineers in understanding and solving the micro scale problems of water management faced by the rural population. Various topics to be covered in the course are identified.

India receives an annual average rainfall of 1170 mm while the global average is 700 mm. Although this appears to be a significant amount, the spatial and temporal distribution of the rainfall is highly uneven. The rainfall at Cherrapunji in the northeastern state of Meghalaya is about 13000 mm, probably the highest in the world, while that at towns like Jaiselmer and Barmer, located in Thar desert near the western boundary, is 150-200 mm. In most of the country, the rainfall is seasonal (monsoonal), lasting for 3-4 months and made of 30-60 rainy days. Half of the annual rainfall occurs in heavy spells of short duration and the total duration of these spells is 20-30 hours. Most of the precipitation during these high intensity storms runs off the surface, with poor ground water recharge and considerable soil erosion (Pisharoty, 1990).

About 80% of the total area of the country (320 million ha) is located in the semi-arid tropical belt where the average temperatures are high and evaporation losses from surface water bodies are considerable. Further 66% of the total area is covered by hard rocks, mostly granites and basalts, which have poor primary permeability and storage capacity. Extensive measurements in a score of basins indicate that only 8 to 10% of the total annual rainfall in hard rock areas is contributed to the aquifers, after percolation through the soil and the unsaturated zone (Athavale, 1991).

The net result of this natural situation and tremendous growth in the population over the preceding 50 years is that of scarcity of drinking water during summer months in practically every city and village irrespective of its location and amount of annual precipitation. At the same time, during the monsoon period, there is surplus of water even in areas of low rainfall and this surplus is lost to the sea, after causing floods in some cases.

Substainable supply of potable water can be achieved in this situation only through effective water management practices at the village level. The Govt. agencies do undertake programs of construction of water supply reservoirs and pipeline distribution systems. However, experience shows that they have not fulfilled the needs of the people. Although the problem of water scarcity in summer months is all pervading, the reasons and, therefore, the solutions of the problem are different for different agroclimatic regions. In addition, several thousand villages have problems of water quality in the form of presence of excess Fluoride/Arsenic/Iron/TDS. The author has developed robust, village (micro) level solutions for these twin problems of quantity and quality and carried out field experiments in each case (Table 1). The basic concept in each of these interventions will be described in oral presentation.

Table 1. Technologies developed for sustainable supply of drinking water to villages in India.

Scarcity Situation		
Area and rainfall range	Problem	Solution
Low rainfall (<400 mm) Thar Desert	Inherent Scarcity	Enhancement of runoff to traditional water harvesting structures
Medium rainfall (400-1000 mm)Southern peninsula Hard rock	Overexploitation (Lowering of water table)	Artificial Recharge: Three different approaches
High rainfall (>2000 mm) Western Coast & Himalaya Foothills including Northeast region	Hilly Terrain (Quick runoff)	In-situ water harvesting
Poor Quality		
Area	Problem	Solution
Coastal regions, Saline aquifer tracts of Gujarat, Rajasthan, Haryana, etc.	Incursion of saline water in fresh water aquifers	Installation of salinity sensing devices on wells
Fluorosis affected areas of Andhra Pradesh, Karnataka, Gujarat, etc.	Fluoride above permissible limit in groundwater	Artificial recharge & recovery of water
West Bengal	Arsenic above permissible limit in groundwater	Artificial recharge through village ponds (Proposed)
Orissa, Tamilnadu, Maharashtra, Andhra Pradesh, etc.	Excess iron in groundwater	Sanitary open wells in place of borewells

Some case studies, where one or two of these technologies have been applied and which involve participation by a large number of villagers are described below. The author is associated with these cases in different capacities. In some cases he has initiated the activity, in some he has corrected the work in mid-course and in some cases he has provided a scientific explanation at the impact of people's water, soil and forest management activities on their environment. Detailed discussion with the villagers and helping them to understand the scientific principles involved in water management has been a common thread in all these activities.

1. Mass movement of artificial recharge of wells and construction of percolation tanks by the villagers of Saurashtra region of Gujarat

The Saurashtra peninsula is covered by basalts (Deccan Trap) and the black cotton soil (vertisol) derived from them. The seasonal rainfall is 500-600 mm. The plateau is drained by several ephemeral streams having steep gradients. The people of the area suffer from scarcity of drinking water in summer months and need more water for crop saving irrigation. Some pockets in the area have excess Fluoride in groundwater and its coastal area aquifers have suffered from seawater incursion due to overexploitation. Swadhyay (literary meaning self study), a voluntary organization, has made the people conscious about the need of water conservation and since 1994 the people have constructed eight hundred percolation tanks and transferred runoff water into more than 100,000 open wells (Patel, 1998). While this laudable effort has provided immediate gains, proper scientific care is not being practiced in recharging the water. I have recently visited the area and explained to various farmers that the water needs to be properly filtered, otherwise the pores in well rocks will be clogged and the permeability will be reduced. Similarly, the bacteria in the organic material may multiply in the course of time. I have suggested that the farmers should collect their runoff water in a farm pond located near the well. This farm pond, having gravel and sand layers at the bottom, would transfer the collected runoff water to the well from the substratum. The water will be cleaned in this process. Any pumping of the well will induce recharge from the farm pond. In short, I have advocated the need of practicing 'INDIRECT' recharge in place of direct pouring of runoff water in the well and suggested a method which is acceptable and adaptable. The farmers have decided to change over to this method of recharge from the coming monsoon season. I have not advised the need of elaborate filtration and chlorination of the recharge water and this would not have been technically feasible at every small farm level and would have decelerated the tempo of their water conservation work.

2. Popular article in farmer's magazine

In May, 1996, I wrote a popular article on various method of "Borewell Recharge" in response to a request from the editor of a farmer's magazine "Adike Pathrike", published in the 'Kannada' language spoken by about 30 million people of the southern state of Karnataka. I do not know this language and, therefore, the editorial staff translated my article into Kannada from my English version. I have recently received a letter from the editor stating that, inspired by my article a farmer has been recharging his well using one of the method suggested in my article. The yield of his borewell has doubled and the quality of water has improved. The editor published this farmer's success story in the October, 1998 issue of the magazine and wanted me to write a scientific explanation of his work for publication in the subsequent issue. This case illustrates that language is not a barrier and technical education of farmers is possible if there is a need and a proper forum.

3. Sustained yield by a water well pump

A farmer in village Marrigudem in Nalgonda district of Andhra Pradesh, drilled a well at a site recommended after geophysical surveys. He struck water and was advised to use a 3 HP motor. However, he wanted to have more water and instead installed a 5 HP motor. This pump would operate for about fifteen seconds and trip off after the water level in the well went below the submersible pump. It will again trip in after a period of 30 seconds or so and this cycle would continue throughout the period of power availability. The exasperated man was offered a solution involving construction of a 200 m3 farm pond near the borewell. He did so in 1997. After the monsoon rains, the pond was filled with runoff water, the pump started working without interruption and the discharge also increased.

The other farmers of the village are now emulating the example of this farmer.

4. Rejuvenation of five rivers

The Tarun Bharat Sangh (Young India Union), a voluntary organization, is actively engaged in soil and water conservatiion activities and in afforestation work in several districts of Rajasthan State over last twenty years. Their main work is in various villages of the north-eastern district of Alwar. They have constructed several hundred check dams, gully plugs, farm bunds and micro percolation tanks in about 600 villages of the area and have banned tree cutting and encouraged afforestation. All this work has been done with participation of the villagers, including the village women, in every stage of decision making, i.e. planning, designing and financial contribution in the form of free labour. The Tarun Bharat Sangh (TBS) mainly worked as a catalyst, provided cement and some other construction material and rendered technical advice. The results have been spectacular. Prior to the intervention by the TBS, the people of the villages were not having enough employment, the crop yield was poor and many wells were going dry. As a result of the several water harvesting and conservation structures, the water table has come up and the wells do not dry even during summer. The crop yield has increased. Additional area has come under cultivation and more fodder is available for cattle. The farm residue is used as fuel in place of forest trees. The average income and health of the people has improved and five ephemeral streams of the area have now become perennial. Shri Rajendra Singh, Secretary of the TBS, invited to visit the area and provide a scientific explanation of changes in the hydrological regime of the area as a result of the water harvesting activity transformation of five rivers from ephemeral to perennial status.

The explanation offered by me is reproduced below:
All the five rivers are located in semi-arid tropical climate area. The average rainfall is about 600 mm and most of the rain (about 80%) occurs during the monsoon months. A large fraction of this rainfall comes down in the form of 3 or 4 storm events. The hilly terrain and the precipitation pattern are conductive to quick runoff of a large percentage of the rainfall. This natural situation was further exacerbated by the deforestation taking place over last 100 years or so. The net result was that most of the incident water was leaving the area during the monsoon months as runoff. The change brought out by the water harvesting structures is explained in Table 2 through a guess-estimate of the water balance of a typical river in the area. It may be noted that these figures are based on experience elsewhere and not on actual experiments or data collection in the area visited.

Table 2. Water balance scenario of a river basin in Alwar district, Rajasthan.

Before intervention		After intervention	
Rainfall	100%	Rainfall	100%
Natural Recharge as a percentage of rainfall	15%	Natural Recharge Artificial Recharge	15% 15%
Evaporation and evapotranspiration From bare soil and plants.	50%	Evaporation and evapotranspiration	60%
Runoff	35%	Runoff – Monsoon period Non-monsoon period (Groundwater component)	10% (20%)

Table-2 shows that the runoff has not been substantially reduced. However, it has been regulated. The static groundwater table in the area had gone below the bottom level of most of the open wells and also below the bed level of the river. As a result, both were remaining dry during the non-monsoon months. The additional recharge, effected by the various water harvesting structures, has raised the water table above both these levels.

Most of the wells have, therefore, water during the summer months. This water is used for irrigation and 33% of this irrigation water goes back to groundwater as a return flow. The area has no surface or canal irrigation. The unutilised groundwater enters the river practically all along its length as effluent seepage. The river has turned perennial mainly because of this regulated distribution of the runoff quantum over the entire year, in place of the situation in which most of the runoff was taking place only during monsoon months. Measurement of the conductivity and turbidity of the river water, during the non-monsoon months, would prove that the water was being contributed by the groundwater. However, at present it's crystal clear appearance and tranquil flow were sufficient to indicate that it was the groundwater, gentling oozing on the river bed, which had made it perennial. The above figures suggest that about 5% of the groundwater is used for irrigation and about 1.5% to 2% of the irrigation is returned to groundwater for leaving the area as effluent seepage in the river draining it. The rest of the irrigation water is returned to atmosphere by way of evapotranspiration (by plants) and evaporation. In my opinion, the residents of the catchment area of the river, are entitled to utilize this small fraction of the rainfall endowment of their watershed, for some improvement in their standard of living.

5. Defluoridation of drinking water

In several states in India, the groundwater has Fluoride concentration above the permissible limit of 1.5 ppm/1. Mottling of teeth and skeletal damage, common symptoms of 'Fluorosis', are developed in people who have been ingesting the water for long periods. The number of Fluorosis affected people is estimated as 2.5 million. Government departments have tried to tackle the problem by installing chemical defluoridation plants on many water supply wells. However, it was found that the villagers are not wanting or able to maintain these plants and almost all of the plants have become defunct. Some international aid agencies have provided domestic defluoridation units to villagers in some districts. However, they found that the kits were not used regularly. An experiment on Artificial Recharge through injection borehole in granitic terrain has been carried out at village Malkapur in Nalgonda district of Andhra Pradesh (Muralidharan & Athavale, 1998). This experiment has shown that the Artificial Recharge technique can be used for augmenting groundwater and reducing Fluoride concentration. However, replication of this method at village level will require some time.

In the light of this experience, the author decided to introduce the scheme of supplying defluoridated drinking water in a primary school in Kunderu village in Nalgonda district in Andhra Pradesh. The science teacher was trained in removing excess Fluoride with the use of Alum. A recent random check proved that the teachers are able to use this simple technique efficiently and bring down the Fluoride level in supply water from 5.5 ppm to 1.4 ppm. The children are now drinking good quality water at least during school hours. The scheme is in operation for one year. Some other advantages are that the children see a simple experiment in chemistry and become aware of the hazards of Fluorosis in early stage. It may be mentioned here that the author had to work on the psyche of the villagers for about a year before the scheme was implemented. This was when the villagers were not required to bear the expenses of implementing the scheme.

A new training course on "Water Harvesting and Conservation"

Based on my experience in interacting with villagers in different agroclimatic zones of India, I feel that the topic of water harvesting and conservation with the village watershed as a unit has several dimensions and all of these are not covered in any of the ongoing engineering or geology of hydrology degree courses in academic institutions, at least in India.

The term water harvesting refers to collection and storage of rain water and also other activities aimed at harvesting surface and groundwater, prevention of losses through evaporation and seepage and all other hydrological studies and engineering interventions, aimed at conservation and efficient utilization of the limited water endowment of a physiographic unit such as a watershed.

The various topics which can come under water harvesting and conservation are:

I Construction of permanent/portable storage structures.
I. Farm ponds, either for supplemental irrigation or for augmentation of ground water.
II. Check dams.
III. Percolation tanks at appropriate sites based on geological consideration, design of percolation tanks.
IV. Reclamation/Revitalization of traditional water arresting structures.
V. Artificial recharge through wells.
VI. Control of evaporation from surface water bodies.
VII. Prevention of seepage losses in appropriate situations.
VIII. Enhancement of runoff through mechanical and chemical treatment in catchment area.
IX. Sub-surface dams to arrest base-flow of groundwater.
X. Soil and water conservation practices comprising contour and terrace bunding.
XI. Control of seawater incursion in coastal aquifers.
XII. Control of transpiration without affecting normal plant growth.

It is felt that a one year interdisciplinary post graduate diploma course, covering the above mentioned topics should be introduced in Universities, to prepare the graduates from engineering geology or hydrology streams, for providing micro-scale solutions for sustainable supply of drinking and irrigation water, through interventions appropriate to the local, geological, topographical and climatic conditions.

References

Athavale, R.N. (1991) Role of groundwater in sustainable crop production in India, (Proc.Int.Conf.. on *Sustainable Land Management*, Napier, New Zealand, Nov. 1991), 241-251.

Muralidharan , D. & Athavale, R.N. (1998) A Base Paper "*Artificial Recharge in India*", National Geophysical Research Institute, Hyderabad, June, 1998.

Patel, A.S. (1998) Artificial Recharge in India, (Proc. Workshop on *Artificial Recharge in India*, Aug.27-29, 1998), National Geophysical Research Institute, Hyderabad. (in print).

Pisharoty, P. R. (1990) Indian Rainfall and Water Conservation, (Proc. All India Sem. on "*Modern techniques of rain water harvesting, water conservation and artificial recharge for drinking water, afforestation, horticulture and agriculture*", Pune, 19-21, Nov. 1990), 50-53,

Water school

L'école de l'eau

Pallano V., F. Doda
A.M.G.A. S.p.A., Italy

Résumé

Comme dans tous les secteurs de la production, dans l'industrie de l'eau aussi le rôle de la programmation/planification devient de plus en plus important : la nécessité de garantir un service de qualité, efficient et efficace impose, en effet, une réorganisation des activités techniques et de gestion.

En particulier le niveau actuel du secteur italien de l'eau apparaît mal adapté par rapport aux paramètres en vigueur dans d'autres pays de l'Union Européenne : les Directives Communautaires, d'une part, visant à améliorer la qualité du service et à protéger le consommateur, et les réglementations nationales, d'autre part, focalisées sur une amélioration du service en fonction de critères non seulement qualitatifs, mais également d'ordre économique, imposent aux entreprises de s'adapter aux nouveaux standards en des temps réduits.

Federgasacqua, la fédération italienne comptant parmi ses membres 303 entreprises opérant dans le secteur de l'eau, travaille depuis quelques temps à la création d'une " Ecole de l'Eau ", une initiative de formation visant à la confrontation et à la synthèse des expériences et du know-how acquis dans les différents pays de l'Union Européenne. Proposant des cours de formation visant à la qualification et l'actualisation des connaissances de figures professionelles spécifiques appartenant au secteur, l'" Ecole de l'Eau " a pour objectif de donner aux entreprises les moyens d'être compétitives sur le marché national en se conformant aux critères imposés par la règlementation relative à la réorganisation des services de l'eau (Loi 36/94).

La loi 36/94 prévoit, en fait, une réorganisation des entreprises gérant les services de l'eau, qui doivent aujourd'hui adopter une approche entrepreneuriale pour atteindre un haut niveau d'efficience du point de vue technique et, surtout, économique. Le dépassement du concept d'eau comme bien inépuisable devra comporter l'adoption de stratégies de gestion novatrices et efficaces dans le nouveau contexte.

L'"Ecole de l'Eau " propose, d'une façon plus générale, une amélioration globale de la culture des opérateurs dans le secteur par le biais de cours dont les programmes sont élaborés en fonction des exigences du marché et du tissu économique de référence.

Dans le cadre du programme communautaire ADAPT – Adaptation de la main d'oeuvre aux mutations industrielles, Federgasacqua a réalisé, en collaboration avec A.M.G.A. S.p.A., une expérience pilote en Ligurie, dans le but d'inaugurer une nouvelle approche de la formation, utilisant la confrontation entre expériences d'organisation et de gestion acquises dans des contextes différents. La formation a été surtout adressée aux cadres moyens et supérieurs des entreprises, intéressés par le changement en cours des aspects touchant la gestion du cycle intégral de l'eau.

Dans la ligne des objectifs de l'Union Européenne, qui se fixe comme but d'expérimenter et de réaliser de nouvelles approches dans les politiques de formation professionnelle, par le biais, notamment, de la création de réseaux transnationaux, Federgasacqua et AMGA ont consolidé, grâce à leur participation à ADAPT, l'expérience acquise dans le cadre de la formation continue qui constitue un instrument permettant une mise à jour constante des connaissances en fonction des mutations et des évolutions du secteur de l'industrie de l'eau, et par conséquent, une hausse du niveau culturel du personnel concerné.

Abstract
As is usually happening in all the production sectors, in the water industry planning is getting more and more important: the need to provide a service meeting all the quality, effectiveness and efficiency standards calls for a re-organisation of technical and management activities.
In particular the water sector level, in Italy, is not in compliance with the parameters followed by other countries in the E.U. Community. Directives, on one hand, focusing on the quality service and the user protection, National Regulations, on the other hand, focusing on a service improvement according to qualitative and economic parameters, are inviting the water companies to adapt to the new standards in the short term.
Federgasacqua, the Italian association gathering 303 members operating in the water sector, has been working for a long time to establish a "Water School", a training initiative aimed at a comparison and synthesis between the experiences and know-how in the different E.U. countries. Therefore, the "Water School", through its training programmes focused on the qualification and updating of specific professional profiles in the water sector, intends to put companies in the position to effectively compete in the national market, according to the criteria provided for by the law related to water services re-organisation (Law 36/94).
In fact, Law 36/94 is aimed at a re-organisation of the entities managing the water services, which have to adopt an entrepreneurial approach to reach a high level of technical and, above all, economic efficiency. The abandoning of the conception of water as an unlimited resource should involve, as a consequence, the adoption of innovative management strategies, which could prove to be effective in this new scenario.
The "Water School" intends, more generally, to spur a global improvement in the culture of operators in the water sector, through a study on processes and contents defined in compliance with economic and market needs.
Within the Community Programme ADAPT - Adaptation of the workforce to industrial changes, Federgasacqua, in co-operation with AMGA S.p.A. has carried out in Liguria a pilot project, aimed at defining new training standards, through the comparison between organisational and management experiences from different contexts. The training project has been mainly addressed to executive and technical staff, which are mostly interested in the ongoing changes concerning the management of the whole water cycle.
With their ADAPT project, in line with E.U. aims, related to the definition and experimentation of new approaches in vocational training policies, Federgasacqua and AMGA have consolidated, also through transnational networking, their experience in the field of continuing training, as a tool to keep in touch with changes and evolutions in the water industry sector, with a consequent increase of culture for the training beneficiaries.

1. Introduction

Planning is becoming more and more important in the water industry, as in all sectors of production: administrative and technical aspects of services must be reorganised to provide a guarantee of quality, effectiveness and efficiency.

The Italian water industry compares particularly poorly with the standard set by other European Union nations: water companies will have to improve their standards in the near future to comply with EU Directives aimed at improving service quality and defending consumers' interests and with national legislation aimed at improving both quality and economic aspects of service.

The Italian water industry is extremely fragmented, involving more than 14,000 administrative bodies, including water supply systems administrators (5,500), municipalities responsible for sewer networks (7,000), and water treatment facilities (2,000). In the drinking water sector, the water supply systems administrators deal with over 11,500 aqueducts supplying the 8,000 municipalities in the nation.

The distribution network is highly problematic: the majority of water supplying systems, covering a total of 150,000 km of pipelines, are not interconnected, a fact which is not only uneconomical but results in unequal service.

In this context, the challenge is to come up with a water management policy that will permit reorganisation and optimisation of the water service, overcoming its current fragmentation and grouping administrative units together where advisable.

New administrative institutions are needed to replace the current inadequate organisation of certain public services, which has direct repercussions on the quality of service, the protection of municipal resources and the efficiency of service, and results in higher costs which are directly or indirectly born by users. This is the case of sewer services and wastewater treatment facilities, which more and more evidently require management on a truly industrial scale, involving the operation of complex plants requiring constant maintenance and updating in the wake of technological progress.

Not only must we discover an economical way of managing the water industry, we must also reorganise the services involved in the various stages of the water cycle (water supply systems, sewers, wastewater treatment facilities) on a unitary basis, assigning management of all these services to a single professionally qualified service provider.

It was to remedy this situation that the government passed law 36/94 in January of 1994, a law which, if implemented even in part, would result in a noticeable improvement in water services. The law requires an integrated water service to be managed by «optimal areas», to be defined by regional administrations and implemented by passing local laws: the integrated water service will be run by a single administrative body which will act as a leader within its territory.
In short, the new organisation of integrated water services requires regional administrations to implement a series of measures in a logical and chronological order, substantially consisting of:

- definition of optimal areas;
- regulation of co-operation between different municipal bodies located within the same optimal area;
- adoption and application of a standard convention regulating relations between municipal bodies located within a single optimal area and their water service provider;
- regulation of the transfer of personnel to the water service provider from the local associations, special companies and other local administrative bodies which have managed water services in the past.

The new water service management will thus achieve a number of objectives: overcoming inadequate administration; putting back together a productive cycle which must not, for functional and legal reasons, be managed in separate stages; creating opportunities for co-operation between network services employing similar plants to permit them to make better use of new technologies; and, finally, granting water service providers the degree of independence that a service must have if it is to be run as a business.

The law requires the optimal areas to be defined within catchment basins or sub-basins. The basic aim behind this new way of dividing the territory is to do away with fragmented management of water services by organising an integrated water service in an area large enough to permit efficient management, «... defined on the basis of physical, demographic, and technical criteria as well as politico-administrative boundaries» (art. 8, paragraph 1).

The service provider must guarantee that good service will be provided within each area, and is «required to achieve economically and financially balanced management» (art. 11, paragraph 2).
Implementation of the Galli Law was begun in January of 1999 with the creation of an optimised territory in the Upper Valdarno area, which includes Arezzo and 36 other municipalities with a total of 350,000 inhabitants.

The 37 municipalities in the Arezzo optimal area decided to put the law into effect for the first time with water services managed by a mixed partnership in which the private shareholder, who subscribes 46% of the capital, worth 67 billion lire, assumes full responsibility for administration and investment.

A call for bids was published in October of 1998 for the selection of the minority shareholder, and the choice was made in January of 1999 by a consortium of businesses including Suez Lyonnaise des Eaux, A.M.G.A. S.p.A., IRIDE, Monte dei Paschi di Siena and the Banca Popolare dell'Etruria e del Lazio.

A joint-stock company is currently being established which will be granted a mandate to provide integrated water services for a twenty-five year period.

2. ADAPT – Water school

Job training is the subject of much talk these days. The European Union wishes to experiment and come up with new approaches to training policies, in part through the creation of international networks which will permit people to share and compare their experiences. Ongoing training or professional development is a topic of particular interest, as it represents a highly effective tool for keeping personnel up to date on the continual changes in various sectors of industry and services and thus increasing the level of education of human resources.

Training is of fundamental importance in the water industry as it takes on its new form: training that confers a new way of thinking about service management and organisation, and which is therefore intended especially for management-level personnel: the people in charge of planning and implementing the activities required to comply with the new legislation.

By taking on an active role in this training, rather than just sitting back and waiting for the results, we not only gain a global view of the problems of the water industry, we also let participants into a network of relationships that is bound to increase a company's potential.

Training, in the traditional sense, is currently undergoing a complete transformation and renewal at all levels: from university and secondary school education to on-the-job training, which in Italy is largely the concern of the Regional and Provincial Administrations, though with public funds, and of course company training. These forms of training were, in the past, separate from one another, but they are now converging toward what could one day be an integrated educational system. Though this aim may be far in the future, it is noteworthy how the various educational systems are attempting to locate a space in which to communicate and exchange ideas with one another, clearly the outcome of a shared need.

In water service management, the new scenario created by the Galli Law requires people to acquire skills and exchange information in ways they never could have foreseen. The law requires the creation of a local administrative body managed through an assembly of the municipalities. Within this assembly, the municipalities themselves shall determine the work to be performed, and on the basis of this, come up with a financing plan, and determine a fee dynamic in relation to which they may set up an investment dynamic. They shall also be required to keep a check on the effectiveness and efficiency of the service provided by the appointed water service provider.

Water service management is being radically changed, and the transformation will require education, above all of management-level personnel in the municipally owned organisations and public bodies which are finding themselves caught up in this complex process of transformation.

This is why the educational system as a whole must plan for ongoing training, for professional development and updating; and this is why there must be dialogue among the various educational systems, to permit mutual recognition of the skills people acquire in various different training contexts.

Federgasacqua, a national federation with 303 members in the water industry, has been working for some time on a project in response to the new requirements: a «Water School» which will provide an opportunity for training, comparison, and synthesis of experiences and know-how acquired in various European Union nations. The «Water School» could provide training programmes for qualification and ongoing development of professionals in jobs specific to the water industry, with the aim of making water companies competitive on the national scene in accordance with the principles of the Galli Law.

The general aim of the «Water School» will be an overall increase in the level of education of human resources working in the water industry, to be achieved through study of particular content and processes identified in response to the requirements of the market and the economic context. The school will promote an entrepreneurial type of water management which aims to guarantee high technical and, above all, economic efficiency. The old idea of unlimited water resources is now outdated, and we must progress beyond it to a new model involving innovative administrative strategies that will be effective in the new scenario.

Another opportunity for action in this area is offered by the community programme referred to as ADAPT – Adaptation of the workforce to industrial change. Under this programme, Federgasacqua, in collaboration with A.M.G.A. S.p.A., is working on a pilot project in Liguria which is intended to usher in a new way of thinking about training, through side-by-side comparison of experiences of administration and management in different contexts. This type of training is prevalently addressed to middle- to high-level management, the people most directly involved in the current changes in administrative and managerial aspects of integrated water cycle management.

Their participation in a number of European Union programmes (Leonardo da Vinci, ADAPT) has given Federgasacqua and A.M.G.A. an opportunity to organise training initiatives with the participation of qualified experts, projects with relevance to the changes in progress in the organisation of work in the water industry: projects which respond to the real needs of companies.

The partners involved in the ADAPT – Water School project have carried out studies in the past aimed at identifying the requirement for employee training in the water industry in various European nations, and have produced a detailed map to use as the basis for development of training plans adaptable to various contexts.

The studies have demonstrated that, though the various nations in the European Union may have different training requirements, they all share common objectives: implementation of an adequate response to European regulations, improvement of the quality of the service they provide, and achievement of customer satisfaction.

To achieve these results, they need to organise their administrative bodies appropriately and identify the types of training that the various service providers will require. A transnational training system and standardised educational systems will be required to achieve this aim. This is why Federgasacqua and AMGA are participating in an international partnership under another European project, COMP-AQUA, co-financed by the Leonardo programme, which aims to develop a system of training credits and a flexible source of ongoing training for the water industry which will include opportunities for comparison and exchange of experiences.

Improving skills transfer in operation and maintenance to address scarce resources and increasing demands on water supplies

Optimisation du transfert de connaissances dans l'exploitation et la maintenance afin de résoudre les problèmes relatifs à la pénurie des ressources et à l'augmentation de la demande en eau

Farley M.
All Water Technology Ltd., UK

Abstract
Operation and Maintenance (O&M) is crucial to the successful management and sustainability of water supply networks and sanitation systems. Inadequate O&M leads to inefficient practice, ineffective services, and waste of precious resources. The WHO Operation & Maintenance Working Group's aims are to address these inadequacies and promote the improvement of O&M in water supply and sanitation services. The Group has been active in preparing a number of tools - guidelines, manuals and training packages. One of the tools is a training and resource package – A Leakage Control Training Package. The package has been tested in several countries and is now being distributed to selected agencies for training trainers. The paper outlines the philosophy of the Group and the principal tools developed. It contains a case study on testing one of the tools in Vietnam and Pakistan. The Group is now considering the its future role in disseminating the tools via regional and national training agencies.

Résumé
L'Exploitation et la Maintenance sont des éléments essentiels pour une bonne gestion et pour assurer la soutenabilité des réseaux d'alimentation en eau et des systèmes d'assainissement. L'utilisation de mauvaises techniques d'exploitation et d'entretien engendre une exploitation et des services inefficaces ainsi qu'un gaspillage de ressources précieuses. Les objectifs du groupe de travail sur l'Exploitation et la Maintenance (le Groupe) de l'Organisation Mondiale de la Santé sont de promouvoir l'optimisation de l'exploitation et de l'entretien dans les services de fourniture d'eau et les systèmes d'assainissement. Le Groupe a joué un rôle actif dans l'élaboration d'un certain nombre d'instruments à savoir des directives, des manuels et du matériel de formation. Un de ces instruments est le dossier de formation pour l'élimination des fuites. Ce dossier a été mis à l'épreuve dans un nombre de pays et il est maintenant distribué à certains organismes pour l'instruction de leur personnel de formation. Cette communication trace les grandes lignes de la philosophie du Groupe ainsi que les principaux instruments conçus. Elle décrit une étude de cas pour mettre un d'instrument à l'épreuve au Vietnam et au Pakistan. Le Groupe envisage maintenant, de distribuer les instruments, dans l'avenir, à travers des organismes de formation régionaux et nationaux.

1. The role of O&M in the water sector

Operation and Maintenance (O&M) is crucial to the successful management and sustainability of water supply networks and sanitation systems, whatever the level of technology, infrastructure, and institutional development. The O&M philosophy applies as much to handpumps as it does to more advanced treatment works and water distribution systems. It requires forward planning and operator understanding at all stages of water supply and sanitation processes. Transfer of knowledge and skills to ensure best practice is therefore crucial to operating and maintaining the system. The key factors and constraints which are directly related to O&M, and to the efficiency and effectiveness of water supply and sanitation services include:

- inadequate data
- insufficient funds and inefficient use of funds
- inappropriate system design
- poor management of water supply facilities
- low profile
- overlapping responsibilities
- political interference
- inadequate policies and legal frameworks

With a mandate to address these inadequacies, the World Health Organisation Operation and Maintenance Working Group (the Group) was launched at the Water Supply and Sanitation Collaborative Council meeting at The Hague in 1988. The Group is composed of specialists, academics and practitioners representing a wide range of global agencies and institutions. The Group's brief was to initiate co-operation between external support agencies and developing countries, enabling them to develop tools and methodologies for improving their O&M procedures. This would in turn optimise the efficiency of their water supply and sanitation services. Since that meeting the Group has met annually, and has addressed the key factors which are directly related to the performance of O&M and to the efficiency of services. Ten years on, the Group is disseminating the results of its activities and tools, and is in a strong position to advise institutions and policy makers on O&M procedures.

2. Improving O&M practices

The Group's aims are to promote the improvement of O&M in water supply and sanitation services, and to raise the level of awareness of its benefits. Its scope of work includes raising the profile of O&M through presentations, conferences and workshops, and by promotional literature, consolidating all the available guideline materials, and improving O&M by developing and applying new ones. It also aims to encourage exchange of information within a network of key players, led by members of the Group, and to guide water sector practitioners in the practical application of the tools.

The Tools

The Group has been active in preparing a number of tools - guidelines, manuals and training packages. Some of these are fully developed and have been tested in various countries, others are still in various stages of preparation. The tools are intended to be used as an integrated package for optimising O&M services. To allow the tools to have widespread use some of them have been translated into French, Spanish and Portuguese. The principal tools are:

Case Studies. Group members have contributed material to form a dossier of 22 case studies describing their experience of different projects and concepts of operation and maintenance in a range of countries. The studies have been compiled into one document, which has been distributed to interested agencies. The document is also available in French and Portuguese.

Status Assessment. Developed as a response to the lack of sufficient guidelines for assessing O&M services in both urban and rural areas, this tool combines a literature and information database with a methodology for assessing O&M status. The tool has been tested by the World Bank. A sanitation module is to be prepared.

A Guide for Managers of Urban Water and Sanitation Systems. This document examines the factors which may prevent systems working efficiently, and provides guidelines and solutions for optimisation. The same principles can be used to extend optimisation to fringe and poor areas. The guide was published in 1994 as an official WHO publication.

Leakage Control Training Package. This training and resource package provides another link in the chain of system optimisation. It adopts a logical and "user-friendly" approach to training water practitioners at a range of levels, from senior managers to leak inspectors. Each module can be varied in content depending on the depth of knowledge required for a particular level of trainee. The package has been tested in Vietnam and Pakistan, and is now being distributed to agencies for training trainers.

Optimising Water Treatment Plants. A practical approach to the improvement of water treatment plant performance, this document summarises the field experience in upgrading and improving a wide range of water treatment plants throughout the developing world. The guidelines show how both capacity and water quality of plants can be improved.

Training Package on O&M Management in Rural Areas. Developed for raising the level of training and for optimising the scarce resources for training activities in developing countries, this package provides hands-on material for conducting a course. The material can be adapted to local situations and makes use of local resource persons. The package has been tested in Namibia, and is available for use by external support agencies or water agencies in developing countries. French and Portuguese versions are also available.

Models of Management Systems. This document evaluates the factors which influence the development of O&M management systems for rural water supply and sanitation facilities in developing countries. It describes models used in eight countries and offers guidance to planners and designers in selecting the most appropriate one. The document is available in French and a Portuguese version is planned.

Linking Technology Choice to O&M. This manual is a practitioner's guide for selecting appropriate technology. The tool aims to inform users of the O&M implications, particularly cost factors, of each technology.

Network Survey. This is a new activity. The Group's aim is to prepare a manual on measurement and monitoring techniques for use in the field, to gain a better understanding of how a network is operating.

Operation and Maintenance of Handpumps. Another new tool, this is a manual to give guidance on the O&M requirements for handpumps, including technology choice, costs, and availability of spare parts etc.

3. Promoting O&M at practitioner level

Having reached a stage where it has successfully developed a number of tools to service the identified needs, the Group is now activating the next stage, which is to promote their application. Promotion is being achieved via an advocacy paper aimed at policy makers, by posters, and by an information sheet for each tool. Actions to disseminate the activities and tools of the Group include development and support of training initiatives and programmes, and providing guidelines for training centres to follow. To demonstrate the use and results of its tools the Group will organise implementation projects, initially at city level, to maximise technical support and to ensure success. Since 1997 The Group has organised its annual meeting to coincide with a national water sector Conference (e.g. New Delhi 1997), so that Group members are able to contribute to presentations and discussions at both venues

The Group's role in future years will reflect the need to:
- continue to achieve results and demonstrate them
- modify existing tools and expand the toolkit
- convince the policy makers of the benefits of O&M

With increasing privatisation and decentralisation, the Group will also concentrate on institutional development and management, and consider how O&M can flourish under new initiatives and privatisation. This needs the close involvement of the beneficiaries, who need to be convinced of the cost savings and improvement of services. The strength of the Group lies in the professional expertise of its members, and their ability to progress activities in their own countries and regions. However, the consolidation of these strengths and the continued success of the Group depends on networking between individuals both within and outside the Group. The Group has formed an O&M network of contacts, professionals and experts who are integral to disseminating and sustaining O&M activities. Focal contacts in training institutions and other organisations, linked to O&M by a project, or as an NGO, consultant or manufacturer, will be key actors in the dissemination of the tools. The training institutions will also act as implementing agencies, providing experts to present training courses and workshops.

4. Dissemination of tools – leakage control case study

Water loss management, including leakage control, is one of the crucial issues to be dealt with if an improvement in the efficiency and the effectiveness of water supply systems is to be achieved. Although the analytical techniques and institutional changes involved in developing a water loss programme are well established, until recently there has been little appropriate material to support training programmes addressing this issue. To fill this gap, WHO commissioned, as one of the Group's tools, the production of a leakage control training and resource package. It is intended to be a key contributor to operation and maintenance (O&M) and demand management programmes in developing countries.

Aims of the Training Package

The Leakage Control Training Package is a set of source material from which trainers can select material, and add their own material, to suit local conditions and network characteristics. The issues addressed in the package are grouped into specific modules, embracing the various aspects of water loss management and leakage control. The modules cover a wide range of techniques and procedures to analyse losses and to carry out leakage control activities. The intention is to select and assemble appropriate modules to suit the target audience, thereby acknowledging specific problems and constraints faced by different water practitioners. One of the features of the package is its flexibility, as it might be necessary to add or adapt modules and to update them periodically, without changing the overall framework of the package. The package contains text material, arranged in subject modules, together with notes for trainers on how to approach each module. A set of 35mm slides and overhead projector slides is included in each package. The initial target audience for the package in each region or country is expected to be a core group of trainers, who will derive their own courses, workshops, and seminars, and who will disseminate the methodologies to other practitioners in the water supply sector.

Organisation of the Modules

Water loss and leakage control activities have been grouped into modules. The criteria for grouping activities take into account the types of professionals involved in their execution (managers, engineers, technicians, etc.) and the characteristics and level of development of the water distribution network. Each module is self-contained so that it can be directed to different types of target populations. Therefore, the modules contain as much information as required for the achievement of their particular objectives, and to provide a high degree of flexibility in their assembly in preparation for training courses.

The training package adopts a logical and "user-friendly" approach to training water practitioners at a range of levels, from senior managers to leak inspectors. Each module can be varied in content depending on the depth of knowledge required for a particular level of trainee. For example, engineers and managers could explore in detail the institutional and financial aspects of leakage control, and would benefit from a cost benefit exercise to select and develop an appropriate policy. Engineers and technicians responsible for managing a system and detecting leaks would benefit from an understanding of these principles, but the main thrust of their programme would be based on those modules with a more practical and technical approach to system management.

An attempt has been made to make the training material as broad as possible, and it includes references to published work from papers, journals, seminar material, guidelines, and some textbooks. The trainer can quickly assimilate the appropriate material from each module, and assemble his/her own form of words for each module. The "textbook approach" to training has been deliberately avoided. Much of the material has been collated and reworked from training course notes and visual aids taken from a range of courses and workshops designed and presented by consultants in countries worldwide. It also gives descriptions of techniques and technology from which trainers can select user-suitable equipment.

Two essential components of the package are:
1) tailoring the course to a particular utility or community's requirements by finding out current system practice, problem areas, strengths and weaknesses etc.
2) involving the trainees in producing an "Action Plan" for their system, based on local knowledge and new skills gained from the course.
3) having a practical demonstration of the available equipment over a range of technology, followed by field demonstration (at a pre-selected and prepared site near to the course venue). All trainees should have the opportunity to handle the equipment and become familiar with procedures (e.g. programming a data logger, measuring a flow profile, locating a leak). This approach requires a little planning beforehand and during the first day of the course. Delegates are required to provide written information about their water supply, e.g. physical details like topography, population and demography, magnitude of losses, local influencing factors etc. as well as a description of the current leakage control policy, if any. At a suitable point during the first day, when the course programme, etc. has been introduced, individual trainees, or a representative of a group, are invited to make a brief presentation on the background and current practice of their water supply department.
This has three purposes:
1) it acts as an icebreaker;
2) it helps to stimulate discussion;
3) it provides local material and experiences.

From this point on, all the points raised can be dealt with in subsequent modules, again with trainee feedback and active participation. Local knowledge thus gained is invaluable for constructing an Action Plan at the end of the course, when the trainees will benefit from group work - comparison of different ideas and views will act as an additional stimulus to discussion. This process can be used anywhere in the world, whatever the network characteristics or the level of infrastructure development. There is always some improvement which trainees can make to their current leakage control procedures (repair of broken equipment, repair of visible leaks etc.)

Testing The Leakage Control Training Package

The package has been tested at several workshops, two sponsored by the World Bank in Hanoi and Ho Chi Minh City, Vietnam, and two by WHO in Lahore, Pakistan and in Haiphong, Vietnam. All the workshops emphasised the value of trainees analysing the strength and weaknesses of their systems, and then selecting and developing workable solutions, tailored to their needs, from the range of technical and economic options outlined in the package. The trainees were able to highlight the key steps for action to progress sustainable solutions.

Valuable lessons were learned from testing this particular tool:
1) the presentation style of the package, though attractive, was too bulky and expensive to produce for easy dissemination. This would be overcome by re-packaging the material in a hard-bound document rather than a ring-binder, and by replacing the slide material with a list of suggested topics from which trainers can compile their own slides.

2) some of the philosophy and technology of leakage control has naturally been superseded since the material was first drafted. The package will be updated with a review of state of the art techniques and equipment for leakage monitoring, detection and location.

3) The package does not contain enough of a 'practical' element, e.g equipment operating methods and procedures. A step by step guide to field operations and using the equipment will be included. These deficiencies are all being corrected in a revised version.

5. Conclusions

The tools developed by the WHO O&M Working Group have been designed to address the key factors which have prevented operation and maintenance procedures being introduced in the water supply sector. With increasing demands on diminishing resources it is essential that systems operate at their full potential. With the support of policy makers and decision makers, water sector practitioners, and the Operation and Maintenance Working Group network, the tools are being implemented to help achieve this goal. Testing and revising the tools is an essential step in their effective use.

Acknowledgements

The author acknowledges the continuing support of the Water Supply and Sanitation Collaborative Council and the World Health Organisation to the Operation and Maintenance Working Group, and the dedication of Group members in developing, promoting, disseminating, and implementing the tools.

Continuing education and lifelong learning for environment - water in Czech Republic

Formation de toute la vie dans le domaine de l´environnement et de l´industrie aquicole en République Tchéque

Kulhavy F.
Czech Republic

Abstract
One of the most significant targets is the necessity to expediently resolve the concrete utilisation all of agreements with European Union Directives regarding water management, in particular with the „EU Council Directive's on Water Management Policies and Water Management Planning". Given the conditions of our location, in which exploitable sources are absolutely prevalent due to the accumulation of rainfall, the requirement to fully coordinate the water management policy concept with the concept of landscape development and protection features the basic condition for the future fulfilment of EU legislation-related requirements. In practice, that is according to The Associated Agreement, The White Book and other documents from water-environmental area, to prepare informative, adductive, legislative and economical conditions for integration of Czech Republic into EU. On the base of thorough analyses of the present legislation in the sphere of management of water resources, protection of environment and landscape, there was a new proposal introduced by the group of scientists, ecologists, technicians and lawyers. From needs of specialist practice, follow necessary of profession Landscape engineer legalisation as successor earlier Cultural technical engineer (Kulturtechnik Ingenieur). This expert would be entrusted with full responsibility for permanently utilised landscape, it means as co-ordinator of all degrees of spatial landscape planning, approvement of relevant territorial plans, control and management of water-economic and hydromeliorative buildings operation, having important part in reach of ecological stability in landscape.
The several examples of education projects, exchange programmes and applications of infosystems at all planning, management, utilisation and protection process of landscape are illustrated by the paper. There are given projects which aim to support the quality and the capacity for innovation of vocational training, which can be developed through transnational exchange programmes.
As an conclusions of this paper, several basic ideas for innovation of our lifelong learning in the water industry, asses the possibility of sharing training programs at European level will be given.

Résumé
Cette période historique se caractérise par un besoin urgent d´accorder la législative tchèque avec celle de l´Union Européenne. Dans le domaine de l'industrie aquicole, il faut l´accorder surtout avec la „Directive du Comité de l´Union Européenne sur la Politique de l´Aménagement des Eaux et de la Planification Aquicole".
Dans nos conditions oú les sources exploitables sont représentées pour la plupart par des

eaux pluviales, le besoin de coordonner la conception de la politique de l´aménagement des eaux avec la conception de l´aménagement du territoire et de la protection de la nature est l´une des conditions indispensables pour remplir au future nos engagements envers les demandes de l´Union Européenne.
Vu la situation actuelle, La Compagnie Tchéque des Ingénieurs-Aménagistes du Territoire a décidé d´imposer la renaissance officielle de la spécialisation „Ingénieur-Aménagiste du Territoire" (ancien „Ingénieur de Technique et de Culture" - Kulturtechnik Ingenieur), qui comportera aussi la formation complémentaire des spécialistes existants dans le domaine de l´aménagement du territoire. Pendant la formation de toute la vie, les spécialistes recevront une formation dans plusieurs domaines, et il seront capables de coordonner les problèmes de l´aménagement des eaux avec l´environnement et avec la planification de l´aménagement du territoire.
Cette contribution traite l´étude de l´état actuel de l´enseignement dans les écoles d´apprentissage, secondaires et supérieures. Elle décrit aussi la publication éditée en 1998 par la Chambre Tchéque des Ingénieurs Autorisés et par l´Union des Entrepreneurs dans l´Industrie de Bâtiments, qui comporte 700 programmes de formation.
Vu le niveau actuel de connaissance professionnel des représentants de la régie, de ceux des bureaux d´études et ceux qui mettent les projets en pratique, la situation dans ce domaine peut _tre améliorée seulement par formation de toute la vie des ces personnes, augmentation de la production de leur travail, élimination de la corruption, et par cela minimalisation des risques de non-stabilité écologique par un travail non-qualifié.
A la fin de la contribution, les trois principes sont indiqués: accorder nos activités éducatives avec celles de la Communauté, réaliser des échanges réciproques des programmes de formation, organiser des séminaires internationaux des spécialistes, ce qui menera á adaptabilité supranationale et á mobilité des spécialistes, reconnaissance de leur qualification,.. etc.

1. Introduction

The present times are characterized by the necessity to expediently resolve our commitment to harmonise Czech law with valid European Union Directives, in particular with the „ EU Council Directive's on Water Management Policy and Water Management Planning". The most essential is preparation and adoption of the amendment of the Water act. In relation to the mentioned EU directive, a process of water management planning and institutional arrangement is initiated, so that the practice and supervision of water management correspond to the state under preparation in the EU. Water resources in our country are replenished, nearly exclusively, by recipitation. Their volume in the long term average amounts annually to 16.7 billion m^3, representing 1 621 m^3 per capita. Fluctuation from dry to wet years ranges between 8 and 19 billion m^3. About 30 to 50 % of water wealth can be considered exploitable.

The present legislature of the Czech Republic concerning landscape/nature care lacks the necessary inter-disciplinary co-ordination. Efforts made by our university representatives, as well as by other professionals, in order to remove existing shortcomings are greatly influenced by EU legislation, along with the declaration of the Conference of Ministers entitled „On Water and Sustainable Development" of March 1998. Therefore, the Czech Association of Landscape Engineers decided to assume a more active approach in asserting the renewal of the „landscape engineering" field (formerly called cultural-technical engineering - kulturtechnik ingenieur), including broadening the professional capacities of current water managers - mainly in terms of lifelong learning - into a multi-disciplinary field dealing with the co-ordination of water management and environmental landscape planning. The relevance of further education for these professionals derives from the following universally-significant facts:

- The necessity to introduce all professionals to European Community Law, as well as to newly-passed legislature;
- One of the most important current tasks is to educate both wide sections of professionals and state officials to make them take an active approach towards the environment. In practice this means that decision-making by directive, as applied so far in state administration, will have to be replaced by professional consulting on the methods of efficient economic management attainable in new conditions, and supplemented - depending on the available capacities - with economic stimulation. Only in extreme situations should the public interest be asserted by way of persuading the people about the universal (cross-societal) purposefulness of a certain measure. Further, it is necessary to guide all employees working at every level of administration and operation to make them feel personally responsible for the quality of handed-over or taken-over work, being aware of the danger of losing their jobs;
- By December 31, 1997 (see the 1997 Report) water management for the Czech Republic has the following parameters:

Total area of Czech Republic	7 886 566 ha - 100.0 %
of which - agriculture land	4 279 712 ha - 54. 3 %
arable land	3 090 609 ha - 39.2 %
grass and pastures	953 267 ha - 12.1 %
other agriculture land	235 836 ha - 3.0 %
forest land	2 631 802 ha - 33.4 %
urban areas	129 619 ha - 1.6 %
water bodies	159 393 ha - 2.0 %
other land	686 040 ha - 8.7 %
drainage area	1 081 500 ha - 13.7 %
irrigation systems	132 401 ha - 1.7 %
the exposure area of soil erosion	1 341 324 ha - 17.0 %
the inclination of arable land from 3° to 7°	43.4 %
over 7° 8.7 %	

Water management infrastructure in 1997:

stream important for water management uses:,	15 289 km, of which regulated 5 200 km
small water streams:	60 711 km, of which regulated 15 900 km,
drainage canals:	12 700 km,
waterways and water locks:	303 km waterways and 62 water locks,
water reservoirs, dams:	216, volume of 3 311 mil.m3,
small water reservoirs and fish ponds	24 000, volume of 625 mil.m3,
weirs:	approx. 1 000,
dikes:	612 km,
total number of employees at landscape water management:	4 187,
share of River Board corporations	3 704,
State Land Reclamation Authority	431,
Forest of the Czech Republic, state enterprise	52,
Drinking water supply systems:	3 500 public water supply systems, 51 000 km of water pipes,
sewerage systems:	approx. 24 000 km of sewers, 630 000 of house branch connections,
waste water treatment plants:	870 municipal WWTP and 1 600 plant WWTP,
hydroelectric plants:	1 131 plants, installed capacity of 2 135 MW.

- It derives from the above-given technical parameters of water management structures and equipment that there is a lack of professionals to ensure proper operation and careful maintenance. Due to the lack of funds the operational and maintenance work is often carried out by cheap labour with no relevant qualifications. Meanwhile, the low quality of such work seriously jeopardizes the environmental stability of the landscape.

2. Present education for environment - water

Water management-oriented courses offered by vocational schools have been cancelled due to adverse economic conditions and changes in the vocational education system. The high-school-type water management education offered by 7 schools as an independent course of study has also been terminated due to those schools' minimum interest in new students. As a consequence, those schools provide professional water management education only as a part of civil engineering and engineering structures courses. Three high schools oriented on industrial studies continue to offer this course of study. The Energy Institute, CZ, has introduced a two-year course of industrial water management for high-school graduates. 20 technically-oriented high schools provide – in terms of environmental education – a course on ecology. Three technical schools for high-school graduates have included these courses in their syllabi: water and waste management, landscape protection, and environment protection and administration.

The system of university education follows up on a long-term tradition and offers all types of degree studies (engineer, bachelor, doctoral, distant, etc. courses). In 1997 university graduates completed their courses of studies in the following fields: water management and water structures (at 4 Universities), water technology (at 1 U), environmental engineering (at 2 U), landscape engineering (at 4 U), technology of environmental protection (at 3 U).
The system of professional education is purposefully complemented, as stated below, with activities undertaken by excellent professionals in relevant professional chambers and societies. (The Czech Chamber of Authorized Engineers, The Czech Chamber of Architects, Czech Association of Landscape Engineers, etc.).

3. Continuing education and lifelong learning

Professional education, along with erudite research, represents a significant part of the basic intangible investments made by each and every developed country, and are a precondition for efficiency, successful exports, and competitiveness in world markets. These conditions can be fulfilled mainly through life-long education of qualified staff at all levels and of all professional specialities. This fact made the EU countries establish various international back-up educational programs (SESAM, SOCRATES, LEONARDO DA VINCI, etc.). This type of education has been incorporated in legislation, and is provided with significant economic support on the part of the state and the EU. The legal duty to ensure the renewal, broadening, and intensifying of professional education in our country is borne by universities and professional chambers and organizations. The universal responsibility for the environment and for the landscape establishes the state's indirect duty to ensure that all activities developed in the landscape are carried out in a professional, economical, and efficient way. Given the professional competence of people working for the state administration, as well as those involved in designing and implementing practical operational processes, the necessary requirements can be achieved only through well-targeted, life-time education of current employees, and by hiring people for vacant jobs strictly in compliance with their required professional capacity. This will enable us to enhance an employees' interest in life-long education, while simultaneously improving their labour productivity, and eliminating the danger of corruption (in the field of subsidized construction works), as well as that of jeopardizing the ecological stability of the landscape resulting from unqualified work.

In 1998 the Czech Chamber of Authorized Engineers, in co-operation with the Union of Business People in Civil Engineering, submitted to the professionals in their field a proposed program for life-long education of construction professionals (the program was called 'Construction Academy'), whose main goal was to offer all employees in the field of civil engineering a survey of further education possibilities, and to create - in cooperation with other participants in that process - a system to guarantee the professional quality of future educational programs. The respective publication features 700 educational programs divided in three qualification categories (higher qualification programs for engineers, architects and generalists; mid-level qualification programs for technicians and other specialists; and basic qualification programs for machinists, craftsmen, etc.), and further in four main types of building activities (for contractors, designers, investors, and administration clerks). Depending on the nature of the educational activities, the programs are split into four groups: preparatory, specialized, up-dating, and development groups.

Based on the introduced analysis of the current status of professional qualifications and the number of employees working in the field of water management and landscape engineering at the level of state administration, project designing institutions, research, implementing companies, and operations proper, the required tendency for life-long education can be defined as follows, considering our accession to the European Community that is under preparation:

- In harmony with the European Council's documents listed in the White Book ("Education and Learning: Towards Learning Society") it will be necessary to interconnect our educational activities with the EU's initiatives.
- It will be necessary to create organizational conditions for the supra-national promotion of co-operation and innovation in the field of continuing education and training among experts from all company types and institutions with the possibility to follow up on previously introduced programs as: FORMAQUA PILOT, MULTI-AQUA, COMPAQUA, etc.
- It will be necessary to promotion of the participation of Czech experts in European and international programmes in all fields related to environment and water.
- It will be necessary to promotion the development, exchange and dissemination of educational and training materials.
- It will be necessary to create conditions for the exchange of experts between our and other European companies and universities.
- It will be necessary to organize international workshops of erudite experts dealing with topical professional issues, and leading to supra-national adaptability and employees' mobility (recognition of their qualification, etc.).
- It will be necessary to extend the scope of distant and post-graduate studies for all qualification categories.
- It will be necessary to prepare expert courses at the level of a vocational school course for water specialists.

References

Ministry of Agriculture of the Czech Republic: Report on State of Water Management in the Czech Republic in 1997, 176 pp.
The Czech Chamber of Authorized Engineers, Union of Business in Civil Engineering: Lifelong Education of Construction Professionals, 1998, 136 pp

Theme E

Open and Distance Learning (ODL)

Enseignement à distance

Theme E : Open and Distance Learning (ODL)

Rapporteur: R.H.J. Sellin (UK)

After welcoming all speakers and participants to the session on ODL the chairman introduced the keynote speaker, Professor Takle from the USA, who spoke to his paper "**An Interdisciplinary Internet course on Global Change for present and future decision makers**". He stressed the importance of the feedback process in this type of course. It was organised in both Blocks and learning units, typically there are 13 learning units in a block. It is currently available in three languages, English, Spanish and Portuguese being targeted at the moment at the whole of the Americas. It is importune with this type of course delivered on the Internet that there is a daily update of the Student Dialogue Page, showing deadlines, scores, etc.
The speaker raised the problem of how to deal effectively with the rapidly developing science basis in a topic such as Climate Change. Also he stressed the importance of students attending a number of physical classes with their tutor present, and also carrying out at least one assignment in class as a performance security measure.

J C Deutsch then gave his paper "**Implementing 3N: a New technology, a New science, a New teaching method**". He illustrated the changing paradigms in teaching philosophy to take account of a new awareness of risk and how this can be handled; how the analysis of such situations are increasingly being run by numerical modellers, and how it is necessary to come to terms with the internet, in particular the use of interactive exercises and the ever increasing problem of on-line assistance.

Eric Waarners, "**Triton: a distance learning experience using modelling as a tool in teaching integrated urban drainage**", described developments in the TRITON course, first launched in 1992. This course contained three elements: (1) Sewer systems, (2) Sewer treatment works, (3) Receiving waters. This is now available on a limited network (Portacom), a predecessor of the Internet. The campus-wide version has been in use by e-mail during 1995-1997. It has been concluded that, overall, the objectives were over ambitious and as a consequence some of the educational aims were lost. A new form: "Integrated urban water quality management" is being prepared in which qualitative aspects only are being considered. An important lesson here has been learnt: **KIS – Keep It Simple**.

A fuller description of Jacques Ganoulis' work "**A New Paradigm**" appears in the annual report of ETNET Water-Environment . He poses the question: "What are the necessary skills required to reach this level of teaching? There are four dimensions to the objectives required here: Technical / Economic / Social / Environmental. He also made reference to TEST-EAU, an internet based assessment scheme for skills in water-based topics. This will be found in Key-Water.

T. Hofmann introduced "**MAMBO – development report of a distance learning system in the field of water and environment**". This hypertext-based product provides for different learner models; it also requires expanded capacity to cope with the writing, storage, mark-up and other novel aspects of the system. He stressed the need to create an effective publishing team comprising Author, Editor and Experts. This in turn needs an efficient management system, such as can be provided by traditional publishers' experience.

C Carbonnel described "**L'Or Bleu**", co-produced by STRASS Productions, an interactive encyclopaedia on CD-Rom treating water from a global viewpoint and in the light of the problems this raises. He stressed that this had been produced with secondary school and

family audiences in mind. It possesses all the familiar navigational and graphical facilities found in this type of product. It can be found on the website: www.our.planet.earth on the European server.
Robert Sellin described the CAL project "**CALWARE**" and in particular the title Fluid Mechanics, at the moment the most advanced of the four titles undertaken in this project. Extended trials had been carried out with a class of first year Civil Engineering students in spite of the uncompleted state of the title. Most students were found to adapt very quickly to this type of teaching but a minority lacked the intellectual maturity to tackle a new and difficult subject in this independent manner. The only obstacle to completing this was the inadequate funding available.

L. Tuhovcak introduced an **ODL** scheme devised and used at the Technical University of Brno – Faculty of Civil Engineering. This covers a wide range of activities in the general area of **Water Management**. Considerable experience has been obtained with this teaching system, supported by the Tempus-Phare programme of the EU, mainly in the postgraduate courses in the Faculty. He commented that water consumption in Moravia has fallen by 30% since the widespread introduction of metering.

J. Delleur explained "**Student, Scholar and Multimedia exchanges in the environment/ water field**". This concentrates on the need for effective co-operation between the United States and Europe in the water area. The current exchange programme is between five Universities in Europe and four in the USA. The lead sites are the VUB in Belgium and Purdue University for the USA. ODL teaching material is now being prepared to extend the impact of the scheme. Language training, if appropriate is an integral part of the project, as is also cultural preparation prior to exchange visits. All the usual problems were met, notably those of student motivation and costs.

R. Verhoeven introduced their paper "Towards rivers where nature and humans meet". This study was set up under the Socrates Pilot project scheme and was concerned with European collaboration between seven partners from six countries to consider river restoration. It took the form of the collection and comparison of typical case studies of integrated water management schemes in each of the countries represented. It proved of great interest to compare how the objectives were set and met in the different countries considered. This resulted in the development of a more global view on river management projects in Europe.

Keynote lecture

An interdisciplinary Internet course on Global Change for present and future decision-makers

Cours interdisciplinaire avec l'Internet sur le changement global destiné aux décideurs d'aujourd'hui et de demain.

Takle E. S., Taber M. R., Fils D.
International Institute of Theoretical and Applied Physics,
Iowa State University, USA

Abstract
We have developed an Internet-based university course addressing issues of global environmental change. The course provides access to recent scientific literature and structured learning activities on a wide range of global environmental issues. An electronic dialog allows on-line discussion organized by topic. A web-based laboratory allows students to test hypotheses and conceptual models by accessing and running a research-quality model of soil-vegetation-atmosphere interactions. Each student has a personalized password-protected electronic portfolio for managing all interaction with the course and the laboratory. A global learning resource network has been established to facilitate multi-directional flow of information and ideas from many countries on global change issues.

Résumé
Nous avons mis au point un cours universitaire basé sur l'Internet traitant du changement global relatif à l'environnement. Le cours permet un accès à la littérature scientifique récente et à des activités d'apprentissage structurées traitant sur une grande gamme des questions globales et environnementales. Un dialogue électronique permet un discours relié à l'informatique classé par thème. Un laboratoire créé à partir du réseau mondial permet aux étudiants de tester des hypothèses et des schémas conceptuels en ayant accès et en exécutant un schéma d'un très haut niveau scientifique des interactions entre le sol, la végétation et l'atmosphère. Chaque étudiant a un portefeuille électronique protégé par un mot de passe personnalisé pour gérer toutes les interactions avec le cours et le laboratoire. Un réseau fournissant des moyens pour l'apprentissage global a été établi pour faciliter l'échange des informations dans plusieurs directions et des idées, provenant de nombreux pays sur les questions du changement global.

1. Introduction

The spherical shell of less than 10 km thickness at the surface of planet Earth is unique in that only here does water exist in abundance in all three phases. Solar illumination of this fragile, water-rich shell has given rise to the only known place in the universe for life as we know it. From the philosophical considerations of the origin of life to the practical need for washing dinner plates, water permeates every element of human existence. Professor G. O. P. Obasi, in an address to the American Meteorological Society Annual Meeting in Dallas Texas (USA) (Obasi, 1999) reported that freshwater resources now have fallen below the 1000 m^3 annual per capita level (a common benchmark for water scarcity) in 22 countries, and that pressure on the world's water resources will continue to increase. Freshwater shortage and water quality will be dominant problems in the next century and may jeopardize all other efforts to secure sustainable development (Obasi, 1999).

Almost 20 years ago the U.S. government intelligence services estimated that there were at least 10 locations in the world where war could break out over dwindling shared water resources (Starr, 1991). Starr goes on to point out that the United Nations International Children's Emergency Fund (UNICEF) estimates that 40,000 children worldwide (mostly on the African continent) are dying every day from hunger or disease caused by lack of water or from contaminated water and that by the end of this century (less than one year), almost 40% of the African population will be at risk of death or disease from water scarcity or contamination.

There is need for authoritative information on rapidly evolving international, national, and local dimensions of water issues and the role of water in the larger issue of global environmental change and global sustainable development. This need requires networks of researchers and educators to transform research results into educational materials appropriate for multiple-target audiences and available over the Internet. We have developed an Internet Global Change course for senior university undergraduates and beginning graduate students covering a broad range of topics in the general area of global environmental change, including a unit on water issues.

2. Access to information

The gravity of global environmental problems, including availability of adequate supplies of safe water, calls for prompt international action. However, not all nations have equally timely access to information relating to the scope and magnitude of changes, of both natural and anthropogenic origin, occurring in our global environment. As a consequence, scientists may not have adequate time to evaluate the impact of changes or the political response to these changes for their own countries. For a number of reasons, global change materials, including materials on water, are not always accessible to instructors in university courses in a timely manner. In some cases, access to research reports is limited due to cost of journals. In other cases, instructors, facing increased pressure to do more instruction with fewer resources, do not have the time to keep abreast of rapid advances outside their own specialty. Because of the closely linked relationships of different components of the earth system, however, graduates of our universities will increasingly be involved in careers that bring them into multi-disciplinary issues of global change. This creates a need for improved communication among students from different disciplines and different geographic regions, particularly for addressing the policy and human dimensions of global environmental change.

We have developed a Global Change course on the Internet for senior undergraduates and beginning graduate students. The course also would be suitable for practitioners and decision-makers seeking an overview of how the water environment is interconnected with all global change issues. Parts of the course have been used for introductory environmental science courses at other universities and colleges. And numerous e-mail messages and Internet requests indicates it is being used by high-school students, writers, government researchers, publishers, and many others.

3. Course description

Objectives of the Global Change course are: (a) to demonstrate the interconnectedness of the earth's environmental system and to explore the scientific evidence for changes in the global environment, (b) to instill in students the value of peer-reviewed literature on global-change issues, and (c) to engage students, by means of the Internet, in dialog among themselves, with outside experts, and with students from other countries on the scientific, economic, social, political, and ethical implications of these global changes.

The course consists of 3 blocks each having 13 individual but interconnected learning units spanning the spectrum of global change issues. Each learning unit consists of a set of objectives, a learning narrative (transcript of a conventional lecture) including images and links to other information sources, a quiz over the learning narrative that is automatically and instantaneously graded and recorded, links to other related sites, a link to the search engine for the Iowa State University Library, a "question to ponder" as a post-classtime discussion starter, and a link to a publicly available post-classtime electronic dialog on the learning unit topic. Some units have interactive on-line experiments for student to complete and report results.

Students manage their interaction with the course through their personal Internet portfolios in which are archived all their electronic submissions, instructor's grades and comments, and responses of other students, faculty, or others to electronic dialog comments. Students' reviews of research papers are posted on the web and linked where appropriate to learning narratives. Cooperative learning is implemented through summaries, created and posted on the web by small groups, of classtime presentations and discussion. These serve as a catalyst to facilitate post-classtime discussion.

The interactive Internet-based electronic dialog provides an organized framework for student discussion on each of the 42 Global Change topics. Students enter questions, comments, newfound information, new websites pertinent to the topic, and responses to other students. Outside experts are invited to enter the dialog by answering student questions and pose questions for students to ponder. Dialog from each class is archived and made available to later classes. This allows future students to take advantage of particular interchanges, especially with outside experts. An ancillary advantage of the student electronic dialog, particularly for international implementations, is that students from several countries are able to dialog with other students on trans-national issues such as water policy. The class environment allows students from several countries to explore the concept of collective and enlightened self-interest related to regional and global environmental problems. These discussions can build understanding and friendship among students who eventually assume positions of leadership in science and government in their respective countries, that may contribute to more effective resolution of future international water issues.

4. On-line laboratory

Learning is enhanced if students have access to interactive tools that allow hypothesis testing and concept evaluation. Some physical processes, such as components of the hydrological cycle, are complex and not easily visualized or conceptualized by simple diagrams or numerical relationships. For instance, the evapotranspiration of water from plants depends on meteorological conditions, soil conditions, and structure and phenological stage of the plants. A complete description of how water is cycled through atmosphere, soil, and plant requires information about the physical and biological processes for the plant, physics of water movement in soils, transport of water vapor in the atmosphere, and how direct and diffuse solar and infrared radiation are distributed within the plant canopy. So-called soil-vegetation-atmosphere-transfer (SVAT) numerical models capture these processes and are now being used as submodels for global and regional climate models for studying climate change (Sellers, et al 1986; Sellers et al, 1996).

A prototype on-line laboratory has been developed to accompany the Global Change course. In our present implementation of this laboratory, students access over the web, through their portfolios, a research-quality SVAT model, SiB2 (Sellers et al, 1986), resident on a computer workstation at our institute. The student performs a prescribed but open-ended set of experiments to examine atmosphere-soil-biosphere-hydrosphere interactions, with input and output (including graphical material) being managed through the online portfolio.

5. Global learning resource network

Water issues are regional and global, and understanding water issues requires information on a variety of meteorologic, hydrologic, geographic, technical, cultural, and political topics. Information originating in a particular country may not be applicable in other countries due to cultural, geographic, or political reasons. Solutions to trans-national water problems require international teams of scientists and decision-makers that are fully informed of the latest research and technical information and who have the mutual understanding and cultural awareness to achieve workable solutions. A goal of the Global Change course is to establish a Global Learning Resource Network of global change information and people as a managed but organic network for multi-directional flow of information and ideas relating to global environmental issues. The first step in forming this network was taken in September and October 1998 when a team of 13 Latin American university faculty members assembled at Iowa State University for a planning meeting. This team developed the topology for dissemination of the Global Change course and began translating the Global Change course to Spanish and Portuguese for coordinated delivery in 16 Central and South American countries belonging to the Inter-American Institute for Global Change Research (IAI). Wide and multi-lingual delivery is expected to begin in Fall 1999. To be truly effective, this global network needs wide representation from both developed and developing countries. Additional participating countries are being sought for this network.

Acknowledgements
Development and implementation of the course and extension to Latin American universities is supported under funding from Iowa State University, UNESCO, the US National Science Foundation, the Global Environmental Facility, and the Inter-American Institute for Global Change Research. Particular acknowledgment for the Latin American extension is given to Prof. Eduardo Banus, Director of the Project RLA/92/G34 - IAI/UNDP/GEF/WMO who provided, not only funding for the Latin American workshop, but perhaps more importantly a vision for the multi-directional network.

References

Obasi, G. O. P., 1999: Hydrology and water resources: A global challenge for WMO. 14th Conference on Hydrology, American Meteorological Society, Dallas, Texas, 10-15 January 1999, 330-332.

Sellers, P. J., Y. Mintz, Y. C. Sud, A. Dalcher, 1986: A simple biosphere model (SiB) for use within general circulation models. *Journal of the Atmospheric Sciences*, 43, 505-531.

Sellers, P. J., D. A. Randall, G. J. Collatz, J. A. Berry, C. B. Field, D. A. Dazlich, C. Zhang, G. D. Collelo, and L. Bounoua, 1996: A revised land surface parameterization (SiB2) for atmospheric GCMs. Part I: Model formulation. *Journal of Climate* 9, 676-705.

Starr, J. R. 1991: Water wars. Foreign Policy 82 (Spring), 17-36.

3N: une Nouvelle technologie, une Nouvelle science, une Nouvelle pédagogie

Implementing 3N : a New technology, a New science, a New teaching method

DEUTSCH J. C., TASSIN B.
Cereve, ENPC, France

Résumé
Depuis 7 ans un cours d'hydrologie urbaine a été ouvert à l'Ecole nationale des ponts et chaussées. Ce cours est l'un des premiers en France qui enseigne les concepts, les outils et les méthodes de cette nouvelle discipline scientifique, qui a émergé aux USA dans les années soixantes. Enseigner une telle discipline nécessite des approches pédagogiques spécifiques, et depuis la création du cours, une pédagogie de projet a été mise en oeuvre. C'est dans le cadre de ces innovations pédagogiques que les nouvelles technologies de l'information sont utilisées depuis 2 années. Elles permettent non seulement la diffusion de supports écrits, mais aussi servent de support à l'évaluation du processus d'apprentissage, à différentes échelles depuis les exercices d'application jusqu'au projet final du cours. Ces outils favorisent l'établissement d'une pédagogie de type " bottom-up " et permettent un rythme d'apprentissage individualisé. Les évaluations réalisées par les étudiants se sont révélées positives.

Abstract
Since 7 years, a course in urban hydrology has been open in the third year of the curriculum of Ecole nationale des ponts et chaussées. This course is one of the first in France whose aim is to teach the concepts, the methods and the tools of a new science : urban hydrology, which has come to light in the USA during the sixties. Teaching such a discipline requires specific pedagogical methods and from the beginning, a project oriented course was developed. To help in this orientation, new information technologies (NIT) are used since 2 years, not only for the diffusion of written materials but also for testing the learning process at various scales, from drill-exercices to project analysis. NIT make easier a bottom-up pedagogical approach and allow the student to be nearer of his own rythm of learning The assessment of these methods by the students sounds very positive.

Depuis 7 ans, l'Ecole Nationale des Ponts et Chaussées a créé un cours sur l'hydrologie urbaine. Il a été un des premiers en France à avoir comme objectif d'enseigner les outils les méthodes et les concepts d'une nouvelle science issue de l'assainissement : l'hydrologie urbaine. Ce cours est destiné aux élèves de troisième année de l'ENPC ainsi qu'aux étudiants du DEA Sciences et Techniques de l'Environnement commun à l'Ecole nationale des ponts et chaussées, l'Ecole nationale du génie rural, des eaux et des forêts et l'Université de Paris-Val de Marne. Il dure 30 heures réparties en tranches de trois heures sur un trimestre.

1. Le contexte scientifique et technique

Selon Eurydice 92 (1997), l'hydrologie urbaine s'intéresse à la partie du cycle de l'eau affectée par l'urbanisation ou affectant le fonctionnement de la ville. Le traitement et la distribution de l'eau potable, même s'ils sont indubitablement liés à la gestion urbaine de l'eau, ne sont cependant pas généralement rattachés au champ de l'hydrologie urbaine. Elle est apparue à la fin des années 1960 aux USA, suite à prise de conscience par un certain nombre de chercheurs de la demande sociale existante à propos de l'hydrologie en milieu urbain. Dans la lancée d'un premier programme de recherche lancé par l'American Society of Civil Engineers en 1967, des premières rencontres internationales ont été initiées à partir de 1972. Depuis, a été créé un comité commun sur l'hydrologie urbaine à l'Association Internationale de Recherches Hydrauliques (AIRH) et à l'Association Internationale pour la Qualité des Eaux (AIQE) qui organise tous les trois ans une conférence internationale (ICUSD).

L'hydrologie urbaine s'est constitué en domaine spécifique pour essayer de résoudre les problèmes particuliers de l'écoulement des eaux dans les espaces urbains :
(a) faible dimension des bassins versants, et par voie de conséquence particularités des pluies critiques (pluies courtes et violentes liés à des phénomènes convectifs)
(b) forte imperméabilisation des sols en zone urbaine qui modifie la nature du ruissellement quantitativement (diminution des pertes à l'écoulement, accélération du mouvement de l'eau) et qualitativement (modification de la nature des polluants entraînés par l'eau)
(c) caractère artificialisé du réseau hydrographique
(d)vulnérabilité des espaces urbains face au risque d'inondation et importance des enjeux financiers, environnementaux et sociaux
(d) modification de la perception des usages de l'eau
(e) évolution rapide de l'occupation des sols limitant les possibilités d'utilisation des méthodes statistiques fondées sur l'observation du passé pour prédire l'avenir
(f) partenariat étroit avec le gestionnaire de l'assainissement.

Parallèlement à l'émergence de l'hydrologie urbaine, et en relation avec le développement des recherches, le métier de l'assainissement s'est complètement transformé dans les trente dernières années. J.F CYR *et al*, (1998).ont décrit les principales modifications qui sont intervenues.

L'ingénieur en assainissement des années 1960 avait comme formation principale le génie civil. La diversification des objectifs fixés au service d'assainissement, et en particulier la prise en compte de la protection du milieu naturel, ont amené les spécialistes à faire appel à une grande variété de disciplines scientifiques pour concevoir, réaliser et gérer des réseaux d'assainissement.

De nouveaux outils scientifiques, méthodologiques et technologiques sont utilisés. La modélisation des phénomènes s'est développé. Parallèlement , des moyens de mesure ont été installés, aussi bien pour la pluviométrie que pour la débitmétrie et l'estimation de la qualité des eaux.

En ce qui concerne les eaux pluviales urbaines, on est passé, dans les années 70, d'un objectif d'évacuation des eaux le plus loin et le plus vite possible des agglomérations, à un objectif de rétention de ces eaux au plus près de leur production. Ceci a conduit au développement, d'une part, des techniques alternatives au réseau d'assainissement, et d'autre part, à la notion de gestion en temps réel des systèmes d'assainissement. Aujourd'hui l'approche normative de la conception des réseaux d'assainissement : protection contre la pluie décennale, est en passe d'être abandonnée pour une approche "gestion du risque" qui peut concerner l'ensemble des acteurs intervenant sur la ville. Pour essayer de faire comprendre les causes de ces transformations, leurs évolutions futures et ainsi de mettre les étudiants en mesure, autant que possible, de s'adapter rapidement à un milieu professionnel, les objectifs pédagogiques du cours ont dû être complètement explicités.

2. Le contexte pédagogique

Les objectifs immédiats ont été précisés de la manière suivante : analyser les différents objectifs d'un système d'assainissement, présenter l'ensemble des interactions entre l'assainissement et les autres techniques urbaines, détailler les récents et nombreux progrès des méthodes de conception et de gestion des ouvrages. Tous ces éléments sont présentés dans le cadre d'une gestion globale des interactions eau-urbanisme.

La trame du cours est basée sur une idéalisation de la démarche d'un ingénieur polyvalent vis à vis des problèmes d'assainissement. Après une introduction consacrée à une évolution de l'assainissement au cours du temps, une première partie est consacrée à la conception des réseaux en insistant sur les problèmes de choix qui peuvent se poser. Dans une deuxième partie on aborde les problèmes d'exploitation et de gestion.

La nécessité de faire passer de multiples messages sur : l'utilisation de connaissances multidisciplinaires, le bien-fondé d'une approche intégrée des problèmes et l'obligation d'avoir une réflexion sur les dernières connaissances développées dans les centres de recherche nous a conduit dès le début à privilégier la réalisation d'un projet final par les étudiants.

Celui-ci est conçu de manière à permettre d'utiliser les outils, les méthodes et les conceptions qui ont été présentées dans le cours et d'expérimenter les adaptations nécessaires à un problème local. Les documents fournis se rapprochent le plus possible de ce dont on peut disposer dans le cadre d'un projet réel. L'idéal, qui a pu être réalisé cette année, est de travailler sur un site réel que les étudiants peuvent visiter.

Ce projet final est le dernier maillon d'une chaîne d'exercices d'application de plus en plus complexes qui doivent faciliter l'acquisition des connaissances. Il permet aussi de montrer que le concept de "la meilleure solution" n'existe pas dans l'absolu, et dépend des objectifs qui sont fixés au départ. Ce qui est jugé, c'est la cohérence du projet, et non pas la solution présentée.

En fait, l'objectif large que nous avons fixé au cours est de mieux faire comprendre, par une mise en pratique partielle, ce qu'est le métier d'ingénieur, compris comme étant la capacité de fournir une ou des solutions à des problèmes posés de manière technique par la demande sociale, en utilisant toutes les ressources de l'état de l'art et de la connaissance du contexte local.

3. Mise en œuvre des objectifs pédagogiques

Les NTIC constituent un outil de choix dans la mise en œuvre pratique des objectifs pédagogiques du cours. En effet, elles permettent :
(a) de compléter les heures de cours entre étudiants et enseignants, qui imposent comme dans le théâtre classique l'unité de lieu et l'unité de temps, par un contact supplémentaire via des outils de communication de type forum ou courrier électronique ;
(b) l'ouverture vers l'environnement extérieur via des liens soit avec des professionnels en prise avec la matière enseignée, soit avec des sites pertinents abordant la matière enseignée;

(c) un rythme d'apprentissage personnalisé. Dans le cas de formation multidisciplinaires, dans lesquelles les étudiants sont souvent de niveau hétérogènes, elles sont la seule manière d'organiser des séances d'application du cours que chaque étudiant suit à son propre rythme ce qui ne peut être envisagé dans une séance de travaux dirigés ;
(d) de mettre à disposition des supports d'information non seulement de type texte ou hypertexte, mais aussi de type image, son ou vidéo ;
L'implémentation des NTIC dans le cadre du cours d'hydrologie urbaine s'est faite principalement au travers de la conception d'un site Internet (http://www.enpc.fr/cergrene/HomePages/tassin/hydurb99/index.html). Par ailleurs, les séances de cours sont complétées par des projections vidéo, en tant que de besoin, qui ne sont pas reliées au site Internet.

Ce site est constitués des éléments suivants :
(a) des supports de cours ;
(b) des exercices d'application de points particuliers du cours ;
(c) un exercice de synthèse sur la conception des réseaux d'assainissement, sous forme d'un jeu de rôle : Vilegout ;
(d) un site d'information et d'échange dédié au projet final du cours.

Supports de cours

Les supports de cours sont disponibles à la fois au format hypertext et sous un format imprimable. Ils sont constitués de notes de cours et de supports des présentations orales. Pour les enseignants, ces formats facilitent la mise à jour des documents. Ils permettent aussi de limiter les impressions, l'usager n'imprimant que les parties des documents dont il a expressément besoin. Enfin, les étudiants passant une part importante de leur scolarité en stage, ces documents qui sont en accès libre, peuvent éventuellement être consultés de l'extérieur en cas de besoin. Cet usage peut aussi intéresser un ingénieur dans le cadre de son activité professionnelle, à l'instar d'une formation continue.

Ces supports de cours sont complétés par une liste de liens vers des sites Internet liés à la thématique de l'hydrologie urbaine. Les sites existant sont majoritairement d'origine nord-américaine et peuvent être classés en 3 catégories :
(a) sites à vocation de recherche. Emanant d'institution de recherche ces sites contiennent des descriptifs de programmes ainsi que des rapports de recherche et des prépublications scientifiques qui peuvent utilement compléter les enseignements et/ou aider les étudiants dans leur projet ;
(b) sites à vocation d'enseignement. Moins nombreux que les précédents, la plupart ne contiennent que peu d'information en dehors de programme d'enseignement. Quelques uns toutefois ouvrent l'accès à des supports de cours ;
(c) sites à vocation d'information. Emanant pour la plupart de collectivités locales, ils exposent les réalisations locales effectuées dans le domaine et constituent une base d'exemple très diversifiée pour les étudiants. On y trouve aussi des consignes à la destinations de la population locale pour minimiser les impacts des rejets d'eau usées ou de pluie sur les milieux récepteurs.

Exercices d'application du cours

Dans sa conception actuelle, le site présente quatre exercices qui ont pour objectif de favoriser l'assimilation du contenu de l'enseignement par l'étudiant. Un effort de réflexion est exigé quant à l'analyse des résultats obtenus. Pour chaque exercice, l'énoncé est exposé sur le site, ainsi qu'une aide (rappel de cours, principe de mise en œuvre de concepts, règles de dimensionnement) à l'exercice. Les réponses sont transmises par courrier électronique au responsable de l'enseignement par l'intermédiaire d'un formulaire. Au delà d'une date limite convenue à l'avance, le formulaire de réponse est remplacé par la solution de l'exercice. Par ailleurs, chaque exercice fait l'objet, en cours magistral, d'un commentaire des principales erreurs rencontrées.

Exercice de synthèse sur la conception des réseaux : jeu de rôle Vilegout
Ce travail de synthèse n'a pas pour objectif de vérifier si les techniques de dimensionnement des différents éléments constitutifs d'un réseau d'assainissement ont bien été acquises. Les exercices d'application du cours doivent y répondre. Il vise à mettre les étudiants face à une situation réelle mais simplifiée, de conception d'un réseau d'assainissement sur une collectivité locale de taille moyenne et les oblige à se poser les questions les plus pertinentes face au problème posé, pour les amener à proposer une solution qui dans la pratique servirait de ligne directrice d'action au bureau d'étude qui serait chargé du travail. Il s'agit donc principalement d'un travail de questionnement de proposition d'hypothèses qu'il s'agit de valider ou d'invalider. Il fonctionne comme un jeu de rôle, dans lequel les enseignants sont les animateurs Il se compose des étapes suivantes (Figure 1):

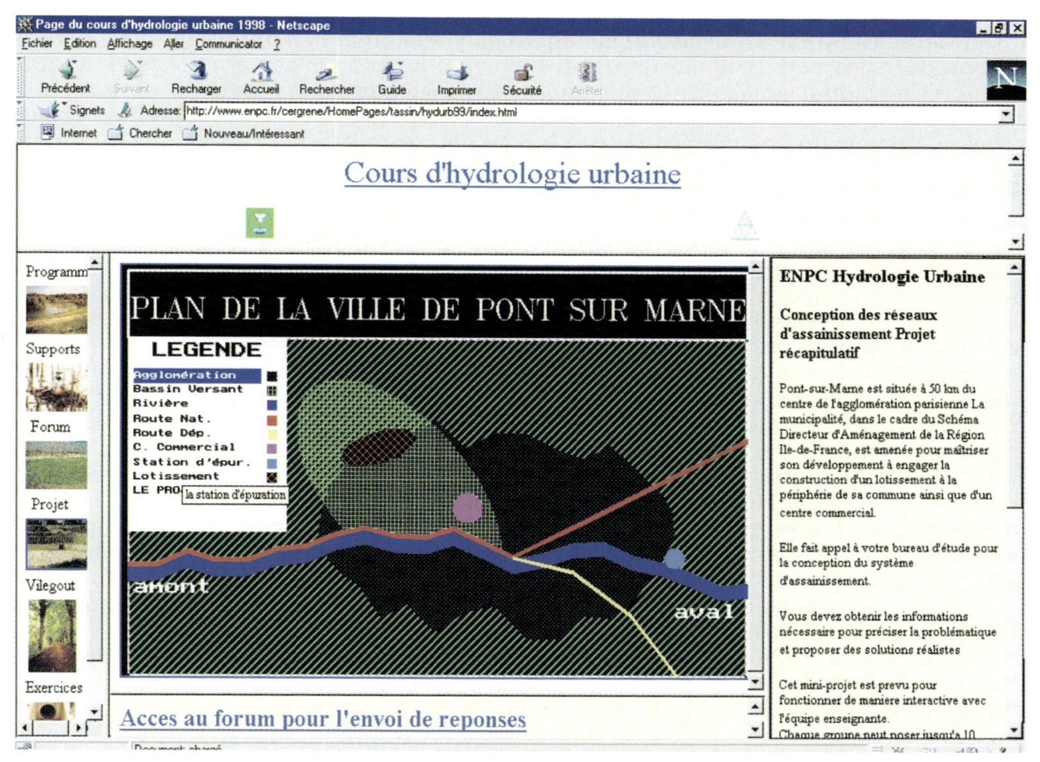

Figure 1. Page d'entrée sur le site du cours d'hydrologie urbaine de l'exercice de synthèse sur la conception des réseaux d'assainissement

(a) prise de connaissance du projet et des données de base, à partir de pages hypertextes;
(b) établissement de l'échange entre étudiants et enseignants. Les étudiants regroupés en groupes de 3 peuvent poser jusqu'à 10 questions au total à l'équipe enseignante sur le projet. Ces questions sont posées au travers d'un forum électronique. Ainsi chaque

groupe a accès aux questions posées par les autres groupes. Chaque question reçoit une réponse sous 24 heures, y compris durant les fins de semaine ;
(c) Proposition d'une solution, postée sur le forum. Comme pour les exercices, les solutions proposées sont ensuite analysées par l'équipe enseignante, en cours magistral.
Le temps laissé aux étudiants pour réaliser le projet, permet une bonne maturation, accélérée par le travail en groupe. De plus, le forum électronique, permet de combiner la formulation écrite, donc la formalisation de la pensée et l'interactivité. Les solutions proposées par les étudiants sont en général très riches et innovantes.

Aide à la conduite du projet final

Le projet final atteint un degré d'intégration supplémentaire par rapport au projet précédent. Sur un cas réel, situé à proximité des locaux de l'ENPC, les étudiants doivent résoudre un problème d'hydrologie urbaine, comportant à la fois des volets touchant à l'aspect quantitatif et des volets touchant à la qualité des eaux et à l'aménagement des rivières urbaines. On insiste au travers de cette variété de facettes du projet sur la nécessaire approche globale des actions d'aménagement touchant à l'environnement. Au cours de ce projet, les aspects quantitatifs sont traités à l'aide d'un logiciel utilisé en milieu professionnel, qui nécessite un investissement en temps important mais permet de proposer des solutions qui dépassent le cadre de projets scolaires. Les NTIC sont aussi utilisées pour améliorer le dialogue entre enseignants et étudiants, au travers là aussi d'un forum électronique, sur lequel les étudiants peuvent questionner l'équipe enseignante au sujet des difficultés qu'ils peuvent rencontrer.

4. Evaluation et conclusion

Le cours d'hydrologie urbaine de l'ENPC teste les NTIC pour la deuxième année en 1999. Il est donc encore nécessairement dans un état expérimental et évolutif. Les évaluations réalisées jusqu'à présent et basées à la fois sur les avis des étudiants et ceux de l'équipe enseignante permettent de dégager les conclusions suivantes :
(a) Les différents éléments constitutifs des exercices permettent un apprentissage facile. Les exercices n'empiètent plus sur les heures de cours magistral qui peuvent être pleinement utilisés pour des cours magistraux ;
(b) la souplesse de fonctionnement des deux projets, qui repose sur l'usage d'un logiciel et des NTIC permet aux étudiants d'exprimer leur créativité et leur capacité d'innovation. Cette créativité est renforcé par la possibilité d'accès via les liens internet à un grand nombre d'étude de cas exposées de manière très explicite ;
(c) l'usage des forum électronique permet de faciliter les échanges entre étudiants et enseignants, de les densifier en dehors des heures d'enseignement sans engendrer de contrainte d'emploi du temps ;
(d) Aujourd'hui, la plupart des étudiants sont très intéressés par cette approche et n'éprouvent pas de difficulté d'utilisation ;
(e) L'approche que nous avons développé, loin d'aller vers des universités virtuelles désincarnées, dans lesquelles l'enseignant reste un être énigmatique, est basée sur l'utilisation de NTIC pour renforcer les liens entre enseignants et étudiants, même si le face à face n'est plus nécessaire. Les différentes niveaux d'évaluation, depuis les exercices d'application jusqu'au projet global, permettent de s'assurer de la bonne assimilation des connaissances et de vérifier que les étudiants peuvent traiter à la fois des problèmes techniques ponctuels et des problèmes beaucoup plus généraux d'hydrologie urbaine à l'échelle d'une ville. . Il est possible dans cette optique d'imaginer un développement de ce concept, qui pourrait aller jusqu'à une réduction, voire une suppression des cours magistraux et qui faciliterait la transition entre milieu scolaire et milieu professionnel.

Références

Eurydice92 (1997) Encyclopédie de l'hydrologie urbaine et de l'assainissement, (coordinateur B. Chocat).Technique et Documentation, Paris.
Cyr, J.F., Marcoux C., Deutsch J.C.& Lavallée P. (1998) L'hydrologie urbaine : nouvelles problématiques, nouvelles approches de solutions. *Revue des sciences de l'eau,* **11** (numéro spécial), 51-60.

Student, scholar and multimedia exchange in environmental / water resources engineering and sciences

J.W. Delleur[1], A. Van der Beken[2], E. R. Blatchley III[1]
[1]. School of Civil Engineering, Purdue University, West Lafayette, IN 47906, USA
[2]. Interuniversity Programme in Water Resources Engineering, Vrije Universiteit Brussel, 1050 Brussels, Belgium

Abstract.
The United States and the European Community have a significant need for engineers and scientists with international experience. This project addresses this need by focusing on the transatlantic relationship between the United States and the European Union. The purpose of this program is to enhance the current practice for transatlantic cooperation in environmental / water resources engineering and science. Important scientific advances have been made in Europe, particularly in the fields of water treatment, environmental hydrology and hydraulics. In the United States, important advances have taken place in wastewater and solid waste treatment. Taking advantage of the best scientific and technological developments on both sides of the Atlantic, this program aims to prepare the students for a career in which they will have to plan, design, operate, maintain and rehabilitate the needed environmental and water resources infrastructure in the foreseeable future. At the same time, the program seeks to prepare individuals to develop a cultural awareness for operating in an international framework. The exchange program is supplemented by a significant distance learning component using multimedia, including the Internet. Two courses are being planned for transatlantic Internet instruction: one at Purdue University and one at the Technological University of Denmark. This paper concludes with a number of academic and implementation issues concerning distance learning using the Internet in the transatlantic framework.

Resumé.
Les Etats Unis d'Amérique et la Communauté Européenne ont un besoin important d'ingénieurs et de scientifiques possédant une expérience internationale. Ce projet s'adresse à ce besoin en focalisant sur les relations transatlantiques entre les Etats Unis et l'Union Européenne. Le but de ce programme est d'améliorer la pratique de l'ingénierie et de la science de l'environnement et des ressources en eau. Des progrès importants ont été faits en Europe, particulièrement dans les sujets du traitement des eaux, de l'hydrologie de l'environnement et de l'hydraulique. Aux Etats Unis, des avancements importants ont eu lieu dans les domaines du traitement des eaux usées et des déchets solides. En profitant des meilleurs développements technologiques et scientifiques de deux côtés de l'Atlantique, ce programme vise à préparer les étudiants pour une carrière dans laquelle ils devront planifier, projeter, opérer et réhabiliter l'infrastructure hydrique et de l'environnement dans un futur prévisible. En même temps le programme cherche à préparer des individus qui développent une connaissance culturelle qui leur permette de travailler dans des structures internationales. Ce programme d'échange est accompagné par une composante importante d'enseignement à distance employant les multimédias incluant l'Internet. Deux cours sont projetés pour enseignement transatlantique par Internet: un à Purdue University et l'autre à l'Université Technologique du Danemark. Cet article termine par un certain nombre de considérations académiques et de mise en oeuvre concernant l'enseignement à distance employant l'Internet dans le cadre transatlantique.

1. Water resources and the environment in the U.S.A. AND E.U.

The field of environmental / water resources engineering and sciences provides many unique challenges. On both sides of the Atlantic the problems of controlling the quality of the environment are paramount. In the early part of this century, nations of the world were concerned with the allocation of water. As we reach the end of this century, environmental issues are predominant. For example, the U.S. and Canada have entered into a number of agreements to reduce the discharge of contaminants into the Great Lakes. The U.S. has built one of the world's largest treatment plants to reduce the pollution of the Colorado River before it enters Mexico.

Similarly, the European nations in the Rhine River basin are collaborating to improve the water quality of this river that serves both potable and industrial water uses. At present the E.U. is concerned with the improvement of the management practices of the water resources and the achievement of a better understanding of the functioning of the environment. Improved infrastructures have to be designed; new standards and new directives have to be issued in a cost effective and unified manner throughout the E.U. The Commission has a key action "Sustainable Management and Quality of Water" in its Fifth Framework Programme RTD (1998-2002) and the "Environment-Water" Task Force of the European Commission has declared that the management of water will be the main issue of the 21st century. A proposal for a Framework Directive on Water Policy is presently being discussed by the E.U. member States.

Important scientific advances have been made in Europe, particularly in the fields of water treatment, environmental hydrology and hydraulics. For example, several European countries have developed and applied water treatment technologies that surpass those available in the U.S. in terms of efficiency and performance. The U.S., on the other hand, leads the world in the development of wastewater treatment processes. Likewise, recent advances have been made in Denmark, France, Germany and the United Kingdom in the theory and modeling of contaminant transport in urban sewer networks. These European nations are clearly leading the United States in this domain. However, the United States is in the lead in the theory and modeling of contaminant transport in the saturated aquifers and in the unsaturated zone, as well as in the techniques of remediation of groundwater contamination.

The continued growth of large metropolises, the increasing number of automotive vehicles and the proliferation of industrial wastes continue to pose challenging problems. The problems of the supply of potable water, treatment of wastewater, disposal of solid wastes and the sustainable management of rivers, lakes and groundwater water resources are some of the many subjects of interest. Competency in these areas is essential to provide an acceptable environmental infrastructure for the twenty-first century. Looking toward the future, recent research findings on global climate change point to a strong probability of increased variability and extremes in precipitation. This would intensify existing problems in water quantity, water quality, sewage treatment, erosion and urban storm water routing.

2. The program

Taking advantage of the best scientific and technological developments on both sides of the Atlantic, this program aims to prepare the students for a career in which they will have to plan, design, operate, maintain and rehabilitate the needed environmental and water resources infrastructure in the foreseeable future.

The Institutions. The European institutions participating in the project include five universities located in Belgium, Denmark, Greece and the United Kingdom and one private research laboratory located in France. They are: the Free University of Brussels (VUB),

Brussels, Belgium (European lead institution); the Technical University of Denmark, Lyngby, Denmark; the Aristotle University of Thessaloniki, Greece; the University College London, London, UK; the University of Lancaster, Lancaster, UK; and the Lyonnaise des Eaux, Le Pecq, France
The USA institutions include four universities; all located in the Midwest and part of the "Big Ten" Universities consortium. They are: Purdue University, West Lafayette, Indiana (US lead institution); the University of Illinois, Urbana, Illinois; the University of Iowa, Iowa City, Iowa; and the University of Minnesota, Minneapolis, Minnesota.

The students. The students targeted for this program are full time Bachelors, Masters, and Ph.D. candidates majoring in water resources / environmental engineering and sciences. The program is thus oriented towards third and fourth year undergraduates and graduate students in any year.

The Schools of Engineering at Purdue have the goal that 10% of its undergraduate student body should have an international academic experience. This would permit graduates from its programs to better compete in surroundings which have become substantially more global. Participating in the proposed project will make this goal more reachable.

Studies and traineeships. Students may participate on this program for either one semester or an academic year. The latter is preferred. Summer research traineeships sessions are offered at University College London and at Lyonnaise des Eaux. Credits earned while taking courses overseas on this program will be applied directly to the student's home institution degree. The programs are conducted in English with the exceptions of *Aristotle University of Thessaloniki*, Greece and *Lyonnaise des Eaux*, France.
Applicants must be in good academic standing and have attained a minimum grade point average of 2.75 / 4.0 for undergraduates and 3.0 / 4.0 for graduate students or equivalent. Students should demonstrate a strong professional and cultural interest in studying abroad and indicate the potential benefit that would be gained from participation in the program.

Language and cultural preparation. The cultural component is an integral part of this program. Previous training in the history, literature and culture of the host country and of Europe or America in general is recommended and language preparation, if applicable, is mandatory. This can be achieved by taking appropriate non-technical elective courses during the one or two semesters before travelling supplemented by at least one course at the host institution. This cultural awareness is deemed essential to prepare the students for a multinational career.

Student exchanges. The proposed student exchange program would make it possible for the students to participate in the most recent advances in environmental / water resources engineering and science on both sides of the Atlantic. The program seeks to develop in a structured way transatlantic student mobility in the field of environmental / water resources engineering and sciences through courses with full academic recognition, research internships and practical training and placement in industry. The program seeks to prepare competent individuals to design, operate, maintain and rehabilitate the needed environmental infrastructure for the foreseeable future. The broadest participation is visualized through the existing networks of universities and enterprises.

Faculty exchanges. The student exchange will be supplemented by faculty exchanges. These will be for seminars or short intensive courses of one to four weeks duration in particular to supplement Internet and other multimedia courses

Multimedia distance learning exchanges. Finally, selected courses will be offered via the Internet. A number of Internet based course offerings are available, for example at the

Technical University of Denmark and at Purdue University. These provide exceptional opportunities for distance learning. This mode of asynchronous learning is particularly well suited to the teaching of science and engineering concepts. In addition, with this mode of education it is possible to overcome the problems of different academic schedules and time differentials that cannot be resolved in synchronous satellite TV courses. The Internet courses can be accompanied by CD-ROMs and complemented by short courses given by the instructor at the receiving institution. This mode may be supplemented by video taped sessions and by short visits of the instructors to the remote institutions.

Duration and Financing. The project has a funded life of three years, from October 1998 to October 2001. The budget is funded approximately equally by the European Commission and by the U.S. Department of Education. The budget includes funding for transatlantic travel of faculty for short courses and seminars.
The budget also includes support for some 80 student mobility grants in the amount of $1,500 or 1260 EUROs apiece. Each institution will have approximately 8 to 10 grants. The grants will be awarded competitively, based on the academic performance, the expected benefits to accrue from the study abroad experience, and the language, historical and cultural preparation planned by the student. The students on this program pay the base fees (e.g., tuition & fees) to their home institutions. For the students receiving the mobility grants, the costs on this program are very similar to those costs associated with education at the student's home institution.
It is important to note that the budget does not include the student tuition fees at the host institution or the fees for distant learning courses at the institution where the course originates. It is assumed that, over the life of the project, there are equal numbers of students travelling in each direction. Each student pays the tuition and fees at the home institution and thus equilibrating the cost of fees. Similarly it is assumed that equal numbers of students will be taking the Internet courses in each direction over the life of the project. This procedure places a constraint on the number and distribution of the students travelling or taking distance learning courses. This balancing requirement places an important limitation, particularly on distance learning, essentially requiring the same number of courses originating in Europe and in the U.S.A.

3. Multimedia courses

Among the many multimedia techniques let us cite CD-ROMS, videotapes, video-conferencing, television and the Internet. The advantage of CD-ROMS and the Internet is that they allow the juxtaposed use of text, sound, still images, full video, dynamic comparisons, software for calculations and graphics all combined in one system. Furthermore, these techniques permit the students to interact, which is necessary for them to learn to think conceptually.
The efforts in connection with this project involve several of these techniques. The Free University Brussels (VUB) has contributed with others to the development of a CD-ROM based course in fluid mechanics that is in its beta version. The participants hope to supplement this course with a CD-ROM collection of pictures of hydraulic structures and other relevant photographs that would illustrate the course. Videotape courses are under consideration. Several courses in Environmental Engineering, such as CE 550 and 551 (Physico-Chemical Processes in Environmental Engineering I and II) using this medium are planned at Purdue University during the next couple of years. Video conferencing is under investigation and its use will strongly depend on the budgetary constraints. So far, the attention has been centered on the use of the Internet. This effort will be discussed in further detail in the following paragraphs.

Distance learning using the Internet. Television video-conferencing provides an attractive learning medium to overcome distance between teacher and students. But in the case of transatlantic courses, there is the additional obstacle of time differentials. There is not only the difference of 6 to 7 hours but also the different academic calendars. The Internet provides an interesting tool for distance learning that liberates the participants of most of the distance and time separation constraints.

Distance learning tries to remove or relax the constraints of time and place in the educational process. Because of the inherent difficulties in establishing an educational relationship between the student and the instructor separated in space and time, it is strongly dependent on technology. This dependence can be extensive because of the importance of the hardware and software and the speed with which these evolve.
In the traditional face to face education, the student-teacher relationship is instantaneous. In the Internet based education, this relationship is media based and asynchronous. The instructional process must therefore be designed for a completely different learning situation. This creates a substantial demand of time and effort in the development of Internet courses.

Internet courses are now available at many universities. For example, the Open University in the UK offers many Internet courses. Over 4000 students connect to the Open University network every day. In the US, the University of Illinois at Urbana-Champaign offers a Master's program in library and information sciences and the University of Phoenix (Arizona) offers four undergraduate and five graduate programs on line. Strong and Harmon (1998) have compared these U.S.A. online degree programs.

There is a wealth of offerings on the World Wide Web, but it is not organized and there is no control on the quality. The Global Distance Education Network (GDENet) provides a collection of documents about distance education particularly related to human development in developing countries (Moore, 1998). The GDEN is reachable from the World Bank web page. The resource section of this site lists journals, books, conferences, and courses on the www, distance education guides and distance teaching universities.

Purdue University. At Purdue University, the development of Internet courses is facilitated by the *Multimedia Instructional Development Center.* As part of the distance learning aspect of this program, Purdue University will attempt to provide several courses via the Internet. Possible Internet course offerings include:
- CE 597N (Civil Engineering). Internet Resources Design and Development. This course provides an understanding of the underlying framework of the Internet and the tools that make it so valuable. Projects are in Environmental / Water resources sciences and Engineering.
- ABE 521 (Agricultural and Biological Engineering). Soil and Water Conservation Management. Through a pro-active participation, students acquire knowledge of the principles and planning methods that enable environmentally responsible uses of land and water resources.
- ABE 526. Watershed System Design. GIS analysis of basin hydrology and water quality modeling.
- ABE 591M. Advanced Soil and Water Conservation Engineering. This course equips students in environmental resources engineering with the understanding of the hydrologic processes and associated design skills.

ABE 526, *Watershed Systems Design*. Among the previous courses ABE 526 was selected for the second semester of the 1998-1999 and 1999-2000 academic years. This is because the subject matter is important in the analysis and management of water resources systems. It focuses on information technologies, especially Geographic Information Systems (GIS) and computer simulation as they apply to the analysis of watersheds. In addition, this course is taught regularly in the second semester of each academic year and its offering is not dependent upon an overseas audience.

ABE 526 is a 3 credit course. The instructor, Professor Bernard A. Engel, is well known for his expertise in the field of Geographical Information Systems. He is a consultant to NASA in this area. The course is aimed at graduate students or senior undergraduates (or equivalent) in a field related to water resources. The course assumes the students have no or little knowledge of GIS.

As water resources are increasingly managed at the watershed level, the goal of this course is to provide an introduction to GIS and hydrologic/water quality models for analysis of water resources issues in primarily rural watersheds. In the first portion, students learn to utilize GIS to analyze hydrologic and water quality issues. In the second portion of the course, hydrologic/water quality models that are interfaced with GIS are used to analyze water issues at the watershed level. The course syllabus is available at http://danpatch.ecn.purdue.edu/~engelb/abe526/
The majority of the course is available via the WWW. Assignments are completed using GIS software and computer models. Students submit assignments electronically. The instructor is available to address questions via e-mail (typically daily reply). Students need access to a WWW browser to complete course materials and arrangements for the acquisition of the software are made with the instructor. It is expected that the instructor would travel overseas to visit the distance classes shortly before the end of the course.
The first offering of this course was expected to be in the Spring of 1999. The course information was transmitted to the European participants in late December 1998. Although a limiting starting date of February 1st 1999 was set, there were no registrations from overseas. U.S. participants were not eligible because non-transatlantic exchanges are not permitted under this program.

It is intended to re-offer this course in the Spring of 2000 which would allow sufficient time for promoting the course. It is also evident that more flexibility is needed in the academic programs so that these distance learning courses can be included in the students' plans of study.

Technical University of Denmark. An Internet course is in preparation at the Technical University of Denmark under the direction of Professor Poul Harremoës and Mr. Eric Warnaars. The Integrated Urban Water Quality Management course (63421) is on the subject of environmental/hydrologic modeling of urban drainage systems including treatment of wastes and impact on receiving waters. A beta version of this Internet course will be available later this year and the final version in 2000. It is intended to include the final version in this transatlantic cooperation program.

The Department of Environmental Science and Engineering at DTU is well known for its flagship course 63350 "*Water Quality Processes*" under the direction of Professor Harremoës. This course is a precursor to the one to be used in this project. In this earlier course, the textbook is available on the Internet. The text describes the relationships associated with the release of pollutants in the aquatic environment. The course on Water Quality Processes was used to test the Internet as a platform for providing support and information to students in parallel with a "traditional" teaching approach. This hybrid course was very successful and received a positive student response. The website was also used to give a course with similar content in Thailand, again taught in a traditional way with the Internet. A team teaching approach was used: the course staff includes three Professors, an assistant and a coordinator.

The Integrated Water Quality Management 63421 objectives are threefold:
1. to impart to the student an understanding of the operational interplay between the three components in the urban drainage system: the sewer system, the treatment plant and the receiving waters,

2. to convey to the student the importance of this interplay for the reduction of pollution released during rain runoff,
3. to teach the student how to utilize an integrated modeling tool for achieving optimized performance.
A modular teaching approach is used, whereby each module builds on the next in increasing complexity. The course is made up of the following seven modules:

1: Analysis of system
2: Nutrient loads
3: Oxygen depletion
4: Sensitivity analysis: treatment plant overflow
5: Effects of new system components
6: Rule based control
7. Competitive optimization

The key element that links the modules and actively involves the students is an integrated software-modeling tool that will be used to illustrate the system dynamics. The students will use the software-modeling tool to complete the exercises at the end of each module. The exercises are based on a case study that is used in all modules. In the final module the student is cast in the role of a consultant vying for a contract to minimize the pollution impact to a system at the least cost. Fellow students are the competing consultants.

4. Student recruitment.

A program brochure, a poster and a web site (***http://www.ecn.purdue.edu/EWRES***) are now available to promote the program. Most partner universities have organized information meetings for the students. One Master's student from the Free University of Brussels (VUB) spent the 1998-1999 academic year at Purdue University. An information meeting was held at Purdue University in mid January 1999 as a result of which one Civil Engineering undergraduate student participated in a research traineeship at Lyonnaise des Eaux during the Summer 1999. A student from the University of Illinois is also scheduled to be taking part in a research traineeship at Lyonnaise des Eaux during the Fall semester 1999. A graduate student from the Technical University of Denmark is spending the Fall 1999 semester at Purdue University. He is taking via the Internet the beta version of the course "Integrated Water Quality Management" originating in Denmark. He is also preparing the material for the Internet course on "Geographical Information and Watershed System Design" that will originate at Purdue University in the Spring of 2000. He could possibly serve as a tutor for this course at DTU in the Spring of 2000. It is anticipated that four other students will participate in the program in the year 2000.

5. Issues to be considered

From the plans of Purdue University and of the Denmark Technological University it is clear that the Internet approach has the potential for greatly magnifying the scope and the impact of technological learning across cultural and geographic distances. However these plans are not without problems.

One important issue is the *transferability* of courses from one program to another across the Atlantic and the flexibility of programs to accept new courses. This lack of flexibility has militated against the movement of U.S. graduate students to Europe and against the movement of first degree students from Europe to travel to the U.S. A second point is the *synchronicity* of coursework so that the courses fit in the students' plans of study at the appropriate time. A third point is the *personal interaction.* As distance education is self-educational, it is often criticized for its lack of face to face interaction. For this reason, this project includes the possibility of one to two week visits of the instructor to the overseas receiving sites, preferably towards the end of the course. A fourth point is the development of a *security system* to limit access to those students who are actually enrolled in the course. A fifth point is the implementation of *in situ* activities at the receiving sites such as technical support and supervision for examinations. A systems approach that considers many of these issues can be found in Elliot (1990) and the recent advances in instructional design and in the development of open

systems of information and learning are discussed, for example, in Khan (1997) and in Mantyla and Gividen (1997).
Another important issue is the operational and production costs. The waiver of the tuition fees for the overseas participants was a new issue at Purdue University that could be solved only to a limited extent through the bilateral agreement requiring equal number of students in each direction across the Atlantic. The cost of production of these Internet and multimedia courses can be quite large. Their development generally requires a team. The subject matter expert needs the support of an instructional designer, a programmer, a graphic artist, and a project manager who keeps everything on track. The role of the teacher is split into several sub-functions performed by specialists. This often requires outside funding. The course at DTU, for example, is being developed with the support of the European Community. As stated by Peters (1993), distance education needs capital investments, a concentration of available resources and a qualified centralized administration.

From previous experience, it is apparent that distance learning involving advanced technologies is more successful with mature and *motivated students*. For this reason, minimum ages of 21 and 23 have been imposed at the Open University and at the University of Phoenix, respectively. In this transatlantic exchange the students are focusing on a well-defined professional subject or are graduate students and the question of maturity and motivation is not expected to be a problem. It has also been shown that the students benefit from a highly interactive environment. Several of these issues and others were pointed out by Strong and Harmon, (1997).

In conclusion, distance learning, and the use of technologies such as the Internet in particular, help create a setting that can enhance international collaboration among scholars and can promote the coordination of educational resources. In the case of this project, distance learning can bridge the Atlantic and the national and linguistic boundaries in addition to the traditional boundaries of time and space. The combined faculty expertise from the cooperating institutions makes it possible to disseminate high quality courses to an expanded number of students on several campuses in several countries. Distance education also has the potential to enhance the social and cultural understanding across the participating countries, thus training the participants to function more effectively in a transatlantic framework. As we emerge in the new millenium and as the new European currency takes its place in the world, the necessity of this intercultural and inter-technical collaboration becomes more urgent.

Acknowledgement. The program described in this article is funded by the European Commission Directorate General XXII, Education, Training and Youth, Brussels, Belgium and by the Fund for the Improvement of Postsecondary Education, U.S. Department of Education, Washington, D.C., U.S.A.

References

Elliot, S., 1990. *Distance Education Systems*, 1990. FAO Economic and Social Development Paper 67, Food and Agriculture Organization of the United Nations, Rome.
Khan, B.H., 1997. *Web-Based Instruction*, Educational Technology Publications, Englewood Cliffs, NJ. 480 pp.
Mantyla, K. and Gividen, J.R., 1997. *Distance Learning: a Step-by-Step Guide for Trainers,* American Society for Training and Development, Alexandria, VA, 179 pp.
Moore, M. G. 1998. The global education network, Editorial, *The American Journal of Distance Education*, v.12, No. 3, p. 1-3.
Peters, O. 1993. Distance education in a post-industrial society, in D. Keegan (ed.): *Theoretical Principles of Distance Education*, Routledge, London, p. 39-58.
Strong, R.W. and Harmon, E.G. 1997. Online graduate degrees: a review of three Internet-based Master's degree offerings, *The American Journal of Distance Education*, No.3, p. 58-70.

Basic knowledge and new skill requirements for water resources management in the information society

J. Ganoulis

Laboratory of Hydraulics,
Aristotle University of Thessaloniki, Greece

Abstract

The use of new information and communication technologies in information and training in the field of water resources management is a new challenge. Distance learning especially, and Internet based education and evaluation modules offer new possibilities for improving skills and acquiring new knowledge.

In order to deal with changes in different existing professions in water resources management and also to meet the needs from the emergence of new jobs, both new skills and basic knowledge are needed. To encourage the acquisition of new knowledge and to validate new skills, testing of basic and professional abilities on the Internet is becoming an important tool.

In this paper, two pilot projects funded by the European Commission – DGXXII are briefly described. The projects named TEST-EAU and TEST-EAU Pro aim to provide tests for the validation of basic and professional knowledge and skills in water resources management.

1. Introduction

Drastic changes in the organisation of every-day work indicate that after the agricultural and industrial revolutions we are now experiencing a third, namely the *information revolution*. With the exponential progress of science and technology, the globalisation of the economy and the upheaval of new communication techniques, our society is rapidly changing. Education and training in the field of water resources is also greatly influenced by this change. Environmental protection and preservation of water resources are very important components for sustainable development together with technical feasibility, economic performance and social equity. Although no unique relationship exists between education and economic growth, it is well recognised that high level education and training are necessary conditions for efficient environmental and water resources management.

However, despite the fact that much progress has been achieved, Europe today faces many challenges, which are related to unemployment, social exclusion and the need to adapt to new professional activities. This is also the case in the water market, where new skills are required for new jobs connected to new technologies.

Markets in the environment-water sector are growing fast all over the world. In Europe, it is estimated that only for water services (water supply + sewage) the market counts for approximately 14 billion of ECU's per year. Together with development of economic activities, professional needs are changing and ask for particular efforts in education and training. Because the current educational system cannot cope with new professional demands, the gap between the professional training system and actual current professional needs is increasing.

This gap may be reduced by the acquisition of new skills and new knowledge through extensive use of new Information and Communication Technology. In today's world, the

use of new information technology for education and training, such as networking and distance learning is an emerging challenge and a new opportunity.
The projects TEST-EAU and TEST-EAU [Pro] are two of the pilot projects which have been recently selected by the European Commission, DGXXII, to validate newly acquired knowledge, skills and professional competence through network technologies (Internet and Intranet services). Aiming to develop original software for the validation of "basic knowledge" in the area of ecological systems interacting with water resources, TEST-EAU started under the DGXXII Socrates Programme.
(http://socrates.civil.auth.gr/test-eau/ and
http://europa.eu.int/en/comm/dg22/tests/liste.html)
It has been recently extended under the Leonardo Programme with the name TEST-EAU [Pro].
(http://www2.oieau.fr/testeaupro/)
TEST-EAU [Pro] includes educational modules and tests for professional competences in the field of environment-water.

2. Description of the projects

2.1 Partners

Main partners in the project are:
- Prof. J. Ganoulis, Co-ordinator, Aristotle University of Thessaloniki, Greece
- Prof. A. Van Der Beken, Contractor, Vrije Universiteit Brussel, Belgium
- Mr. G. Neveu, Principal Partner, Office International de l'Eau, Limoges, France
- Mrs. M. Couchoud, Principal Partner, CEDEX, Madrid, Spain
- Mrs. L. Diaco, Principal Partner, FEDERGASACQUA, Rome, Italy and
- Prof. C. Thirriot, Advisor, Institut de Mécanique des Fluides, Toulouse, France.

L. Rorris, AUT, GR is responsible for computer programming. The projects benefit for input and scientific support from the European Thematic Network ETNET-Environment/Water and from TECHWARE TECHnology for Water Resources (Co-ordinator, Prof. A. Van Der Beken).

2.2 Objectives and structure

TEST-EAU aims to evaluate and validate the basic knowledge on environment / water that any responsible European citizen should have. TEST-EAU Pro deals with more specialised professional knowledge and skills. Several case studies from particular water related situations are also under development.
In TEST-EAU questions are structured in *three different modules of tests* (qualitative, quantitative and reflection), which are available on the INTERNET. The questions are about environmental issues and conflicts arising from interactions between man, water, and ecosystems. In order to stimulate further reading, questions are presented positively, providing an explanation of the context for any specific question, an on-line glossary, correction of false answers and links to other WEB sites in order to obtain more information. The general structure of TEST-EAU is shown in Fig. 1.

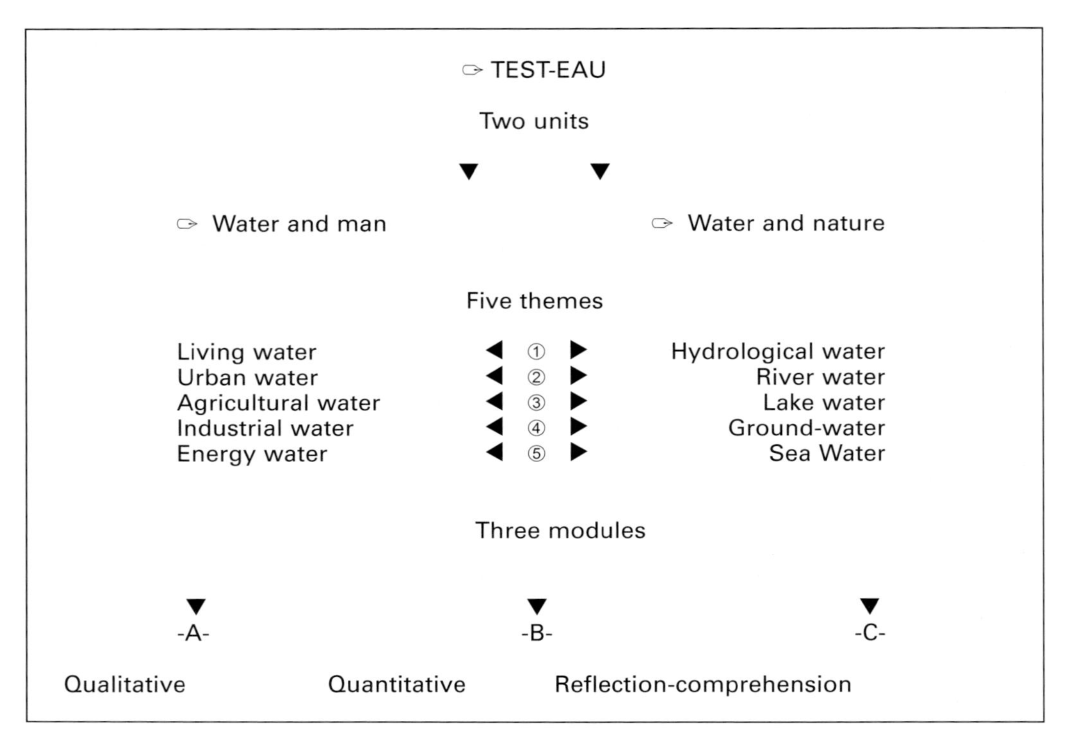

Figure 1 General structure of TEST-EAU

Each of the three modules (qualitative, quantitative and reflection/comprehension), consists of two units:

- Water and Man,

and

- Water and Nature.

For each unit five themes have been retained (Fig.1) which are:

biological, urban, agricultural and energy water for the unit 1

and

hydrological, river, lake, sea and ground water for the unit 2.

Questions are designed to demonstrate personal skills while avoiding duplication. They are progressive, interactive and user friendly.

2.3 Target groups

Potential users of the test include the following

- ❑ Professional associations in water distribution and wastewater collection
- ❑ Local authorities and companies in water supply and wastewater treatment
- ❑ Associations for environmental protection
- ❑ Students and teachers in secondary education
- ❑ Adults in professional training

2.4 Type of questions

The following types of questions are actually included in the tests
- ❑ MCQ: Multiple Choice Questions
- ❑ True or False
- ❑ Target objects on images
- ❑ Drag and drop
- ❑ Fill in blank spaces

The description of different case studies and questions related to them are under development in TEST-EAU [Pro].
For every test, a set of questions is randomly selected from a large data base, located on different servers (three-tier architecture, suitable for multilingual applications). Instead of scrolling the screen, questions are presented on an intuitive graphic interface containing navigation bottoms, multimedia presentations, a glossary, and links to other WEB sites.
In order to target different groups of users, the contents of the tests will be structured in three different levels: 1) elementary, 2) intermediate and 3) advanced

3. Software and Interaction

« TEST-EAU » is cross-platform independent and may run in any computer system that can support a fourth generation Web-Browser such as Intel/Win95-Intel/WinNT running Netscape's Communicator >= 4.0 or MS IE. >= 4.0, and all Unix systems supported by Netscape i.e. Unix, HP-UX, AIX, Sun Solaris, SunOs, Intel/Linux and last but not least Macintosh Power Pc/68k)
Different programming languages have been used in parts of the system, such as:
- ■ Microsoft Visual Basic
- ■ HTML, Dynamic HTML
- ■ C++
- ■ Java 1.1

A combination of HTML, JavaScript and Java has been finally adopted. If we define as a client the PC or terminal used for testing, TEST-EAU and TEST-EAUPro are based on the new model of a thin client: this means that the software is located mainly on the server and is downloaded interactively any time the test is taken (Fig.2).

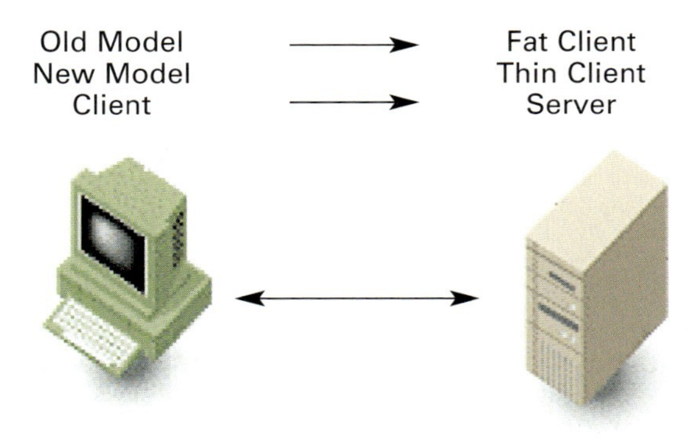

Figure 2 Client-server interaction.

3.1 Client and server models

The main characteristics of a fat or thin client model may be summarised as follows:
Fat Client

- Native application is developed on the client machine.
- Different versions should be provided for different clients.
- Distributing and maintenance problems.

Thin Client

- Client application is common for all different computing platforms.
- Application runs across networks.
- Distribution and maintenance very easy (central to the server).

The thin client model has been adopted for TEST-EAU and TEST-EAU Pro.

3.2 Client - server communication

The server side software has the following characteristics

- A Java daemon (TestEau daemon) communicates with the RDBMS Data Base.
- Three-Tier Architecture is used. (Figure 3).
- Java is great for communication with DataBases with the use of JDBC (natively or via ODBC)
- Data Base connectivity is transparent for the programmer.
- The server side can be easily ported to different architectures (Unix, WinNT).

Since the communication uses TCP/IP, TEST-EAU and TEST-EAUPro can be deployed over LANs or the Internet.

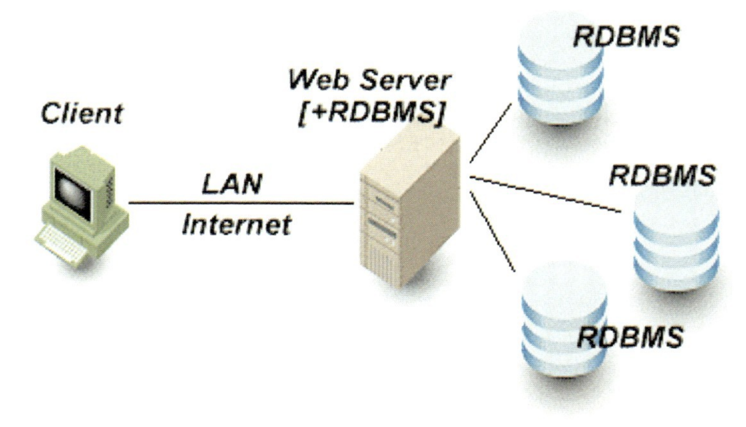

Figure 3 Internet structure of TEST-EAU..

4. Conclusions

New information and communication technologies may be used to improve education and training in Water Resources Management. Acquisition of new knowledge and the development of new professional skills may be enhanced by means of tests and validation modules available on the Internet.
The main properties of two pilot projects for validating basic knowledge and professional skills (TEST-EAU and TEST-EAU ^{Pro}) are summarised in this paper. The author and partners of the projects wish to acknowledge the financial and logistic support of the DGXXII-European Commission.

MAMBO – Development report of a distance learning system in the field of water and environment

Hofmann T., Schmidt-Tjarksen T.
Bauhaus-Universität Weimar, Weiterbildendes Studium Wasser und Umwelt, Germany

Abstract
The creation and publishing of multimedia enhanced distance learning material differs from the task of doing so for a paper based version. The introduction of new structures like hypermedia in contrast to conventional, linear layout of information, leads to expanded requirements in writing, storage and markup of content. The described development of such a working environment shows that many constraints known from conventional publishing processes have to be applied, but also extended to allow for efficient workflows.

Résumé
La création et la publication du matériau utilisé pour l'enseignement à distance se distingue des exigences (de procédé de la rédaction d'une version imprimée) à une version imprimée. Cettes versions ont une disposition linéaire des information.
L'introduction des structures nouvelles comme hypermedia demande des formes étendues de l'écrit, de la mise en mémoire et de la marque (distinction) de contenu.
Ce developpement du champ d'activité (des environs du travail) montre que beaucoup des restrictions du procès conventionel de la publication doivent non seulement appliquer, mais aussi élargir pour permettre un capacité de travail (une opération de travail; un processus de travail) efficace.

1. Introduction

Distance learning in Weimar
For the last three years, the Bauhaus Unversity of Weimar has been offering distance learning courses in the field of water and environment. Students may choose from 14 courses offered in Weimar and an additional 16 courses resulting from the cooperation with the University of Hannover. Thus, a student is able to assemble an individual curriculum for his or her studies. It is planned to end these studies with a masters degree, the permit procedure for which is currently in work.

The organisation of the courses follows a conventional scheme. The students receive a printed chapter of a course every two weeks, containing the course material and exercises, which are sent back for correction.

Questions on the material and the exercises are answered by the responsible trainer of each course, who receives and answeres the questions via phone, fax or e-mail.
Each course ends with a one-week meeting, during which additional lessons, exercises, excursions and a final exam take place.

HSP III and Mambo
Since January 1st, 1998, we, the authors of this paper, are making up the working group „Multimedia in Distance Learning", funded with means from the German HSP III funds. It is our job to implement at least one of the 14 courses offered in Weimar into a multimedia-enhanced (mm-enhanced) environment, and to perform an evaluation on the final product.
The aim of our work is to show if and how the use of mm elements in such an environment can add quality to the learning process in comparison to conventional paper based distance learning.

The result will be a combination of programs which have to be installed by the user on his or her machine at home, which will allow to access the course material from a CD.

2. Why Multimedia and interactive elements

Depending on whom we ask, the attitude towards using „new technologies" in teaching in general and distance learning especially is biased.

On the one hand we find enthusiasm, mostly combined with a certain naivite, towards the use of animations, simulations and the like. The bottom line rounds up to the belief that adding mm and interactive elements to conventional content always adds quality.

For the time being, we are somewhat less confident in the possibilities of mm. Still, positive reactions of students to development work including mm elements lead to believe that its usage may have some impact onto the learning process.

Furthermore, new technologies like the simulation of wastewater treatment processes offer new possibilities to show details of a very complex topic, which can hardly be done otherwise, especially for distance learning.

The impact of such implementations should be investigated more thoroughly.

On the other hand it is claimed that excessive usage of mm waters the learning effect or is simply too expensive. Both arguments are important, as they show that there is still not enough experience at hand. The first is, at least partly, disproved by, e.g., Koubek et al. (1998), the second simply depends on factors like experience, availability of tools, and models of distribution of mm elements.

It is our aim to focus on questions like the ones mentioned and to find adequate answers, as we consider the results to be valuable not only for the complex of distance learning, but also in the learning community in general.

We regard it as a benefit that we are working closely together with the group realizing the conventional distance learnig courses. This allows for access to experiences in the field of coaching the students, but also in the field of creating the course material.

The goal of this work is to implement a mm distance learning software environment using modern technology, which supports distributed working, which is usable by the average user, and leads to a better learning for the same price. Furthermore, the result of this work will be evaluated against the results achieved with the conventional courses.

3. Hypertext and multimedia

Elements used in a teaching environment
It is necessary at this point to give an overview of the elements which will be used in such a learning environment. The following list results from both views, a logical and technical, when

looking at content and technology available, and includes:
a) Text
b) Equations
c) Figures
d) Animations
e) Interactive elements.
Using these elements, the next thing we have to do is bring them in an order which provides a context suited for teaching.

Linking of elements
We do so using a term and a technology known as linking, which is exactly the same as known from web-browsers. The easiest way to do linking is to put all elements into a sequential order, which leads to a „previous" and a „next" link for all elements but the first and last. All elements are thus put into a relation to each other. Using such a sequential manner of linking leads to a structure we know from books with pages.

Hypertext and value of links
Such sequential structures are easy to create and maintain, but suffer from the following disadvantages for our purpose:
a) They are boring
b) They are incomplete if you want to map complex circumstances, since they lack the chance to reference to related information contained in other elements

The solution to these problems is rather simple: Allow for multiple links from one element to others. Systems like these are often referred to as Hypertext, or Hypermedia if the elements are not constrained to be text. For our intended use, Hypermedia offers the following advantages:
a) The mapping of complex information can be done more adequately
b) This system may lead to a saving of time and resources. Once that elements from other authors are made accessible for linked references, they can be reused from others, if they are known or can be found.

There are two important points to a learning system set up in such a concept:
a) More and more of the net information contained in the system is shifted from the pure elements towards the links, since they open up a new context in which the elements information can be seen in.
b) Such a system is really shitty to set up and maintain.

We will tell you more about why this prooves to be difficult in the next part.

4. Aspects of authoring and publishing

What is a book?
In our context, this is a rather simple question. A book is a sequence of text and figures, which for a long time in the history of writing has been written and composed by a single person (author), who also published these papers himself.
The result of his work was a book of a certain quality, which mainly depended on his own work, and in which the information usually was arranged in a sequential manner.

Publishing as it is
This process has evolved over the years to a distributed system with one or more authors, experts responsible for figures, photographs, reproduction or printing and a publisher, responsible for the structure and formatting of the book as well as controlling.
The result of this teamwork process is again a book, the quality of which depends now on complying with stringent conditions set up by the publisher. These conditions mainly concern structure, layout and organization of workflow.

The differences to the former process are the distribution of tasks, which lead to dependencies between the single steps of the workflow showing more clearly and result in a higher need for communication and feedback.

Creation and Publishing of Multimedia Content in a Hypermedia environment
The following description is a preliminary attempt to set up roles in the team of authors, experts and publisher(s) stemming from our first experiences in our work. It is in no way to be seen as a final, proofed definition.

In this environment, authors not only provide for the writing of text, they also propose links according to their knowledge of the topics they are writing about. Ideally, these propositions are not only made for catchwords, but specific other elements they can refer to.

Experts not only have to work on the same tasks as before, but now also contribute to user interface, mm objects, interactive elements and the IT-part of publishing.

The publisher has to focus even more on questions concerning the structure of the content he is providing. Still, the layout of not only the content, but also the user interface and navigation elements is very important to ensure ease of use. The task of controlling the publishing is extended towards the quality assurance as seen in the software business.

The result of such a collection and publication can, ideally, be manifold. Web sites, computer based training courses, videos and books can be the resulting output of the publishing processes, provided that enough information is present to turn the hyperstructure into something linear suitable for books and videos.

The quality of the resulting publication depends even more on the strict compliance with stringent conditions, which is basic for the chance to reuse an object in different contexts.

Differences and requirements
Differences to publishing as we know it are manifold. We will try to list just the most important ones:
a) The system is more complex, with all obvious consequences
b) The system is more powerful in terms of showing complex connections, thus allowing for different points of view for a single question or problem.
c) The system may, given there are enough different approaches provided in the structure and/or user interface, offer better access for different learner models on the students side.
d) The work flow is very new to all members of the process – authors, experts, publishers and users.
e) New constraints which are, up to now, not at all clear from the beginning have to be met rigorously by all contributors, who, lacking experience, cannot estimate dependencies of other work in progress.
f) As a result from the hypermedia structure, the possibility to insert automatic links to index or glossary entries and the need for navigation in different forms we end up with many links.

All these points listed above make it clear for the experienced computer user, that we need a new working environment, which has to be designed in a manner to help out with most of the problems resulting from them.

5. CMS – another three letter acronym

What is a CMS and why do we need it
A content management system (CMS) is the term used for a software solution designed to manage content. Since we haven't found a definition of this term which is more precise than that, we would like to describe with this term a system which allows us to import, store, man-

age and publish content for multiple purposes. A detailed description of such a system follows below.

A short calculation of links

In a hypermedia environment, links and their management are essential. To explain why we believe that a CMS is indispensable, let us do a very rough assumption on the number of links to be maintained.

In a first try, 1000 pages of course material have been split up into roughly 400 text elements, attached to which are about 170 figures, about 60 tables and some 80 equations. These numbers will change depending on the granularity of the text elements and the nature of the content.

Assuming that there are four links for navigational purposes and four reference links to other elements for each text (which will probably be not sufficient), we get 3200 links to maintain only for the text interface we want to implement.

In addition to that we want to supply a nonlinear, secondary graphical user interface allowing for another structure of the same content provided in the paper version. This structure is made up of a tree with at least 25 000 links only for navigation in the tree (four levels, 12, 16, 12, 12 elements), most of which can and have to be set up automatically, probably by inheriting in an object oriented environment. This does not include additional reference links to elements.

We believe that now it is clear why we want to have such a CMS.

6. Design of a CMS environment suitable for our needs

The layout of a CMS suitable for our needs is schematically shown in Fig. 1. There are four main fields of functionality allocated around a central processor, right now implemented in a scripting language environment called Frontier, serving as a connecting and controlling unit. This system is being assembled, programmed and tested against our needs right now.

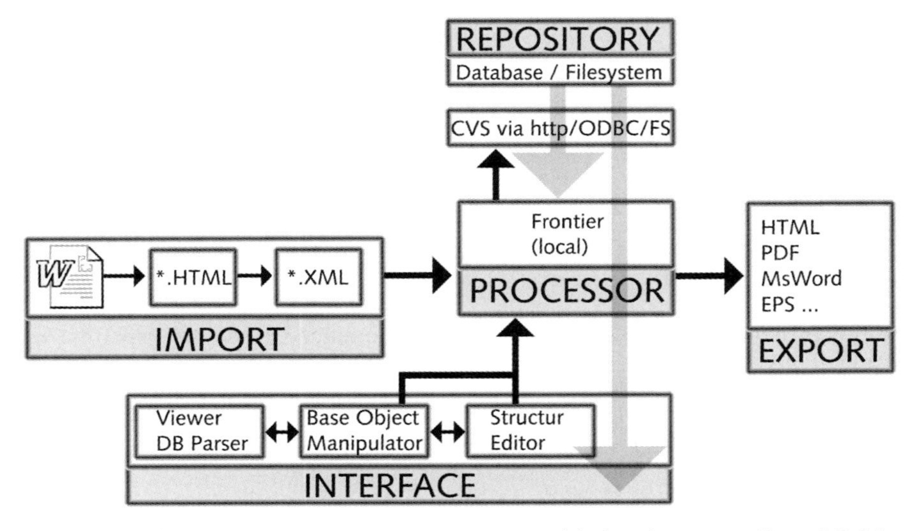

Figure 1. Layout of a content management system usable in a hypermedia publishing environment

The import section covers our need to get large amounts of content into the system, like when importing a whole course for the first time. Due to the original format of the documents (Winword 7.0), different steps have to be taken to accomplish the task of converting, splitting and importing the content into the system without losing too much information.

Once the elements are imported in the system, an interface has to provide for the task of verification and edition of imported data and metainformation. Almost more important than that is the ability to create, view and edit hyperstructures by inserting and maintaining links. This process, if going beyond automatic referencing against indexes or glossaries, can only be done by authors or editors familiar with the content.

Both kinds of information, elements as well as structures, have to be stored in a repository, possibly controlled by a concurrent versions system (CVS), which mainly takes care of versioning and history of file changes.
The fourth and last field concerns the publishing process. For the time being we are at developing the functionality to publish for three formats, Winword, to be able to communicate with authors, Framemaker for internal publishing purposes and HTML, which is the format we use for our CD distribution.

The processor is the central unit. Actions triggered from or performed in one of the four connected areas are fed into the processor where they are controlled, validated and, if possible and allowed, carried out. Examples of such actions are to allow for editing of text for certain users, setting up and maintaining of dependencies of elements within structures, or provide search capabilities for authors looking for reusable elements.

7. What we do, then

As aforementioned, we try to develop structures which allow us to import large amounts of differently formatted text, figures, equations and tables and to describe, store and reuse them adequately. This will eventually lead to a set of guidelines according to which existing course material will have to be reformatted.

Resulting from this step we will have elements which have to be checked for correctness, completeness and quality, and to which we will have to add metainformation which allows for searching, finding and managing the elements.

Additionally, we create a new number of interactive or mm elements to expand the possibilities of mediating course information to distant students. These elements are imported into the CMS mainly like elements stemming from Word files.

The possibility to not only add references, but also create whole new structures of accessing the information stored in the course material opens a new field of distance teaching. The results of our work will undergo an evaluation against the conventional paper based course material.

References

A.Koubek, S. Lo, E. Meisterhofer, R. Posch (1998) Lernen mit Multimedia, FH-Joanneum Arbeitspapier 98-001, http://zml.fh-joanneum.at/arbeitspapier1.html

L'OR BLEU© - une encyclopédie interactive sur CD-Rom sur le thème de l'Eau traité à l'échelle planétaire et dans l'ensemble de sa problématique

Marleix A., Carbonnel C.

Coproduit par STRASS Productions et les éditions de l'UNESCO et conçu en collaboration avec de nombreux scientifiques internationaux, le CD-Rom L'Or Bleu est la première encyclopédie interactive consacrée à l'EAU et à son rôle dans notre environnement.
Par d'immenses voyages à travers le monde et le temps, L'Or Bleu propose de découvrir l'eau sous tous ses aspects : sciences, nature, histoire, société, et enjeux.

Les auteurs de L'Or Bleu, et l'équipe qui a collaboré à cette réalisation, souhaitent faire prendre conscience des enjeux primordiaux que représente l'eau pour l'humanité. Ils montrent les liens permanents que nous avons avec l'eau, depuis la nuit des temps, et ils nous proposent de fantastiques voyages à travers la planète, et à travers l'histoire, à la quête de l'Or Bleu..

Comme tous les titres qui viendront enrichir la collection Notre Planète Terre, L'Or Bleu concilie rigueur scientifique et réalisation interactive attrayante et ludique.
Les concepteurs de ce CD-Rom ont en effet ciblé tous les publics, de l'enfant de 8 ans à la curiosité débordante, jusqu'à l'adulte qui souhaite se distraire tout en s'informant.

Un choix de deux modes de navigation facilite l'accès à l'information : l'utilisateur peut parcourir librement et de manière visuelle l'intégralité du CD-Rom, représentée par une gigantesque fresque interactive, ou rechercher un thème précis à l'aide d'une base de données thématique. La richesse du titre, complétée par une connexion Internet, en fait également un outil d'enseignement particulièrement adapté aux collégiens et lycéens.

L'Or Bleu sortira au premier semestre 99 en version Française et en version Anglaise dans le courant de l'année 1999. Des accords de distribution sont en cours de discussion, pour réaliser différentes versions localisées (Espagnol, Italien, Allemand).

Points forts :

- un CD-Rom accessible à tous;
- des informations supervisées par des experts internationaux;
- un double système de navigation très pratique : un mode intuitif (la fresque) et une base de données thématique;
- une richesse graphique remarquable;
- plus de 600 photos, 200 graphiques, schémas et cartes, des animations 2D et des

séquences en images de synthèse 3D, près d'une heure de commentaires et plusieurs séquences vidéo plein écran, une base de données indexée par thème et un glossaire accessible par hypertexte à tout moment, possibilité d'exporter l'ensemble des textes (près de 300 pages).

BLUE GOLD - an interactive encyclopaedic CD-Rom on the subject of Water dealt with on a world scale and in all its aspects

Co-produced by STRASS Productions and UNESCO and designed with the cooperation of eminent international scientists, BLUE GOLD is the first interactive encyclopaedia on CD-Rom devoted to water and to its role in our environment.
Using extensive travel in time and space, BLUE GOLD allows users to discover water in all its aspects - scientific, natural, history, social - and appreciate all its implications.

The authors of BLUE GOLD and all the team who contributed to the production, want to raise people's awareness of water's fundamental issues. They demonstrate the continuous links that we have with water, ever since the beginning of time, and they offer us amazing journeys around the planet and in history, in a quest for Water.

Like all the titles that will be added to the Our Planet Earth series, BLUE GOLD is a CD-Rom which combines a scientific approach with an attractive and fun interactive product.
The CD-Rom's designers have targeted all publics, from 8-year-olds full of curiosity to adults who want to be informed and entertained.

There is a choice of two navigation modes making access to information easier : the user can run freely through the whole CD-Rom visually, as represented by a huge interactive fresco, or search for a specific topic using a subject database. The richness of the publication, including an Internet connection, also makes it an educational tool which is especially suitable for secondary school students.

BLUE GOLD will be published in the first semester 1999 in French and in English afew months later. Distribution agreements are being negotiated to localise the title in different foreign versions (Spanish, Italian, German).

Key features :

- one CD-Rom accessible to everyone;
- information supervised by international experts;
- practical double navigation system : an intuitive mode (the fresco) and a subject database;
- outstanding graphics;
- more than 600 photographs, 200 charts, diagrams and maps, two-D animation and clips of three-D computer-generated images, one hour of commentary and several full screen video clips, a database indexed by subject and a glossary accessible via hypertext at any time, exportation of the whole text.

Experience with CAL as a teaching aid in Fluid Mechanics

Expérience avec EAO comme support pédagogique dans la Mécanique des Fluides

Sellin[1] R. H. J., Davis[2] J. P.
[1]Emeritus Professor and [2]Reader,
Department of Civil Engineering, University of Bristol, UK

Abstract
This paper reports on the development and trialling of a multimedia Computer Aided Learning title designed as an alternative to a standard lecture course in Fluid Mechanics for first year Civil Engineering undergraduate students. The implementation attempted to improve the overall learning experience, to improve the quality of interaction between tutor and students while at the same time seeking to significantly reduce staff input time. The results suggest that this can be made to work well given the necessary support framework especially in terms of assessments.

Résumé
Cet exposé concerne le développement et l'essai d'un programme multimédia d'Enseignement Assisté sur Ordinateur servant d'alternative au cour magistral en Mécanique des Fluides pendant la première année des étudiants en Génie Civile. Cette application a essayé d'améliorer globalement l'expérience d'apprentissage, d'améliorer la qualité de l'interaction entre le professeur et les étudiants tout en cherchant à réduire de manière considérable le temps d'enseignement dispensé par les professeurs. Les résultats montrent qu'avec un peu plus d'ajustement cette application pourrait être couronnée de succès et atteindre ces objectifs.

1. Introduction

The training needs of the European water Industry at all levels are considerable. There are particular needs for in-career training, updating and training in specialist areas. Modern methods utilising the latest forms of computer-based technology and allowing a distance learning approach offer a solution to some of these requirements. CALWARE is a project initially funded in 1993 by the EC COMETT programme under the auspices of TECHWARE, a European non-profit making association for training and information transfer for the water industry. This ambitious programme was to be the work of some 20 partners drawn mainly from Universities in the European Union (see Figure 1).

Its objective was to produce four multimedia CAL titles for the European water resources sector as shown in Figure 1. These were selected following a training needs analysis in the water sector. The first title, Fluid Mechanics, has been largely finished and used in this state in full scale. trials. The other three titles are at various stage of production but not yet at the trialling stage.

This paper discusses the lessons learnt, both in the production of the Fluid Mechanics title (it amounts to over twenty hours of Computer Assisted Learning Material), and also from the experience gained in the course of its trials.

Although the main target group for CALWARE is mid-career hydraulic and water engineers, some of the titles have application to undergraduate courses. These factors have been borne in mind at the design stage.

2. Learning principles and the type of CAL used

The computer can be used in the learning process in different ways depending on the objectives of the material. If the purpose is simply to raise awareness or supply reference material then a hypertext system is appropriate. Other approaches are needed for teaching motor skills. In the context of this project a basically linear learning programme was chosen. This has been implemented in a modular form to allow the user as much freedom as possible to look at individual topics within the title in isolation.

The development of the philosophy of the CALWARE material has been guided by three main issues, following Bork (1991). The aim has been to produce material which will have high levels of Interest, Interaction and Individualisation.

Interest

For training packages to be successful, that is for them to be generally accepted and widely used, two main hurdles have to be over come. These hurdles are the acceptance of the material both by the learners and also by the trainers. There is a notorious unwillingness among trainers to accept anything which isn't done exactly the way they would have done it themselves, the "not invented here" syndrome.

Title	Partner	Country
Fluid Mechanics	University of Bristol	UK)
	University of Limerick	(IRL)
	University of Calabria	(I)
	Labtegnos	(I)
	Free University of Brussels - VUB	B)
Open Channel Flow	University of Leuven	(B)
	University of Glasgow	(UK)
	Travers Morgan	(UK)
	Hydrodata	(B)
	LHCN Liege	(B)
Pressurised Pipe Systems	University of Valencia	(E)
	Egevesa	(E)
	AMGA Genova	(I)
Urban Drainage	TU Denmark	(DK)
	HR Wallingford	(UK)
	Free University of Brussels - VUB	(B)

Figure 1. CALWARE Titles and Partners

The problem should be eased if there is a breadth of consultation between partners in development and significant trialling of prototype products. Difficulties with both trainers and learners can be eased further however if the material is found to be really stimulating.

The teaching of Hydraulics and Water Resources relies heavily on pictorial and graphical material, and is therefore ideally suited to be adapted to a multimedia approach. This dictated the use of both video and audio.

Interaction

The CAL material must have a high level of interaction with the learner in order to maintain interest and to ensure that the learning process is both effective and efficient. Bork (1991) refers to both the degree and quality of interaction as being important. Conventional hypertext systems may provide a high degree of interaction but the quality of this interaction is very low. They expose the learner to a lot of material, but there is little evidence to show that the material is being absorbed in this situation.

Interactivity has so far been accomplished in three distinct ways handled within a "training overlay" which is described later.

Interactive problem solving

Methods have been developed to use tried and tested problems where the most common wrong answers are known. These answers can be trapped and the most appropriate feedback given. The user cannot get access to the worked solution until he or she has crossed a number of interactive hurdles.

It is awkward to ask users to type in equations on a computer so where it is necessary to get an equation from the user, he or she is asked to assemble the equation from a group of available parameter groups on the screen using the drag and drop approach. The resulting equation is then compared with the acceptable correct forms by using an algorithm which assigns a number to each group in such a way that the acceptable correct combinations correspond to a set of unique numbers.

Requests for predictions. As part of the learning process the user is frequently asked to express, often graphically with an interactive animation, what is going to happen next.

Simulations. Existing simulations of hydraulic systems are used as and when appropriate.

Hypertext links. Although this is a lower level of interactivity, hotspots are provided to enable the user to move around in order to look up information as well as checking on terms and concepts which may have been forgotten. This leads on to the final point.

Individualisation

Conventional training courses provide material at a fixed pace, but individual users learn at different rates. Computer assisted learning provides an opportunity to correct this mismatch. However there are other important ways in which the system should be tailored to the needs of the individual user.

It is felt that intelligent tutoring systems which model the behaviour of the user are not yet mature enough for practical implementation on a large scale project. Ideally the CAL program should have a metasystem capable of detecting, either implicitly or explicitly, what knowledge the user has already. Otherwise the user may become too quickly discouraged by lack of progress or bored by the trivia they have to go through.

The material, though in a mainly linear tutorial form, has levels above and below the main route which are accessible through the use of hotspots. The lower level holds prerequisite concepts and other backups to the main material. The upper level provides advanced material which is supplementary to the main learning objectives of the title, but which can provide increased interest and motivation for the more advanced user.

Risks and choices

The decision to go for a highly interactive multimedia approach has consequent risks attached to it. These risks include the immaturity of the PC video technology at that time and the cost and acceptability of the delivery system. Some of these issues are discussed by Davis (1994).

The development methodology had also to be decided. The choice was between using an authoring package , which reduced the work load but imposed restrictions, and using a high level language with greater flexibility.

It was decided to use the authoring language approach with a highly structured shell or template to speed the production, ease the project management and material consistency,

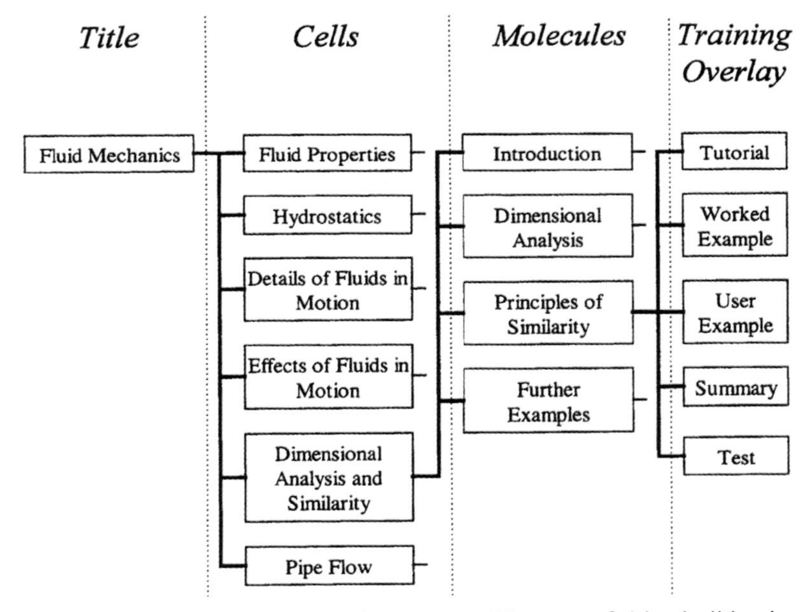

and allow the author to concentrate on the content. The use of this shell is also discussed in more detail in Davis (1994). The package Authorware Professional was chosen because of its facilities which included built4n assessment procedures.
Figure 2. Example of the CAL WARE Structure - Fluid Mechanics Title

3. The CALWARE shell

Once the authoring package has been chosen one of the partners was tasked with producing a shell or template. This embodies the structure of the package, the look of the backgrounds of the screens at all hierarchical levels, the navigation systems and all the buttons necessary for the navigation and look up facilities. The use of a shell like this has two main consequences.

The first is that the domain experts have only to put their material into the shell without having to worry about screen design and all the other navigational issues, thus speeding up production. The other consequence is that it imposes a significant restriction on the freedom of the domain expert in the way he or she presents that material. It was felt that this restriction would have to be accepted in the interests of meeting the project objectives.

The structure of the shell divides the material up into three levels, the cells, molecules, and the working level, as shown in Figure 2. Further, the working level has superimposed on it a Training Overlay. This specifies the way in which the material will be delivered.

The tutorial section for each molecule is required to take less than twenty minutes to complete before the user is given industrially based worked examples and exercises to rehearse the ideas and techniques.

4. The implementation

The target group for the trial was a mixed first year intake of the University of Bristol Civil Engineering and Engineering Mathematics Departments - 75 students in all. This mixture enabled us to compare the reactions of the 'computer friendly' Engineering Mathematicians with a cohort of Civil Engineering students, some of whom sometimes exhibit computer phobia. Thus the overall group contained both 'freaks' as well as novices who had never done much computing before.

Implementation strategy

The constraints for the implementation were as follows. The timetabled contact hours for the students to be exposed to the CAL material were to be the same as in a previous year when the standard lecture approach was used. The input of support staff time, that is from postgraduate demonstrators, was also to be the same as previously. Only 15 multimedia computers were available which entailed restrictions in timetabling. The implementation was based on the model outlined below.

Existing conventional teaching methods for a group of 60 students

The conventional approach used:
1. Twenty hours lectures given by the tutor (professor) - conveying both principles and techniques.
2. Twenty hours examples classes - practising the use of the material, staffed by tutor and postgraduate demonstrators.

A total of 40 tutor contract hours.

New approach using CALWARE

The new approach is based on a mix of various forms of student-tutor interaction, giving the computer the task of information transfer, while improving the quality of student-tutor interaction for reinforcement.

The new approach uses:
1.Students working in pairs at timetabled sessions at multimedia computers - mixed tutorial and practice. These computers are also available in 24 hours access mode. Postgraduate demonstrators available during timetabled sessions.

1. Five hours of lectures through the year to motivate, put in context, provide overview and summaries.
2. Glass divided into 4 groups of 15 which meet with lecturer 6 times during the year for seminars using the 'Socratic dialogue' approach.That is the students are closely questioned about their understanding of the material, with the opportunity to discuss the issues involved in the various concepts.
This adds up to a total of 29 contact hours for the tutor, a saving of nearly 30%.

This new approach is intended to produce the following benefits:
1. A more stimulating learning environment for the student.
2. A facility which allows the student to go over the material as often as required, at the student's own convenience.
3. Higher quality contact with the lecturer through more interesting lectures and closer, more intense, small group seminar sessions.
4. A more enjoyable teaching experience for staff, having higher quality interaction with the students.
5.A reduction in staff time of nearly 30%.
At the end of the contract period 80% of the Fluid Mechanics material had been converted into CAL form and mounted on a CD-ROM platform, one title had been abandoned due to inadequate progress, and the remaining three titles were partially completed though not taken to the CD-ROM stage.

5. Assessment of the CAL project

The assessment of the success of the implementation was carried out by the following means:
(a) User response through questionnaires covenng both usability of the system and learnability of the material.
(b) A log book was kept to register user comments on a day to day basis.
(c) Direct observation during selected sessions throughout the course to obtain views on:

1. initial reactions to the material
2. ease of use or frustrations experienced
3. effectiveness of working in pairs - does an intelligent dialogue develop successfully in most cases?
4. progress tests
5. a traditional exam at the end of the year was used in this first evaluation year to measure 'success'. This is important as there is no real control group being used. It is therefore only possible to make comparison with the 'success' rate of the previous year's group.

The results of the questionnaires will be presented first and afterwards discussed in the light of the general observations made during the progress of the trial.
Scores for each question were given on a scale of 1 to 5, expressing their agreement or not with the statement. A score of 5 meant that 'the user strongly agreed with the statement' whereas a score of 1 indicated 'strong disagreement'.

This kind of scoring system has been used for many years for student feedback on courses in the Faculty of Engineering at Bristol. Experience suggests that with a set of returns over twenty then any score over 4 indicates a very satisfactory agreement with the statement, while a score lower than 2 represents a significant and widely shared disagreement that requires serious attention. The results from the nine sections will now be discussed in turn.

Overall organisation of the course

This is in many ways the key section for this trial as it deals with the implementation system adopted rather than with the quality and appropriateness of the material presented. Studentsfelt that the balance between time given to computer delivery and lectures was not right. They reported an excellent response to the lectures and even more so to the seminars (scoring 4.11). Unfortunately the students did not understand that the overview style of the lectures and the interactive style of the seminars was only possible because the bulk of the information transfer task had been given to the computers.

Two main points of dissatisfaction were identified, the first related to the quality of postgraduate support available (score 1.63) and the second related to self-discipline and the setting of targets (apparent from the free text feedback). The training given to the demonstrators attending the timetabled computer sessions can obviously be addressed and this should result in an improved response here in the future. It is intended in future to have a linked local News group on the network to provide extra support. The second failing was that we did not realise how much guidance the students needed to help them work out how much material they ought to cover each time they came to the computer. Several responses indicated how a student felt secure if he or she had attended a lecture, written down what they perceived they were meant to write down and gone away again irrespective of whether they had actually learnt anything then or not. Whereas working at the computer they were not really sure if they had done 'what they were supposed to have done'. Future implementations must therefore provide more frequent and clearer targets as well as clearer indications through tests and interactive work that they had achieved the objectives for the section.

The overall response to the course was however very encouraging. The material itself provided no obstacles, as will be seen from the following sections.

Organisation of Information and Content

The highest scoring statement in this section were: the information is logically organised (3.84) and the content develops logically (3.87). The objectives were clear (3.66) and the examples related to the objectives (3.45). This illustrates the importance of clearly specifying the behavioural objectives at the start of each section - a key feature of CALWARE.

The lowest scores were obtained for: the feedback in the user examples is at an appropriate level (2.47). The interactive examples have two levels of help system. The first is given through the use of hint buttons. The second is given by traps which recognise common mistakes in numerical answers given by the user and provide appropriate comment. The response to this statement shows how difficult it is to pitch help at the right level to suit every user, and more work is needed in this area.

Observations

One of the novel aspects of this trial was to investigate the effectiveness of the students working in pairs. This was not as effective as had been hoped partly because there was not enough space around the computers for two people to work together well, and partly because this mode of working was not enforced from the start. The progress tests showed little difference from the marks obtained in the progress test of the previous years group with one difference. The CAL group's marks showed a slightly increased spread between individual students.

Other aspects of the questionnaire

Other sections of the questionnaire deal with more technical aspects of the CAL system and the results of these are reported in full in Davis (1996).

6. Conclusions

taking into account the difficulties caused by the incomplete nature of the CAL part of the pourse~ the professors concerned were nevertheless of the opinion that a higher degree of academic maturity would be required in future target student groups for this kind of material. Alternatively, the appropriate level of self discipline could be imposed by careful study plans, targets and progress testing. This last component is seen as the rnost important element in the scheme. In fact it is recommended that future developments should be assessment led rather than having the assessment element as an add-on feature. The development of these features adds yet more to the initial costs of developing such systems.

The material itself and its style were however very successful. The training overlay and the attendant navigational style was found to be natural and intuitive and is recommended as a model for future developments. The navigational facilities did in fact pre-figure the style that is now commonplace in web based systems. This all leads to the conclusion that future systems may have to emerge gradually on web sites as the tutorial and multimedia material is collected and loaded year by year so that what starts out as a subject support site can gradually become a computer aided learning site by degrees.

References

Bork, A. (1991). Learning in the TwentyAirst Century: Interactive Multimedia Technology. In Proc. Int. Conf. on Computer Aided Learning and Instruction in Science and Engineering (CALISCE'91). Lausanne.

Davis, J. P. (1994). Multi-media CAL: The reality. In Hydroinformatics '94, Verwey, Minns, Babovic & Maksimovic (eds). Balkema, Rotterdam.

Davis, J. P. (1996). Large scale implementation of computer assisted learning. In Hydroinformatics '96, Muller (ed). Balkema, Rotterdam.

Using of new information's technologies in open and distance learning

Tuhovcák, L., Stara, V., Valkovic, P.
Institute of Municipal Water Management, Faculty of civil engineering, Technical University of Brno, Czech Republic

Abstract
The structure of daily and distance learning in the Technical University of Brno - Faculty of Civil Engineering, the main postgraduate courses for waterworks staff, the methods of co-operation between water industry and universities and international co-operation is present in the first part. The practical knowledge of producing the multimedia materials as new educational instruments, recommendation of the methods and special software packages for this production, exploitation of this media with the Internet possibilities in the educational context and results of the Institute of Municipal Water Management in this field are presented in the second part of this paper.

1. Introduction

The education system of our university has a long-time tradition. The Czech Technical University in Brno was found 100 years ago in 1899. Its actual name is Technical University of Brno and it has nine faculties located in Brno and Zlín. The Faculty of Civil Engineering with more than 3 000 students is the biggest faculty of the university at this moment. Students can study at six main undergraduate study fields
Overground Structures
Material Engineering
Structures and Transportation Engineering
Water Management and Water Structures
Geodesy and Cartography
Economics and Management of the Building Industry

and four postgraduate courses

Theory of Structures
Material Engineering
Water Management and Water Structures
Geodesy and Cartography

The new changed political, social, economic and sociable conditions of our society after 1989, fair interest and the pressure of public to complete the full university education forced faculties to organize also the distance form of study for undergraduate students.

2. Open and Distance Learning

Faculty of Civil Engineering has organized the distance learning from academic year 1994/95. The students of this type of study must do the entrance exam in the Mathematics and Physics. The students have six years study plan as a standard but they can modify it to their individual conditions on nine years till. The study is finished in diploma project and official National final examination. They will receive the same title *Ing.* like daily students. The basic forms of distance study are

individual study with materials (prospectus, literature, manuals, video, software packages and etc.)
individual consultation through phone, fax, e-mail
individual non obligatory attendance consultation organized in the faculty campus four times per each semester on Friday afternoon and Saturday
obligatory attendance consultation each semester

This form of study is very difficult for the students. More than 60 % students stop their study during the first two years of study.

2.1 Short courses and international cooperation

The Institute of Municipal Water Management (IMWM) is one of the institutes which are responsible for the study field Water Management and Water Structures. IMVM provides also professional training courses especially in water distribution systems, urban drainage, sewerage and wastewater treatment. The courses are aimed at staff of water-works companies which require updating and re-training as part of their career development. The training is normally provided in the form of short courses (1 - 5 days) and can be held on campus of faculty or organized on an 'in-house' basis for particular companies. The IMWM participated also in TEMPUS program supported by EU. The Join European Project S_JEP 09109-95 *Continuing Education Centre for Water Industry - CECWI* was running during years 1995-1998. The partners of this project from Czech Republic were

Czech Technical University (Coordinator, contractor of the project)	Department of Hydraulics and Hydrology
Technical University of Brno	Institute of Municipal Water Management
University of Agriculture	Department of Water Resources
Union of Water Management	

and from EU

Polytechnical University of Valencia	Hydraulic and Environmental Eng. Department
Heriot-Watt University Edinburgh	Department of Civil and Offshore Engineering
University of Porto	Department of Civil Engineering

The CECWI was established at the Department of Hydraulics and Hydrology of the Czech Technical University in Prague. Special lecture room with didactical equipment was constituted also in Technical University of Brno as a local office of CECWI. The CECWI organized each academic year 4 short courses (2-3 days) for staff and management of Czech water industry. For example in academic year 1997/98 CECWI organized these courses:

Catchment Management plan
(February 1998, 3 days, 3 lecturer from EC)
Means of Operative Hydrology in Forecasts and during Controlling the Water Resources Systems - the Role of Hydroinformatics
(March 1998, 3 days, 1 lecture from EC)
Management and Leakage Control of Water Supply Systems
(June 1998, 3 days, 2 lecturers from EC)
Soil and Groundwater Pollution and Remediation
(June 1998, 3 days, 1 lecture from EC)

More than twenty experts from Czech water industry took the participation at each course. The materials with ISBN code were prepared also for each course.

3. New informations technologies in learning

The IMWM develops using of new information technology for education and preparation of study materials during last 5 years. We are interested mainly in the Internet and multimedia exploitation.

3.1 Internet

The IMWM sets up the first special www server with problems of water in the Czech republic. Its name is WATER (*http://water.fce.vutbr.cz*) and it was established in spring 1995 (Fig. 1). There is a link to information systems of faculty with basic information of prospectus, courses, staff, research and etc. The students can monitor also their study results through INTRANET. The students can find on the server WATER also the most interesting www addresses in water sector. They are separated in chapter "Other resources" into five main categories:

Czech government institutions
International non government institutions and associations
Research instituions and universities
Water Industry, companies
Libraries

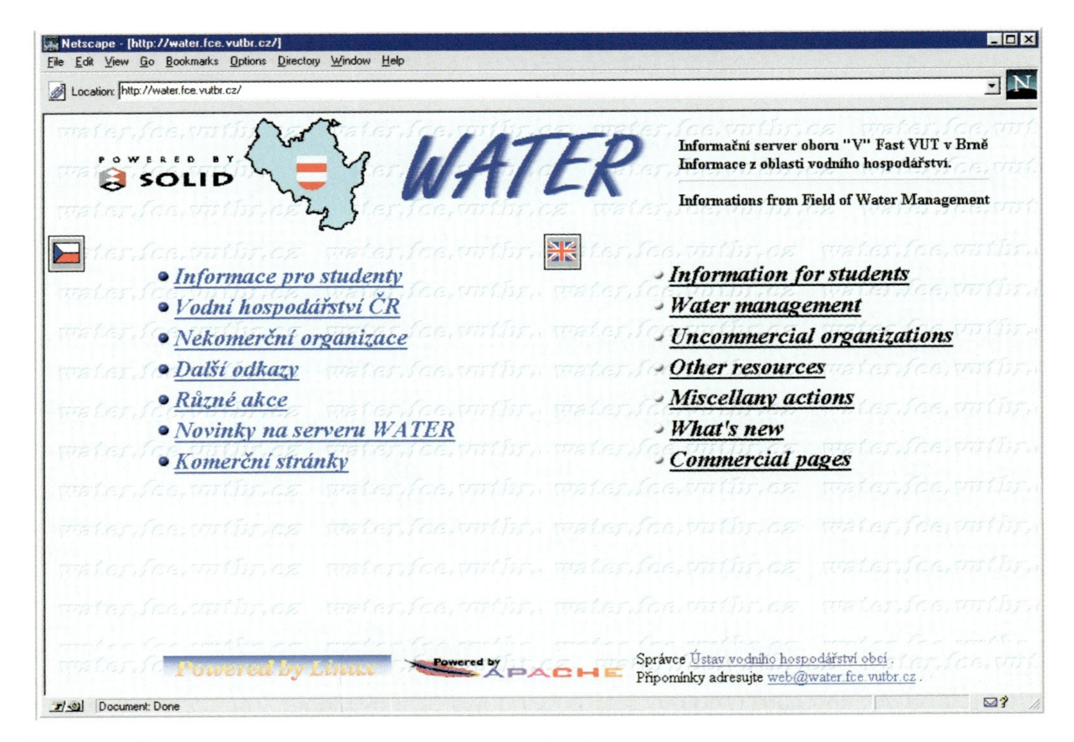

Figure 1. The entry screen of www server WATER

There are also materials which students need for the study in some courses. For example there is also a database of pumps with possibility to choose the pump for the individual conditions or exercises - Fig. 2. The database is powered by MiniSQL.

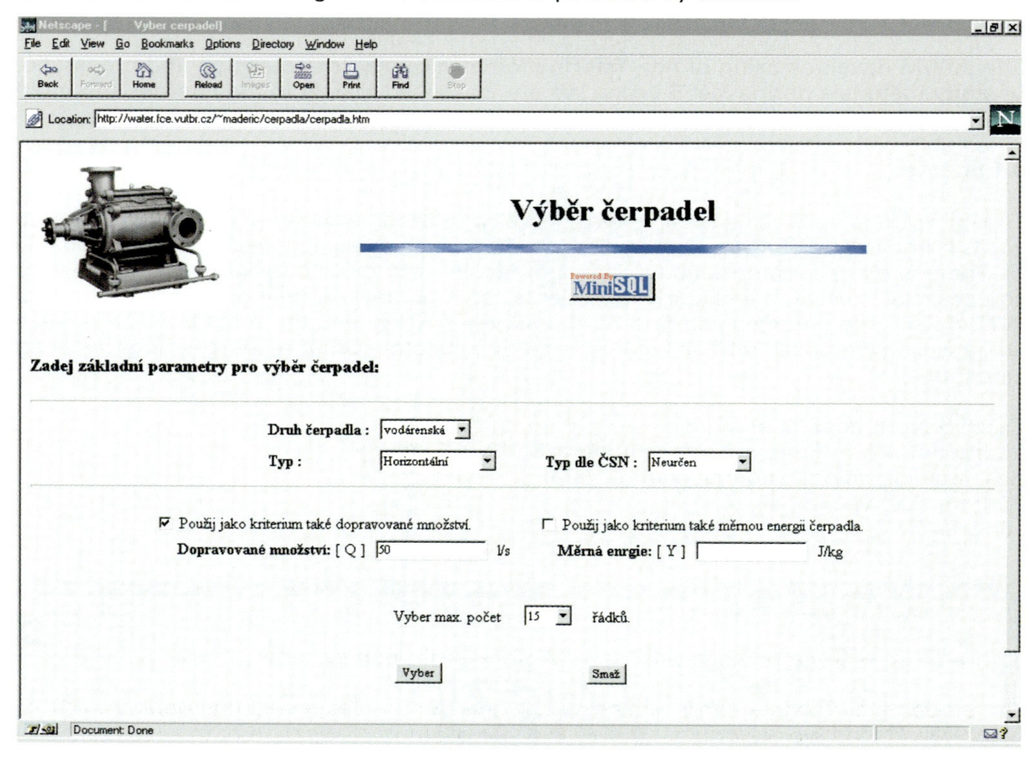

Figure 2. The possibility to choose the pump from database of pumps

3.2 Multimedia

We started to develop also the multimedia as a new educational materials approximately two years ago. We sought the relative simple and cheep software for preparing the multimedia materials. We tested several different product and the main criteria were possibility used the prepared materials also in HTML format, simple preparation with using WORD format and without complicated way of programming and friendly to end users without special software for browsing. After evaluation procedure we have chosen the Zoner Context software package. It is special software for designing of electronic book and catalog publishing. This software includes its own advanced text and hypertext *Zoner Context Editor* which combines a wide variety of data types - text, tables, imagines, drawings, databases, animation, audio and video.

Zoner Context Compiler prepares the final output files for end users. They must use a freeware viewer *Zoner Context Reader* for reading the final electronic book or other materials. Zoner software package need a minimal hardware configuration (486 processor, 8 MB of RAM and VGA), software requirements is Windows 3.1x or higher. The software exist in the Czech and also in English version. More information you can obtain in http://www.zoner.com .Last year we prepared first version of electronic book "Water distribution systems" (Fig.3) with this software as diploma project of one of our students.

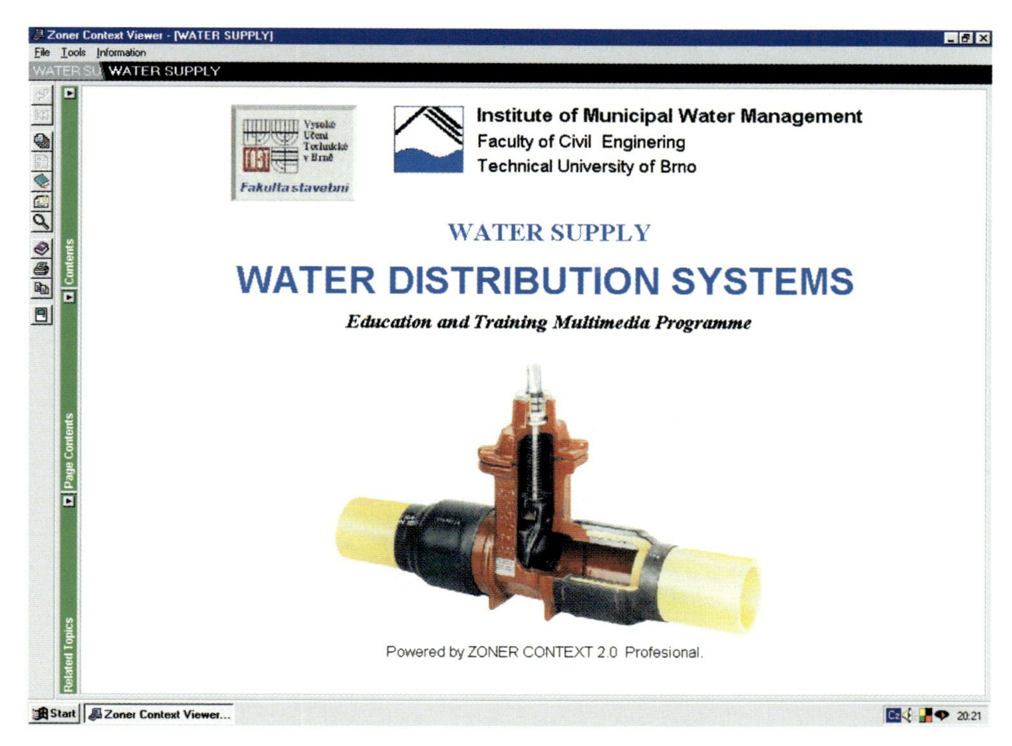

Figure 3. Title screen of electronic book "Water distribution systems"

Our students and workers from three waterworks companies tested this book. After their recommendation we are preparing now the first commercial version of this book in Czech language with the financial support of firms HAWLE, Krohne and HOBAS. This version will contain the text, photos, graphs, tables, equations, animation sequences, audio comments, video sequences, several separate software products, some catalogs of products and final evaluation test of knowledge. We plan that main user of this study materials will be our daily and distance students, managers and workers of water industry, designers and all people of this professional interest.

4. Conclusion

We obtain the first knowledge with these technologies. We are sure that these technologies represent the near perspective in preparing of study materials for all students included open and distance learning. We would like to contact our partner in the other countries with the same interest. We see the potential cooperation with partners under programme as Leonardo da Vinci and other programs of EU. We hope we will have possibility to exchange our opinion and knowledge with other universities and special training institutions.

References

Ganoulis,L., Rorris, L., Van der Beken, A., Couttier, S., Neveu, G., Thirriot, C., Couchoud, M. & Diaco, L. (1998) Use of new technology for the education and training in the field of environment/water: the European Project "TEST-EAU". In: *European Water Policy* (Proc. Venice Inter. Conf., May 1998), 193-200.

Tempus (1998), Structural Joint European Project S_JEP 09109-95 Continuing Education Centre for Water Industry. Final Report and Statement of Expenditure, Department of Hydraulics & Hydrology, Czech Technical University in Prague, Czech Republic.

Zoner Context 3 (1999), Users Manual, Zoner Software, Brno, Czech Republic

TRITON: A distance learning experience using modelling as a tool in teaching integrated urban drainage

Eric Warnaars, Poul Haremoës, Egon Loke
Department of Environmental Science and Engineering
Technical University of Denmark, DK-2800 Lyngby, Denmark

Abstract
This paper focuses on the experience gained at the department in developing and running a distance-based course that used software modelling tools to educate university students and industry professionals in the urban drainage field. The content of the course deals with the integration of the three components in the urban runoff system, namely: the sewer system, the treatment plant, and the receiving waters. Adequate understanding of the interplay between these three components is only possible through use of models and computer simulations. Therefore the core of the original training course was based on the use of commercially available computer simulation models, used by students to complete a series of exercises based on a case study. The course ran successfully for five years despite certain drawbacks and valuable lessons were learned. Finally, the content and concept of the original course was combined with the lessons learned to create a new distance-based course.

Résumé
Ce papier porte sur l'expérience acquise dans le développement et le fonctionnement d'un cours à distance qui utilise des logiciels de modélisation dans l'éducation des étudiants et des professionnels dans le domaine de l'hydrologie urbaine. Le contenu du cours concerne l'intégration des trois composantes du système d'écoulement urbain, à savoir: le système des égouts, la station d'épuration et les récepteurs. La compréhension adéquate de l'interaction entre ces trois composantes est possible uniquement par l'utilisation des modèles et des simulations sur l'ordinateur. Par conséquent, le fondement du cours original a été basé sur l'utilisation des modèles de simulation commerciaux disponibles, utilisés par les étudiants pour compléter une série d'exercices basés sur une étude de cas. Le cours a fonctionné avec succès pendant cinq ans malgré certains inconvénients et des leçons précieuses ont été apprises. Finalement, le contenu et le concept du cours original ont été réunis aux leçons apprises afin de créer un nouveau cours à distance.

Introduction

Traditionally, university education in the engineering sciences has been based on a system of classroom lectures that end in an evaluation that takes the form of a single examination often in combination with exercises and/or reports that must be submitted. Lecture times are fixed in time and place and this is where the majority of the understanding is imparted to the learners.

The abatement of pollution problems in the aquatic environment has now become a key aspect in the training of civil engineers in the water sector and has triggered a need for the retraining of practising engineers. In particular, urban areas are the cause of problems owing to the large quantities of runoff water during rain events that enter the sewer systems and cause overflows into receiving waters. Research made it increasingly clear that to adequately manage receiving waters all contributing sources needed to be considered. This was a clear change of paradigm that required new courses to be developed.

However, developing such a highly specialised course in a newly emerging area meant that the number of students and industry professionals available at any one place was very small. In addition, the research results that made up the core of the material that needed to be taught was dispersed among different organisations. Thus, to collect as many learners as possible into this new field and to integrate the knowledge base, a proposal was put together to create a distance learning based course. The distance element liberated the course from the constraints of place and time, which was an important consideration for industry professionals (i.e. outside working hours).

Background

The Department of Environmental Science and Engineering (IMT) at the Technical University of Denmark (DTU) decided to seek funding from the COMETT II programme of the European Community to develop a course that would embrace a holistic approach to the management of urban waters considering both the quantitative and qualitative aspects.

Since the topic for this course was relevant throughout Europe it was felt appropriate to seek funding on a European scale and to involve experts across Europe to contribute their knowledge to the teaching of this topic. To sustain this Europe-wide dimension it was additionally important that the course be accessible to students (and professionals) throughout Europe. Hence a distance-based learning approach was decided upon.

The EC-programme COMETT II was adopted December 1988 for a period of five years starting on January 1, 1990. The IMT project that was submitted to the Commission was asked to merge with another project co-ordinated by the Greek Company: Action Business Centre and the result was a pilot project called TRITON. The first course resulting from this project was the computer conferencing training course on 'Integrated Urban Runoff'.

The integrated concept

The basic concept behind the course is that the three components in the urban runoff system form a unity:

- The sewer system
- The treatment plant
- The receiving waters

This unity is best designed and operated by taking the interplay between the three components into account. However, this is in sharp contrast to the practise of the past and for the most part the present as well. In the extreme case, the sewer system is designed and operated by civil engineers with fixed rules as to the permissible discharge to the receiving waters and the load on the treatment plant during rain.

In turn, the treatment plant is designed and operated by mechanical engineers (or chemical engineers) with fixed (or poorly defined) discharge rules during rain events. The receiving waters are supervised by natural scientists, for the most part made up of biologists.

Recently, discharges have also come to be regulated by issuing financial levies or penalties, and this brings the economist into the picture.
This mixture of operating and regulatory agencies and the tendency to use fixed rules or simple levies or penalties does not facilitate optimal solutions. However, by increasing the knowledge of the totality of the system; by better understanding the significant interplay between the components; and by improving computer programmes with which to simulate the performance of the whole system, provides the basis for better design and operation of urban water systems to decrease the pollution impact on aquatic systems.

The original TRITON course

The course content that was required to form a basic understanding of the system components was no different to that which had been taught in each of the three component-areas. Established courses existed on sewer systems, wastewater treatment plants, and water pollution. What was required was a way in which to join these three established areas into a single course.

The only way in which the real dynamics of the interplay between the sewer system, wastewater treatment plant and receiving waters can be conveyed to the learner (and brought to life) is through the use of computer simulations that model the total system. Using computer models as part of the course will involve the active participation of the learner in preparing, running, and analysing the results of the models. This active participation is, in turn, a proven way to convey understanding to the learner.

Thus, the core of the training course was based on the use of computer programmes, which have been developed by industry professionals to such a degree that the complexity of the system can be simulated.

The topics covered in the Course were divided into two parts. The first part concentrated on the quantitative (hydrological) aspects of modelling the integrated system. The second part concentrated on the qualitative (pollution) impacts on the receiving waters. Each part was built up out of 7-modules, with text material that presented the basic theory followed by exercises that applied the theory by using the simulation model to solve the questions. The exercises were all based on the same case study, which was a catchment area in greater Copenhagen. The catchment and all the relevant parameters and data were presented to the students in the first module. From there each module built on the previous increasing the level of complexity of the system until the last module where the student was asked to optimise the system.

The key elements in Parts I and II were:

- Calibration of model parameters
- Evaluation of uncertainties in the simulation results
- Impact on receiving waters from urban runoff pollution
- Dynamic control of sewer systems
- Evaluation of alternatives based on the integrated concept

As mentioned previously, the course gained a high level of quality and a European Dimension by involving several experts from different European Countries. This meant that different authors throughout Europe wrote the text that made up the basic course material. In addition, the main author of the module actually carried out the correction of the exercise associated with that module.

In conducting the exercises the students made use of commercial software models. These models, together with the hardware locks and course material was sent by post to each of the participants; or given to them if they were able to travel to attend the initial meeting held at the start of the course in Denmark at IMT-DTU. The purpose of this meeting was to give everyone an opportunity to meet before returning to their home organisation. The opportunity to meet in person was felt important to entice participants to make use of the communication and discussion facilities available. It was also the opportunity to explain how to use the communication and discussion facilities and how to install the model software and where to place the data files.

The learners were expected to submit their responses to the exercises according to a fixed timetable. This was one module per week. The answers to the questions were found by running the computer simulation and submitting these to the teacher assigned to that module. The teacher would then have one week to review and respond to the answers. The teacher would inform the course co-ordinator when a student successfully completed a module. The co-ordinator recorded the progress of each of the participants and at the end of the course issued a certificate. Each of the two parts of the course could be taken separately. Although if Part I was not taken then proof of equivalent education on that subject was required before it was possible to register for Part II alone.

Experience

It should be borne in mind that this course was originally developed and run in a distance learning mode before the advent of the Internet. Today we have become very familiar with the concept of e-mail and other forms of distance communication, such as on-line chat sessions, discussion groups and electronic mail-bases often with archived discussion strings available on web-sites.

Originally, as the course was run for the first time in September 1992 through to 1994 a costly mainframe conference programme, called Portacom, was used. The experience of using this system was that it was cumbersome and not inviting enough to the participants to entice them to use the message boards and discussion rooms that were created in the system. The main use was simply to post answers to the exercises and communicate with the teachers and co-ordinator. This can be equated to the way in which e-mail is used today, only at a significantly greater cost.

Under the COMETT II programme the courses officially stopped in 1994 as no request was made for further funding. The course averaged 15 students from across Europe annually, including both university students and professionals.

From 1995 to 1997 the course was integrated as a normal course at DTU (Course 63421, Integrated Urban Runoff). The course was targeted at 4th year students at IMT-DTU. The decision was taken to continue to run the course in a distance-mode as the concept of the electronic classroom was (in 1995) a hot topic. At this point in time e-mail was becoming well established and this replaced the Portacom system that had by now become obsolete and prohibitively expensive.

In these three years, all participating students were from DTU and a special computer-lab was made available to the students to enable them to complete the course and to better protect the commercial software. Some of the teachers, however, were still external and so the European dimension was maintained.

The limited number of computers available in the computer lab meant that students were placed into groups of two to complete the course. On average there were some 12 students taking the course over these three years.

In 1998 the decision was taken to restructure and update the course. The rational for this decision was based on the key lessons learned from the experience of running the course for 5 years.

Lessons Learned

In year one of the course the feedback from all participants was that the course was too demanding and could not be completed within the allotted time. In fact there were serious delays and teachers found themselves marking exercises weeks after the due date. This trend continued even when students were placed in groups. It seemed to take students almost twice the allotted time to complete the modules and associated exercises. Thus the exercises were simplified in 1996 and again in 1997.

The commercial software programmes that were used were far to detailed and cumbersome to use. The 1992 TRITON-course was developed on the basis of a combination of two simulation programmes, one for the sewer system and one for the receiving waters. However, the commercial programme turned out to be increasingly unfit to be used as a teaching tool. In fact, the complexity of the commercial programme became an obstacle to conveying basic understanding. Thus, as opposed to spending time gaining insight into the interplay of the system the student was learning how to use the commercial model and lost the overview.

One of the other aspects that surfaced regarding the model was the displeasure of the owners of the commercial software regarding cases of unauthorised software use. This was the reason for installing the software on machines in a computer lab and no longer allowing the software to be installed elsewhere.

Overall the course turned out to be over-ambitious in terms of the material that had to be completed and the commercial software proved too unwieldy to achieve the educational objectives of the course. Nonetheless, the experience gained in running this course in a distance learning mode is invaluable.

Future

The advances in Internet technology and its widespread acceptance and use as an information resource open the way to re-establishing the TRITON course as an international course; although with some significant modifications, based on the lessons learned.

In 1998 a proposal was submitted to received financial support from the European Commission, Directorate General XXII – Education, Training and Youth, under the SOCRATES-programme for transnational cooperation projects under the action "Open and Distance Learning" (ODL). This proposal was successful and funding was provided initially for one-year for the ETNET.UWREM-project (European Team for New Methods of Education and Training in Urban Water Resources Management). The project started on September 1st, 1998.
This new course focuses on the pollution aspects and not on the hydraulic. The number of modules and exercises has been reduced to 7-modules. The teaching objectives of the course are as follows:

- Provide the student with a basic understanding of the features of interplay between the three components in the urban drainage system: the sewer system, the treatment plant and the receiving waters.
- Provide the student with an understanding of the function of each facility in relation to the whole system; including design of new facilities, like structures of best management practise, overflow treatment.
- Illustrate to the student new approaches to cost-effective performance, like real time control.

The course is intended primarily for university students aiming at an MSc-degree in environmental engineering. The course is a superstructure on top of the traditional courses, recommended for the last semester before the MSc-thesis.
The tools applied for distance learning are:

- A simulation model available as a computer programme, downloaded from the homepage of IMT, DTU, with a learning friendly interface suitable for distance learning.
- A result presentation programme with animation of results, so as to facilitate the understanding of the interplay between the components in the urban drainage system.

- Bi-weekly exercises, consisting of problem-sets that require the learner to use the simulation model and results-animation tool in formulating the answers.
- A textbook providing the core information to understanding the concepts in each module before starting the exercises and user manuals on use of the programmes.
- On-line chat sessions between the student and the supervisor, with support and explanations to the students during the problems solving of the exercises.

The model and the interface will be as simple as is necessary to handle the simulations. The model is simple in order to run fast and not to confuse the key issues with the mechanics of model sophistication. Integration forms the key issue: How do the different components interact. This cannot be understood with the complication of the most sophisticated approaches. The choice is further motivated by the fact that uncertainties are more associated with the lack of knowledge of the processes in the sewer system, the treatment plant and the receiving waters than with the refined detail of the hydrodynamics of the hydraulic flow phenomena. The standardised unit hydrograph (corrected to fit kinematic wave properties) is used for flow simulation. The event mean concentration is used to calculate pollution mass transfer. That is more than sufficient for the purpose of this course. The graduated engineers may use more sophisticated models in practise when the practical circumstances justify the higher degree of sophistication.

Conclusion

The original course was at the cutting edge of technology. The dynamic nature of the subject matter meant that using a computer simulation model was a very appropriate way to teach students, when combined with a specific case study. However the original course was too ambitious and the timeframe for completion was impossible. In addition the model was too detailed and learners became lost in the mechanics of the model, losing the overall insight into the dynamics. Thus, a complete revision of the course occurred reducing the volume of work and creating a dedicated model tailored to the educational need of the course. The key element that prevails in both versions is applying an interactive approach to learning whereby the student uses a simulation model to gain an understanding of the cause-effect relationship of his actions on the system and so better comprehend the complex dynamics involved. This interactive element (the computer simulation), a well-defined case study, clear exercises and a timetable for submitting exercises makes this course suitable to run in a distance mode.

Towards rivers where nature and human meet, through a European case studies-module

Huygens M., Verhoeven R.
Hydraulics Laboratory - University of Gent, Belgium

Abstract
In the framework of the Socrates European Module Pilot projects an international collaboration between seven European partners was set up to produce an example of possible approaches to river restoration. The basic idea of the module "River design and environmental protection in Europe" was to collect some illustrative case-studies on integrated water management and environmental development, as they independently are developped by the respective international partners. Each participant explores an example of a national project within a global schedule of approach which was commonly agreed before. All seven participants (Spain, France, Sweden, Denmark, Norway, Wallonia and Flanders) show a typical project on integrated river water management, oriented to local problems, achievements or targets. It is very interesting to collect and discuss this broad spectrum of actions in the framework of river restoration. From there, one gets a more common (European) view on the global river management and the respective approaches in some countries.

Résumé
Dans le cadre de projets Européens du programme Socrates, une collaboration internationale entre sept partenaires est élaborée pour la création d'une module éducative sur l'idée de restauration des rivières et les approches typiques dans les pays Européens. L'idée fondamentale de cette module "Aménagement et protection environnementale des rivières en Europe", c'est de collectionner des projets illustratives sur l'aménagement intégrale et le développement environnemental de rivières dans les pays partenaires respectives. C'est ainsi qu'apparut l'idée de favoriser la création de plusieurs équipes multidisciplinaires dans les différentes universités pour préparer une documentation qui puisse servir à une analyse comparative de la façon d'affronter le dessin des espaces fluviaux dans quelques pays de l' Union Européenne. Chaque partenaire décrit son exemple nationale dans un cadre commun où on avait définie les sujets à appliquer sur la base d'une table des matières. De là, les sept institutions (de l' Espagne, de la France, de la Suède, du Danmark, de la Norvège, de la Wallonie et de la Flandre) ont représenté leur propre projet d'aménagement intégral d' une rivière représentative ; à la fois d' une manière globale mais aussi en attirant l' attention sur les problèmes locales et les solutions spécifiques. De cette façon, ce livre puisse être utile comme matériel d' enseignement dans tous ces cas où on prétend accroître les possibilités d' une compréhension réciproque entre spécialistes ou étudiants des différentes disciplines indiqués.

1. International context

One of the principal achievements of the United Nations Conference on the Environment and Development (otherwise known as the "Earth Summit") held in Rio de Janeiro in 1992 was the signing of the convention on Biological Diversity by many countries, including the

United States of America and virtually all European States. To prevent a decline in productivity and the quality of life the Earth needs to be managed to conserve biodiversity, and biological resources have to be used sustainably. Each signatory to the Convention has to prepare national plans for the conservation and sustainable use of biodiversity under Article 6, and is required to integrate these plans into other sectors. Within Europe, for example, the conservation of biodiversity has become a prominent issue since 1992, where a high level of public awareness and concern has led to the need to complement a protectionist approach with new policies designed to create or restore nature and wildlife. The European Directive on the Conservation of Natural Habitats and Wild Fauna and Flora states that a coherent European ecological network be established under the title Natura 2000 (European Union 1992) to maintain or, where appropriate, restore. Such concepts are leading directly to the promotion of actions outside strictly protected area and are introducing the principles of sustainability for all forms of natural resource management (Stanners and Bordeau, 1995).

With the growing recognition of biodiversity issues worldwide, there is a clear need to consider both the protection of existing resources ans the restauration of damaged ones to sustain ecosystems. Goals for river management should include improving the physical condition of watercourses to provide a diversity of habitats, reducing the self-purification capacity of rivers.

Within this context the purpose of the book is to develop guiding principles for use by river managers in striving towards sustainable river channel restoration projects. The approach focuses on sustainability (and therefore recoverability) and views the system as interconnected (i.e. reach, floodplain and catchment). Sustainability requires re-establishment of biophysical processes and interactions that confer persistence and resilience in the face of normal environmental perturbations over the longer term. This concluding chapter reviews some of the challenges and critical issues and suggests solutions that are required to accelerate progress.

2. Origin and implementation of the work

A new research team on river design and planning was created at the University of Cantabria (Spain) during 1994 by putting together the resources of five scientific areas: ecology, physical geography, regional geography, hydraulic engineering and city and regional planning, and by taking advantage that rivers are of common interest for all five disciplines. In the mean time, the most recent studies on river interventions started to suggest the use of multidisciplinary approaches and that interventions should not be restricted to the river channel but should also include land uses and activities near the river channel. Thus, a discipline that had been considered the exclusive concern of hydraulic engineering was developing towards a multidisciplinary situation.

In due course, the European Union countries had started to incorporate multidisciplinary approaches, on a greater or lesser extent, evolving from very different administrative and physical situations. But even taking into account that rivers and fluvial territories are an integrated system not divided by disciplines with interrelations between hydraulic, ecological, land use, etc. aspects, in most of these countries knowledge on rivers was still transmitted to students through unidisciplinary views.

As a result of these, it was considered interesting to promote the creation of multidisciplinary teams in several universities, in order to elaborate a joint material to be used for a European comparison of the extent that design of fluvial areas was done taking into consideration not only hydraulic criteria, but also environmental (ecological and geomorphologic) as well as land use criteria. This material would also be useful for university graduate level courses open to students from several disciplines.

This project was launched in 1995 to match the deadline for European Module projects under the European Union SOCRATES Programme, in its experimental phase. Two of ten priority areas of this call for projects were Design in Civil Engineering and Environmental Protection. The project we have referred to matched perfectly with this call for projects and became specific with multidisciplinary teams from Belgium, Denmark, France, Norway, Spain, Sweden and the United Kingdom. The proposal "River Channel Design and Environmental Protection in Europe" coordinated by the University of Cantabria was approved, the work started in September 1996 and it was finished in June 1998.

The project basis consists of the analysis of several river transformations and actions, one river per country (with some exceptions); a common table of contents was defined for all countries. This table of contents was defined to include all the issues that are raised in most river interventions. The table of contents for each case study includes three major fields. First, the existing information and the description of the river characteristics, second, the administrative framework under which actions are undertaken, and third, the objectives and techniques that are used. New knowledge is required to intervene on rivers in a compatible way with the environment, but in many cases it is enough to consider jointly the already existing knowledge of different disciplines. The aspects that have been analysed in each case study are going to be described and justified below, in particular their usefulness to understand the requirements for multidisciplinar and integrated fluvial design and planning.

The table of contents for the first section, which embrases the river characteristics and the existing information about them, includes the following aspects:

- The natural river characteristics in relation to climate, relief, hydrology, fauna and flora.
- The river, floodplain, and basin transformations.
- The pressures excerted upon rivers and their public image.
- The existing information on river corridors, floodplains and basins (liquid and solid flows, flows in channel and floodplains, flora and fauna, water quality and uses, etc.).

Human interventions on rivers and their corresponding technical studies may depend, in our opinion, upon the four groups of aspects indicated above: their characteristics, their degree of transformation, the existing information, and the pressures that are considered relevant by public opinion. The lack of information on any of these aspects means that it will not be taken into account in the technical design process, but also the way rivers are conceived in each country.

The table of contents' second section, which embraces the administrative framework under which actions are undertaken, covers the following aspects:

- The legal and regulatory framework.
- The institutions responsible for river management and their relations to those responsible for land use planning and for the protection of nature.
- The policies and the administrative and public participation procedures used when acting on rivers.

Each one of these three aspects is also indicative of the way rivers are conceived in each country. The emphasis is paid first, to the existing legislation and whether it refers only to hydraulic issues or to a more comprehensive fluvial understanding, second, to the type of river administration, being it a basin specific administration or an administration responsible for many other issues and which divides the river in several portions, and third, to the existing relations between the administration responsible for rivers and those responsible for land use planning and protection of nature. These three aspects highlight to what degree rivers are considered under a very narrow hydraulic and channel vision or under a more broad ecological and territorial understanding.

The table of contents' third section, which embraces the policies and techniques used for river interventions, covers the following aspects.

- The priority objectives for river interventions (i.e. supply of water, flood protection, environmental protection, etc.).
- The techniques used for river interventions.
- The technical studies, the way and degree the river channel and its adjoining land are taken into account , and the consideration given to scientific disciplines (hydraulics, ecology, geomorphology, land use planning, etc.).
- The evaluation procedures used by the technical studies.

This third section seeks to clarify to what degree rivers are considered complex realities by analysing the pursued objectives, the techniques used, and the technical studies.

3. Structures and teaching use of the book

The book is organized in three parts: the first one is of a general character, while the second one is composed of the river case study of each country. Finally the third one is a comparative analysis of the indicated case studies.

The work contained in this book is the result of a cooperative effort involving eight universities and several non-university institutions and of the work of thirty-three university faculty members and around twenty technical advisers. The work attempts to focus attention on the environmentally compatible fluvial developments that are taking place in several European countries. Emphasis is also given to the need for a multidisciplinary approach to fluvial management in which town and country planning plays an important role. The project has been mainly financed by the Socrates-Erasmus program of the European Commission. But it would have not been viable without complementary financial support from university and departmental funds and without the unselfishness dedication of time by the people. The river case studies presented in the book have only been possible because the institutions responsible for river management have provided information, materials, maps, etc.

All the texts are in three languages: english, french and spanish, but all the tables and maps are only in one language (english); this is why at the end of the book there is a list of technical words in the three languages.

The first chapter, as it has already been said, is an introduction to the study itself and to the general situation of rivers in Europe. It highlights the different situation of European rivers in relation to the impacts caused by the use of fluvial resources, by the use of land near them, and by engineering constructions. It will become clear later on that European rivers have a quite different present situation if compared by country or regionally, and this fact would have to be reflected in the priorities considered in river management and design in each country.

Chapter two seeks to present a theoretical framework to approach river design in a compatible way with nature. Based on the arguments pointed out below and during the implementation of the study, it was decided that this theoretical framework would be composed of river restoration general principles. One of the objectives of this study is to understand how to transform fluvial areas in an environmental compatible way. This is done in three different ways: first, by reducing the impacts on rivers that are in good natural conditions but cause problems for humans; second, by maintaining or enhancing the environmental, hydraulic, geomorphologic, and land use characteristics of rivers that have a good environmental situation; and third, by enhancing or restoring rivers that are deteriorated. In our opinion, the issues that have to be considered in all the above three mentioned situations appear as well in most river restoration projects. This is why chapter two reproduces the text from Brookes and Shields (1996) "Towards an Approach to sustainable Projects" first published in the book "*River Channel restoration. Guiding Principles for Sustainable Projects*" of the same authors.

The chapters that compose the second part of the book describe the rivers selected as case studies for each country.

* *"Planning of fluvial areas in northern Spain: the river Saja"* has been done by the University of Cantabria team. The Saja basin is a relevant example of the river situation of the Spanish cantabrian watershed, in which river corridors play a major role in the settlement and communication patterns and where the main impacts happen in the lower section of rivers.
* *"Planning of the river Isère in France: a centenarian evolution"* has been done by the team of the Joseph Fourrier University and the National Polytechnique Institute both from Grenoble. This case highlights the transformation process of the Isère river and how nowadays interventions are related to or a consequence of previous actions exerted on the river.
* *"The Mark-Vliet basin, a cross boundary study of the Netherlands and Flanders"* - "Managing small streams in Flanders, Belgium: the Gonde-Molenbeek river" have been done by two teams from the University of Gent. The two most relevant arguments of the first one are the situation caused by different water standards in the two countries and the importance of sediments as carriers of pollution. The second case study underlines the addition of small impacts importance and the need for adequate methods and technologies for small streams, often different from those used for big rivers and in many cases derived from or induced by public and/or local community involvement.
* *"Planning of the river Ourthe in Wallonia, Belgium"* has been done by the University of Liege. The risks of camping and second home sites recently installed on floodplains, the importance given to river continuity for salmons, and the different administrations responsible for and legal systems applicable to navigable and non navigable rivers are highlighted in this case study.
* *"A case study of the Swedish basins: Helgean"*, that has been produces by the University of Lund, indicates that agricultural land and pollution, public access to nature, and reduction of nutrients load into the Baltic Sea are key issues to understand the way rivers are treated and managed in Sweden today.
* *"A restoration project for de river Skjern: Denmark"*, done by Alborg University, presents how an old channelisation project of the river Skjern lower section and its parallel land reclamation project for agriculture, typical example of river interventions undertaken during the sixties, is now almost totally retransformed through a restoration project, through specific means, technologies and institutional framework.
* *"Hydro-electric use and fluvial planning in Norway: The river Orkla"* has been produced by four institutions located in Trondheim, the Norwegian University of Science and Technology, SINTEF, the Norwegian Institute for Natural Sciences and the Norwegian Administration for Hydraulic Resources and Energy. This case study calls our attention to the great number of technical regulations produced by the Norwegian administration in order to design better river interventions, to the experience of multiuse planning of rivers and related land, and finally to the preservation of some Norwegian rivers from energy production.

European partners of the project

The book's third and final part is a comparative analysis of the case studies. The comparison has a similar structure to the one followed for each case study.

This book is addressed to persons which already have a basic knowledge of rivers. Basic fluvial ecology, geomorphology, hydraulics or land use planning are not included neither the objective of this book, and thus the reader should not expect to find in it a systematic explanation of river dynamics or engineering technologies. Taking this into account, the book may be useful as an interdisciplinary and communication learning material between specialists or students of the different disciplines already indicated.

4. Future developments

During the development of the module, a preliminar lecture series was already given at the University of Santander (Spain) to point out the possibilities and shortcomings in this teaching project. An internal evaluation was used to finalize the module-lectures notes of the book. Currently, the Swedish partner from the University of Lund (Department of Water Resources Engineering - Prof. R. Larsson) has already applied this book-module as part of his courses.
A local educative website to find out about the course was set up for an international student group at http://aqua.tvrl.lth.se/rolf/flodrest/RIVER_HOME.htm, with all relevant information and links to associated partners and their activities. Actually, the final editorial actions are finished and a hard copy of the book is available for all European students.

Besides the intensified scientific contacts between all European partners in the project, the result of this collaboration will help to understand and explore a more universal river design and environmental management in European countries. By teaching this module, young people will get a basic knowledge to support a further development towards inviting European rivers in the nearby future.

References

River design and environmental protection in Europe (1998), Universidad de Cantabria - Scientific Editor J.M. de Urena, ISBN 84-8102-202-0, Graficas Calima Santander, Spain.

River channel restoration: guiding principles for sustainable projects (1996), A. Brookes and F.D. Jr. Shields, John Wiley & Sons, Chichester, UK.

Directive on the conservation of natural habitats and wild fauna and flora (92/43/EEC) (1992), European Union, Brussels, Belgium.

Theme F

Postgraduate education and research training

Enseignement post-diplome et formation par la recherche

Theme F: Postgraduate education and research training

Rapporteur: J. Ganoulis

Greece

Postgraduate education and research training in Hydraulics and Water Resources Engineering is a major concern of Universities and research institutions. Trends in formulating new curricula in this domain reflect actual changes and developments in the Water Resources Engineering profession. This has become even clearer from the most interesting papers presented in this session.

There is no doubt that water resources currently face multiple threats leading to the so-called 'water crisis'. On the one hand, economic growth in different sectors, such as tourism, industry and agriculture, has resulted in a significant increase in water demand and brought about increased social conflicts. On the other hand, global environmental change and hydrologic uncertainties, together with the increasing pollution of rivers, lakes and groundwater, have decreased the availability of water in different sectors.

In his keynote lecture Prof. H. KOBUS described the activities of the IAHR European Graduate School of Hydraulics, which also are a specific project of the European Thematic network on Education and Training for Environment-Water (ETNET. Environment-Water). He stressed that a unified approach for research training and professional education should be under taken.

A different philosophy is emerging in the engineering profession, aiming to take into account not only technical and economic reliability, but also environmental risks and social equity. The formulation of new methodologies integrating technical, economic, environmental and social issues is a challenge towards a new paradigm for sustainable water resources management. The above question is a specific project of ETNET. Environment-Water, entitled ***'A European Paradigm for Integrated Water Management'*** (SP-E).

This shift is also reflected in the postgraduate curricula presented by Prof. BARGAOUI (Tunisia). Prof. THEVENOT (France) presented a common 1-year postgraduate curriculum of three French Institutions (Paris VI, ENPC, and ENGREF) on Environmental Science and Technology, containing interdisciplinary modules on system analysis, modelling and decision making. The European experience from two Research Training Networks (WET and SWAMIEE) was summarised by Dr. SHIEL (EU). The role of new information technologies in postgraduate training was emphasised by Dr. MOLKENTHIN (Germany), who described the programme "Hydroinformatics" under the IAHR European School of Hydraulics.

The main conclusions of this session may be summarized as follows:

1) There is a need for a unified approach in postgraduate education and research training.

2) A interdisciplinary methodology integrating environmental and social aspects is needed in postgraduate education and training.

3) New tools emerging from new information technologies should be developed and considered as an integrated part of postgraduate education and research training.

Keynote lecture

IAHR European Graduate School of Hydraulics: Towards a European Education and Training Program

Kobus H., Lensing H. J.
Institut für Wasserbau, Universität Stuttgart, Germany

Abstract
The IAHR European Graduate School of Hydraulics (IAHR-EGH) is part of the European Thematic Network of Education and Training for Environment-Water (ETNET. ENVIRONMENT-WATER) and a main activity of the IAHR Committee on Education and Professional Development (CEPD) [4].
IAHR-EGH is a network of more than 50 universities and research institutions from 18 European countries [6] offering short courses directed at graduate students, post-graduates and professionals. A course program spanning interdisciplinary fields in hydraulic engineering on a high academic level with appropriate quality control is presented. Successful completion, proven in a final examination, is acknowledged by an IAHR certificate. In order to classify the course program, the courses are differentiated into two categories: design oriented courses directed mainly at MSc students and professionals from industry, and research oriented courses directed mainly at PhD students and staff from universities and research institutions. Additionally, technical training courses can complement the course program. In the long run the various courses will be arranged to complement each other to form specialized segments of higher education oriented towards hydroscience, water engineering and water resources management.
The IAHR-EGH had a successful start. The first six pilot courses 1997 were attended by 120 participants from several European countries. In 1998, 15 courses were carried out. For 1999, 20 courses are announced by the network partners.
Based on the experience of the courses 1997 and 1998 and the developed modules of an international education and training network the recommended steps towards establishment of a coordinated course program and a complete framework for IAHR-EGH concerning quality control of the courses, IAHR-EGH visibility and structure is presented.

1. Introduction

Water is indispensable for life and for the development of our societies and has become a coveted economic good, whose management will be one of the main global issues of the 21st century. Water related problems such as water shortages, floods and water pollution are not limited to specific regions or countries. The development and realisation of sustainable solutions are called for on an international scale. Global needs that require specialized education and training have been recognized by UNESCO and by water-related international organizations such as IAHR [1,2,3,4].

For coping with water related problems, an increasing number of well qualified professionals is required. New techniques and research results must be transferred into practice immediately. Furthermore, a need for qualified and specialised training of research students oriented at interdisciplinary collaboration and holistic management exists. Since formal education programs in Europe are traditionally oriented at the respective national systems, there is a specific need for continuing education and training courses on a European level.

2. Scope and objectives

The following general agreement of the IAHR-EGH network partners was a first important milestone towards implementation of the network: IAHR-EGH should function as an umbrella for the benefit of all partners, channeling and coordinating individual activities with a minimum of administration, maintaining a maximum of individual freedom and responsibility for the various course organizers, but providing quality assurance for the activities. Regardless of the coordination by IAHR-EGH, full responsibility for content, finances and arrangements for the course lies with the course organizers.

The individual courses offered by the network partners under the umbrella of IAHR-EGH are normally taught in English, and they are directed at graduate students, post graduates, scientists and professionals. The IAHR-EGH program spans interdisciplinary fields in the water-environment domain on a high academic level. In order to classify the course program the courses have been differentiated into three main categories:
Category D: **Design oriented courses** directed mainly at MSc students and professionals from industry
Category R: **Research and development oriented courses** directed mainly at PhD students and staff from universities and research institutions
Category T: **Training courses**, which aim at providing certain skills and methods for the participants, can supplement the course program.

In the long run the various courses will be arranged to complement each other to form specialized segments of higher education oriented towards hydroscience and water engineering such as:

- fluid mechanics and dynamics,
- river hydraulics and sediment transport,
- hydraulic structures,
- coastal hydraulic engineering,
- groundwater resources: management, contamination and remediation,
- environmental hydraulics, water quality and ecosystems,
- hydroinformatics and water resources management.

3. Development of a continuing education and training network

For the process of initiating an internationally oriented continuing education and training, a certain organizational structure was developed by IAHR-EGH during the three year period 1997-1999. For the evolution of the program, the broad range of courses to be considered was grouped into main group topics, and an international steering group was established (table 1). The steering group members take care of the coordination and harmonization of course contents, titles and dates of courses and take initiatives towards a representative (reasonably "complete") set of courses and inspire the "inationalization" of courses. The system is open for further main group topics, depending on the development of the IAHR-EGH course program.

Table 1. Main group topics and members of the IAHR-EGH steering group

Main group topic	steering group members
IAHR-EGH network coordination	Prof. H. Kobus Universität Stuttgart, Germany Dr. H.J. Lensing Universität Stuttgart, Germany
Rivers and sediment transport	Prof. A. Ervine University of Glasgow, UK Prof. A. Armanini Università Degli Studi di Trento, Italy
Hydraulic structures and hydropower	Prof. E. Kalkani National Technical University of Athens, Greece Prof. F. Nestmann University of Karlsruhe, Germany
Groundwater	Prof. R. Helmig Technical University of Braunschweig, Germany Prof. S. Troisi University of Calabria, Italy
Environmental hydraulics	Prof. G. Jirka Universität Karlsruhe, Germany Dr. G. Perrusquia Chalmers University of Technology, Sweden
Hydroinformatics	Prof. R.A. Falconer University of Wales, Cardiff, UK Dr. F. Molkenthin Brandenburgische TU Cottbus, Germany

For academic recognition of IAHR-EGH courses, the definition of prerequisites, a detailed course description, satisfactory completion of a final examination and the declaration of credits are necessary. The implementation of a common credit system allows for the transfer of credits and supports the mobility of students. Furthermore, the credit system facilitates the comparison of courses and could eventually form the basis for a European degree system.
The European Community Course Credit Transfer System (ECTS) is being implemented for academic recognition of the courses. An assessment of the quality of work will be done according to the ECTS grading scale if this is requested by the participants. The ECTS system facilitates institutions recognition of learning achievements of students through the use of commonly understood measurements - credits and grades - and also provides the means for interpretation of national systems of higher education.
The goal of IAHR-EGH is to offer courses at a high academic level and to provide quality assurance for these activities. Guidelines for the definition of quality control criteria were a clear definition to avoid an excessive administration and a comprehensible control procedure for the course organizers. For the realisation of IAHR-EGH courses the following quality criteria have been defined:

Only institutions offering PhD programs in the water field can become active partners.

Lecturers of IAHR-EGH courses must be Professors at an active institution who have supervised several PhD candidates and have major publications (e.g. one paper per year in refereed journals and significant input to International Conferences) or lecturers with an equivalent qualification and experience in consulting.

The qualification must be met by the host institution and the main lecturer of the course. In case of instructor teams, supplementary lecturers with different backgrounds may also be involved.

The quality control on the level of the course is also very important. Therefore, all courses undergo a "post mortem" evaluation by the participants and the lecturers, which is used for improvement of the courses as well as an important element of quality control.

The quality control is carried out by the IAHR-EGH steering group, which also considers the course notes as well as the post mortem evaluation of the courses for their decision. The development of unified quality control and recognition procedures is an important step in offering a European dimension to the educational system.

Course announcements and certificates of the individual IAHR-EGH courses have a common logo and layout including standard reference texts for the course program and ETNET.ENVIRONMENT-WATER. In the newsletter of the International Association for Hydraulic Research, a regular page on "News from IAHR-EGH" is being published. A flyer, a post card and a poster have been prepared for the network partners and advertisement at conferences.

4. IAHR-EGH course program

The direct start of pilot courses and the parallel implementation and evolution of a framework for IAHR-EGH were the main activities in the three year period 1997-1999. This double track development of IAHR-EGH has several advantages. The experiences from the courses provide additional information for the evolution of the network, and in turn the course program enforces the development of a framework for IAHR-EGH taking care of the needs of the active institutions, the needs of the lecturer and the needs of the participants. New conceptualized criteria, procedures and network structure modules can be directly tested with respect to practicability in order to establish an effective network structure. The dual track procedure has lead directly to a practice oriented and operative program and network.

Table 2 and 3 give an overview of the courses which were offered in 1997 and 1998. The number of participants lay between 4 and 90 with a mean between 15 and 25. Most of the courses had a duration of 3 to 5 days. The duration of the final examination varies from 30 minutes to 2 hours.

Table 2. Overview of the pilot courses 1997

Pilot courses 1997	Nr. of participants
University of Stuttgart, Germany Numerical Modelling: Multi Phase Flow and Transport Processes	22
University of Stuttgart, Germany Concepts of Geostatistics and Stochastic Modelling	
Technical University of Athens, Greece Hydropower Engineering	4
University of Erlangen-Nürnberg, Germany Principles of Turbulence Modelling	57
University of Trento, Italy Regulation of Small Rivers and Torrents	16
University of Glasgow, UK River Flood Hydraulics and Sediment Transport	17

Table 3: Overview of IAHR-EGH courses 1998

Courses 1998	Nr. of participants
Oxford Brookes University, UK River Hydraulics and Hydrology	15
Oxford Brookes University, UK Coastal Engineering and Management	22
University of Stuttgart, Germany Numerical Modelling: Multiphase Flow and Transport Processes in the Subsurface	33
University of Erlangen-Nürnberg, Germany Einführung in die Strömungsmechanik	25
Oxford Brookes University, UK Urban Drainage and Network Simulation Modelling	28
University of Erlangen-Nürnberg, Germany Hitzdraht und Laser-Doppler-Anemometrie	39
University of Erlangen-Nürnberg, Germany Numerische Strömungsmechanik	92
Oxford Brookes University, UK Sediment Transport: Theory and Practice	11
University of Calabria, Italy Experimental Data and Validation of Groundwater Mathematical Model for Reclamation of Polluted Sites	12
University of Trento, Italy Regulation of Small Rivers and Torrents	15
Technical University of Cottbus, Germany Hydroinformatics Systems	18
University of Porto, Portugal Disposal of Effluents in Aquatic Environment	34
University of Glasgow, UK River Flood Hydraulics and Sediment Transport	9
University of Stuttgart, Germany Risk Analysis for Water Resources Engineering	21
Chalmers University of Technology, Sweden Ocean Wave Energy	6

Especially the research oriented courses, directed mainly at PhD students and staff from universities and research institutions, were attended by participants from several European countries. Main goal of the participants were the introduction to new scientific topics or new methods and the development of scientific contacts. The recognition of the courses for degree programs or for continuing professional development (CPD) programs were only relevant for participants from countries with formal CPD programs. The high number of successful IAHR-EGH courses underlines that there is a market for continuing education and training on a high academic level in Europe.

Table 4. IAHR-EGH course program 1999

Oxford Brookes University, UK
    RIVER HYDROLOGY AND HYDRAULICS
Technical University of Braunschweig, Germany
    SUBSURFACE MICROBIAL PROCESSES: MASS TRANSPORT, REACTIONS AND BIOREMEDIATION
Oxford Brookes University, UK
    COASTAL ENGINEERING AND MANAGEMENT

University of Erlangen-Nürnberg, Germany
TURBULENCE AND TURBULENCE MODELLING
Oxford Brookes University, UK
URBAN DRAINAGE AND NETWORK SIMULATION MODELLING
Chalmers University of Technology, Sweden
NON-STATIONARY FLOW IN OPEN CHANNELS AND SEWERS
Czech Technical University, Czech Republic
MODELLING FLOW AND CONTAMINANT TRANSPORT IN THE SUBSURFACE WITH EMPHASIS ON MULTIPHASE FLOW
Oxford Brookes University, UK
SEDIMENT TRANSPORT THEORY AND PRACTICE
Technical University of Karlsruhe, Germany
ENVIRONMENTAL FLUIDS MECHANICS
Imperial College of Science, Tech.and Medicine, UK
GIS IN URBAN WATER SYSTEMS
University of Sheffield, UK
VISUAL MODFLOW: THE PROVEN STANDARD FOR 3-D GROUNDWATER FLOW AND CONTAMINANT TRANSPORT
University of Calabria, Italy
REMEDIATION OF POLLUTED SITES: DATA SAMPLING AND GROUNDWATER MODEL
National Technical University of Athens, Greece
HYDROPOWER ENGINEERING
Technical University of Cottbus, Germany
HYDROINFORMATICS SYSTEMS
University of Porto, Portugal
DISPOSAL OF EFFLUENTS IN AQUATIC ENVIRONMENT
The University of Glasgow, UK
FLOOD HYDRAULICS AND SEDIMENT TRANSPORT
Technical University Berlin, Germany
PROTECTION OF GROUNDWATER WITH RESPECT TO HANDLING SUBSTANCES HAZARDOUS TO WATER AND SOIL
University of Trento, Italy
REGULATION OF SMALL RIVERS AND TORRENTS
University of Stuttgart, Germany
FUSSY MODELLING AND APPLICATIONS IN WATER RESOURCES

Table 4 gives an overview of the 1999 IAHR-EGH course program. Currently, 19 courses have been announced by the network partners. Further details about the courses as well as the IAHR-EGH network are presented on the IAHR-EGH home page at:
http://www.uni-stuttgart.de/UNIuser/iws/IAHR/home.html
In order to ensure current information for network partners and interested parties, the IAHR-EGH home page is regulary updated.

5. Perspectives of IAHR-EGH

Based on the experience of the course program 97/98, the existing network structure and the implemented modules presented above, it is intended to continue the developments towards a fully integrated IAHR-EGH course program. In a first step towards a coordinated course program, some course organizers join forces in the development of common courses with a lecturer team. This may lead to sequences of offering the course alternating between two or more locations. Some of the course organizers have already involved an international group of lecturers. These initiatives improve the scientific contacts and allow for courses enriched by actual results of recent research projects. Furthermore, alternating locations enhance the number of potential participants by reducing travel and accommodation costs for the participants. Especially professional companies prefer "in house" courses because they are less expensive to run. For instance, R. Helmig has repeated his course "Numerical Modelling of Multiphase Flow and Transport Processes in the Subsurface" in 1997 at the Bundesanstalt für Geowissenschaften und Rohstoffe, Germany.

In a further step, the development of coherent sets of courses for individual IAHR-EGH main topics such as Hydroinformatics, Groundwater, or River Engineering is envisaged. Based on core courses of collaborating institutes, the steering group members aim at developing coherent sets of courses for the main topics. The following general guidelines are agreed for the envisaged course programs:

At the present level of development, the parallel evolution of coordinated course programs for the main topics seems preverable over a total catch - all program. For each main topic, the coordination of the course program can be based on thematically linked courses and standing collaborations.

The individual main group programs should aim at a high academic level of the courses fullfilling the IAHR-EGH quality criteria, international teams of instructors and an international audience.

Further education forms such as "young scientists workshops" will be considered in the course programs to support an active participation of young scientifically oriented postgraduates. The combination of several education forms represents a more comprehensive education of young scientists and offers the participants the opportunity to introduce own interests and activities without the limitations of conventional symposia and conferences.

For the successful development of coherent programs, it is intended to group courses to "Summer Schools" or "Training Camps" - perhaps at annually changing locations. These activities would represent a consequent continuation of the envisioned combination of several education forms. A further relevant education form represents ODL methods. In 1999 a first distance learning course will be performed by the IAHR-EGH hydroinformatics group [5], and it is foreseen that ODL methods may develop into key element of IAHR-EGH, since they can effectively reduce time needs and costs of the participants.

In the long run, the various program segments with thematically linked and coordinated courses can serve as modules of water related Master and PhD programs. For instance, some IAHR-EGH courses were considered as electives for the students of an internationally oriented Master of Science Program "Water Recourses Engineering and Management (WAREM)" at the University of Stuttgart. In the long run, the consideration of coordinated sets of IAHR-EGH courses in curricula of Master and PhD programs of collaborating institutions can broaden the individual course programs and improve the quality of education. However, a fundamental condition for inclusion in curricula is continuity in form of a regular repetition of the courses. A long term vision - depending of the general development of European educational systems - could be a fully integrated IAHR-EGH course program which could serve as basis for a European degree granted by several network partners from European universities and research institutions.

References

1 H. Kobus, E. Plate, H.W. Shen & A. Szöllosi-Nagy (1994): Education of Hydraulic Engineers., IAHR/UNESCO Panel Report, Journal of Hydraulic Research, Vol. 32(2), 41-54
2 H. Kobus (1995): IAHR European Graduate School of Hydraulics., in Proceedings of the Euroworkshop on Evaluation and Assessment of Present and Future Continuing Education

and Training (CET) for Water Recources Engineering., Brussels, 20.-22. April 1995, published by Techware, Brussels, July 1995, 117-125
3 J. McNown & H. Kobus (1996): Education in Hydraulics: From the nineties to the zeros., in Issues and Directions in Hydraulics, Hsg.: T. Nakato & R. Ettema, Balkema, Rotterdam
4 H.J. Lensing & H. Kobus (1997): IAHR European Graduate School of Hydraulics: Basic Network Concept, Developments and Perspectives., Contribution to the Seminar S4 on Continuing Education and Training of the XXVII. IAHR Congress, San Francisco, USA, 13-22
5 F. Molkenthin & R. Falconer (1999): Hydroinformatics and the Learning Society - A task inside the IAHR European Graduate School of Hydraulics., in Proceedings of the International Symposium The Learning Society and the Environment, Paris, June 2-4, 1999
6 IAHR European Graduate School of Hydraulics, List of Network Partners
Home page: http://www.uni-stuttgart.de/UNIuser/iws/IAHR/home.html

Une expérience en cours: la formation doctorale en modélisation en hydraulique et environnement à l' ENIT

Bargaoui Z., Rais S.
ENIT, Tunisie

1. Introduction

La société cognitive se trouve être l'un des pôles du « triangle des cognitions ». En effet, selon Ganascia (1996), les études relatives à la connaissance se distribuent sur trois pôles : le sujet psychique, le sujet physiologique et la société. Si le premier pôle est relatif à l'examen des capacités d'acquisition et de mobilisation des connaissances, à l'étude des phénomènes de compréhension et de mémorisation du langage, le second pôle contribue à établir le lien entre le système nerveux et l'activité cognitive. Etant donné que la connaissance ne relève pas seulement d'individus isolés mais d'êtres sociaux, les échanges entre individus et les moyens de communication sont de la plus haute importance. Ainsi, le rôle du troisième pôle est-il de rapporter la connaissance à sa circulation au sein des individus : son échange et sa transmission y sont étudiés. Le langage, la société, l'économie caractérisent ce troisième pôle.

Les psychologues et les neurobiologistes s'intéressent au champ de la connaissance au singulier. Les linguistes, les sémanticiens, les anthropologues et les sociologues couvrent les champs des connaissances au pluriel (en rapport avec le contenu de la connaissance). Les formateurs (enseignants) contribuent à la transmission des connaissance mais en tant que chercheurs, les universitaires sont en même temps confrontés chaque jour aux problèmes d'accès à la connaissance : connaissance des mécanismes et des processus en place dans les phénomènes qu'ils étudient, conception des expérimentations permettant d'y accéder.

Dans les sciences de l'eau et plus particulièrement dans le domaine de la modélisation hydrologique, les approches élaborées dans le cadre des sciences cognitives ont été souvent appliquées à la connaissance des phénomènes hydrologiques. Nous pensons en particulier à l'approche comportementaliste dans laquelle le sujet est vu « comme une boîte aveugle » qui est analogue à celle de la boîte noire en modélisation hydrologique. Dans cette approche, « le sujet pensant et réfléchissant est évacué au profit d'un substitut abstrait et opaque, capable seulement de répondre, par des réactions quantifiables, à des injonctions quantifiables, sans que les images, les motivations, la volonté, ...ne soient jamais présentes. Tout le travail du psychologue consiste alors à engendrer des lois d'expérience pour relier entrées et sorties » (Ganascia ,1996). Il est frappant de constater à quel point cette description de la démarche du psychologue ressemble à celle du modélisateur lorsqu'il adopte une approche de relation entrée-sortie ou de boite noire. Cette approche interdit « la prise en considération de consignes explicites ou d'éléments contextuels autres que ceux réglés par l'expérience ou induits par elle ». En effet, c'est le jeu des entrées et sorties du modèle hydrologique qui en fixe alors les paramètres.

De même, l'approche cognitive dans laquelle « toute action est interprétée en termes de fonctions et qui permet de considérer les mécanismes (inhibition, activation, compensation) » se traduit en modélisation hydrologique par le recours aux modèles à base physique qui, contrairement aux premiers, décrivent les phénomènes en place. L'inventaire des fonctions cognitives, leur décomposition en fonctions élémentaires dont les déterminants sont parfaitement élucidés et validés expérimentalement, trouve son homologue dans la séparation et le traitement des divers processus mis en jeu (évaporation, infiltration, ...) ainsi que dans la discrétisation spatiale.
Enfin, tout comme pour les sciences cognitives, la modélisation hydrologique se construit « à partir d'une masse de données empiriques, de protocoles d'expérimentation, de constructions mathématiques, de simulations informatiques ».

Le pont étant ainsi jeté, nous pouvons passer maintenant à la présentation de cette expérience du DEA de Modélisation en Hydraulique et Environnement MHE mise en place à l'ENIT. Nous commencons par présenter la situation de la formation dans le tissu universitaire et post-universitaire : les sources de l'input, la place de l'équipe d'encadrement dans la Recherche en Tunisie. Dans un second volet, nous examinons la demande en formation : son importance, ses spécificités, son évolution, et son issue. Enfin, l'ancrage économique et sectoriel est analysé du point de vue de l'offre de postes, de l'interdisciplinarité de la formation et du réseau scientifique et économique dans lequel elle s'insert à l'échelle nationale et internationale.

2. Situation dans le tissu universitaire et post-universitaire

2.1. Evaluation de l'input

Le DEA recrute principalement parmi les ingénieurs du Génie civil, Génie rural, Génie minier, Génie de l'environnement, Génie industriel et Génie chimique ainsi que parmi les ingénieurs géologues. Les titulaires d'une maîtrise en Sciences physiques ou Mathématiques figurent également parmi ceux pouvant postuler à cette formation. A combien est évaluée cette population ?
En Tunisie (MES, 1998 , a), le nombre d'étudiants ayant obtenu le diplôme d'ingénieur en juin 1998 représentait environ un millier. Les titulaires d'une maîtrise atteignaient dans le même temps un effectif de 1522 personnes. Sur les trois universités de : Tunis II (Université des Sciences et Techniques et de Médecine de Tunis), du Centre et du Sud, la répartition des effectifs est indiquée en tableau 1. Nous précisons que les ingénieurs qui suivaient l'ancien régime étaient de deux catégories : ingénieurs des Travaux de l'Etat (4 ans après le baccalauréat) et Ingénieurs Diplômés (6 ans après le bac) alors que ceux du nouveau régime font 5 ans de manière uniforme (et ce, depuis 1996-97).

Tableau 1. Répartition des diplômés par université

Diplôme	Tunis II	Centre (Sousse)	Sud (Sfax)	Total
Maîtrise Math, phys, nat	933	353	236	1522
Ingénieur ancien régime	191	49	37	277
Ingénieur nouveau régime	382	82	246	710
Total ingénieurs	573	131	283	987

Source : Bureau des études de la planification et de la programmation, MES, 1997-98.

Pour avoir le total des gradués susceptibles d'être intéressés par le suivi de cette formation, on peut ajouter à ces chiffres, le nombre d'équivalences au diplôme national d'ingénieur qui ont pu être accordées la même année par le Ministère de l'Enseignement Supérieur MES et qui sont au nombre de 142.

Les étudiants du troisième cycle en Sciences et Techniques sont au nombre de 861 (MES, 1998) alors qu'ils sont 89 en Sciences Agronomiques. Considérant que nous avons en moyenne 25 étudiants inscrits par an, nous en concluons qu'environ 3% des étudiants de troisième cycle en Sciences et Techniques et en Sciences Agronomiques sont inscrits au DEA de Modélisation en Hydraulique et environnement de l'ENIT.

2.2. Ancrage de l'équipe d'encadrement

La formation doctorale est proposée par le laboratoire d'Hydraulique de l'ENIT, qui est constitué d'un groupe d'une douzaine d'enseignants-chercheurs permanents de l'ENIT, auxquels se sont associés des enseignants-chercheurs d'autres établissements universitaires ainsi que des chercheurs d'établissements de Recherche. Dans l'ensemble, ce personnel encadre environ 15 à vingt étudiants en DEA et Thèse de Doctorat par an.

Ce laboratoire a été fondé en 1980, année au cours de laquelle de jeunes enseignants-chercheurs ont été recrutés. D'autres les ont suivis pendant quatre à cinq années encore. Actuellement, on peut distinguer au sein de ce laboratoire quatre équipes travaillant sur les thèmes de recherche suivants : Mécanique des fluides et procédés de l'eau - Hydrogéologie - Hydrologie de surface - Hydraulique fluviale et maritime et mettant en commun leurs compétences et leurs ressources. Comment se présentent ces équipes au sein du tissu de la recherche reconnu par le MES?

Parmi les équipes faisant partie du Répertoire de la Recherche Universitaire 1997-98 (MES, 1998,b), nous en avons relevé 11 en Sciences de la Terre, 15 en Sciences et Techniques de l'Ingénieur (STI), 3 en Chimie, 5 en Physique et enfin 4 en Sciences de la Vie dont les thèmes de recherche ont une interférence avec les thèmes affichés par le laboratoire d'Hydraulique (soit un total de 38 équipes). C'est en rapport avec ces équipes que se brossera la situation du laboratoire d'Hydraulique.

En Sciences de la Terre, il s'agit des équipes de recherche couvrant les thèmes suivants : Sédimentologie/ géomorphologie / pédologie /Hydrogéologie et géothermie / Climatologie / Géochimie. En Sciences et Techniques de l'Ingénieur (STI), Physique, Chimie, nous les avons classé en : Mécanique des fluides / Mécanique des fluides et thermique / Traitement des eaux usées / Transfert de chaleur et de masse / chimie appliquée à l'environnement. Enfin, en Sciences de la Vie, ce sont des équipes travaillant en Hydro-bio-écologie qui ont été ciblées. Les équipes du laboratoire d'hydraulique sont répertoriées au sein du thème Sciences et Techniques de l'Ingénieur. Leur effectif total en 1997-98 était de 32 chercheurs et la taille moyenne d'une équipe était de 8 chercheurs, ce qui, compte tenu des effectifs affichés par les autres équipes semble tout à fait représentatif du tissu tunisien. En effet, le tableau 2 montre qu'en moyenne, les équipes qui nous intéressent et qui se sont présentées en 1997-98 au financement de la Direction Générale de la Recherche Scientifique et Technique (DGRST) du MES sont en moyenne constituées de 8 membres. La plus grosse équipe rencontrée compte 23 personnes.

Tableau 2 : Effectifs des équipes dans les trois thèmes génériques

Domaine	Effectif moyen en chercheurs	Min-Max
Sciences de la Terre	10	4 - 14
STI, Physique, Chimie	7	2 - 23
Sciences de la vie*	7	3 - 9
Laboratoire d'hydraulique	8	6- 9

(*) appliquées à l'environnement

Le potentiel de chercheurs dans ces différentes équipes peut être subdivisé en chercheurs permanents (de l'établissement de tutelle ainsi que des checheurs associés) et chercheurs contractuels qui sont principalement les doctorants. Le laboratoire d'Hydraulique compte 13 permanents et 19 doctorants soit un taux d'encadrement de 1.5 permanent pour un étudiant en formation. Comment se présente ce taux parmi les équipes définies plus haut?

Tableau 3. Répartition des ressources humaines entre encadreurs et doctorants

Domaine	Permanents	doctorants	Taux d'encadrement
Sciences de la Terre	3	6	2
STI	3.9	3.8	1
Physique	8	6.8	1.2
Chimie	3.6	2.3	1.6
Sciences de la vie*	2.2	4.3	0.5
Total	18.5	23.2	-

(*) appliquées à l'environnement

Selon le tableau 3, le taux d'encadrement de l'équipe dépasse largement celui d'une équipe de STI mais reste bien inférieur à celui d'une équipe en Sciences de la Terre. Pour les Sciences de la Vie appliquées à l'environnement, il nous semble que la faiblesse de ce taux, de même que le petit nombre de permanents qui lui sont associés, alliés à une demande relativement forte de la part des doctorants tient son origine dans la nouveauté de ce thème dans les facultés.

Au total, le nombre de permanents dans les 38 équipes examinées correspond à 336 chercheurs dont 3.9% de l'effectif est affilié au le laboratoire d'Hydraulique. Le nombre de doctorants au sein de ces équipes atteint 185 dont 10.2% travaille dans le laboratoire d'Hydraulique. Nous devons finalement noter que ces équipes comptent généralement un seul professeur dans l'effectif des permanents. Une équipe en compte 4 et une autre trois alors que cinq équipes (13%) n'en comptent aucun.

Quant à la répartition géographique de ces équipes, nous en trouvons neuf qui se situent à Sfax (Université du Sud) avec 52 chercheurs (15% de l'effectif total), cinq comprenant un total de 44 chercheurs dans la région de Monastir (Université du Centre) et trois équipes totalisant 22 chercheurs à Gabès (Université du Sud). Le reste des équipes appartient à l'université Tunis II (tableau 4)

Tableau 4. Répartition géographique des équipes et du potentiel humain

Université	Nombre d'équipes*	potentiel humain
Tunis II	21	218
du Sud à Sfax	9	52
du Sud à Gabès	3	22
du Centre à Monastir	5	44

(*) relativement au domaine considéré qui est Hydraulique et Environnement.

Ainsi, rapporté au potentiel humain de l'université d'appartenance, celui du laboratoire d'Hydraulique de l'ENIT constitue finalement 15% de l'effectif, avec 19% des doctorants (le nombre de doctorants étant de 125).

Il faut cependant insister sur le fait que c'est l'association des trois axes : Mécanique des fluides, Hydrologie continentale, procédés de l'eau qui a permis de faire émerger ce groupe dans le tissu de la recherche. En effet, considérées séparément, les quatre équipes ne constituent plus une masse critique. En Mécanique des fluides, nous trouvons alors que les équipes de la faculté des Sciences de Tunis (19 chercheurs) et de l'ENIM de Monastir (17 chercheurs, en Transfert de chaleur et de masse) constituent les deux pôles en Tunisie. De même, dans le domaine des sciences de la Terre auquel nous associerons l'Hydrologie continentale, l'essentiel des chercheurs du MES se trouve à la faculté des Sciences de Tunis. Un important lot de chercheurs se situe également dans les écoles agronomiques et du Génie rural, non comprises dans le répertoire qui nous sert de support.

3. Demande en formation

3.1. Demande potentielle

Faisant l'hypothèse que 60% des étudiants diplômés des maîtrises de Mathématiques, Physiques et Sciences Naturelles ont leur diplôme en Mathématiques et Physiques et que 25% des ingénieurs diplômés et ayant obtenu l'équivalence sortent des filières de Génie Civil, Génie Rural, ...nous obtenons un potentiel de demandeurs pour cette formation et pour les formations équivalentes organisées dans d'autres institutions, qui serait de l'ordre de: .6*1522+.25*987+.25*142=1160.

Pour avoir une idée des formations équivalentes et adressées à la même population que celle que nous visons, nous nous sommes limitées à l'université Tunis II, université d'appartenance de l'ENIT. Nous avons répertorié les DEA suivants : Géologie et écologie générale (faculté des sciences de Tunis), Sciences de l'environnement (faculté des sciences de Bizerte), Transfert de chaleur et de masse (faculté des sciences de Tunis).

Les diplômés de l'Université Tunis II représentent 60% des diplômés tunisiens. Leur nombre correspondrait à 0.6*933+.25*459=675 diplômés potentiellement candidats. Ici, le chiffre 459 se rapporte aux ingénieurs de l'ENIT (ancien et nouveau régime), aux ingénieurs de L'Ecole Nationale Polytechnique de Tunis EPT ainsi qu'aux ingénieurs en Agronomie et Agroalimentaire. En considérant seulement le nouveau régime pour les ingénieurs, le nombre de candidats potentiels tombe à 358, ce qui conduit à 650 diplômés ciblés. Le tableau 5 donne la répartition des diplômés de l'université Tunis II selon les filières de recrutement.

Tableau 5. Répartition des diplômés selon les Génies à l'Université Tunis II

ENIT	277
Géologie	18
Ecole Polytechnique	31
Agronomie et Agroalimentaire	151

3.2. Etablissements dont est issue la demande

La demande en formation provient principalement d'ingénieurs diplômés, et très faiblement de maîtrisards. A l'Université Tunis II, nous avons compté dix institutions (guide de l'université, 1998) à partir desquels pourrait se manifester une demande en formation. Il s'agit en premier lieu de l'ENIT (1) puis des facultés des Sciences de Tunis FST (2) et de Bizerte FSB (3), de l'Ecole Supérieure des Sciences et Techniques de Tunis ESSTT (4), de l'Ecole Supérieure des Industries agro-alimentaires de Tunis ESIAT (5), de l'Ecole Nationale Polytechnique de Tunis EPT (6) et enfin des écoles d'Agronomie et de Génie rural à savoir : l'Institut National Agronomique de Tunis INAT (7), l'Ecole Supérieure des Ingénieurs de l'Equipement Rural de Medjez El Bab ESIER (8), l'Ecole Supérieure d'Agriculture de Mograne ESAM (9) et de l'Ecole Supérieure d'Agriculture du Kef ESAK (10). Toujours en 1997-98, l'effectif des étudiants issus des formations dispensées dans ces établissements et possédant les prérequis pour suivre le DEA MHE va d'une trentaine (EPT) à 600 (FST).

Plus généralement, le tableau 6 présente une liste de 16 établissements (facultés des Sciences, écoles de Génie, écoles d'Agriculture) contenant les demandeurs potentiels. Nous avons enregistré au moins une demande depuis 1991-92, année de création de la formation de la part de quatorze d'entre eux. Les formations initiales dont provient le potentiel de demandes sont ainsi les suivantes, par établissement (tableau 6).

Tableau 6. Formations initiales support de la demande potentielle

Numéro	Etablissement	Formation d'origine
1	ENIT	ingénieur en Génie Civil et industriel
2	FST	maîtrise de Physique, de mathématiques, ingénieur diplômé en Géosciences
3	FSB	maîtrise de Physique, de mathématiques
4	ESSTT	maîtrise de Génie Civil
5	ESIAT	ingénieur Génie alimentaire
6	EPT	ingénieur polyvalent
7	INAT	ingénieur du Génie rural eaux et forêts, en Économie et développement (gestion des ressources naturelles)
8	ESIER	ingénieur en Hydraulique et aménagements
9	ESAM	ingénieur en agro-économie
10	ESAK	ingénieur agronome
11	ENIM	ingénieur en Génie énergétique
12	ENIS	ingénieur en géoressources et environnement
13	FSM	maîtrise de Physique, de mathématiques
14	FSS	maîtrise de Physique, de mathématiques
15	ISTIM	ingénieur génie minier
16	ESAMa	ingénieur agronome

Parmi ces formations, certaines sont anciennes (Génie Civil à l'ENIT, Génie Rural à l'INAT, maitrises à la FST) alors que d'autres sont un peu plus récentes ou ont disparu.

3.3. Demande effective

Le tableau 7 permet de suivre le nombre de candidats qui ont soumis un dossier d'inscription. Parmi eux, les diplômés de Tunisie proviennent de 14 établissements parmi les 16 cités plus haut. Aucun candidat de l'ESAM ni de l'ESAK ne s'est présenté jusqu'ici. La répartition des candidats selon l'institution d'origine (tunisienne ou étrangère) ainsi que la diversité de ces institutions sont également présentées dans ce tableau.

Tableau 7. Répartition des candidats selon l'origine et diversité des institutions de la demande

Année	Nombre de demandes			Nombre d'institutions		
Provenance*	Tunisie	Etranger	Total	Tunisie	Etranger	
1991-92	80	2	82	9	2	1M+1D
1992-93	39	5	44	6	5	3M+ 2D
1993-94	31	7	38	7	7	5M+1A+1D
1994-95	19	8	27	7	6	7M+1A
1995-96	35	6	41	7	4	7M
1996-97	57	12	69	7	6	2 A +8M+2D

Année	Nombre de demandes			Nombre d'institutions		
	Tunisie	Etranger	Total	Tunisie	Etranger	Provenance*
1997-98	37	9	46	8	4	4A+5M
1998-99	43	10	53	10	5	8A+2M
Total	343	59	402	-	-	31M+16A+6D

(*) : M=Maghreb, A=Afrique sub-saharienne, D=divers

L'année du lancement semble avoir été une année singulière. C'est ainsi qu'avec 80 candidats de Tunisie et seulement 33 admis à s'inscrire, elle aura contribué à infléchir la demande dans le sens de la rencontre de l'offre. Depuis, la moyenne du nombre de candidats à l'inscription se situe autour de 37. Ce chiffre ne représente qu'environ 6% du nombre de candidats potentiels en considérant les ingénieurs nouveau régime et les maîtrisards concernés (37/650).

Le tableau 8 donne le nombre d'inscrits I par an. Comparativement au nombre de demandes exprimées, nous avons en moyenne sur les huit années de fonctionnement 2.2 demandes pour une place (402/180).

3.4. Structure de la demande

Le nombre de candidats étrangers est insignifiant au départ mais constitue maintenant environ 20% de la demande. La diversité des établissements d'origine est assez importante (de 6 à 10 établissements différents). Enfin, les candidats maghrébins majoritaires au début parmi les étrangers se voient maintenant relayés par les candidats d'Afrique subsaharienne. Il faut toutefois noter que les candidatures d'étudiants étrangers ont peu de chances d'aboutir à mois qu'elles ne passent pas par le biais de la coopération bilatérale qui est seule habilitée selon la réglementation en vigueur à autoriser l'inscription des étudiants étrangers. C'est ainsi que compte tenu de ces contraintes ainsi que de l'inexistence d'un système de bourses, le nombre de candidats étrangers qui ont effectivement suivi cette formation est de seulement 9.

3.5. Évaluation des taux de réussite

La formation a lieu théoriquement en deux ans. Elle peut se dérouler au maximum sur trois ans. Pendant sa première année d'étude, l'étudiant suit un certain nombre de modules d'enseignement et passe un examen semestriel. A l'expérience, on se rend compte qu'il se manifeste un taux d'abandon relativement important en cours de première année. En effet, le nombre effectif de candidats E, c'est à dire le nombre de ceux qui finalement passent les examens, est bien plus réduit que celui des inscrits. La quantité : 100- E/I constitue une mesure du taux d'abandon (tableau 8). L'instauration d'un droit d'inscription de 200 $US depuis 1996-97 semble avoir contribué à le diminuer.

Tableau 8. Répartition des étudiants

Année	Nombre d'inscrits I	Candidats effectifs E	Nombre d'admis A	E/I (%	A/E (%)
1991-92	25	13	6	52	46
1992-93	19	5	2	26	40
1993-94	25	6	3	24	50
1994-95	13	8	4	62	50
1995-96	15	7	4	47	57
1996-97	19	13	5	68	38
1997-98	38	12	9	32	75
1998-99	26	21	-	81	-
Total	180	85	33*		

* jusqu'à 1996-97 ce nombre est de 24

Le taux de réussite en première année est défini comme le rapport des étudiants ayant réussi la première année d'étude à celui des inscrits effectifs. Selon le tableau 8, ce taux tend à augmenter au fil des ans. Bien entendu le contexte économique est pour beaucoup aussi bien pour fixer l'importance du nombre de demandes d'inscription que dans la décision de l'étudiant de se maintenir en formation. Pour nombre d'ingénieurs diplômés, l'inscription au DEA ne présente en effet qu'une alternative au chômage ou au stage d'insertion professionnelle, passage presque obligé pour la majorité d'entre eux. Les premiers mois de l'année scolaire nous apparaissent comme une période où se manifeste la détermination de l'étudiant à poursuivre un troisième cycle, généralement dans des conditions matérielles difficiles, caractérisées par l'absence de financement par l'université. Par ailleurs, nous pensons que la réduction de la durée de la formation initiale des ingénieurs (5 ans au lieu de 6 en ancien régime) peut constituer une motivation pour la poursuite de leurs études.

La première année du DEA est suivie d'une ou de deux années d'élaboration d'un mémoire de DEA. Nous assistons également à un taux d'abandon non négligeable à l'issue de la réussite en première année. N'étant généralement pas financés par des contrats industriels, les étudiants poursuivent leurs travaux de mémoire en comptant sur leurs financements propres. Aussi lorsqu'une offre d'emploi intéressante se présente en cours de route, il arrive que l'étudiant abandonne la réalisation de son mémoire. Ainsi, 24 étudiants ont réussi la première année d'études jusqu' en 1996-97 alors que seuls 18 d'entre eux ont soutenu leur mémoire (un dix neuvième va bientôt les rejoindre). Le taux de passage de la première année à l'obtention finale du diplôme est ainsi de 79%.

4. Ancrage économique et sectoriel

4.1. Offres de postes

Cette offre se situe nécessairement dans le cadre général des recrutements effectués par le MES et des offres d'emploi adressées aux diplômés des formations scientifiques et techniques. Il se trouve que malgré le doublement des effectifs dans l'enseignement supérieur en dix ans (120 000 étudiants en 1996), l'offre générale de postes d'enseignement reste relativement restreinte. En 1995, 168 enseignants ont été recrutés pour le grade d'assistant, 61 en concours externe pour les maitres-assistants et 50 pour les maîtres de conférence. Au total 445 postes ont été pourvus, toutes formations confondues (Banque Mondiale, 1998). Évidemment, cette situation générale n'est pas en faveur des titulaires de notre formation pour les deux raisons suivantes : d'une part, les filières classiques restent dominantes, ce qui constitue un biais en défaveur des formations scientifiques et techniques (Banque Mondiale, 1998). D'autre part, ce constat prévaut également au sein des formations scientifiques et techniques : les maîtrises classiques restent dominantes dans les facultés et les filières de Génie classiques dominantes dans les écoles d'ingénieurs. Le système modulaire mis en place dans les formations d'ingénieur depuis trois ans n'a pas encore donné lieu à des formations à la carte ni à des formations où la spécialisation est poussée. De toute manière, si faible que soit le recrutement, les facultés continuent à recruter traditionnellement parmi les doctorants en Sciences Physiques ou de la Terre. En même temps, l'offre des écoles de Génie reste très mince étant donné le faible nombre de leurs étudiants en spécialisation.

Quant au milieu industriel, il ne reconnaît pas explicitement le plus de qualification apporté par un DEA dans le cas d'un ingénieur diplômé. Même l'opération de mise à niveau menée actuellement dans le pays vise essentiellement le recrutement des ingénieurs diplômés au sein des entreprises en vue d'une augmentation significative de leur taux d'encadrement. Pour être plus concrètes, nous allons suivre le devenir des 18 diplômés du DEA.

Parmi ceux-ci huit ont continué une carrière de chercheur en s'inscrivant à une thèse à l'ENIT (trois) ou à l'étranger (Canada, France ou pays d'origine). Quatre ont continué leur carrière professionnelle d'origine, avec amélioration de leur statut. Trois ont trouvé un emploi ou exercent à titre privé, mais sans rapport direct avec la formation doctorale. Un se destine à continuer en thèse et travaille actuellement en tant que chercheur contractuel. Enfin, nous n'avons pas de nouvelles du dernier. Nous constatons ainsi, que pratiquement la moitié des titulaires du diplôme continuent le cursus universitaire, ce qui est encourageant.

Il faudrait signaler ici qu'il y a un changement institutionnel qui s'opère actuellement dans le domaine de l'organisation de la Recherche en Tunisie. Il va dans le sens d'une plus grande reconnaissance des laboratoires : subventions de recherche plus substantielles permettant la reconnaissance des chercheurs doctorants ainsi qu'une gestion plus autonome de la recherche. Nous voulons espérer que les nouvelles conditions d'organisation et de financement de la Recherche contribueront au renforcement des formations doctorales « marginales » comme la nôtre (marginale du point de vue de l'importance de la demande mais pas du secteur auquel elle s'adresse).

4.2. Exploration des besoins dans le domaine de la recherche

A l'échelle de l'ENIT, il n'y a pas d'instance d'exploration de ces besoins. Il n'y a pas non plus de Direction de la Recherche. Cependant, les équipes en charge du DEA sont en réseau avec des institutions, entreprises, bureaux d'ingénieurs-conseils, ce qui leur permet d'explorer indirectement les besoins en recherche du milieu économique. En quelque sorte un circuit informel s'est installé pour la recherche des sujets d'encadrement et pour leur financement éventuel (en réalité peu présent), à la faveur des contacts et projets entrepris avant l'institution du DEA. C'est ainsi que nos interlocuteurs institutionnels privilégiés sont:

- pour l'Hydrologie et la gestion de l'eau : Au Ministère de l'Agriculture, principalement la Direction des Ressources en eaux, la Direction des Etudes et Grands Travaux Hydrauliques, la Direction de Conservation en eaux et sols, la SONEDE, société d'exploitation des eaux (qui intervient également au titre de la qualité de l'eau), la SECADENORD société d'exploitation du Canal Medjerda-Cap Bon, ainsi que les directions au sein des commissariats régionaux de Développement, qui sont les institutions décentralisées au niveau des régions pour le Ministère de l'Agriculture.

- pour les applications de la mécanique des fluides à l'analyse des écosystèmes lacustres et côtiers, à l'hydrodynamique, à la pollution hydrique et aux procédés de l'eau, notre principal vis à vis est le Ministère de l'Environnement avec ses multiples agences : l'Office national de l'assainissement, l'Agence nationale de protection de l'environnement, l'agence de protection du littoral, le Centre international des technologies de l'eau, la Société de promotion du lac de Tunis, etc...

Les sujets de DEA soutenus ou actuellement engagés proviennent pour 51% de propositions soumises de la part de nos partenaires économiques. Pour le reste, le sujet proposé à l'initiative d'un directeur de recherche a obtenu l'accord d'une institution qui a appuyé sa réalisation. L'appui rencontré se situe rarement au plan financier mais concerne surtout l'acquisition de l'information (plans et résultats d'expériences, mise à disposition de banques de données, et de documents internes, visites de sites, ..).

4.3. Coopération interdisciplinaire et internationale

L'organisation du DEA au sein d'un même laboratoire ne doit pas masquer le caractère interdisciplinaire de la formation. En effet, celle-ci est donnée par des enseignants spécialisés dans les domaines qu'elle est supposée couvrir à savoir : la Mécanique des fluides fondamentale, ses applications aux procédés de l'eau et à la modélisation hydrodynamique d'une part et l'Hydrologie des systèmes continentaux avec considération des problèmes de gestion des ressources en eau. Ce caractère interdisciplinaire apparaît nettement au niveau des enseignements en accord avec l'esprit général de la formation qui voudrait associer la Mécanique des fluides et l'Hydrologie. Une tentative de séparation de la formation en deux options : Mécanique des fluides/Hydrologie- ressources en eaux, n'a pas abouti faute d'un nombre suffisant de candidats. Ses chances d'aboutir nous semblent encore minces aujourd'hui tant la demande en formation est faible. Toutefois à l'occasion des mémoires de DEA et des quelques thèses qui leur ont fait suite, la discipline a chaque fois repris le dessus, ce qui semble logique au niveau d'un DEA. Le département Génie industriel de l'Enit a contribué depuis maintenant trois ans à l'encadrement au sein du DEA. Mais, cette interdisciplinarité entre les équipes de l'ENIT ne doit pas faire oublier qu'un effort certain reste à consentir pour associer des spécialistes en Sciences de la vie : chimistes, hydrobiologistes, biologistes ainsi que des économistes travaillant dans le domaine de l'eau.

Au niveau international, nous avons établi en Tunisie des liens avec des institutions de recherche tels que l'ORSTOM. Plusieurs projets de mémoire de DEA ont été définis et cofinancés par cette institution et continuent à l'être. Cette coopération s'est également manifestée avec des équipes en France (Mécanique des fluides, Génie des procédés), au Canada (Hydrologie et gestion de l'eau), ainsi qu'en Allemagne dans une moindre mesure. Un projet de renforcement institutionnel canadien en Tunisie est actuellement en cours et a parmi ses objectifs de créer et de raffermir le réseau universitaire du laboratoire d'Hydraulique.

5. Conclusion

Le triangle (laboratoire en charge de la formation- enseignants, demande de formation-chercheurs, ancrage de la formation- secteur économique) a été examiné dans divers aspects. Il apparaît que la pérennité de cette formation doctorale, qui est objectivement interdisciplinaire, dépend fortement du cadre socio-économique dans lequel elle se situe.

Son input, essentiellement des ingénieurs diplômés, varie avec le contexte du marché de l'emploi offert aux ingénieurs. En l'absence du financement de la recherche, le parcours de l'étudiant inscrit apparaît comme un parcours du combattant. De ce point de vue, ce DEA est très précaire. Heureusement, en contre-poids, dans un cadre institutionnel informel, cette formation bénéficie de l'appui de partenaires économiques intéressés par le résultat des recherches mais qui sont eux-mêmes incapables, de part leur organisation institutionnelle, de participer au financement de la recherche.

Bibliographie

Banque Mondiale (1998). L'enseignement supérieur tunisien :Enjeux et avenir. Les rapports économiques de la banque mondiale.

Ganascia JB. 1996. Les sciences cognitives. Éditions Dominos. Flammarion.

Guide de l'université (1998). Université des Sciences, des Techniques et de Médecine de Tunis. Année universitaire 1998-99

MES, 1998,a. Publication de la direction de la prospective, des statistiques et de l'informatique. 1997-98.

MES. 1998,b. Répertoire de la Recherche Universitaire 1997-98

Hydroinformatics and the Learning Society - A task inside the IAHR European Graduate School of Hydraulics

Molkenthin F.[1], Falconer R.[2]
[1] Brandenburg University of Technology at Cottbus, Germany
[2] Cardiff University, United Kingdom

Abstract
The IAHR-EGH »Hydroinformatics" programme offers a combination of courses for academic education and application oriented training. Theses courses provide European students and engineers with state-of-the-art knowledge and engineering expertise transfer in hydroinformatics applications as well as in the development of hydroinformatics systems for study, research and practice. Modern Internet/WWW techniques, distance learning and collaborative engineering will open up a new dimension for education in hydroinformatics across Europe (and especially in Eastern Europe) independent from the location. In this way the EGH courses on Hydroinformatics satisfy the needs for the learning society in Europe in terms of international continuous education and training in this future oriented and important field.

1. European Graduate School of Hydraulics of the IAHR

The European Graduate School of Hydraulics of the IAHR (International Association for Hydraulic Research) [1] is a partner of the European Thematic Network on Education and Training ETNET.ENVIRONMENT-WATER, sponsored by the SOCRATES/ERASMUS programme of the European Commission. Under its umbrella several European educational institutions offer short courses of a high academic level in hydraulics and related fields. The courses offered are grouped under five major themes [4]. This contribution presents the activities under the »Hydroinformatics" programme.

2. Hydroinformatics

Hydroinformatics is a central discipline within hydro-engineering. As a part of its core, hydroinformatics concerns the modelling of information contained or generated by simulation of physical, biological and chemical processes and the impact of hydraulic structures and their operation on the aquatic environment, as well as the working processes of engineers in consultancy and administration. Hydroinformatics encompasses methods and techniques to design, implement and apply software systems (so-called hydroinformatics systems) relevant to hydro-engineering. The methods and techniques applied originate from computer science, engineering science, mathematics and physics, as well as chemistry, biology and sociology. Hydroinformatics systems are modular software systems, designed and developed to support engineering design and planning procedures, as well as to support working processes (such as the real time operation of a water treatment works). They contain and integrate in an open and flexible manner different software components (e.g. databases, analysis, numerical simulation, information management, GIS, visualisation, knowledge bases, documentation, conferencing and so on) for the relevant application fields in hydro-engineering.

2.1. Evolution of Hydroinformatics

The roots of hydroinformatics stem for over thirty years, when the available capacities of computer was sufficient for numerical solutions of hydro-environmental problems. Computer applications have significantly changed over the past few decades and have significantly influenced hydraulic modelling. From early times hydraulic models have been used as research and application tools of numerical solutions to the governing partial differential equations of hydraulics, with making these tools much more practical for a wider range of problems. The appearance of the object oriented paradigm lead to a widening of this view towards the use of hydroinformatics tools. Its foundation has been highlighted by ABBOTT [2] and others. Knowledge based methods, decision support, uncertainty analysis and new methods such as genetic algorithms, fuzzy logic and neural networks, have entered the field in the last few years. Besides these recent methods, also the working process of hydraulic engineering is being revolutionised by the progress in computer and network technology [3]. The recent progress in hydroinformatics has lead to large complex hydroinformatics systems, which integrate all of the relevant tasks of engineering in a modern, network based interdisciplinary environment.

The speed of the process has dramatically increased since 1990 by the upcoming Internet and the evolution by the general availability of low cost PCs and workstations. The Engineer of today - especially those who are actively involved in the water management - have to follow this evolution. This demands an ongoing and continuous education and training in the field.

2.2. Importance of Hydroinformatics for Society

Water is a fundamental requirement for human life and a non increasing resource. Water supply to humanity all over the world, with its different climatic and geological regions, demands for complex engineering management and planning. These technical infrastructure requirements (such as irrigation and urban drainage systems, pump stations, river regulation, navigation channels, harbours, dikes, dams etc.) have to be designed, planned, constructed and managed under ecological, economic, cultural and social requirements and conditions, defined by society and governed by the natural environment. These tasks can only be achieved with a deep understanding of the interaction of the nature and the engineering structures under environmental aspects. Hydroinformatics provides methods, techniques, models and systems to simulate these natural processes and the interaction with the human intervention, and to manage and operate the water resource structures. Therefore the application of hydroinformatics systems has a wide importance in hydro-engineering of the natural environment and in a global view for the necessary satisfaction of society, with water and infrastructure for protection of human life.

2.3. Education and Training in Hydroinformatics

Computer based modelling of nature and human interventions demands a distinctive understanding of the natural and structural behaviour of processes, as well as knowledge of, experience and of applied techniques and methods of the different hydroinformatics software components. The growing complexity and the increased interdisciplinarity of engineering projects requires a wide academic education and practical training of modern-day engineer in the actual state of science and practise. The »first" academic education in the national education systems (i.e. school and study) cannot provide the necessary knowledge and experience for life long engineering practise and the ability to operate in a global market, including transnational cooperations. The dramatic ongoing progress in hydroscience, computer science and engineering and the corresponding changes in the working processes have increased the necessity to create an international (i.e. European) education and a continuous training in hydroinformatics at a high level.

These developments require universities to play a more active role in creating an international framework of education and training for researchers and consultant companies etc. A more comprehensive post-graduated education and training for small and medium enterprises (SMEs) in Europe will strengthen the competition in applied research and development and engineering, essential within a general move towards a »knowledge and learning based society". In addressing this task the EGH offers, under the topic of »hydroinformatics", courses which cover in their syllabi the research and design of hydroinformatics systems for academic purposes and on the other hand training courses for professional specialisation at the level of applying such systems. In this way the IAHR-EGH courses on »Hydroinformatics" help the »Learning Society" in Europe to follow the ongoing progress in research and practice within the »Water-Environment» domain.

3. Hydroinformatics in the EGH

The EGH courses on hydroinformatics aim to transfer key knowledge, ability and experience to enable participants to develop, adapt and apply the corresponding software tools. The structure and content of the courses is dynamically adapted to reflect the ongoing progress in information and communication technology (ICT), as well as the change in working processes in engineering, management and sustainable development.

3.1. Course Concept

The EGH courses are classified into three types of courses. Research oriented courses are directed mainly towards PhD students and staff from universities and research institutions as well as interested professionals with a high academic education. Design oriented courses are directed primarily towards MSc students, as well as interested professionals require a more academic education in the field. Under an appropriate quality control of these two types of courses, European students will qualify for academic recognition by credit points (ECTS). Training courses are addressing application oriented expertise, mainly for professionals as continuous education and practical training. Training courses are offered by universities, consulting companies and commercial software provider giving professionals and students an introduction and training in the application of these systems.

As a basic discipline hydroinformatics and the complementary courses focus on the basic techniques, methods, models, processes, systems and applications which are used for research and engineering in a wide sense in the field of hydraulics and hydrology, such as flow and transport processes in rivers, estuaries and coastal waters, groundwater, urban drainage and water resources.

The course structure is based on a number of regular core courses, which include fundamental and general aspects of hydroinformatics and hydroinformatics systems (e.g. »Hydroinformatics Systems" [5]), and a number of irregular courses which focus on special aspects of hydroinformatics (e.g. »GIS in Urban Water Systems", »Urban Hydroinformatics", »WWW based Collaborative Engineering in Hydroscience"). The first group cover the core content of this discipline, while the second group encompass a flexible course programme to consider the ongoing progress and diversity in hydroscience and ICT-application.

The EGH courses on hydroinformatics are »European courses". Lecturers and participants come from various European countries. For instance the regular core course in 1999, entitled »Hydroinformatics Systems", will be held in close cooperation with lecturers from 5 European countries (D,F,GB,LT,NL). Experience from the course held in 1998 (with 18 participants from 11 European countries) showed that there was a demand for such European based education. To support the European education concept the lecturers and the courses will move between the European countries over the years.

3.2. IC-Technology for Distance Learning in Hydroinformatics

The Internet and the WWW provide a modern ICT framework for offering a new dimension to education and training in Europe. Hydroinformatics as a bridge between hydro-engineering and computer science, provides a means of using a modern network technology for education and training. In 1999 a course, entitled »WWW based Collaborative Engineering in Hydroscience" will be run as an experiment for distance education. The course will be held simultaneously at two different locations: at Delft, in The Netherlands, and at Cottbus, in Germany. Participants will be given a selected task as part as of a hydroinformatics project to be solved within groups, residing at both sites. Using the Internet lecturers and students will stay online and in contact with one another to solve the given problem in a collaboration framework. Students from other locations (i.e. potentially over Europe) might follow the course over the Internet/WWW. This approach will open up a new dimension to EGH courses in the future with courses being held in parallel, at several locations at the same time. This will enable to portfolio of courses to be extended with remote exercises and supervision over long periods.

3.3. Prerequisites in the education system

This contribution has shown the importance of European education and training courses in hydroinformatics for the learning society. This extension and change in the role of European universities towards an international education of a high academic standard has to be reflected in the structure of the universities as education institutions. Lecturers and students are no longer fixed to a single university in their activities. This will demand for more flexibility and mobility for both groups. It can be expected that regular students will receive typically 10-20 % and more of their education from outside of their home university. Lecturers will offer courses at several locations and will teach a significant number of participants (i.e. students and practitioners) from outside via the Internet. Practitioners will spend up to 10 % of their regular working time contributing towards a continuous education. This evolution in education has been highlighted by politicians and industrials, however, the practical support for such initiatives is often insufficient. The European Commission supports this development in a leading role by broad activities of its programmes and frameworks. Nevertheless, our experience of obtaining the necessary support (e.g. travel costs) for such courses are variable and disappointing, mostly highlighted by considerable effort in term of applications and little success. Special support for students and lecturers from East European countries is an actual problem, due to the economic difficulties in those countries. We hope that this will change in the near future and that the demand of the »Learning Society" in education and training for »Water - Environment" can be satisfied more by Europe for Europe.

References

1 H.J. Lensing & H. Kobus (1997): IAHR European Graduate School of Hydraulics: Basic Network Concept, Developments and Perspectives. Contribution to the Seminar on Continuing Education and Training of the XXVII. IAHR Congress, San Francisco, USA, 1997
2 Abbott,M.B.: *Hydroinformatics: A Copernican revolution in hydraulics* in IAHR Journal of Hydraulic Research, Volume 32, 'Hydroinformatics', 1994

3 Molkenthin, F.; Holz, K.-P. Working Process in a Virtual InstituteProceedings Int. Conf. Hydroinformatics 98, Copenhagen/Denmark, pp. 941-948 A.A.Balkema, Rotterdam 1998
4 IAHR European Graduate School of Hydraulics
Homepage: http://www.uni-stuttgart.de/UNIuser/iws/IAHR/home.html
5 EGH-Course Hydroinformatics Systems
Homepage: http://www.bauinf.tu-cottbus.de/EGH

18 ans d'existence d'un diplôme multidisciplinaire, le Diplôme d'Etudes Approfondies en Sciences et Techniques de l'Environnement : la majorité ?

18 years for an interdisciplinary one-year postgraduate degree on environmental science and technology : the full age ?

Daniel Thévenot et Bruno Tassin
Cereve, Université Paris XII-Val de Marne, 61 avenue du Général de Gaulle, 94010 Créteil Cedex, France.

Résumé.
En 1980 deux établissements d'enseignement supérieur, une université et l'École Nationale des Ponts et Chaussées, rejointe ultérieurement par l'École Nationale du Génie Rural des Eaux et des Forêts, ont uni leurs moyens pour organiser le DEA Sciences et Techniques de l'Environnement (STE). Ce diplôme associe 200 h de cours et un stage de 6 à 9 mois dans une équipe de recherche. Durant ses 18 années d'existence, il apparaît trois périodes : (a) de 1980 à 1988, peu de cours obligatoires et de nombreux modules optionnels sont proposés ; (b) de 1989 à 1997, les cours optionnels sont moins nombreux et regroupés en trois filières ou domaines ; (c) depuis 1998 le diplôme propose trois cours obligatoires et trois optionnels, choisis parmi les neuf offerts. Le DEA STE est consacré à l'environnement et au développement durable à l'échelle des bassins versants urbains. L'évolution de son programme montre comment un ensemble de modules monodisciplinaires a été remplacé par une nouvelle approche environnementale associant analyse de systèmes, modélisation et analyse de choix de projets.

Abstract.
In 1980, two institutes of higher education – one university and a civil engineering school (ENPC), followed by an agriculture and forestry engineering school (ENGREF) - decided to join their efforts in order to organise a 1-year post graduate degree: "Diplôme d'Études Approfondies en Sciences et Techniques de l'Environnement" (DEA STE). The programme includes 200 hours of teaching and a 6 to 9 month placement in a research laboratory. Three major periods appear in the degree evolution: (a) from 1980 to 1988, a few major courses and numerous elective ones were proposed; (b) from 1989 to 1997, the number of elective course was dramatically reduced and organised in three major fields; (c) within the current period the degree associates 3 major and 3 elective courses, chosen within 9 possibilities. This degree is devoted to environment and sustainable development, focusing on anthropic effects on water and soil compartments of catchment basins. The programme evolution illustrates how an addition of monodisciplinar modules was replaced by a new pedagogical approach based on system analysis, modelling and analysis of strategic decisions.

Pourquoi un tel diplôme multidisciplinaire ?

Constatant l'importance des problèmes environnementaux dans la société et les besoins de connaissance scientifique en ce domaine, deux établissements d'enseignement supérieur, l'Université Paris XII-Val de Marne (UPVM) et une école d'ingénieurs, l'École Nationale des Ponts et Chaussées (ENPC), rejointe ultérieurement par l'École Nationale du Génie Rural des Eaux et des Forêts (ENGREF), ont décidé de mettre en commun leurs moyens pédagogiques et scientifiques, afin de proposer, en 1980, un diplôme de troisième cycle, le Diplôme d'Études Approfondies en Sciences et Techniques de l'Environnement (DEA STE). Ce diplôme est consacré à l'environnement et au développement durable, en se situant à mésoéchelle, c'est à dire à une échelle régionale ou de bassin versant rural ou urbain. Une telle échelle permet, en effet, de mettre en évidence l'influence des activités humaines sur les hydrosystèmes et les sols. Au contraire de la plupart des diplômes de DEA qui se focalisent sur une seule discipline afin d'en permettre l'approfondissement, le DEA STE, compte tenu de son champ d'application, a toujours pris en compte les approches multidisciplinaires en associant les sciences naturelles (physique, mécanique, hydraulique, géologie, chimie, écologie, microbiologie...) et sociales (économie, gestion, urbanisme, aménagement, politique...).

Evolution historique du DEA STE

Pendant ses 18 années d'existence, l'approche pédagogique et l'organisation du programme du DEA STE ont subi de nombreuses modifications par amélioration de la prise en compte de leur caractère multidisciplinaire. Néanmoins les caractéristiques principales du diplôme, liées à son statut de DEA, ont été maintenues : une formation orientée vers la recherche, comprenant 200 h de cours et un stage de 6 à 9 mois, au sein d'une équipe de recherche, généralement associée à l'enseignement. L'effectif des promotions est resté compris entre 20 et 40 étudiants.
Trois grandes étapes jalonnent l'histoire du DEA STE.

Les origines : un fonctionnement à la carte !
Pendant la première période, de 1980 à 1988, le noyau du DEA STE s'est constitué autour de quelques cours obligatoires. Une trentaine de modules optionnels, la plupart ouverts dans d'autres filières ou diplômes, ont été proposés aux étudiants. Chacun de ces modules était généralement centré sur une seule discipline. L'ensemble de ceux-ci, riche et foisonnant, associait par exemple la chimie de l'atmosphère et la faune sauvage.

Les trois filières du DEA STE : des ensembles cohérents !
Pendant la seconde période, de 1988 à 1997, le nombre de cours optionnels a été considérablement restreint et organisé en trois larges champs d'application : les sciences et techniques de l'eau, les sciences et techniques des sols, et l'économie et la gestion de l'environnement. Il est apparu que de plus en plus d'étudiants étaient intéressés par un mélange de modules issus de deux ou trois filières différentes. Ainsi eau et sol, ou eau et gestion environnementale sont des domaines qui interagissent et l'existence de frontières nettes entre les trois filières est apparue trop restrictive.

Aujourd'hui : un itinéraire personnalisé pour chaque étudiant !
Comme à son origine, le DEA STE associe, depuis 1998, un tronc commun (comprenant trois cours obligatoires), et trois cours optionnels : ceux-ci sont maintenant en nombre restreint et choisis parmi neuf possibilités (Tableau 1). Cette organisation permet à la fois de délivrer un message commun, faisant la spécificité du diplôme, et d'adapter de façon souple le programme des enseignements à la fois aux connaissances initiales de chaque étudiant et à son projet de recherche. Celui-ci se déroule d'octobre à juin ou septembre, en parallèle avec les enseignements.

Tableau 1. Organisation pédagogique du DEA STE

Tronc commun : 3 modules obligatoires (80 h)	Modules optionnels : 3 modules parmi 9 (72 à 93 h)
Interfaces et environnement : cycles et processus biogéochimiques	Écologie des sols
Mesures et Environnement	Gestion de l'eau
Économie de l'Environnement	Gestion des risques et environnement
	Gestion et valorisation des déchets
	Hydroécologie et gestion des systèmes aquatiques
	Hydrologie quantitative
	Hydrologie urbaine
	Politique et droit et de l'environnement
	Sols et matériaux : des interfaces dans l'environnement

La procédure d'admission au DEA STE présente une originalité qui constitue un défi pour les étudiants. Ils doivent, souvent plusieurs mois avant de présenter leur candidature, prendre contact avec les équipes d'accueil qui ont des activités de recherche dans leur domaine de compétence et d'intérêt. Après échange d'informations réciproques et accord, ils doivent construire avec leur futur maître de stage un projet de recherche suffisamment structuré et justifié pour pouvoir être présenté et défendu oralement lors de leur entretien d'admission. Une telle démarche garanti la forte motivation des candidats pour leur choix du DEA STE et de leur projet de recherche, leur permettant de commencer leur travail de recherche dès le mois d'octobre. Elle permet aussi aux responsables pédagogiques d'assurer la cohérence, pour chaque étudiant, entre choix de modules optionnels et thème de recherche personnel. Cette procédure est particulièrement difficile à réaliser pour les étudiants étrangers, surtout depuis la mise en place des nouvelles procédures de délivrance des visas. Cependant, la proportion d'étudiants étrangers admis au DEA STE reste comprise entre 8 et 33% (valeurs des 5 dernières années).
analyse de l'évolution pédagogique

La multidisciplinarité par la quantité
L'ouverture d'une formation à l'environnement, impliquant à la fois une université et une grande école d'ingénieur, était considérée au début des années 80 comme un acte particulièrement innovant, créant un lien entre la recherche scientifique de pointe et l'art de l'ingénieur. Néanmoins, la culture et les principes pédagogiques de l'équipe enseignante imposaient des formes d'enseignement très classiques, basées sur des approches monodisciplinaires. Ainsi la nécessaire multidisciplinarité, absente à l'intérieur de chaque module, était remplacée par une multiplication du nombre de module. L'étudiant acquérait sa culture " d'honnête homme de l'environnement " au travers d'un empilement de savoirs disjoints, avec une forte base de chimie et de sciences de l'ingénieur. Un certain nombre de contraintes se sont rapidement révélées.

(a) Choix des modules optionnels. Il était impossible à un étudiant de suivre l'ensemble des modules, pour de simples questions de volume horaire. Deux stratégies de choix de modules ont été mises en œuvre par les étudiants. La première consistait à approfondir les domaines correspondant aux compétences acquises en formation initiale. Ceci n'apportait évidemment pas l'ouverture souhaitée dans cette formation. L'autre stratégie, consistait à papillonner dans des disciplines dont les connaissances de base n'étaient pas maîtrisées. Le gain pour l'étudiant restait alors très superficiel, puisqu'il n'avait pas les moyens, sauf investissement important, d'acquérir les compétences qui auraient été nécessaires pour bénéficier pleinement des enseignements reçus.
(b) Hétérogénéité de la population étudiante. L'appellation " environnement " et la volonté sous-jacente de multidisciplinarité constitue un appel à un recrutement très hétérogène en fonction des parcours initiaux. Ainsi dans les premières années, le DEA STE voyait se côtoyer médecins, ingénieurs, économistes, chimistes. Cependant, la pédagogie mise en œuvre ne permettait pas d'exploiter cette diversité. Seuls les savoirs minimaux, loin des derniers développements de la recherche, pouvaient être transmis par ce système " top-down ".
(c) Organisation matérielle des modules optionnels. D'un point de vue pratique, le nombre moyen d'étudiant par cours restait très faible. Ainsi, chaque année, un certain nombre de modules ont dus être fermés ou se dérouler devant un très petit effectif.

La multidisciplinarité par la méthode
L'ensemble des innovations pédagogiques qui sont apparues tout au long des 18 années d'existence du DEA STE apparaît a posteriori comme mu par un unique ressort, à savoir développer une véritable formation multidisciplinaire en environnement. L'objectif n'est plus alors de chercher à couvrir le plus grand nombre de champs scientifiques. Au contraire, sur un très petit nombre de champs, les étudiants vont être amenés à découvrir aux travers des enseignements et de leur expérience de la recherche les spécificités des méthodes d'étude scientifiques de l'environnement. Charge à eux, par la suite d'utiliser cette interaction entre un champ et une méthode comme un paradigme pour pouvoir ensuite la décliner dans d'autres champs scientifiques.
Dans cette optique, le DEA STE se caractérise par : (a) un équilibre entre les disciplines des sciences de la nature et celles relevant des sciences de l'homme et de la société, (b) le développement de l'approche systémique. De nombreux enseignements (processus biogéochimiques, hydrologie urbaine, hydrologie générale, gestion des risques,...) sont basés sur une analyse de systèmes complexes et des interactions existantes entre différents éléments de ces systèmes. En particulier ces interactions peuvent être d'ordre physique, physico-chimique, biologique pour ce qui concerne les systèmes naturels, mais aussi intégrer des dimensions économiques et sociales, dès que les activités humaines deviennent un élément incontournable de leur fonctionnement. Cette analyse systémique permet, en outre, de mettre en évidence les différentes échelles de temps et d'espace relatives aux fonctionnements de ces systèmes et d'aborder les difficiles problèmes d'intégration d'échelle. Dans cette voie l'outil de modélisation est amené à jouer un rôle majeur. Il est donc aussi intégré dans les enseignements, en tant qu'outil, mais ne fait pas l'objet d'un enseignement spécifique. Ces évolutions du DEA STE sont tout à fait conformes au constat dressé lors de la conférence " Engineering education and training for sustainable dévelopment " (WFEO et al., 1997) qui prônait, dans le domaine de l'environnement, la multidisciplinarité par l'approche systémique et la modélisation.
Au-delà de l'organisation des enseignements, les méthodes pédagogiques mises en œuvre à l'intérieur des modules intègrent la multidisciplinarité. En particulier, plus que l'approche quantitative du savoir, c'est l'attitude générale de l'étudiant et sa capacité d'approche d'un problème environnemental que l'on cherche à évaluer. Dans cette optique les étudiants sont amenés à réaliser un travail collectif : analyse ou conception d'un programme de recherche en environnement, étude de cas à l'échelle d'un secteur de ville en hydrologie urbaine ou en gestion des déchets, qui nécessitent une application des connaissances reçues durant les phases d'enseignement, non pas de manière ponctuelle et

restrictive, mais au contraire de manière intégrée. Ces travaux sont menés en équipes formées d'étudiants de formations initiales différentes. Une mise en commun de cultures s'effectue alors au sein de chaque groupe, qui non seulement permet d'enrichir ces travaux, mais surtout permet l'acquérir l'expérience du travail en groupe dans un contexte pluriculturel.
Enfin, le travail de recherche en laboratoire, qui s'étale sur la totalité de la durée de la formation suit cette même démarche. Les équipes d'accueil de doctorants proposent maintenant des sujets de recherche permettant de mettre en œuvre les méthodes étudiées dans les enseignements.

Forces et faiblesses

Dans le contexte de l'enseignement supérieur

Après 18 années d'évolution, le DEA STE considère avoir réellement progressé sur la mise en place d'une formation multidisciplinaire dans le domaine de l'environnement. L'organisation du cursus, les contenus des modules, les méthodes pédagogiques employées, l'organisation du stage de recherche, aux travers de leurs évolutions, le montrent.
Toutefois cette originalité et cet effort de mise en œuvre pratique de la multidisciplinarité se heurtent aux schémas qui restent très classiques des instances académiques. Si elles affirment souhaiter le développement de la multidisciplinarité, ces dernières restent le plus souvent dans une logique monodisciplinaire. Cette monodisciplinarité se retrouve dans la structuration administrative des instances d'évaluation des formations de l'enseignement supérieur ce qui rend très difficile l'évaluation d'une formation qui a fait de la multidisciplinarité son objectif.
Par contre la qualité des travaux scientifiques menés dans la continuité du DEA, au cours de la formation doctorale en sciences et techniques de l'environnement, peut être facilement évaluée au travers du critère très officiel des publications scientifiques dans des revues à comité de lecture. En moyenne chaque travail de thèse donne lieu à au moins deux publications, ce qui confirme la qualité des travaux menés et des approches scientifiques suivies.
Là encore, toutefois, une critique intrinsèque à l'approche scientifique choisie apparaît fréquemment. En effet les jurys de thèse sont souvent constitués par des experts scientifiques d'un domaine particulier, qui mettent le plus souvent en avant les avancées supplémentaires qui auraient pu être effectués dans leur discipline, mais occultent les passerelles établies entre des disciplines qui souvent ne se parlent guère, comme l'écologie et le droit, la chimie et la mécanique des fluides, la microbiologie et la dynamique des particules.
En plagiant Einstein, on pourrait dire " la multidisciplinarité ne triomphera jamais, ce sont ses détracteurs qui mourront ".

Dans le fonctionnement interne de la formation

L'organisation d'une formation, selon les principes précédemment exposés pose des problèmes de forme très particuliers.
Le premier est de s'assurer de la cohérence du cheminement individuel de chaque étudiant dans l'ensemble des modules optionnels. Elle est assurée, lors de l'admission, après la présentation par le candidat de son projet de sujet de recherche, par le candidat, par une discussion avec les responsables pédagogiques de la formation.
Par ailleurs, chaque module étant devenu multidisciplinaire, il est nécessaire afin d'atteindre un niveau scientifique satisfaisant, de faire intervenir non plus un mais plusieurs intervenants par module, eux même préparés à cette approche, et d'organiser efficacement leur coordination.

Conclusion

En conclusion, nous pensons que de tels enseignements multidisciplinaires sont nécessaires pour donner à chaque étudiant la possibilité d'étendre son propre domaine de connaissance et de prendre en compte l'environnement et le développement durable. Ce programme a été apprécié par 18 promotions successives d'étudiants qui, une fois diplômés, ont des activités professionnelles très différentes : enseignement et recherche universitaire, mais aussi recherche finalisée ou expertise, gestion, études d'impact... Malgré les difficultés actuelles d'emploi des jeunes, les diplômés du DEA STE n'ont pas eu trop de mal à trouver du travail car ils ont, au cours de leur enseignement et projet de recherche, travaillé avec de nombreux professionnels. Ainsi, depuis dix ans, le DEA STE des Journées Sciences et Techniques de l'Environnement, donnant la parole à des doctorants, des enseignants, des chercheurs et des professionnels français et européens, sur un thème transversal d'intérêt général : eaux pluviales urbaines, gestion intégrée des milieux aquatiques, gestion durable des milieux périurbains, mesure et environnement... Nous pensons donc que les difficultés rencontrées pour définir et gérer des enseignements multidisciplinaires sont bien compensées par l'intérêt d'une préparation large, sinon exhaustive, aux défis environnementaux.

Remerciements

Nous remercions les 18 promotions d'étudiants, qui ont suivi cette formation, qui nous ont accompagné tout au long de cette expérience, ainsi que l'ensemble des équipes enseignantes successives, qui ont participé parfois avec vigueur et toujours avec enthousiasme à l'évolution de cette formation.

References

WFEO, UNEP, WBCSD and ENPC (1998): Joint conference on engineering education and training for sustainable development,24-26 September 1997, Final report 23 p.

European Commission Research Training Networks in the field of Water Technologies

Réseaux de formation par la recherche dans le domaine des technologies de l'eau

Shiel J.
Directorate General for Science, Research and Development, European Commission, Brussels, Belgium

Abstract
The Research Training Networks allow researchers from five or more research teams in at least three countries to join their efforts in a common research project and to constitute, in this manner, groups capable of performing research of higher quality. and in so doing to promote the training and mobility of young researchers.
The WET (Wetland Ecology and Technology) network started on 01/09/1996 and will run for 42 months. It is co-ordinated by the Christian-Albrechts Universität Kiel. Partners include teams from Denmark, Italy, the Netherlands, Portugal and Sweden. The network is performing research related to wetland rehabilitation and reconstruction.
The TMR Network SWAMIEE (Sediment and Water Movement In Industrialised Estuarine Environments) project started on 01/01/1998, runs for 36 months and is co-ordinated by the University of Southampton. Partners include teams from France, Portugal, Italy, Ireland, the United Kingdom and Denmark. The network is performing research related to the effective management of industrialised estuarine environments.

Résumé
Les Réseaux de formation par la Recherche permettent à des chercheurs provenant de cinq équipes de recherche dans au moins trois pays d'unir leurs efforts au sein d'un projet de recherche commun et de constituer, de cette manière, des groupes capables d'effectuer une recherche de qualité supérieure afin de faciliter la formation et la mobilité des jeunes chercheurs.
Le réseau WET (l'écologie et la technologie des zones humides a commencé le 01/09/96 et est a une durée de 42 mois. Il est coordonné par Christian-Albrechts Universitaet, Kiel. Les partenaires incluent des équipes du Danemark, d'Italie, des Pays-Bas, du Portugal et de Suède. Le réseau étudie la réadaptation et la reconstruction des zones humides.
Le réseau SWAIMEE (le dépôt et le mouvement dans les environnement d'estuaires industrialisés) a commencé le 01/01/98 pour un durée de 36 mois et est coordonné par l'université de Southampton. Les partenaires incluent des équipes de France, du Portugal, du Royaume Uni et du Danemark. Le réseau étudie la gestion efficace des environnements d'estuaires industrialisés.

Earlier this year the European Commission launched the Improving Human Potential Programme. The aims of the programme, which has a total budget of 1280 million euro, are centred on two main areas of activity:
- the improvement of the human research potential and
- the strengthening of the socio-economic knowledge base.

The programme will be implemented through a series of five actions:
- The support of training and mobility of researchers (Research Training Networks and Marie Curie Fellowships worth a total of 858 million euro).
- Enhancement of access to research infrastructures (182 million euro).
- Promotion of scientific and technological excellence (50 million euro).
- Improvement of the socio-economic knowledge base (165 million euro).
- Support for the development of scientific and technology policies in Europe (25 million euro).

The Research Training Networks'activity is open to all fields of scientific research that contribute to the Community's objectives in RTD. Research teams from a number of European countries are encouraged to work together on a top quality joint research project and in so doing to promote the training and mobility of young researchers (i.e. aged 35 years or less at a pre- or post-doctoral level). The Programme follows on from the earlier Training and Mobility of Researchers (TMR 1994-1998) and Human Capital and Mobility (HCM 1992-96) Programmes.

A total of 249 Research Training Networks are currently being supported under the Training and Mobility of Researchers Programme (TMR 1994 – 1998). The networks involve on average eight partners from five countries and will provide training for some 6000 young researchers. The Networks cover a wide range of scientific disciplines including chemistry, earth sciences, engineering, mathematics, information science, physics and economic, social and human sciences The average duration of a TMR network contract is 3.5 years and the average funding is 1.6 million euro.

The primary objective of Research Training Networks' is to promote training through research at both the pre- and post-doctoral level within the context of high quality transnational, collaborative research projects. Community support is awarded to help researchers collaborate on experiments and to reinforce research teams of the network through the temporary appointment of young researchers from another EU country or an Associated state. Support is also provided to cover the networking costs and the costs of co-ordinating the collaborative research project. It is the task of each network to distribute the research responsibilities between its research teams and to co-ordinate the research operations so that cooperation and communication is as open and efficient as possible. Each Network ensures the diffusion of its principal research results through the publication of papers, brochures and overview articles. Arrangements are made, wherever possible to establish regular dialogue with industrial laboratories, particularly from SMEs that could exploit the research findings or finance an extension of the research towards new objectives. As far as possible, SMEs from less-favoured regions are associated in this dialogue and encouraged to integrate with transnational research teams. The Networks are supported for three to four years.

In the field of water technologies, there are currently two relevant TMR Network projects:

The Wetland Ecology and Technology (WET) project is co-ordinated by the Christian-Albrechts Universitaet Kiel and involves partners teams from Denmark, Italy, the Netherlands, Portugal and Sweden. The project began in September 1996 and will run for 42 months. The Network is performing research related to wetland rehabilitation and reconstruction.

Its aims include the development of general guidelines and evaluation tools for wetland rehabilitation and reconstruction by:
- Monitoring of and Experimenting on natural wetland ecosystems including: collection of data on processes effective in a wetland, comparison of collected data for a common monitoring strategy; organisation and integration of the collected data by an Ecological Information System (EIS).

- Modelling of natural wetlands including: preparation of a general review of existing wetland models, elaboration of a conceptual framework model, linkage of models to EIS and GIS.
- Designing reconstruction and management of wetland systems including: elaboration of an integrated concept of management strategies and reconstruction designs, elaboration of science-supported guidelines for wetland reconstruction and management.

- In the first two years of the project WET network set important milestones leading to the development of science-supported guidelines and evaluation tools for wetland rehabilitation and management. These included:
- development of an expert system and integration of data sets of vegetation and abiotic parameters from various temperate wetland sites as a base for the development of management concepts on the landscape level.
- testing the application of wetlands to remove nitrate, heavy metals and growth promoters from manure.
- network workshop in Padua to exchange state of the art knowledge on wetland modelling and to intensify interdisciplinary training within the network.

WET participants are monitoring a series of wetland sites in Europe. Each team is contributing to the development of scientific guidelines for wetland reconstruction and management.

- The Ecosystem Research Centre at Kiel University (DE) is undertaking the co-ordination, the provision of data and models and the maintenance of EIS as a common information base for wetland design and management;
- The Royal Danish School of Pharmacy is responsible for modelling and ecological engineering approaches, particularly for non-point source pollution;
- The University of Copenhagen (DK)is undertaking modelling and measurement of changes in nutrient dynamics in a newly constructed wetland;
- The University of Padua (IT) is carrying out modelling and measurement of the development of structure and processes in a reconstructed wetland;
- The University of Utrecht (NL) is modelling the hysteresis processes in nutrient availability in soil- and ground water and vegetation responses;
- The University of Coimbra (PT) is modelling and monitoring structure and processes in a natural wetland;
- The University of Lund (SE) is evaluating alternative design and management of wetlands for the reduction of nitrogen transport to sea;

The principal partners for collaboration and implementation of the results of WET are public authorities and consultancy enterprises. The majority of WET partners have entered into such co-operation. These contacts will be intensified in the future. Guidelines and assessment instruments developed by WET indicating location and design of the appropriate wetland for a proposed management target such as maximizing pollution abatement in the surface water and / or protection / restoration of valuable wild life will provide a scientific basis for public authorities and consultancy enterprises for rehabilitation and reconstruction of wetlands in different regions of Europe.

WET is funding 360 person-months of employment of young researchers. They are being trained in wetland ecology and technology with emphasis on interdisciplinary, co-operative teamwork on a European scale. The training programme includes research activities as well as a series of intensive theoretical and practical courses. Subjects of the courses represent a wide range of expertise in the field of wetland research. Having successfully finished their training, participants will be able to cover a wide spectrum of positions in the field of environmental research planning and management within public offices and engineering companies.

Some of the scientific highlights to date include:
- Demonstration of the role of macrodetrital nutrient transport in coastal waters.
- Description of the coupling of denitrification and pyrite oxidation in freshwater under various hydrological conditions.
- Scientific attendance of reconstruction and establishment of a new freshwater wetland in the Po Delta.
- dentification of biogeochemical processes in tidal pools regarding the nutrient dynamics in estuarine systems and nutrient export from such systems.
- Development of a conceptual model for 1) simulating ecological responses of naturally or anthropogenically introduced changes/disturbances in a number of abiotic and biotic parameters in shallow lakes/wetlands, and for 2) generating ideas about how to manipulate wetland ecosystems .
- From a study conducted in Sweden it can be concluded that plant species diversity in agricultural landscapes can be increased by wetland construction. In constructed wetlands, habitat diversity and management seem to be the most important factors for species diversity.
- In a study on bacterial decomposition of macrophyte detritus in wetlands and the effects of solar radiation, very different patterns in accessibility of dissolved organic matter were observed, depending on the location along the eutrophic gradient, and most likely reflecting huge variations in the microbial assemblage and the nature of organic matter from different sources. Some results indicate that the activity of heterotrophic nanoflagellates was inhibited by solar radiation.

The SWAMIEE Network is co-ordinated by the Department of Oceanography of the University of Southampton. Participating teams come from the University of Bordeaux (FR), CNRS (Geology Laboratory, Rouen, FR), University College Dublin (IE) University of the Algarve (PT), Oxford University (UK), Plymouth University (UK), Danish Hydraulic Institute (DK) and University of Ferrara (IT).
The Sediment and Water Movement in Industrial Estuarine Environments (SWAMIEE Network) is co-ordinated by the School of Ocean and Earth Science of the University of Southampton. Participating teams come from the University of Bordeaux (FR), CNRS (Geology Laboratory, Rouen, FR), University College Dublin (IE) University of the Algarve (PT), Oxford University (UK), Plymouth University (UK), Danish Hydraulic Institute (DK) and University of Ferrara (IT). The project began in January 1998 and will run for 36 months.

Since historical times, estuaries have been the focal points of extensive human activity (eg. Maritime transport, agricultural, residential and industrial development). These marginal marine environments have been put under excessive strain in terms of sediment contamination, coastline evolution and saltmarsh development. There is an apparent need for a co-ordinated approach to the effective management of these systems based upon an improved understanding of the various physical processes operating within them.

The key research objectives of this network are to:
- further the scientific understanding, at specific industrialised and non-industrialised locations, of the movement of water and sediment, at a variety of time scales, through an integrated programme of established and innovative field and laboratory work;
- integrate the various data sets (geological, geomorphological, physicochemical and sedimentological) into predictive (forecasting-hindcasting) numerical models, for use in the establishment of regional patterns of water and sediment movement and their sensitivity to changes in the controlling mechanisms;
- transfer experience and expertise between European research groups working in different research disciplines and to foster the development of an applied approach to those concerned with environmental problems;

- disseminate the results of the research to the various end-users of estuarine environments. Particular emphasis will be placed upon the preparation of protocols for cost-effective applied research, to assist in policy formulation and implementation.

The first twelve months of this programme has inevitably focused on getting nine post-doctoral students in position and starting their preparatory training in the key skills required for the successful completion of this programme. The nine institutions have been subdivided into three broad, but interlinked, groupings: cohesives (Oxford, Plymouth, Rouen, Dublin, Bordeaux), non-cohesives (Southampton, Algarve, Ferrara, Bordeaux) and modelling (DHI). Initial fieldwork is being undertaken on a variety of European estuaries (e.g. Southampton Water, Teignmouth and the Severn Estuary (UK); the Seine and the Loire (FR); and the Guadiana (PT). The results from these diverse environments will be compared and contrasted with the prinicpal field site of the Gironde Estuary (FR). Fieldwork in this site will involve all nine institutions collaborating on two cruises this summer, late June (cohesives) and late September (non-cohesives). In addition the results from these cruise will not only attempt to solve a variety of process dominated problems but will also provide valuable input data for 1D, 2D and 3D models.

In addition to the training of the researchers in the individual labs a total of three workshops are being planned for all post-doctoral students in the three key areas (cohesives, non-cohesives and modelling). The first of these is being planned for late April in Southampton. The work from this Network will be disseminated to the academic community via internationally refereed publications and to the wider maritime community via the Internet and summary reports.

Useful Internet addresses:
For information on the Improving Human Potential Programme:
http://www.cordis.lu/improving/home.html

For information on networks running under the Training and Mobility of Researchers Programme: http://www.cordis.lu/tmr

Theme G

Training mobility and internationalization

Mobilité et internationalisation

Theme G : Training mobility and internationalization

Rapporteur: J. Bogardi (UNESCO)

The session approached the theme from different starting points. Prof Musy in the keynote lecture put the question of postgraduate education into a European perspective calling for increased co-ordination and modular structures of the programmes which would facilitate the conception and implementation of high standard postgraduate education.

Inter-university co-operation under European umbrella is believed to have the necessary political support and a viable chance of recognition of joint "European" degrees. An Italian example, presented by M. Guidici, G. Paravicini and G. Ponzini, proved that international inter-university mobility enhances the level of a study programme even in a national context. It prepares the students better for the professional challenges ahead. Mobility programmes enhance the targeted harmonization of the European higher educational structures relying basically on a two level educational system.

J. Rodda, in reviewing the role of the International Association of Hydrological Sciences (IAHS), a scientific NGO addressing also educational, training and professional development issues related to water, highlights an entirely different approach. International co-operation within IAHS (since 1922) does not only influence science, but both implicitly and explicitly education also. Reference books and conference proceedings of IAHS are primary carriers of new scientific results into practice.

Three contributions by E. Wietsma-Lacka, S. Ignar, and J.P. Carbonnel and R. Drobot deal with the experience obtained through the successful implementation of the EU-supported TEMPUS Programme. This programme did not only open possibilities for Central European educational institutions to keep or even improve their scientific and educational standard and equipment, but it gave them the much needed international exposure. No other programme has proven the educational rewards of academic mobility and international co-operation so decisively than the TEMPUS Programme. This well conceived assistance scheme for university partnerships became a source of mutual benefits. TEMPUS has also proven the importance of student and staff mobility as the most efficient means to foster mutual understanding and friendship, declared political goals of the donor.

The session: "Training mobility and internationalization" presented basically only success stories. The case studies showed that both mobilities and internationalization of higher education can be best implemented within the framework of medium or long-term multiple university partnerships: networks. Networking seems to be the key elements to prepare educational institutions for the challenges ahead, to improve teaching skills, curricula and research programmes of the participating universities and teaching staff and, finally, to guarantee a "better product": the graduate.

Networking means also sharing of resources in both monetary and intellectual terms. The shortest summary of this session could be three words: networking, networking, networking.
It is felt however that the demonstrated proof of the potential of networking does not substitute reaping its long-term rewards. Partnerships, the interaction and co-ordination between networks at a higher level, as suggested by the proposed GOUTTE of Water initiative could turn the European success stories into sustainable solutions at the global scale.

It is felt that networking is not only the contribution of institutions of higher learning to join forces to prepare the next generation of professionals for the challenges of the "looming water crisis", but it can be seen as one of the most pronounced realizations of the "learning society" itself. During the ensuing General Forum discussions the role of education, training and networking was analyzed and suggested that a coherent view of the sector, those who educate and those to be educated, should be conceived via an internet-based consultation and dialogue among the stakeholders, as a Water-Education-Training (WET) Vision within the ongoing World Water Vision consultation process. This sector vision should then be presented during the forthcoming 2nd World Water Forum, 17-22 March 2000 in The Hague, The Netherlands.

Keynote lecture

REFORME : Une " Euroformation " dans le domaine des ressources en eau

Musy A.
Ecole polytechnique fédérale de Lausanne, Suisse

Résumé
La nécessité de former des spécialistes en gestion des eaux dans le monde entier n'est plus à démontrer, ni celle d'ailleurs de traiter une telle problématique aux échelles régionales et transfrontalières. Des efforts substantiels sont encore à réaliser en matière de formation et surtout de postformation dans le domaine de la gestion intégrale et intégrée des ressources en eau. Ces efforts doivent cependant être coordonnés, non seulement en terme pédagogique et thématique, mais également au niveau socio-économique afin de mieux appréhender les facteurs locaux ou régionaux agissant sur un phénomène plus global. Actuellement, les formations postuniversitaires sont organisées de manière individuelle, par Institution, et ne sont pas vraiment coordonnées. Or, la mise en commun d'expériences et de connaissances ne peut être que profitable à la fois aux bénéficiaires et aux enseignants. L'organisation modulaire de la formation faciliterait de tels échanges et conduirait, pour chaque module, à une offre élargie en provenance de diverses Institutions. Ceci permettrait à la fois une diversité pédagogique, linguistique et culturelle évidente. Par le biais de convention de partenariats inter-institutionnels, le titre délivré en fin de cours serait cosigné et pourrait comprendre un label "européen" si certaines conditions de mobilité étaient respectées. Celles-ci peuvent être facilitées, grâce à des programmes d'enseignement à distance, que les systèmes modernes de communication nous autorisent à considérer actuellement. Un programme original intitulé REFORME (REseau de FORMation dans le domaine des ressources en Eau) pourrait être mis en place à cet effet. Ses buts et composantes essentielles font l'objet de la présente communication.

Abstract
The need for a high standard education of water resources specialists worldwide is unanimously accepted. The regional and multi-national scale of water resources problems is also well recognized. Substantial efforts are still required in graduated and post-graduated education in the field of full-scale and integrated water resources management. Coordination has to be improved not only under the pedagogic and thematic points of view but also at the social and economical levels. This approach ensures that local and regional factors will be taken into account at a global scale. Currently, post-graduated education is organized on an individual basis, per single Institution. No efficient coordination does exist. However, common knowledge is of benefit to both, students and teaching staff. A modular organization of post-graduated education will definitively contribute to the spreading of such knowledge. A higher quality offer can be foreseen from the joint effort of several Institutions. The resulting pedagogic, linguistic and cultural diversity can also be seen as a significant return from this multi-institutional venture. Through inter-Institution agreements, the delivered diploma could be co-signed and credited with an

"European" label provided some mobility conditions are fulfilled. The latter could be encouraged with distance learning that new communication technologies can now make attractive. An original program entitled REFORME (Réseau de FORMation dans le domaine des ressources en Eau; Network for Post-Graduated Studies in the Field of Water Resources Management), could be implemented for the above mentioned purposes. This paper describes the main components of the proposed network.

1. Le constat

Les ressources en eau de notre planète, aujourd'hui plus que jamais, doivent être gérées de manière très efficace si l'on souhaite vraiment relever le défi de l'adéquation de l'offre à la demande dans une perspective de développement durable. La connaissance des divers éléments d'une telle gestion exige encore des travaux de recherche importants, notamment pour décrire et simuler le comportement d'un système ou pour planifier et maintenir en opération les divers aménagements. La complexité des nouvelles approches est grandissante, surtout en raison de l'imbrication des composants entre eux et de la prise en compte d'éléments connexes et qui concernent plus spécialement les aspects sociaux, économiques et environnementaux. De nouveaux principes doivent être considérés, de nouveaux concepts élaborés, des outils adaptés sont à développer. Or ceux–ci dépendent non seulement du lieu géographique, aux diverses échelles considérées, dans lequel ces réflexions doivent être entreprises, mais aussi du contexte socioculturel sous-jacent. Ce dernier joue un rôle capital, notamment dans l'acceptation et la maintenance des procédures et des techniques de gestion, garantie essentielle de pérennité des travaux exécutés. Même si, d'un point de vue technique, un aménagement spécifique peut être envisagé pour satisfaire un besoin particulier, sa création et son insertion dans un milieu donné peut poser problème dans des régions à caractère fort différent. Un puits de pompage par exemple ne se réalise pas de la même manière ou sur les mêmes bases dans une région à économie forte ou faible. Ces différences sont dues à toutes sortes de facteurs, dont certains d'origine culturelle. Elles doivent être comprises afin de mieux en tenir compte lors de la planification et la réalisation de travaux d'aménagement pour la gestion des ressources en eau.

Afin de faire connaître ces nouveaux développements et applications, de nombreux programmes de formation ont été mis en place dans divers contextes institutionnels et dans plusieurs pays. Ces programmes sont soit intégrés dans des cursus universitaires précis, celle des ingénieurs du génie civil ou du génie rural par exemple, soit dans des cours ad hoc de postformation, organisés par ces mêmes institutions ou par d'autres. Pour les postformations, les participants à ces programmes proviennent d'horizons, tant géographiques que professionnels et culturels fort divers et consacrent plus d'un an en général (une année universitaire de cours et d'exercices, suivie d'un travail de recherche de 4 à 6 mois) pour approfondir leurs connaissances en la matière. Leurs efforts intellectuels sont importants tout comme les sacrifices d'ordre familial et financier consentis. En suivant une formation complète de ce type dans une seule et même institution, ces jeunes bénéficient d'un apport scientifique et technique important certes, mais principalement orienté vers les besoins d'une société localement concernée par des besoins particuliers (hydrologie et ressources en eau en contexte alpin, hydrologie karstique et prospection/protection des eaux en zones essentiellement calcaires, hydrologie et gestion forestières en région à fortes couvertures arbustives, etc.) et non généraux. Il s'agit là d'une vision intéressante mais partielle des aspects globaux d'une gestion "universelle" des ressources en eau qui mériterait d'être élargie. Le moyen utilisé actuellement pour contourner cette difficulté consiste, d'une part à inviter dans un cours de postformation des enseignants étrangers ou ayant une grande expérience professionnelle dans d'autres contextes et situations géographiques et, d'autre part, à suggérer aux étudiants de réaliser leur travail de mémoire de fin de cycle dans d'autres régions. Cette manière de procéder donne quelques satisfactions mais ne saurait remplacer l'apport d'une formation décentralisée.

Enfin, force est de constater qu'aujourd'hui, les coûts de formation sont élevés et que les finances dévolues à leur organisation sont en stagnation, voire en régression. Il en découle donc d'inévitables restructurations accompagnées parfois de formulations de nouvelles règles et contraintes à respecter en matière de participation estudiantine et professorale, afin de mieux optimiser d'un point de vue socio-politique ces divers enseignements. Par ailleurs, les jeunes scientifiques diplômés, intéressés par la gestion des eaux n'ont plus, sans ressources financières additionnelles, autant de flexibilité et de possibilité qu'auparavant pour consacrer une partie importante de leur temps à la post-formation après leurs études. Ils préfèrent souvent et dans le meilleurs des cas, étaler dans le temps leur participation à des cours de postformation ou alors la restreindre à quelques aspects uniquement, même si cette formule ne conduit pas à un titre universitaire. Par contre, la mobilité les intéresse toujours, voire toujours plus car elle permet de répondre à leur volonté d'ouverture et contribue directement à la valorisation de leurs actions lors de la recherche d'un emploi.

Pour toutes ses bonnes raisons, il est temps de repenser la postformation en tenant compte des souhaits des uns et des contraintes des autres, mais surtout de l'intérêt général consistant dans l'application d'une gestion durable des ressources en eau, quelle que soit la région dans laquelle on la réalise. Le projet REFORME présenté dans cette communication est orienté dans cet esprit.

2. Le Cursus

La postformation en matière de gestion des ressources en eau peut être organisée de diverses manières.

Elle se base en premier lieu sur des *connaissances préalables indispensables* dans les sciences de base que les étudiants doivent posséder, notamment en mathématique, physique et statistique. Une description des chapitres essentiels de ces matières, mise à la disposition des étudiants potentiels, entre autre par le biais d'Internet, devrait permettre à ces derniers de situer leur niveau et bagages scientifiques avant de s'engager dans un tel cycle de formation post universitaire.

Des *connaissances complémentaires* dans le domaine des eaux doivent être également requises afin de pouvoir développer de nouvelles matières ou d'en approfondir d'autres dans le cadre d'enseignements spécifiques de postformation. Il s'agit notamment de cours concernant l'hydrologie générale et des sciences apparentées telles l'hydrogéologie et l'hydrométéorologie générales, et d'autres disciplines comme l'hydraulique, la géologie, la pédologie, vu sous un angle assez général. Le choix de ces domaines est fonction des disciplines d'application. Ces enseignements complémentaires peuvent être offerts aux étudiants soit directement dans le cadre d'une formation spécifique organisée par une Institution, soit par le biais d'Internet grâce à des cours développés à cet effet.

Des *connaissances approfondies* dans différentes matières relatives à la gestion des eaux peuvent être alors proposées aux étudiants dans le cadre de cours post universitaires. L'éventail de ces matières devrait être aussi large que possible, laissant un réel choix aux participants, fonction de leur motivation et de leur intérêt. Un regroupement de certaines disciplines enseignées pourrait conduire l'étudiant vers une spécialisation et un titre universitaire précisant le domaine choisi (par exemple, spécialisation "gestion des eaux" ou "hydrogéologie", "aménagements" ou encore "zones alpines, forestières ou semi-aride", etc.).

En fin de cycle un *mémoire de recherche* devrait être imposé pour toute formation délivrant un diplôme. Celui-ci pourrait s'effectuer dans divers contextes institutionnels ou privés et dans diverses régions du monde.

Ce principe de formation n'est pas nouveau et de telles organisations ont déjà lieu dans diverses universités ou grandes Ecoles, par exemple à l'EPFL de Lausanne et à l'Université de Neuchâtel, dans le cadre de la formation interuniversitaire en hydrologie et en hydrogéologie. Mais de telles structures sont en général trop rigides (pas ou peu d'options prévues), trop lourdes (trop de matières enseignées), trop longues (durée de 16 à 18 mois), trop localisées (les aspects locaux et régionaux sont présentés en règle générale). De telles formations intéressent par ailleurs une "clientèle" locale professionnelle restreinte et coûtent relativement chères tant au niveau des étudiants qu'à celui des Institutions qui les organisent. Pour pallier ces inconvénients, quelques adaptations sont réalisées mais ne donnent pas entière satisfaction. Une réorganisation de ces structures mérite d'être abordée dans un autre esprit.

3. L'organisation

Selon le principe de cursus évoqué ci-dessus, un cycle postgrade universitaire dans le domaine des ressources en eau pourrait être organisé de la manière suivante :
Les *connaissances préalables* à tout cycle de postformation doivent être acquises pour tous les étudiants, quelles que soient leurs origines universitaires. Selon une description générale des chapitres nécessaires à connaître dans ces matières, un **auto contrôle** de connaissance peut être suggéré aux étudiants potentiels par le biais d'Internet. Ce test facultatif, lorsque répété plus tard en cas de doute dans une Institution, permettra à cette dernière de récuser sans autre excuse la personne qui ne satisfait pas aux exigences minimales imposées. Celles-ci peuvent être différentes, compte tenu des objectifs finals des spécialisations offertes et fonction des exigences académiques des Institutions co-organisatrices.

Les *connaissances complémentaires* peuvent être acquises soit par des enseignements spécifiques, organisés dans le cadre d'un cursus normal de formation universitaire (formation d'un ingénieur du génie rural de l'EPFL par exemple), soit par des enseignements dédiés adressés uniquement aux étudiants en postformation (cours ad hoc), soit encore au moyen d'enseignements assistés par ordinateur et à distance, par le biais de cours préparés dans cet esprit. Dans le cadre d'un projet de recherche européen, des travaux sont en plein développement dans divers endroits, notamment à l'EPFL de Lausanne (programme ARIADNE – cours Hydrologie générale).

Les disciplines correspondant aux *approfondissements* font l'objet des enseignements **strictement réservés à la postformation**. Elles sont groupées par grande thématique (modélisation hydrologique, gestion des ressources en eau, hydrologie des zones urbanisées, aménagements hydro-agricoles, aménagements de cours d'eau, aspects législatifs, économie des eaux, etc.). Une thématique est elle-même organisée en modules de 25-30 heures (cours, exercices ou projet). A la fin de chaque module, soit après une ou plusieurs semaines, un contrôle de connaissance est réalisé et donne droit à des crédits de type ECTS. Ces crédits, donc les enseignements qui correspondent, sont reconnus et acceptés par l'ensemble des Institutions co-organisatrices de ces postformations.

La *spécialisation (connaissances complémentaires + approfondissements)* est déterminée par un ensemble de thématiques, donc par un ensemble de modules, dont la totalité des heures se situe entre 400 et 600 heures de contact. Ces modules peuvent être organisés soit séquentiellement, avec quelques éventuelles contraintes de modules prérequis (exemple : modélisation hydrologique et hydraulique urbaine), soit en parallèle pour des matières suffisamment distinctes (exemple : modélisation hydrologique et économie des eaux). Cette spécialisation peut s'étaler sur plus d'une année académique, fonction de "l'offre" en modules et de leur positionnement dans le temps.

Lorsqu'un nombre suffisant de crédits est obtenu, l'étudiant peut alors entreprendre son *travail personnel* de fin de cycle. Ce mémoire peut être organisé au besoin en avance et parallèlement au suivi de certains modules mais ne pourra être comptabilisé tant que le nombre de crédits de l'ensemble des modules d'une spécialisation ne sera atteint.

L'organisation inter-institutionnelle est importante et peut se résumer à l'aide de la figure 1.

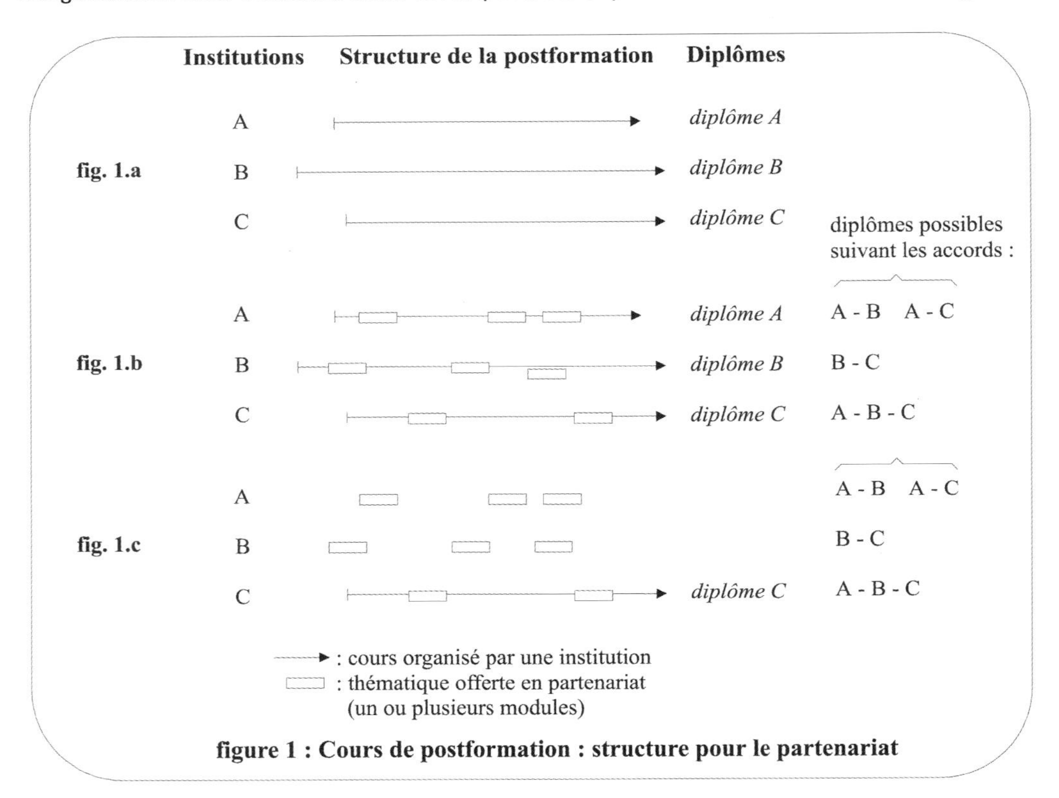

figure 1 : Cours de postformation : structure pour le partenariat

Un groupe d'Institutions (de 2 à 3 selon l'exemple fig.1) s'accordent pour mettre en place une postformation en matière de ressources en eau, orientée vers une problématique déterminée (hydrologie et aménagement des eaux en zones alpines par exemple). Cette postformation est organisée selon plusieurs thématiques spécialisées dont au moins quatre (par exemple) sont requises pour débuter un travail de recherche. Chaque institution offre de manière concertée, un certain nombre de modules dans un ou plusieurs domaines scientifiques et techniques. Les participants sont des étudiants régulièrement inscrits à ce cycle de postformation mais également d'autres professionnels intéressés par une matière uniquement, donc par un seul et unique module. Lorsque le nombre de crédits est suffisant, l'étudiant aborde alors un travail de recherche auprès d'une institution déterminée. La défense de son mémoire passée avec succès, cet étudiant reçoit alors un titre universitaire de l'institution où il a réalisé son travail final, cosigné par les toutes les Institutions partenaires de la spécialisation. Plusieurs combinaisons inter-institutionnelles peuvent être envisagées :

Selon le schéma fig.1a, chaque institution organise par elle-même l'entier de la postformation. Il n'y donc pas de coordination inter-institutionnelle et le titre délivré émane de chaque Institution, conformément à sa propre réglementation interne. Ce cas prévaut aujourd'hui dans la plupart des Ecoles ou Universités offrant une postformation dans le domaine des ressources en eau.

Dans le cas fig.1b, les Institutions A, B, C organisent leur propre postformation par modules spécifiques. Certains cours sont offerts dans une deuxième filière de postformation inter-institutionnelle. L'étudiant ayant suivi 4 thématiques dans 3 Institutions différentes par exemple peut prétendre à un titre cosigné par les Institutions A,B,C. Il peut également rester à un seul endroit et suivre l'entier de la postformation organisée par une seule Institution. En cas de succès, il recevra dans ce cas un titre émanant uniquement de l'Institution organisatrice. Ce dernier cas ne diffère pas de celui présenté en fig.1a.

Le cas de la fig.1c est un cas particulier du précédent. Une ou plusieurs Institutions (A, B selon le schéma présenté ci-dessus) n'offrent des modules que dans le cadre d'une postformation inter-institutionnelle prévue. Il n'y a donc pour elles pas de formation diplômante unique en dehors de la postformation inter-institutionelle (ce qui n'est pas le cas de l'Institution C). Les étudiants non inscrits en C ne pourront donc prétendre qu'au titre inter-universitaire A,B,C.

4. Le réseau

L'organisation d'une telle structure de formation est toutefois conditionnée par la mise sur pied d'un réseau inter-institutionnel. Celui-ci doit être basé sur une volonté réciproque de concentrer des efforts et de coordonner des actions afin d'offrir une postformation conjointe dans le domaine des ressources en eau. Pour atteindre cet objectif, il faut que chaque Institution se sente concernée par quelques aspects de la problématique "eau" d'une part, et soit d'accord, d'autre part, d'associer son nom à celui d'autres partenaires pour délivrer un diplôme inter-universitaire commun. Ce dernier aspect implique nécessairement une relation de confiance car chaque institution doit s'assurer auprès de chacun de ses partenaires non seulement d'un engagement pédagogique réciproque mais aussi d'une qualité d'enseignement et de contrôle de connaissances au moins égale à celle qu'elle s'impose. Les programmes et les méthodes d'enseignement actuels doivent ainsi être adaptés, tout comme les mentalités prévalant dans certains systèmes académiques. Quelques prérogatives doivent être abandonnées au profit d'une internationalisation des compétences et des titres et d'une valorisation plus globale de l'enseignement. Ces relations privilégiées prennent du temps pour être formalisées mais ne sont pas impossibles à réaliser. Afin de gagner immédiatement en efficacité, il serait possible d'utiliser dans ce but les réseaux inter-universitaires existants, le réseau "CLUSTER" par exemple, qui lie 11 Institutions européennes renommées pour certains aspects d'enseignement et de recherche ou d'autres réseaux, tels TIME ou ETNET. Un label supra national, du type "européen", pourrait être également accordé si une commission ad hoc en acceptait les principes. Ce point mérite cependant encore quelques réflexions avant d'être concrétisé.

Par de tels accords, on verrait naître ainsi un nouveau réseau intitulé "REFORME" (REseau de FORMation dans le domaine des ressources en Eau) qui associerait à la fois les compétences de plusieurs institutions et les possibilités de développer en commun un cycle de postformation dans le domaine des ressources en eau. Celui-ci pourra être vraiment élargi à de nombreuses orientations et applications, tant au niveau thématique que géographique. Par conséquent, il sera susceptible d'intéresser de nombreux jeunes scientifiques et professionnels, soucieux de développer de manière durable les principes d'une utilisation rationnelle des eaux et des systèmes qui les concernent.

5. Les avantages

Ils se situent à plusieurs niveaux :
- possibilité offerte aux étudiants de choisir une filière de formation selon leur choix et en fonction de leur motivation et de leurs intérêts,
- occasions, pour des scientifiques ou techniciens, d'améliorer leurs connaissances dans certaines matières uniquement en ne suivant que certains modules,
- ouverture, pour les participants, vers d'autres régions à économie et culture différentes,
- meilleure qualité des formations car thématiques abordées à l'endroit où elles sont perçues et vécues,
- économie d'ordre financier substantielle pour les Institutions n'offrant que certains modules et non plus une postformation complète,
- approfondissements garantis des connaissances par modules, puisque spécialisées,
- concertation indispensable entre spécialistes de diverses régions pour offrir un "menu" adéquat et circonstancié de formation,
- délivrance d'un titre inter-institutionnel, offrant de meilleurs arguments d'embauche aux étudiants, surtout si celui-ci avait de surcroît un "label" général particulier, label "européen" par exemple,
- mise en réseau d'Institutions dans une thématique définie.

6. Les difficultés

Elles sont essentiellement d'ordre administratif et organisationnel :
Au niveau administratif, il convient de coordonner les secrétariats académiques des Institutions partenaires afin de s'accorder sur les niveaux de formation préalablement requis et respectés dans chaque module ainsi que pour le mémoire de recherche. La connaissance réciproque de chaque partenaire et surtout la confiance que ceux-ci s'accordent mutuellement restent la meilleure garantie de succès dans ces négociations, pas toujours faciles à finaliser.

Sur un plan logistique, la difficulté réside surtout dans l'établissement d'un calendrier des actions que chaque Institution doit établir et respecter. Là encore, cette difficulté n'est pas impossible à surmonter mais nécessite de nombreuses discussions.

S'agissant enfin des aspects financiers, il convient de faire remarquer également que la mobilité imposée par ce système coûtera plus cher qu'actuellement aux étudiants, mais que ce surcoût sera largement compensé par l'expérience extraordinaire que vivront les étudiants dans diverses régions.

7. En guise de conclusion

La nécessité de former des scientifiques dans le domaine des ressources en eau n'est plus à démontrer. Elle nécessite cependant et à la fois une "globalisation" des connaissances et un haut degré de spécialisation. Cette ouverture requiert au niveau de l'enseignement une combinaison de disciplines très variées liées à des applications dans divers contextes hydro-climatiques. Seule une mise en réseau de compétences peut répondre à de telles demandes. Et si, de surcroît, la structure de la postformation résultante est bien réfléchie, une offre complémentaire intéressante sera disponible pour des personnes soucieuses de parfaire leurs connaissances dans un domaine thématique particulier, sans titre académique à la clé.

Le projet REFORME présenté ci-dessus permet de répondre à de telles exigences en renforçant la collaboration internationale dans ce domaine, tout en diminuant les coûts impartis, notamment pour chaque Institution partenaire. Ce projet peut présenter un intérêt évident aussi bien pour la collectivité scientifique internationale que pour les institutions universitaires concernées. Il n'attend finalement que de bons parrains pour être défendu et lancé rapidement. L'auteur de cette communication se réjouit par avance de recevoir des propositions concrètes de collaborations dans ce sens et d'envisager des partenariats aussi intéressants qu'efficaces afin de pouvoir mieux relever un défi essentiel de nos sociétés, celui de résoudre à terme et à satisfaction le problème de la gestion intégrale, intégrée et pacifique de nos ressources en eau.

Promoting international co-operation in learning—the role of the International Association of Hydrological Sciences

L'encouragement de la coopération internationale en savoir—le rôle de l'Association Internationale des Sciences Hydrologiques

Rodda J. C.
International Association of Hydrological Sciences,
Institute of Hydrology, UK

Abstract
This paper presents a summary of the activities of the Association, a water orientated non governmental organisation, in relation to the learning society. It looks at the history of IAHS, its current challenges and future aims.

Résumé
Cet article présente un résumé des fonctions de l'association, une organisation non gouvernementale qui adresse les problèmes de l'eau, en relation de la société cognitive. L'article considère l'histoire de l'association, les défis présents et les buts futurs.

1. Introduction

As the oldest of the non profit making international non governmental organisations dealing with hydrology and water resources, the International Association of Hydrological Sciences (IAHS) has made a unique contribution to the world of education and learning and its related areas of economic activity, since it was established in 1922. During these nearly 80 years, the world at large has experienced many far reaching changes, a considerable number impacting on the water environment and on society generally. Some of these changes have been reflected in the activities of the Association and have altered the pattern of its endeavours, many in relation to learning. These changes have not only affected the science of hydrology, but they have also altered the structure of IAHS and how the Association functions. Furthermore, they have become more intrusive over the last decade or so, particularly as economic pressures have built up. Now there are new challenges arising, some scientific, some technological and some due to socio-economic factors, while fresh ones can be foreseen on the horizon which may be potent in causing further alterations to the Association.

2. Mission statement

Many organisations have adopted mission statements over the last 5 to 10 years, statements which purport to encapsulate what that particular body is attempting to do. The IAHS mission statement appears in the initial part of its statutes (IAHS 1996), the basic rules of the Association which govern the way it operates:

> To promote the study of Hydrology as an aspect of the earth sciences and of water resources
> o to study the hydrological cycle on the Earth and the waters of the continents; the surface and groundwaters, snow and ice, including their physical chemical and biological processes, their relation to climate and to other physical and geographical feature as well as the interrelations between them;
> o to study erosion and sedimentation and their relation to the hydrological cycle;
> to examine the hydrological aspects of the use and management of water resources and their change under the influence of man's activities;
> o to provide a firm scientific basis for the optimal utilisation of water resources systems, including the transfer of knowledge on planning, engineering, management and economic aspects of applied hydrology,
> To provide for discussion, comparison and publication of research results.
> To initiate, facilitate and co-ordinate research into, and investigation of those hydrological problems which require international cooperation.

Naturally this statement has faults, which may be common to those adopted by similar professional scientific bodies and organizations. For example, it tries to be all embracing and not to be too specific, it incorporates some outmoded ideas, it needs to be brought up to date and there is always the question of whether or not such a mission can be accomplished. The discovery of significant quantities of water on the Moon in March 1998 and the possibility of water on other moons suggests that the term 'earth sciences' may no longer be appropriate to IAHS. In a similar vein, the reference to 'man's activities' may deserve a change to an equivalent gender free phrase. These and other alterations will, no doubt, be incorporated into the next statutes of the Association, when a revised version is set before a future IAHS General Assembly for its approval. While such amendments can be significant, perhaps more important to discussions here are the references to the learning society, such as to the gaining of knowledge, to the transfer of knowledge and to the provision of a forum for the discussion, comparison and publication of research results. To initiate, facilitate and co-ordinate research is another phrase worthy of note in the current context.

3. History

An outline of the history of the Association illustrates its past contribution to the world of learning and it provides a platform for an understanding of what the future might bring. When they drew up a history of IAHS, Volker and Colenbrander (1995) identified four phases in the development of the Association and the discussion here will follow that classification.

Phase 1: The first participation of hydrology in international scientific cooperation, 1922-1939

After discussions at the 1922 Rome General Assembly of the International Union for Geodesy and Geophysics (IUGG), an International Branch of Scientific Hydrology was established. The term 'scientific' was used in order to distinguish the Branch from those who used rods and twigs to find water and from those interested in the commercial exploitation of mineral waters. A Commission for Glaciers was formed as part of the Branch at the Madrid Assembly in 1924 and at the Prague Assembly in 1927, the Commission absorbed the International Glacier Commission which had been established in 1894. The first of the 'Red Books' was published for the Madrid Assembly, to be followed by similar volumes for the succeeding 20 assemblies. At Madrid, a Commission on Statistics was set up with the tasks of trying to bring some uniformity into the publication of hydrological data and to make inventories of national water resources. At this time hydrologists from some 25 countries were involved in the work of the Branch, which in 1933 became the International Association of Scientific Hydrology (IASH) at the Lisbon Assembly. During the 1930s the commission structure emerged within the Association, when commissions for potamology, limnology, instruments and measurements and subterranean waters were established alongside the existing bodies. At the time of the Washington Assembly in 1939, the Association had produced 26 publications in the form of Red Books.

Phase 2: Recognition and expansion of IASH, 1948-1970

Post war growth saw both hydrology and IASH expand, because of the increasing demand to control and manage water resources through the construction of hydraulic works, because of the need to provide forecasts of floods and to design defences against them and because of a number of other needs. The assemblies in Oslo (1948), Brussels (1951) and Rome (1954) saw an increasing number of volumes published by the Association and then in 1956, IASH convened its first symposium outside an IUGG assembly, in order to meet the demand for more frequent meetings. This symposium was held at Dijon to commemorate the 100th anniversary of the publication by Darcy in 1856 of the law on groundwater flow. In 1956, Professor Leon Tison, who had become Secretary General of the Association in 1948, started the publication of a Quarterly Bulletin, carrying both news of the Association and scientific papers. He also handled the publication of the Red Books, editing the papers, getting them published by a local printer and mailing them himself. More volumes of the Red Books appeared containing scientific papers for symposia and for the general assemblies, then in 1959 Publication No. 50 was produced. For the 1963 Berkley General Assembly, the Association published 5 volumes of proceedings containing more than 2000 pages covering the topics of; snow and ice, evaporation, surface waters. subterranean waters, erosion, precipitation and soil moisture.

In the late 1950s, IASH started to co-operate with the United Nations and its agencies with interests in water, particularly with UNESCO and the World Meteorological Organization (WMO). In 1961, Professor Tison became the first Vice President of what was later to become WMO's Commission for Hydrology (WMO 1986) and when UNESCO launched the International Hydrological Decade in 1965, Professor Tison was reputed to have been one of the three fathers! The Association helped UNESCO formulate programmes and it contributed to the activities of IHD working groups. One of the benefits of the this co-operation was the joint convening of symposia and the publishing of the proceedings, first with UNESCO in 1967 for the Symposium on Fractured Rocks and then in 1972, with WMO, for the Symposium on the Distribution of Precipitation in Mountainous Areas. In 1967 a new series of co-edited publications on Snow and Ice was started with UNESCO, dealing primarily with the fluctuation of glaciers.

Phase 3: Transformation of IASH into IAHS and further growth, 1971-1981

When Professor Tison was forced by ill health to resign from the position of Secretary General, it became obvious that because of the increasing work load and also because of the new thrusts in hydrology and new ideas about the grouping of water sciences, the Association needed a change. Consequently a revised set of Statutes and Bye-laws was brought into force at the Moscow General Assembly in 1971, including a change in the name to the International Association of Hydrological Sciences. This change was to reflect the broadening of the approach, while the number of Commissions was increased in response to this growth in the science.

A new management structure was devised to cope with the increasing number of activities and the growing work load. A Treasurer and an Editor were appointed to operate alongside the Secretary General, to look after the specific tasks of finance and publishing. Annual meetings of the Bureau of the Association were instituted at this time and more frequent meetings were held between the President and Secretary General. An editorial office was established at the Institute of Hydrology, Wallingford, in 1971 and another milestone was reached when Publication No. 100 appeared later that year— the Proceedings of the Symposium on Mathematical Models. The IAHS Bulletin became the Hydrological Sciences Bulletin in 1972, containing only scientific papers, but to provide news of the Association, the Secretary General started to publish a Newsletter, which was circulated initially to IAHS National Committees and to members of Commissions and Committees.

When the IHD came to and end in 1974, the Association was one of the bodies which helped in the establishment of the permanent International Hydrological Programme (IHP) in UNESCO, a programme which was launched in 1975. The Association took part in the Conference to mark the end of the Decade (UNESCO/WMO 1974), the Conference which became the first of the 6-yearly Joint Conferences on Hydrology convened by UNESCO and WMO. During this period the Association adopted a policy of pre-publishing Red Books prior to the symposium concerned, the cost of one copy of the proceedings being contained within each registration fee.

Phase 4: IAHS in a changing environment, 1981-1995

It could be argued that the pace of science accelerated during the 1980s, as environmental change and particularly climate change became more prominent on the research agenda. The development of bigger and better global circulation models seemed a dominant trend, but for hydrologists, their scale and their inability to simulate the hydrological cycle with any degree of verisimilitude and the difficulties of incorporating land-atmosphere relations in them were considerable impediments to progress. The increased power of computers and the advent of the PC, provided opportunities in hydrology to model and simulate hydrological systems that had not been available previously. These opportunities were stimulated by the development of new sensors on the ground and on satellites, to the extent that remote sensing became an important tool for hydrologists.

In response to these developments, IAHS started its own series of Scientific Assemblies , the first being held in Exeter in 1982. The Assembly attracted over 500 participants and the proceedings were published in six volumes with a total of nearly 1800 pages. For the Hamburg IUGG General Assembly in 1983, in contrast to the 1970s, the associations in IUGG were able to arrange their own programmes within the IUGG Assembly and the IAHS attendance and programme benefited considerably as a result. Later during this period the number of Committees in IAHS was increased to three (Table 1) and the joint IAHS/WMO GEWEX Committee was established. Both of the actions resulted from the widening of the role of the Association in response to the scientific challenges arising, the purpose of the GEWEX Committee, in particular, being to establish a role for hydrologists in the burgeoning field of large-scale hydrometeorological field experiments.

Table 1. The Commission and Committee Structure of IAHS

International Commission on Surface Water	ICSW
International Commission on Ground Water	ICGW
International Commission on Continental Erosion	ICCE
International Commission on Snow and Ice	ICSI
International Commission on Water Quality	ICWQ
International Commission on Water Resources Systems	ICWRS
International Committee on Remote Sensing & Data Transmission	ICRSDT
International Committee on Atmosphere-Soil-Vegetation Relations	ICASVR
International Committee on Tracers	ICT

In 1981 the Association, with UNESCO and WMO, started the annual award of a silver medal know as the International Hydrology Prize to hydrologists who had made an outstanding contribution to the science. The Prize was first awarded to Professor Tison for the excellence of his contribution to Association and for his expertise in research and skills in teaching. The Prize has been awarded every year since to hydrologists who have distinguished themselves and brought credit to the science. The Hamburg Assembly agreed to establish an annual award to mark the contributions of young scientists to the Association. The award, which is in the form of a certificate and money, is known as the Tison Award. It is made on the basis of the best paper or papers presented in one of the publications of the Association. The Hamburg Assembly also set up the Hydrology 2000 Working Group, a group of young hydrologists who reported to the Vancouver Assembly in 1987 on probable state and shape of hydrology in the year 2000AD. This volume makes interesting reading as the year in question nears.

In 1983 the name of the Bulletin was changed to the Hydrological Sciences Journal and it became a bimonthly in 1988. The proceedings of the 1984 symposium on the Hydrochemical Balances of Freshwater Systems appeared as IAHS Publication No.150, only 13 years having elapsed since Publication No.100 had been published, compared to 37 years between Numbers 1 and 50. The Newsletter, which had been published by the Secretary General since the 1970s, was put on a more formal basis towards the end of the 1980s, being produced three times a year with a circulation of up to 1000 copies. Two new series of publications were launched in 1989; IAHS Information in Chinese and the Blue Books, a series of monographs written at the invitation of the Association. IAHS Information is a magazine which appears 4 times a year containing selected reviews, reports and papers culled from IAHS sources and which is circulated widely in China. The Blue Books are a series of Special Publications, the initial one having the title Hydrological Phenomena in Geosphere-Biosphere Interactions. Specialised symposia continued to be organised by the Commissions and Committees of the Association, independent of General Assemblies and Scientific Assemblies, to the extent that in the early 1990s, 7 or 8 Red Books appeared in most years. Indeed in 1991 IAHS Publication No. 200 appeared—the proceedings of the Fourth International Symposium on Land Subsidence. UNESCO provided financial support to this publication as a contribution to the IHP, while a number of other Red Books received financial support from a range of bodies and sources during the 1990s. Since 1992 there have been a series of colloquia, known as the George Kovacs Colloquia which have been organised jointly by IAHS and UNESCO for the two or three days directly before the meetings of the Intergovernmental Council of the IHP. The Association continues to work closely with both UNESCO and WMO. IAHS is represented at Council and Commission meetings and in the subsidiary bodies of both. UNESCO and WMO are represented at the meetings of the IAHS Bureau and play influential parts in the affairs of the Association.

Probably the most important IAHS initiative in support of the transfer of knowledge to developing countries was started in 1991, when the Association's Task Force for Developing Countries (TFDC) commenced the mailing of Red Books and, in some cases, the Hydrological Sciences Journal, to libraries in nations in need. Over 70 addresses were selected covering many countries in all continents, on the understanding that the publications are made available to hydrologists and other scientists outside the recipient organisation. The list of addresses has been revised a number of times subsequently, but the main thrust of this important initiative is continuing.

4. For the future

As the Millennium approaches, like many other scientific and professional bodies IAHS is examining its role and considering its future. Where will the new scientific initiatives occur? What will be the impediments to progress? How can the small international community of hydrologists influence events or, at least, be carried along in the strongest part of the current of progress?

The contribution of hydrology to the learning environment has been strongly influenced by technological progress during the last 20 years or so. Easy access to computers and the rise of remote sensing, together with the information revolution have transformed the science in manner that would be unimaginable to the founders of IAHS. These innovations must be part of the accelerating pattern of Schumpter waves (Economist 1999) which characterise the bursts of innovation which started with the Industrial Revolution in the 18th Century. Their consequences and side effects may radically influence the way the Association operates during the coming decades. Video- conferencing, electronic publishing, satellite based systems such as for global positioning, biotechnology, a base on the Moon and other advances may have serious impacts on hydrology and on IAHS activities. However they may be overshadowed by the coming water crisis and the continuing need for hydrology to react to it.

References

UNESCO/WMO, (1974) ***Records of the International Conference on the Results of the International Hydrological Decade and on Future Programmes in Hydrology***, I Final Report, Paris, 2 -13 September 1974, 112.

WMO, (1986) Silver Jubilee of the WMO Commission for Hydrology, *Technical Reports to the Commission for Hydrology*
No. 22, 65.

IAHS, (1996) IAHS Handbook 1995-1999, International Association of Hydrological Sciences, 77

Volker,A., and Colenbrander, H.J., (1995) History of the International Association of Hydrological Sciences, IUGG
Chronicle, Jan/Feb, 13-22.

Economist (1999) Catch the wave, Innovation in Industry Survey *The Economist*, February 20-26, 7-8.

Teaching subsurface hydrology for the «Laurea» degree course in Physics at the University of Milano

L'enseignement de l'hydrologie souterraine dans l'Ecole de Physique de l'Université de Milano

GIUDICI M.*, PARRAVICINI G., PONZINI G.***

*Università degli Studi di Milano, Dipartimento di Scienze della Terra, Sezione di Geofisica, Italy.

**Università degli Studi di Milano, Dipartimento di Fisica, Sezione di Fisica Teorica, Italy.

Abstract

Groundwater hydrology is taught to students of the «Laurea» degree course (roughly equivalent to the British MS) in Physics at the University of Milano. We present the current status of this initiative, the importance of students' mobility through Europe and some future perspectives.

Résumé

L'hydrologie souteraine est enseignée aux etudiants de l'Ecole de Physique de l'Université de Milano (I). On present ici le status de cette initiative, l'importance de la mobilité des etudiants dans les Universitées Europeennes et quelques perspectives pour le future.

1. Introduction

Problems of subsurface hydrology are usually faced by specialists with a degree in Civil, Hydraulic and Environmental Engineering or in Engineering Geology. However Physicists and Geophysicists can give valuable contributions. Firstly, environment is so complex that the management of natural resources involves a great number of different disciplines; a definite answer to environmental problems cannot be given by scientists and technicians within a single expertise. Second, the background of physicists makes them particularly able in modelling, in experiment design and implementation, in the analysis and interpretation of data with reference to mathematical models of transport phenomena as well. This background allows physicists to analyse a great mess of data, to infer and apply the basic physical laws governing transport phenomena, even for very complex natural systems such as sediments and rocks where subsurface water flow takes place.
For these reasons we teach basic notions of groundwater hydrology to the students of the «Laurea» degree course in Physics at the University of Milano (I). In this communication we report the current status of teaching subsurface hydrology in this course, we discuss the importance of a co-operation between different European Institutions for students' mobility and we finally analyse some of the perspectives of this effort for the next years.

2. The «Laurea» degree course in physics

The curriculum for the «Laurea» degree in Physics is divided in three propedeutic years and a final year with specialistic course units. At the University of Milano, the speciality in «Environmental and Earth Physics» is activated at the fourth year. Students can attend the course units «Dynamics of Fluids» (lecturer: G. Parravicini), «Earth Physics» (lecturer: G. Ponzini) and «Laboratory of Earth Physics» (lecturer: M. Giudici), which introduce some basic concepts of subsurface hydrology among other topics. In particular elements of continuum mechanics, perfect and viscous fluid motion and heat transport are taught in «Dynamics of Fluids». «Earth Physics» covers several transport phenomena in geophysics, including water flow and contaminant transport in porous media; emphasis is given to modelling and to experimental and mathematical techniques for the identification of the phenomenological parameters. «Laboratory of Earth Physics» is devoted to the application of geophysical survey methods, e.g., magnetic and geoelectrical prospecting, to environmental - mainly hydrogeological - problems; it includes field data acquisition, processing and interpretation. The course units «Earth Physics» and «Dynamics of fluids» consist in 80 hours of classroom lectures each, whereas «Laboratory of Earth Physics» consists in 120 hours of tutored activity, including introductory lectures (approximately 40 hours), data acquisition in the field, data processing and interpretation in computer laboratory (Giudici et al., 1997a).

Before receiving the «Laurea» degree, students write a thesis, in which they tackle a research theme - which might be quite advanced or standard according to the student's capability. The time required for the preparation of the thesis is normally one year; the thesis can be prepared at the University or at some other public or private Institutions. Since 1995 we have supervised 23 theses on subsurface hydrology; four of these are still to be discussed and the preparation of nine theses has required a stage in other Italian or European Institutions.

The subjects of these theses are related to our research topics which include inverse modelling and identification of physical model parameters, use of geoelectrical prospecting for reconstructing hydrogeological schemes, modelling the aquifer of the town of Milano and others. Some of these theses resulted into presentation to national and international conferences and publication on scientific journals (see, for instance, Giudici et al., 1995, 1997, 1998).

Students choosing the speciality in «Environmental and Earth Physics» can also attend course units on the physics of the Earth's interior («Solid Earth Physics», «Seismology»), atmospheric and environmental physics («Physics of the Atmosphere», «Environmental Physics», «Laboratory of Environmental Physics») and can study problems in these fields for their degree's thesis.

3. The added value of students' mobility in Europe

We took advantage of the Erasmus/Socrates Programme to have students preparing a large part of their thesis at other collaborating Universities. In the first phase of the Erasmus Programme this has been possible because one of us (G. Ponzini) was a member of the Interuniversity Co-operation Programme «Hydraulics and Environmental Engineering 1065/06», whereas in the last years we have inserted students' mobility in the institutional contract of our University.

In particular five of our students completed 4 to 6-month-long visits at Institutions of countries of the European Union within the Erasmus/Socrates programme, whereas one student completed a 3-month-long stage at the Scuola Universitaria Professionale della

Svizzera Italiana at Bellinzona (CH). Two of our students prepared the whole thesis at the JRC of the European Union at Ispra (Italy). We print in Table 1 the complete list of collaborating institutes.

Table I List of European Institutions which hosted students from our University in the field of groundwater hydrology

Number of students	Institutions	Responsibles
1	Universität Hannover (D)	Prof R. Mull
2	Université Paris VI (F)	Prof G. de Marsily and Prof F. Delay
1	LEPT-ENSAM Bordeaux (F)	Dr D. Bernard
1	ENSEEHIT Toulouse (F)	Dr D. Houi
1	SUPSI, Bellinzona (CH)	Dr G. Beatrizotti
2	JRC, Ispra (I)	Dr H. Nguyen

We stress that students' mobility through Europe is very important for several reasons.

a) The students improve their knowledge in subsurface hydrology learning from qualified scientists actively working on particular research arguments which are not among the principal research topics at our University.

b) They can use experimental facilities that are not available at our University; in particular some of them can deal with field applications that require specific instruments and equipment.

c) They work in Institutions whose organisation is different from that of our University; this experience can be very useful for their future in the working world.

d) They co-operate with researchers with different expertise, thus improving their ability to discuss environmental problems in an integrated way.

Moreover we gave lectures at some short courses organised by the above mentioned Interuniversity Co-operation Programme (Ponzini, 1994; Giudici, 1994).

On the other hand no student spent one or more semesters abroad to attend course units; in fact this poses a lot of trouble due to the heterogeneity of the higher education systems in Europe.

4. Future perspectives

In the next years Italy is going to reorganise the University system and the structure of the courses, following the guidelines of the «Joint declaration on harmonisation of the architecture of the European higher education system by the four Ministers in charge for France, Germany, Italy and the United Kingdom» signed at the Sorbonne, Paris, on 25 May 1998. This reform will imply a two-level degree system, followed by doctorate and other post-graduate specialisation courses. It is foreseen that the course for the «Laurea» degree in Physics will split into a first-level degree course and one or more second-level degree courses. The present «Laurea» degree will correspond to second-level degrees (5-year-long course, 300 credits).

The implementation of the new architecture is under debate and construction. Despite the uncertainty about the final details, we can nevertheless anticipate that the present course units of geophysics and dynamics of fluids will be split in several course units, each of which will require less study hours to be completed. In such a way we hope to obtain the following results.

a) To offer more flexible curricula and to allow the students to attend these units at different course years.

b) To facilitate the definition of credits for course units and above all to harmonise better our course units with the European Credit Transfer System; this will promote students' mobility not only for the preparation of their dissertation, but also for attending course units or whole semesters abroad.

c) To co-ordinate these course units better with other course units of environmental, atmospheric and Earth physics offered not only in the degree course in Physics, but also, e.g., in the degree course in Geological Sciences.

d) To promote post-graduate courses at our University or co-operate with the post-graduate courses active at other Institutions, members of the European Thematic Network ETNET.Environment-Water.

References

Giudici, M. (1994) Some practical considerations about the identification of aquifer transmissivities. «Lecture notes» of the «European graduate course» on «Groundwater contamination: from theory to practice», Universität Hannover (D).

Giudici, M., Morossi, G., Parravicini, G. & Ponzini, G. (1995) A new method for the identification of distributed transmissivities. Water Resour. Res. 31, 1969-1988.

Giudici, M., Ortuani, B., Parravicini, G. & Ponzini, G. (1997a) Identification of groundwater flow model parameters with the differential system method. In: Proceedings of IAMG'97 The Third Annual Conference of the International Association for Mathematical Geology (ed. by V. Pawlovsky-Glahn), 796-800. International Center for Numerical Methods in Engineering (CIMNE), Barcelona (E).

Giudici, M. and the students of the Laboratory of Earth Physics (1997b) Teaching geophysics: a «case study». In: Proceedings - 3rd meeting - Environmental and Engineering Geophysics, 463-465. Environmental and Engineering Geophysical Society, European Section, Århus (DK).

Giudici, M., Delay, F., Marsily, G. de, Parravicini, G., Ponzini, G. & Rosazza, A. (1998) Discrete stability of the Differential System Method evaluated with geostatistical techniques. Stochastic Hydrol. and Hydraul. 12, 191-204.

Ponzini, G. (1994) A forward and an inverse problem in the continuous and discrete cases. «Lecture notes» of the «European graduate course» on «Groundwater contamination: from theory to practice», Universität Hannover (D).

Past and On-going TEMPUS Projects - experiences, achievements and effects

Projets terminés TEMPUS – expériences, succès et effets.

Wietsma-Lacke E. D.
Department of Environmental Sciences,
Wageningen Agricultural Universities, The Netherlands

Abstract
This paper provides an overview of selected TEMPUS PHARE and TACIS Projects dealing with education on Water-Environment aspects and implemented in the period 1991-1999. Project cooperation involved 12 EU, 5 Central and Eastern European and 1 overseas countries. The specific feature of the projects is direct involvement of numerous academic staff members on the departments' level. The main projects activities i.e. student and staff mobilities, short intensive courses, development of curricula and teaching laboratories are presented together with achieved outcomes. Next to this the experiences of contracting institution i.e. Wageningen Agricultural University will be presented in the area of project management, project acquisition and financial coordination. The financial figures and budget structure of presented projects are given.
The development of international network between higher education sector, enterprises and local authorities is discusses. The effort is made to show the positive and negative effects of few-year cooperation between EU and Central European universities and how the TEMPUS supports the transformation in Central and Eastern Europe.

Résumé
Dans cet article on a présenté la revue de projets sélectionnés de programme TEMPUS et TACIS qui concernent les aspects d'éducation aux sujets : l'Eau et l'Environnement. Ces projets étaient réalisés dans le période 1991-1999. A coopération participaient 12 pays de EU, 5 d'Europe Centrale et d'Est et Etats-Unis. Le trait spécifique de ces projets est la participation directe de travailleurs académique au niveau de département.
Les activités principales de projets, c'est-à-dire : l'échange d'étudiants et de travailleurs, les courts cours, l'élaboration de programmes d'études, l'agrandissement de laboratoires didactiques, on a présenté ensemble avec des résultats atteints.
Ensuite on a présenté des expériences de l'institution de contrat, c'est-à-dire de l' Université d'Agriculture à Wageningen, dans la domaine de disposition de projets, de préparation d'eux et coordination financier.
Données sur finances et la structure de budget de projets présentés, sont montrés.
On a discuté le développement de réseau international entre le secteur d'éducation supérieure, des entreprises et l' administration locale.
On a pris épreuve de présenter des effets positifs et négatifs de coopération de quelques années entre des universités de EU et de l'Europe Centrale et façon auquel TEMPUS appuyait transformations à l'Europe Centrale et d'Est.

1. Introduction

The TEMPUS Programme was adopted by the Council of Ministers of the European Community in 1990 for the academic years 1990/91 - 1993/94 (TEMPUS I), extended in 1993 for the academic years 1994/95-1997/98 (TEMPUS II) and in 1996 for the academic years 1998/99 - 1999/2000 (TEMPUS II bis).

The main aim of the TEMPUS Programme defined for its first phase i.e. to support the reform of higher education in the countries of Central and Eastern Europe and the former Soviet Union republics, has been extended for its second phase to cover also support for growing interaction and integration with partners in the European Union.

The present paper gives an overview of the evolution of TEMPUS using the examples of selected projects being contracted by Sub-Department of Water Resources, Wageningen Agricultural University. The following two types of projects are presented:
(a) Joint European Projects (JEPs) supporting structural changes in universities of the eligible partners through the cooperation with EU partners. The maximum duration of the projects is three years.

(b) The Complementary Measures/Compact Measures (CMEs) supporting the institutional development of faculties /universities, dissemination of TEMPUs outcomes and feasibility studies.

2. Overview of the selected tempus projects

2.1. Results

From 1991 Sub-Department of Water Resources, Wageningen Agricultural University has been involved in 18 TEMPUS projects, out of which in 15 projects it was the contractor. The projects aim at education on water-environment aspects. The main beneficiaries are institutions from Poland, Hungary, Slovakia, Czech Republic and Russian Federation. The main objectives of these projects aim at the following:

(a) development and/or update of new curricula
(b) development of new PhD programmes
(c) introducing new technologies and techniques (GIS, Internet services, GPS, remote sensing, CAD, multimedia)
(d) introducing distance learning
(e) development of continuing education and retraining schemes
(f) introduction of ECTS (European Credit Transfer System) and internal quality assurance systems

The above mentioned objectives were/are achieved through the following activities:

(a) short visits or retraining visits of academic staff
(b) student training:
    (i) following regular lectures and practical training
    (ii) thesis writing
(c) updating teaching laboratories, computer and office facilities
(d) short intensive courses / PhD seminars
(e) upgrading libraries
(f) development of new laboratories and didactic units
(g) development of inter-faculty study programmes
(h) establishment of centres of excellence.

The projects results expressed in numbers are shown in Table 1 and Table 2. In Table 3 some titles of the short intensive courses are listed.

Table 1. Results of TEMPUS Joint European Projects of which Sub-Department of water Resources is contractor expressed in numbers

TEMPUS JEP	Period	Total budget in ECU	No. of partners	No of short cour-ses/ Semi-nars	No. of staff mob.	No. of . student mob	Equip-ment budget in ECU
EWA-Ring I	1991-1994	1.338.992	25 (16)	7	225	161	216.023
ECEE I	1992-1995	657.200	15 (10)	4	155	56	207.466
SWARP II	1994-1997	776.650	21 (10)	5	234	48	197.099
ICER II	1994-1997	713.500	17 (11)	6	92	70	125.284
PANSED II	1995-1998	746.750	25 (7)	7	182	31	216.673
DATE II	1995-1998	371.850	9 (5)	2	64	5	120.223
STEEM II	1996-1999	681.750	20 (9)	3*	150*	20*	237.212*
ELALM II	1996-1999	339.540	9 (4)	3*	105*	5*	95.948*
INQA II-UM	1998-2000	209.130	7 (3)	1**	96**	-	41.720**
DECES II-IB	1998-2000	239.554	8 (4)	20*	60*	-	26.800**

() - number of partners from Central and Eastern Europe
UM – University Management
IB – Institutional Building
* expected results by August 1999
** expected results by December 2000

Table 2. Results of TEMPUS Complementary Measures/Compact Measures Projects expressed in numbers

TEMPUS CME	Period	Total Budget	No. of Partners	No. of staff Mob.	Equipment Budget
EWARAB	1995-1996	35.000	6 (2)	15	-
CME 1054	1995-1996	45.000	4 (1)	25	-
CME 1064	1995-1006	47.500	5 (1)	31	4.986
CME 2058	1996-1997	35.100	11 (8)	16	-
CME 3053	1998-1999	47.600	4 (1)	21	-

() – number of partners from Central and Eastern Europe

Table 3. Examples of the short intensive courses implemented during the TEMPUS projects

TEMPUS JEP	Title of the course	Target group
EWA-Ring I	Decision Making for Environmentally Oriented Water Management	MSc- and PhD-students
	Hydrometrical Field Training	undergraduate and graduate students
	Optimisation of Water Resources	MSc and PhD students, young researchers
SWARP II	GIS Application	MSc- and PhD-students, young assistants
PANSED II	Integrated Water Resources Management in Rural Areas	young staff and PhD-students
	Internet Application	graduates and young researchers
STEEM II	Modelling Water Flow and Solute transport in Heterogeneous Soil-Plant Atmosphere Systems	MSc- and PhD-students, young researchers
	Modelling of Environmental Systems	PhD-students, young researchers
DECES II-IB	Environmental Impact Assessment (EIA procedures as tools for sustainable development and integrated water resources management)	employees of national and local administration, those working in associations,trade unions, enterprises
	Public Participation and Assess to Information in Environmental Decision Making Processes	employees of national and local administration, those working in associations, trade unions, enterprises

2.2.Management

After eight years of experiences it can be stated that all project contracted by Sub-Department of Water Resources have been successful. The success of the projects is due to:

(a) personal involvement and dedication of staff on the departmental level,

(b) strong financial support from European Commission.

It is clear that one cannot go without the other. Although the projects have been contracted by Wageningen Agricultural University who was the legal responsible for the projects achievements, the eligible partners were always involved in project management as coordinators. The coordinators were responsible for the project merits. On the other hand the good relation between the contractor and coordinator is essential during the problem solving.

The main problems during the project implementations were caused by culture and economic differences between EU and eligible partners. The language was not experienced as an obstacle.

2.3. Evolution

The development of TEMPUS objectives and priorities is a reflection of changes and transformation in EU and Central and Eastern Europe.

European network developed during TEMPUS projects 1991- 1999

Figure 1. European network developed during 8 years of cooperation within the TEMPUS projects

General the TEMPUS projects which were implemented during the phase of TEMPUS I e.g. EWA-Ring and ECEE were regional ones i.e. were involving institutions from several eligible countries. The projects involved only higher education institutions. Very significant achievement (next to the above mentioned) of these projects was creation and consolidation of truly European forum of university departments cooperating in teaching and training in the area of environment, water and agricultural soils. The development of the network laid the foundation of future cooperation and sustainability. The Figure 1. shows the dimension of the European network developed during 8 years of cooperation.

In the TEMPUS II projects e.g. SWARP, ICER, PANSED, STEEM and ELALM the key partners are universities, but other institutions and organizations also participate, including enterprises, public administration agencies, economic organizations and professional associations. The involvement of non-academic institutions and organizations assure that restructured study programmes prepare better the graduates for the actual labour market and its specific requirements in the area of environmental development and protection.

Making use of the achievements of the previous stages a special type of JEP has been anticipated within the framework of TEMPUS II bis i.e. Institutional Building (IB). Project DECES is an example of such. The aim of TEMPUS IB project is to develop and deliver short-cycle training courses for national administrators, professional associations, the semi-public sector and local administrators who are involved in the implementation of the pre-accession strategy. The project consortium consists of universities which provide theoretical support and non-academic institutions such as professional associations, national and local administrations and NGOs who are responsible for both providing training and recruitment of the trainees. Within the project consortium the twinning between EU and Associated countries institutions is required. Generally saying the training developed during the TEMPUS project aim on transfer of the acquis communataire. The aim of TEMPUS IB projects is to ensure that higher level education plays a fundamental role in providing employees of national and local administration and those working in associations, trade unions, enterprise etc. with the tools and expertise necessary for European integration.

3. Recapitulation

So far the education in the area of Water-Environment in Central and Eastern Europe benefited significantly from TEMPUS programme. Not only new curricula and technologies have been introduced. The new activities can be implemented using both the 8-year experience and its contact network in the countries concerned.

Moreover looking from Human resources development and capacity building TEMPUS has indeed been launched to support national polices for reforming higher education sector in Central and East Europe and in this has been as an introduction to the Community programmes in the field of education and training. As such it has indeed three main stages in its evolution:

(a) the first concentrated on university structure and curriculum development,
(b) the second concentrated on universities and their immediate environment,
(c) the third concentrated on universities in their institutional and national context.

To this logical evolution can be added a further element which concerns the adaptation of universities to the new global context of education, which is characterised by life-long learning and by increasing social and economic relevance of education and training.

Life-long learning emphasises the need for continuous adjustment of knowledge and expertise to an ever changing human environment. New economic and social imperatives, on the other hand, require that the education should be seen as a training towards citizenship.

The cooperation between the educational institutions and their environment, represented by various economic and social partners and public authorities, is getting more and more crucial for all of them. In other word, the active marketing of the new social and economic role of the higher education institutions seems to be necessary in any occasion.

Benefits from participation in TEMPUS projects

Les profits de participation aux projects TEMPUS

Ignar S.
Faculty of Land Reclamation and Environmental Engineering,
Warsaw Agricultural University, Poland

Abstract
This paper describes experiences gained from participation in European Union assistance TEMPUS PHARE Programme. This Programme started in year 1990 and it was aimed at higher education system in Central and Eastern Europe. Experiences are shown from a point of view of beneficiary partner, as author's home institution, Faculty of Land Reclamation and Environmental Engineering of the Warsaw Agricultural University has participated in several TEMPUS projects, in all three phases of the Programme. Changes of successive project objectives and goals according to modification of TEMPUS rules and aims and Polish national priorities are described. Direct and indirect benefits from participation in the Programme are shown with further perspectives for pan-European co-operation.

Résumé
Cet article décrit des expériences gagnées pendant la participation au programme d' aide d' Union Européenne TEMPUS PHARE. Le programme a commencé à 1990. Son but était le système d' éducation superiéure à l' Europe Centrale et d' Est. On a présenté ces expériences de point de vue de partnaire qui profite du programme :Département d' Amélioration et de Génie d' Environnement d' Université d' agriculture de Varsovie, qui a participé à toutes trois phase de quelques projects de TEMPUS. On a décrit les changements de tâches et de buts, qui sont conformés aux changements de principes et de tâches de TEMPUS. Ces profits direct et indirects de participation au programme on a présenté ensemble avec les perspecives coopération ultérieur.

1. Introduction

The TEMPUS Programme has been launched in the academic year 1990/91 on the basis of the decision of the Council of Ministers of the European Community on 7th May 1990. It is a part of the EC broad assistance programme called PHARE, which was aimed at support of socio-economic transformations in the Central and Eastern Europe countries. The main goal of TEMPUS was to support the reform of higher education. Three phases of the programme related to the changes in objectives and goals were defined during its implementation: phase I (1990/91 – 1993/94), phase II (1994/95 – 1997/98) and phase II bis (1998/99 – 2000/01). During 10 years of TEMPUS project realisation its objectives and goals have substantially changed. TEMPUS priorities have shown movement from the strict area-based approach to the structural approach. During phase I the key PHARE Programme areas were defined (i.e. areas which should primarily be developed to support the economic, social and political transformations) including agriculture and environment.

Later structural higher education reform issues such as development of university – industry co-operation and development of bachelor-degree courses were introduced ending with more general "European development" of higher education institutions and pre-accession strategy.

2. Achievements of the Faculty of Land Reclamation and Environmental Engineering

Faculty of Land Reclamation and Environmental Engineering of the Warsaw Agricultural University (WAU) participated in total in 7 JEP type projects. First two of them (EWA-RING and ECEE) were devoted to upgrading and development of selected study courses in the area of environment and water. The main implemented activities were: staff and student mobilities, short intensive courses and upgrading of teaching laboratories. Existing course programmes on Hydrology, Hydraulics and Water Resources Management were upgraded. Due to great number of partner universities in both project consorties basis for future international co-operation was established. These projects also allowed Polish partners to get acquainted with new teaching methods and study materials.

Objective of the next JEP 7862-SWARP implemented in 1994-1997 was aimed at the restructuralization of the curricula at 8 Polish universities including the creation of a 4-year Ph.D. course at WAU in the area of Soil and Water Protection and at the development of Geographical Information System training centre at the WAU. Advanced GIS teaching laboratory was developed with modern computer equipment and up to date software. New unit called Laboratory of Environmental Information Systems was set up within the Faculty structure. The following staff members are actually employed within the Laboratory: one professor, one assistant professor, two assistants, one technician and 5 Ph.D. students. Staff of this unit was trained by the means of mobility scheme and short intensive courses. Study course on GIS was developed and incorporated into study programme.

Project JEP 9206-PANSED realised in the period of 1995-98 lead to the creation of new inter-faculty didactic institution at WAU providing education in the field of Environmental Modelling. New subject related study courses were developed and introduced into curriculum. One of them was "Introduction to Internet". It was supported by development of Internet educational laboratory allowing for access to all computer network services. Pilot lecture on GIS was developed as Distance Learning module and it was made accessible through WAU computer network server. Full scale module (15 hours) Distance Learning course on Remote Sensing is being developed jointly with Wageningen Agriculture University as a PANSED follow-up activity supported from the budget of Dutch Ministry of Education. Furthermore short intensive course on Integrated Water Resources Management in Rural Areas was introduced for M.Sc. and Ph.D. students.

Introduction of new interfaculty study programme on Environmental Protection was a general objective of the next JEP 11463-STEEM implemented in the years 1996-1999. As a complementary development of teaching laboratories the equipment for field environmental data acquisition was completed. Two sets of equipment were purchased: Global Positioning System (GPS) for spatial data recording with the use of satellite navigation system and water quality measurement set. This equipment upgraded capabilities for recording of spatial field data to be processed and presented by GIS software. Three short intensive courses (Modelling of Environmental Systems, Modelling Water Flow and Solute Transport in Heterogeneous Soil–Plant–Atmosphere Systems, Multicriterial Decision Support Systems in Environmental Management) were implemented and the teaching materials from these courses will be introduced into the study programme.

Within all TEMPUS JEPs from the phase I and phase II several student mobilities to EU universities are implemented from undergraduate students (following the lectures at EU universities), M.Sc. students (preparing the M.Sc. thesis) and Ph.D. students (conducting project research).

Two newly approved JEPs belong to two main types of projects foreseen for phase II bis of TEMPUS programme. JEP 13374-INQA is a University Management project facilitating introduction of internal quality assurance system and European Credit Transfer System and JEP 13305-DECES is an Institutional Building project. The objective of this project is the development of training centre on EU pre-accession strategy. This centre will conduct the training courses for governmental and local administration on EU regulations in the area od environmental protection and agriculture.

3. Conclusions

Coming to a conclusions it is possible to say that TEMPUS PHARE programme, which has been set up to support the restructuring of higher education system in CEE countries is fully accomplishing its objectives as it was shown using example of achievements of the Faculty of Land Reclamation and Environmental Engineering. Owing to enormous budget and proper arrangement it influences positively most of the elements of teaching activities of the Faculty.

The particular benefits are:
Curricula development. Due to intensive staff mobility scheme it was possible to upgrade of Faculty curriculum. It has started with upgrading of single existing courses: Water Resources Management, Hydrology and Hydraulics. Further, new M.Sc. specialisation was developed (Water Resources Protection) and new 5 years study programme on Environmental Protection.
Creation of new educational laboratories. Three new teaching laboratories were developed (Geographical Information Systems, Internet Services and Water Quality Evaluation). Modern Laboratory equipment was purchased from TEMPUS budget. EU partners consulted selection of equipment items and provided staff retraining.
Completion of field equipment sets. Equipment for Global Positioning System for spatial data recording was purchased making possible to arrange field practical training in modern satellite positioning techniques. Also field equipment for Water quality measurement was completed.
Enrichment of library resources. About one hundred books were purchased for the Faculty library providing substantial upgrade of library resources.
Modernisation of teaching methods. Participation in TEMPUS project allowed for transfer of know-how on modern teaching methods. New equipment (LCD panels, videoprojectors and multimedia computers) were introduced together with new techniques (CD-ROM presentation, Distance Learning and Video-conferences).

Participation of the Faculty of Land Reclamation and Environmental Engineering in above mentioned TEMPUS project has immense impact on undergoing Faculty development in response to new emerging challenging issues for Polish community. It will allow for proper level training of graduates in response to changing labour market needs. It will also facilitate further co-operation with EU universities in forthcoming educational (SOCRATES) and research (5th Framework Programme) programmes.

Formation post-universitaire en Roumanie "Gestion et protection de la ressource en eau"

Graduate programme in Romania "Management and protection of water resources"

Carbonnel J. P.[1], Drobot R.[2]
[1] Université Pierre et Marie Curie Paris 6, France
[2] Université Technique de Construction de Bucarest, Roumanie

Résumé
Dans le cadre du programme européen TEMPUS, une expérience de formation pluridisciplinaire post-universitaire, de type DEA et DESS, dans le domaine Eau et Environnement est présentée. Ce cursus de 6 années (1992-1998) regroupait des enseignants de 4 pays : Roumanie, France, Belgique et Italie.
Après les événements de 1990, la Roumanie s'est trouvée dépourvue de cadres dans le domaine de l'environnement, préoccupation et spécialistes étant inexistants précédemment . Un grand besoin se faisait donc sentir aussi bien au niveau ministériel qu'au niveau des Agences spécialisées et des bureaux d'études commençant à apparaître, besoin augmenté par les nouvelles normes environnementales qui allaient être mises en place et par l'intégration européenne qu'elles supposaient.
Le cursus mis en place s'est, dans un premier temps, inspiré du cursus équivalent français et en particulier a adjoint aux enseignements classiques en hydrologie, des modules de chimie, de biologie et de droit de l'environnement. Durant six années ce cursus a fonctionné avec la participation de 10 enseignants d'Europe de l'Ouest, à raison d'une semaine de cours et de TD en français pour chacun. Des enseignants suisses sont venus rapidement se joindre au groupe TEMPUS. Des professeurs roumains ont pu, par ailleurs, effectuer des stages pédagogiques à l'étranger.
Chaque année le programme a formé de 15 à 20 spécialistes dont le tiers a bénéficié d'un stage de 3 à 4 mois dans des laboratoires d'Europe de l'Ouest pour y effectuer leur mémoire de fin d'étude. Quinze de ces étudiants ont poursuivi leur recherche par une thèse en Roumanie ou en co-tutelle. A ce jour, 24 manuels pédagogiques en roumain, français ou anglais ont pu être édités en 500 exemplaires grâce au projet.
Le cursus est maintenant intégré dans l'enseignement universitaire roumain en tant que « Ecole des Etudes Académiques post-universitaires ». Il vient d'être soutenu par un projet de la Banque Mondiale afin de développer des thèses en co-tutelle pour les meilleurs étudiants de ces six dernières années.

Abstract
A multidisciplinary graduate course in the field of Water and the Environment, within the framework of the European TEMPUS programme is described. This programme has been in operation for 6 years (1992-1998) and enjoyed the participation of teachers from 4 countries: Romania, France, Belgium and Italy.
After the events in 1990, Romania lacked the necessary professionals in the environmental field, a concern and specialisation to which no attention had been given at the time.

A great need was therefore felt both at ministerial level and by the Specialised Agencies as well as by the emerging consulting firms, a need which was further increased by the new environmental standards demanded by European integration.
The programme was initially inspired by its French equivalent and added modules of chemistry, biology and environmental law to traditional courses in hydrology. It has been in operation for six years with the participation of ten West-European instructors each of whom has given one week of lectures and laboratory sessions in French. Swiss teachers rapidly joined the TEMPUS group. Moreover, Romanian faculty staff have obtained teaching internships abroad.
Every year, 15 to 20 specialists have graduated from the programme and a third of them have spent 3 to 4 months in West-European laboratories during which they have written their final dissertations. 15 of these students have continued their research to obtain a Doctorate at Romanian universities or in cooperation with other universities. So far, 24 text books in Romanian, French or English have been published in 500 copies with the help of this project.
The programme has now been integrated into the Romanian university syllabus under the name of « School of graduate academic studies ». It has recently received the backing of a World Bank project with the aim of pursuing joint-authority Ph. D. dissertations for the best students of the last six years.

1. Cadre et objectifs du cursus

Lors des premières Rencontres Hydrologiques Franco-Roumaines qui se sont tenues à Paris en septembre 1991, on s'est très vite aperçu du décalage des préoccupations scientifiques des chercheurs roumains par rapport à leurs collègues français, en particulier dans les domaines de l'environnement, de l'hydro-écologie, de l'hydrochimie... L'approche des scientifiques roumains était, à l'époque, exclusivement orientée vers l'hydrologie quantitative, ignorant presque totalement les processus chimiques et biologiques qui accompagnent les phénomènes physiques. C'est pourquoi très rapidement l'idée d'un transfert de connaissance s'est imposé, aidé en cela par une forte volonté politique des responsables roumains, en particulier du jeune ministère roumain de l'environnement.

L'occasion a été fournie par le programme TEMPUS dont c'était la première année de fonctionnement. Ce programme, destiné à aider à la réforme de l'enseignement supérieur des pays de l'Est, permettait non seulement l'acquisition d'équipements pédagogiques, pour avoir une base logistique comparable à celle des universités d'Europe de l'Ouest, mais aussi des mobilités de personnels et d'étudiants entre les différents partenaires.

Deux programmes TEMPUS se sont déroulés autour du même objectif : créer un vivier d'enseignants, d'ingénieurs et d'étudiants roumains autour de connaissances et d'outils communs, ce groupe devant être susceptible de développer un dialogue scientifique et technique constructif avec leurs homologues européens ; cet objectif ayant pour finalités, d'une part de permettre la création d'un cycle d'enseignement nouveau en Roumanie, d'autre part de favoriser l'intégration européenne des structures et législations roumaines dans le domaine de l'environnement.

Ces deux programmes ont pris les formes suivantes :
« Sciences de l'eau et environnement », 1992-1995, de type DEA français, orienté vers la formation de futurs docteurs destinés à prendre la relève dans l'université roumaine ;
« Gestion et Protection de la Ressource en Eau », 1995-1998, qui correspondait plus à un cursus de type DESS français, destiné à promouvoir des spécialistes dans ce domaine au sein des instituts de recherche, de l'administration ou des bureaux d'études.

Au cours de ces six années, le monde industriel et administratif roumain a participé au développement de ces programmes par son implication dans la formation de ses cadres et en participant à diverses manifestations dont les Rencontres Hydrologiques Franco-Roumaines qui se tiennent tous les deux ans (Paris 1991, Tulcea 1993, Montpellier 1995, Suceava 1997). Par ailleurs, grâce au soutien de la coopération bilatérale française, de nombreux stages en France ont permis à des administratifs et ingénieurs roumains de se familiariser avec les problématiques environnementales et leurs implications tant techniques qu'institutionnelles, ce qui a largement contribué au soutien du programme TEMPUS par ces administrations et instituts spécialisés.

Trois pays de l'Union Européenne (France, Belgique et Italie) et la Suisse ont activement participé à ces formations, en assurant chaque année dix missions didactiques d'une semaine chacune. Les domaines enseignés étaient ceux pour lesquels les compétences des enseignants roumains étaient absentes ou réduites : écologie, modélisation du transport des polluants, modélisation des phénomènes chimiques et biologiques, physique du milieu non-saturé, utilisation des isotopes...

Alors qu'au début du cursus le monde du travail (ingénieurs, administratifs) était majoritairement représenté parmi les étudiants, peu à peu, un plus grand nombre de jeunes gens sortants de l'université s'est intéressé à ce cycle d'étude et est devenu la part la plus importante des étudiants.

La diversité des formations initiales des étudiants des deux programmes TEMPUS, provenant aussi bien de diverses écoles d'hydraulique ou d'ingénieur que des universités (géologie, géographie, aménagement du territoire...), a nécessité la création d'une période de mise à niveau des connaissances, associée à une période de mise à niveau linguistique. Cette façon de procéder a favorisé la sélection des candidats et développé un sentiment de corps qui a été très profitable au déroulement du cursus et à ses suites.

On notera que le premier programme fut piloté par les partenaires français alors que le second le fut directement par les partenaires roumains bénéficiaires ce qui permit une plus grande autonomie du programme vis à vis des objectifs pédagogiques de l'Enseignement Supérieur roumain et, pour le partenaire roumain, d'acquérir une connaissance approfondie des mécanismes de gestion des programmes communautaires.

2. Evolution du cursus

Le cursus a connu une évolution continue pour mieux s'adapter aux besoins exprimés par le monde industriel roumain.
Dans le cadre du premier programme TEMPUS (JEP 3801/92-95), après un tronc commun trois filières étaient prévues : - eaux de surface – eaux souterraines – géotechnique et stabilité des versants. En constatant que la troisième filière présentait peu d'un intérêt pour les partenaires industriels et les étudiants, elle a été intégrée, à partir de l'année 1993-94, dans la problématique des eaux souterraines.
En considérant que les problèmes du milieu ne permettaient pas une distinction nette entre eaux de surface et eaux souterraines, on a uniquement maintenu, dans le deuxième programme TEMPUS (S-JEP 09781 :95-98), une filière concernant la modélisation mathématique avec un certain nombre de disciplines obligatoires pour tous les étudiants.

Cette évolution a, en réalité, suivi de très près l'évolution des besoins qui était parallèle à une certaine évolution des mentalités et à un contexte économique en rénovation rapide. Elle a aussi permis de faire évoluer progressivement le monde de l'enseignement afin de l'amener à changer son mode de fonctionnement traditionnel. Enfin elle a permis la formation des enseignants à une nouvelle pédagogie et la création d'une infrastructure pédagogique nouvelle, illustrée principalement par la mise en place de cellules informatiques et la publication d'un grand nombre de documents pédagogiques nouveaux.

Au départ cantonné uniquement à Bucarest (Université Technique de Construction et Université de Bucarest), le cursus s'est peu à peu ouvert aux universités de province avec lesquels un partenariat s'est créé. Ce sont principalement les universités de Iasi, Timisoara et Cluj qui ont bénéficié de cette ouverture. Là aussi cela a permis l'installation d'un équipement pédagogique minimal et un échange d'étudiants et d'enseignants ouvrant la porte à un cursus national adapté aux besoins du pays.
Depuis 1998, la poursuite du cursus au niveau national roumain, à la suite des deux programmes TEMPUS, s'est effectuée avec l'introduction une nouvelle dimension. En particulier, l'utilisation des SIG comme outil de valorisation de l'information à distribution spatiale présente un intérêt nouveau. A la spécificité du cursus – la modélisation mathématique en hydrologie et hydrogéologie – on a donc ajouté, après le tronc commun, une deuxième filière intéressant plus particulièrement l'utilisation des SIG dans le domaine des sciences de l'eau et de l'environnement.

3. Réalisations des programmes de formation

Le programme TEMPUS de 1992 à 1998 a formé environ 100 spécialistes roumains dans le domaine Eau-Environnement parmi lesquels une trentaine ont effectué un stage de plusieurs mois dans les laboratoires et organismes spécialisés de l'Union Européenne. Une partie de ces derniers a continué en thèse soit dans le cadre de l'enseignement supérieur roumain soit en co-tutelle, principalement en France et en Belgique.
Une soixantaine de missions pédagogiques d'une semaine chacune ont eu lieu de l'Europe de l'Ouest vers la Roumanie ; elles totalisent environ 1500 heures de cours effectuées en français dans le cadre du cursus, auxquelles ils faut ajouter la venue en Roumanie de spécialistes français et belges pour les sessions d'examen et de soutenance de mémoires.

Des activités de recherche-développement, sur budget français, ont soutenu le programme. Ce sont :
les Rencontres Hydrologiques franco-roumaines auxquelles les étudiants ont participé sous forme de présentations de posters,
des séminaires spécialisés (gestion de l'eau, systèmes d'alerte de crues, système des Agences de Bassin...) ouverts à un large public, destinés à former les cadres roumains et à initier les étudiants aux réalisations européennes dans le domaine de la protection et de la gestion de la ressource en eau,
la mise en place d'un projet de recherche en vue de la modélisation globale du fonctionnement de la rivière Arges alimentant en eau la ville de Bucarest, dans le cadre duquel plusieurs mémoires de fin d'étude des étudiants TEMPUS ont été effectués.
L'équipement bureautique (ordinateurs, photocopieuses, scanners, relieuses...) dans un premier temps des deux universités de Bucarest UTCB et Université de Bucarest) puis des universités de province (Iasi, Cluj et Timisoara) a permis le développement des enseignements de modélisation et l'accès des laboratoires participants à l'utilisation de logiciels de plus en plus performants. De plus il a permis de relier la recherche des enseignants à leur activité pédagogique.
Enfin la réalisation de 29 livres didactiques, distribués gratuitement aux étudiants ainsi qu'aux enseignants et responsables des instituts de recherche, a permis la diffusion des cours du cursus à un large public intéressé par les problématiques développées au sein du programme.

4. Conclusions

Sur l'expérience pédagogique dont nous venons de retracer les grandes lignes, un certain nombre de commentaires peuvent être faits :
Un projet de cursus post-universitaire de spécialisation, pour réussir, doit s'appuyer sur un **besoin** socio-économique et une **volonté** du monde économique. Ce fut le cas dans le projet présenté ici et nous tenons à souligner le soutien important que, dans un premier temps, le ministère de l'environnement roumain et la Régie Autonome des Eaux Roumaines lui apportèrent. Ce soutien eut pour effet d'entraîner d'autres instituts et bureaux d'étude à recycler leurs cadres, ce qui alimenta durant 3 à 4 ans le cursus en étudiants motivés et de bon niveau.
Parallèlement les enseignants ont été attirés par le cursus et ont pu s'y investir grâce à l'attrait représenté par l'équipement informatique qui leur permettait de développer leur propre recherche et par les stages à l'étranger qui leur permettaient un ouverture sur les laboratoires extérieurs.
Enfin le cursus, après une première phase de trois ans, entièrement gérée par l'équipe française, a été confié pour la deuxième phase aux partenaires roumains tant au niveau didactique qu'à celui de la gestion financière. Cette nouvelle responsabilité a permis le passage rapide au nouveau cursus purement roumain qui fait maintenant suite aux deux projets TEMPUS.

C'est l'ensemble de ces conditions qui a fait le succès de notre expérience. Nous pouvons avancer que c'est aussi grâce à l'intégration du cursus dans un ensemble d'actions coordonnées autour du thème Eau-Environnement que ce succès a été possible.

"GOUTTE of Water": a Proposed Global Network of Water – Related University Chairs.

GOUTTE of Water": proposition d'un réseau global de Chaires d'Université dans le domain de l'eau.

Bogardi, J. J.
UNESCO, Division of Water Sciences, Section on Sustainable Water Resources Development and Management

Abstract
The experience gained in European context with disciplinary and interdisciplinary networking of university departments (chairs) is very positive. The concept works well, both in networks based on parity (like the ERASMUS programme) and in networks, conceived to channel assistance to target universities (like the TEMPUS programmes).

The paper elaborates the idea to explore networking in global context, linking university institutions both to reflect and to capitalise the ongoing globalisation processes in university education. The suggested network should be based on interacting cells of manageable size and scope, thus creating an embedded (hierarchical) structure. It is proposed to combine, preferably within the collaborative cells themselves, both parity-based and assistance-oriented features of partnerships. Thus the definition of the cells is rather disciplinary or cultural (language of communication, etc) than geographic.

The envisaged framework "GOUTTE of Water" indicates that the partnership would focus on water as an integrating element. By sharing a "Drop of Water" future generations of professionals should learn that water is the schoolmaster of civilisation, a potent agent of co-operation and peace, rather than that of wars. Consequently the network "GOUTTE of Water" can certainly be seen also as a component of the "Culture of Peace" initiative of UNESCO. Its long-term political and strategical importance is undeniable. The framework could be implemented as a natural resource-oriented partnership within the UNESCO UNITWIN (university twinning partnership) programme, incorporating the present and future water-related UNESCO chairs.

The acronym "GOUTTE" stands for Global Organization of Universities for Teaching, Training and Ethics. The French/English name of the suggested network symbolizes the inherent strong intercultural characteristics, while the focus: Teaching, Training and Ethics underlines that the traditional tasks of a university should be interlinked with and used for promoting "New Water Ethics", a novel paradigm and its translation into new approaches in conceiving and conducting water resources development and management.

The necessity and merits of interuniversity partnerships are beyond debate Contrary to their desirability, financing and sustainability are far from being secured. "GOUTTE of

Water" will be proposed to become a substantial component of the educational efforts within the International Hydrological Programme (IHP) of UNESCO.

Résumé
L'expérience acquise en Europe dans le contexte disciplinaire et interdisciplinaire des réseaux des chaires universitaires est très positive. Le concept fonctionne bien, à la fois pour les réseaux fondés sur la parité (comme le programme ERASMUS) et ceux conçus pour canaliser l'aide aux universités cibles (comme les programmes TEMPUS).

Le présent article élabore l'idée du travail en réseau et de son exploration dans un contexte global, poussant les institutions universitaires à la fois vers la réflexion et vers la capitalisation des processus de globalisation en cours dans l'enseignement universitaire. Le réseau suggéré devrait comporter des cellules de taille et d'envergure maîtrisables. Il a été proposé de combiner, de préférence à l'intérieur des cellules mêmes, les deux aspects de parité et d'assistance des partenariats. La définition d'une cellule est donc plutôt disciplinaire ou culturelle (langage de communication, etc.) que géographique.

Le cadre de travail envisagé de "GOUTTE of Water" indique que le partenariat s'intéresserait à l'eau en tant qu'élément d'intégration. En partageant une "goutte d'eau" les professionnels de la génération future devraient apprendre que l'eau est le maître d'école de la civilisation, un agent essentiel pour la coopération et la paix. En conséquence, le réseau "GOUTTE of Water" peut certainement être considéré comme une composante de la "Culture de la Paix", une initiative de l'UNESCO. Le cadre de l'action, orienté vers les ressources, pourrait être mis en œuvre au sein du programme UNITWIN (jumelage des universités) de l'UNESCO incorporant les chaires déjà existantes et à venir, liées à l'eau.

L'acronyme "GOUTTE" veut dire "Global Organization of Universities for Teaching, Training and Ethics. Le nom français/anglais du réseau suggéré symbolise les caractéristiques interculturelles inhérentes, alors que "Teachering, Training and Ethics" souligne le fait que le rôle traditionnel d'une université devrait être lié et utilisé pour promouvoir une "nouvelle éthique de l'eau", paradigme nouveau, et son transfert dans une approche nouvelle de conception et d'orientation du développement et de la gestion des ressources en eau.

La nécessité et le mérite des partenariats interuniversitaires sont indiscutables. Le réseau "GOUTTE of Water" sera proposé pour devenir une composante substantielle des efforts d'Education du Programme Hydrologique International (PHI) de l'UNESCO.

1. Introduction

Disciplinary and interdisciplinary university networking has been very successfully applied in the implementation of the higher educational policy of the European Union. The concept works well for networks based on mutual benefit and parity (like ERASMUS programme) and also in networks conceived basically to assist Central European and CIS universities though transitional phases (such as the TEMPUS PHARE and TACIS programmes). At global scale, UNESCO has recently launched its UNITWIN programme, together with the creation of UNESCO Chairs at participating universities. All these networks proved to be quite successful in the area of water and water-related disciplines. Interesting to note in this regard that the co-organizer of this symposium, ETNET-Environment-Water, is the only environmentally oriented thematic network that has been accepted for funding by the European Union.

This paper outlines the framework of a global water oriented inter-university network: "GOUTTE of Water", a Global Organisation of Universities for Teaching, Training and

Ethics of Water. The network, which is conceived as a UNITWIN-Network itself would function as an "umbrella" organization. It is expected to combine both parity-based and assistance-oriented partnerships. Under the common principles expressed by "GOUTTE of Water" it is expected that the global network will be based on cells, which themselves can be UNITWIN Networks, preferably combining partners from developing and developed universities. Cells are expected to be focused on common interest (for example arid zone hydrology) or interdisciplinary context. The recently created first UNITWIN Network for Water, The Réseau Méditerranéen de l'Eau, is a typical example of a potential cell in "GOUTTE of Water". Practical considerations suggest that cells are formed to use one common language of communication, while the global network might rely on several cells using different languages. This feature, which focuses on water as an integrating factor is underlined by the bilingual abbreviation of the network's name: Drop of Water.

2. Justification

Education is widely regarded as the key to achieve sustainability. In face of the predicted water crisis, expected to occur in the coming decades the need for increased and concentrated efforts in the area of water and related fields is self-explanatory.

Investment into education and university networks is thus a direct investment into our future and peace! But education is also very unique. While it is the basis of sustainability, it will never become self-sustainable. The financial success of a few private university programmes offering academic degrees in business does not change the overall picture. Education is an investment, but the benefits do not occur on the educational accounts. These are political, economic, social, personal, spread over the entire world and shared by all of us.

This general statement is more than true for water related education. It is believed that these principles and statements are generally accepted. But still, can we expect them to be realized?

Water has undoubtedly moved into the focus of the international political agenda, and yet, it has to be concluded that this political water awareness is not at all reflected in respective educational efforts. Many of the water-related programmes, decades-long successful educational training courses are hardly surviving, or had to be closed down as funding for scholarships from governments, but also from bilateral and multilateral donor agencies, has drastically been cut. To make things worse, funding for universities (and especially for the no-high-tech water programmes) are being reduced substantially, even in the rich "Northern" countries. Irrespective of the above mentioned positive balance of the European inter-university co-operation programmes even their funding has considerable decreased, and in case of the TEMPUS Programme the scope has been shifted from academic co-operation to an assistance programme of university administrations.

Beyond the immediate negative consequences we come dangerously close to a vicious downward spiral. Underfunded university programmes do not project good job opportunities for their graduates. Consequently they do not attract the best candidates. We loose the most aspiring students in a "disciplinary braindrain", just when we would need them most to master the "looming water crisis". Admittedly these thoughts are not optimistic, but realizing negative tendencies and exposing the potential dangers are the first steps towards rectification.

What can the academic world do to sustain its standard, irrespective of the vagaries of national or international funding? The first step should certainly be the best utilization of available human and material resources. Pooling these resources and sharing them in a mutually beneficial way should not only alleviate the worst effects of "funding stress" but also help universities to shed the negative image of counterproductive rivalry. Networking, a politically encouraged, yet voluntary from a co-operation can be regarded as the most promising way to succeed the well-financed, but short-lived academic exchange programmes. It is obligatory to note, however that these large scale exchange programmes were only supported in Europe. Nevertheless the achievements are worth to be analysed.

These academic exchange programmes did not only fulfil, or even surpass the political expectations, but they have lead to an unprecedented friendship and co-ordination of activities among professors of European universities. Interesting to note that water orientated networks were and still are among the first, most active, largest and best organized ones.

While their scope has been defined as strictly educational, networks stimulated research as well. In fact, the biggest benefits have been gained by postgraduate and especially PhD students, being suddenly able to rely on the advice and experience of professors far away.

On the contrary to simplistic expectations, networking does not lead to uniformity and academic stagnation. Research and teaching based on co-operation do not eliminate curiosity and competition, the fundamental features of academic life. However networks help to avoid duplication. They lead – on a long-term basis – through resource and human capacity sharing to budget savings. An aspect that could more than offset concurrent costs for electronic equipment, communication and travel. In well established networks, participating members may focus on their own speciality and rely on partners' capacity in other fields, or networking universities may offer joint educational programmes and modules no one ever dreamed to develop alone.

Educational networking is conceivable at different levels, but the above described features, in addition to the problem of communication (language), clearly render them to be most effective at university and postgraduate degree programme levels.

The positive aspects of university networking, which were and are so evident in Europe are also valid for other regional co-operation frameworks. The UNITWIN Network Programme of UNESCO transplants the same concepts into a global context. Synchronized with the UNESCO Chair Programme it offers the best potential to propagate the principles of UNESCO worldwide, through higher educational activities.

3. The concept

It is envisaged to conceive "GOUTTE of Water" as a large, global, water-oriented umbrella organisation of universities and university institutes active in TEACHING and TRAINING and willing to contribute to shape a "New Water Ethics" in academia and, subsequently, in future practice. The ethical dimension of the partnership is to be understood as the explicit requirement to provide also moral leadership in forming and educating professionals and scientists sensitised towards the accepted principles of sustainability, environment-consciousness, equity, culture of peace in water sharing, etc.

"GOUTTE of Water" is basically a forum, where collaborating entities and their programmes can be discussed, compared and concerted. It should function, next to state-of-the-art communication through annual or biannual conventions providing global exchange opportunities for ideas, concepts and results. At present, "GOUTTE of Water" is still at an early stage of conception. Thus, symposia, meetings, like the present "The Learning Society and the Water-Environment", should be used to obtain further stimulating recommendations. It is the aim of this paper to capture the imagination of those present. Further details of "GOUTTE of Water" are rather examples, than final rules of the network.

4. Organisational features

"GOUTTE of Water" is conceived also as a "follow-up" organisation, where (former) geographical limited partnerships can continue their co-operation and extend it to partnerships on a global scale.

Ongoing geographically limited (European) partnerships could be regarded as cells of "GOUTTE of Water". Within this scheme these geographically limited partnerships could be extended.

Within the cells (defined by geography, discipline, language, etc.), the co-operation between universities in developed and developing countries should be stimulated.

"GOUTTE of Water" should attempt to incorporate existing water-related UNESCO Chairs and their respective universities in the network. In fact, UNESCO Chairs should be active in forming cells.

"GOUTTE of Water" refers to water and water-related scientific disciplines in the broadest sense. It is also understood that the framework will rather be based on active cells respecting practical limitations due to interdisciplinary and size than to work in all activities together.

In order to emphasise the co-ordinating role, "GOUTTE of Water" should act as a forum for the exchange of ideas, information and results among the active cells. Joint symposia and seminars are the most likely forum of this co-operation.

As far as possible the "GOUTTE of Water Network" should be self-governing (this is clearly a requirement for the participating cells), but the global network can also be envisaged as one being managed by an institution of academic or post-academic education, closely associated with IHP UNESCO.

"GOUTTE of Water" should be a UNITWIN project, thus belonging to the Education Sector, while its water feature places it into the respective theme of the International Hydrological Programme, within the Science Sector of UNESCO. Thus "GOUTTE of Water" would be an intersectoral UNESCO project.

5. Academic features

The network should have its focus on the ethical issues related to water. These globally valid principles are essential parts of the curriculum of the future generation of professionals and scientists. Ethics is however bound to the general prevailing value system. Consequently, while principles are global, ethical values and ethics may have regional, cultural differences. These could be reflected, and co-ordinated, in the different cells of "GOUTTE of Water". The "New Water Ethics" is to become a practical guide in our approach to water. This implies both research and educational needs.

Emanating from the strong association of "GOUTTE of Water" with the science-orientated components of the International Hydrological Programme (IHP) of UNESCO it is the foreseen that the teaching component of "GOUTTE of Water" should be focused on MSc and PhD level education. Both levels fit better in an international context than the first-degree (BSc) level.

As far as training is concerned the network should focus on the principle of subsidiarity, thus preferably stimulating the training of trainers to disseminate the knowledge acquired in the respective national environment. Training components thus should include didactic techniques, public awareness raising (ethical issues) together with the other features of containing continuing education and training (CE+T). Again the CE+T and train the trainer components are expected to be rather at postgraduate level in an international context.

At the present, conceptual stage, several potential academic activities could be mentioned. Mobilities are certainly form key elements of co-operation. In order to match the needs of each cell, they should retain substantial freedom to define ways and means of their co-operation. "GOUTTE of Water" should not be limited to education alone. Research, specially research education, should be part of its scope. Stimulating certain activities could be achieved by an academic policy to define areas of eventual financial support.

"GOUTTE of Water" cells are preferably formed under the participation of at least one water-related UNESCO Chair, thus enabling a direct injection of academic concepts, principles and practices promoted, encouraged or developed by UNESCO or related institutions.

"GOUTTE of Water" could, at a later stage, take over tasks like advisory function in accreditation, degree comparisons, programme reviews and quality assessment on behalf of national governments, international organizations and its own partners.

While it is primarily university-based, "GOUTTE of Water" should remain flexible to carry out its mandate. Any additional partners (research institutions, colleges, administrations, enterprises) could be considered if their contribution enhances the chances to achieve the basic objectives of "GOUTTE of Water" without jeopardizing its standard.

6. Closing remarks

"GOUTTE of Water" is conceived to provide a coherent approach for UNESCO IHP in the area of university partnerships in the broad area of water related academic ventures.

The present paper aims to open the debate on this concept within the academic community of water-related disciplines and soliciting advice of those experienced in inter-university co-operation.

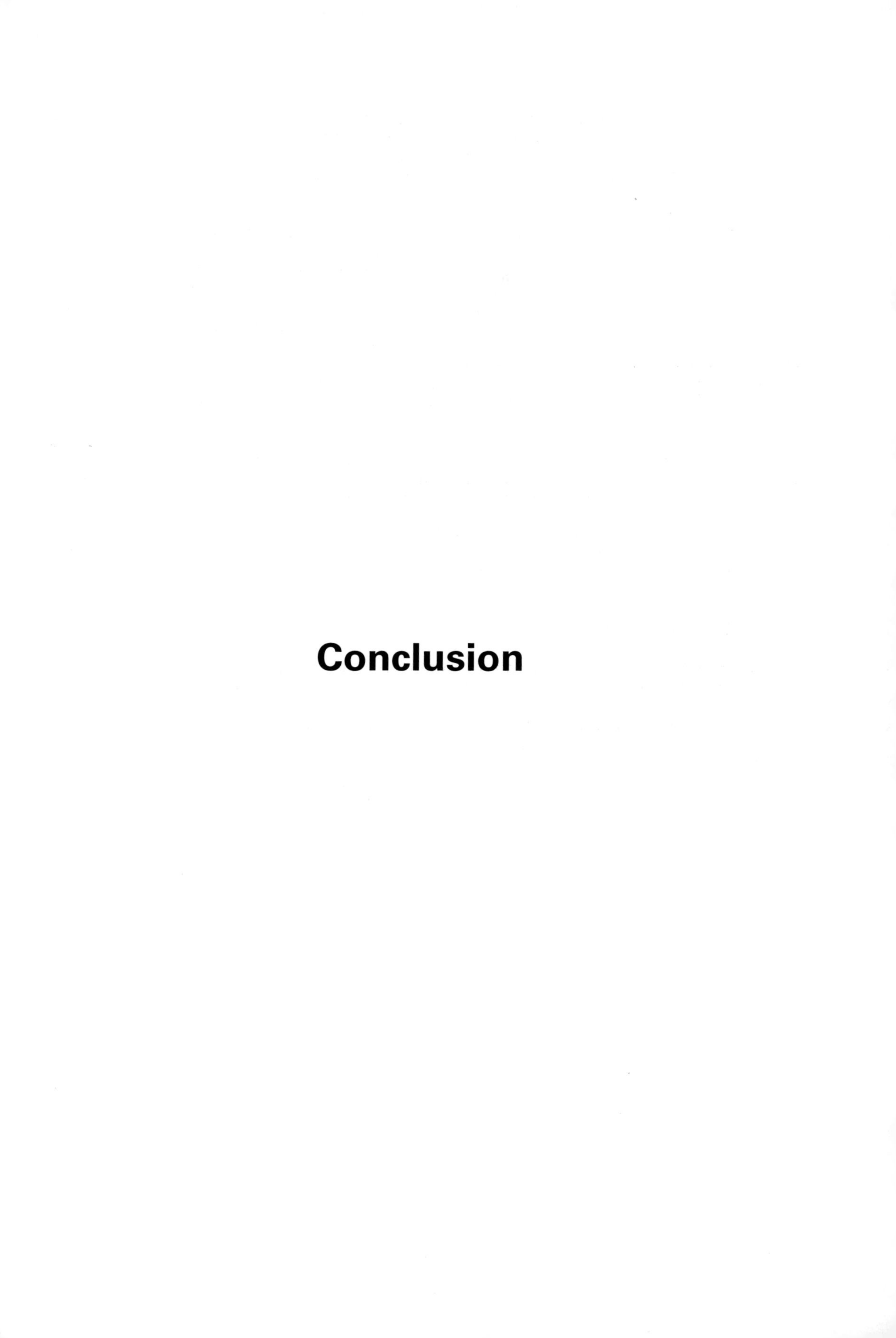

Conclusion

Chairman's Conclusion of the International Symposium

To conclude this symposium, I would like to take a historical perspective.

In a recent conversation, a long-time friend, Dr. R.N. Athavale from the National Geophysical Research Institute in Hyderabad in India, told me about the recent discovery of an ancient river in Northern India, the Vedic Sarasvati*. According to the ancient Indian Mythology, there were three major rivers in India : the Ganges, the Indus and the Sarasvati. These rivers were reported in the famous Rigveda, the oldest texts available to mankind. But in present days, only the two former rivers were still known. River Sarasvati was lost, and considered rather as a myth than a reality. It is only through archaeological and historical research, coupled with paleohydrology, that it was possible to rediscover the lost river, beneath the sands of the Rajasthan desert : the lost river actually existed, and permitted a very prominent civilisation to flourish along its banks : the Early Harappan and Harappan culture, dating back to the fourth millennium BC. The loss of the Sarasvati River is believed to be due to two simultaneous mechanisms : a climate change towards desertification, which started 4000 years BC, and tectonic movements, which diverted the course of the lost river to become a tributary of the Indus. And then these civilisations were lost, and their remains were slowly buried beneath the dunes of the new desert.

Today's world is facing, I believe, a similar problem than the Harappan culture. The danger of the lack of water is a tremendous threat to many countries and civilisations. While climate change is a real concern, and has for instance been observed in particular in the African Sahel during the late sixties, and may be a consequence of the global warming due to the greenhouse gases, the major source of threat to the water resources is the tremendous increase in the world's population, even if the rate of increase seems to have slowed down in recent years. Will we be able to fight the quest for water, and avoid being buried beneath the sand of the desert ? Can we learn, can we teach our hydrologists the science for harnessing our water resources ?

We have tried to address, during these three days, these growing problems, with the viewpoint of a better and more effective approach to teaching and learning in the field of Water-Environment. More than two hundred of us, coming from fifty-one countries, have contributed with their knowledge and experience to the seven themes of the symposium, the conclusions of which have just been summarized.

Let us, for a moment, look back again in history. We learn that the problems of harnessing our water resources have been addressed in essentially two ways by the ancient hydrologists. The first way is to divert and transport the water from where it is available to where it is needed, we could call this the engineering approach. The Egyptians, the Romans, among others, were extremely successful water engineers, and some of the waterworks they have built, like aqueduct or siphons, are still visible today. The first artesian wells, to collect water from deep underground, were probably built by the Arabs in the Sahara in the 9th century. The first wooden suspended channels on the flanks of vertical cliffs in mountainous areas were perhaps built by the Swiss in the Alps, at the same time. This engineering approach has resulted in the numerous dams and canals which have been built and are being built on all the major rivers of the world, and which have resulted in a major change of the ecological functioning of many rivers and wetlands of the world, leading also to intense changes of the transport of matter from the land to the oceans, the consequences of which are only slowly understood.

The second approach to the harnessing of water resources may perhaps be called the ecological approach. It can be best described by the day-to-day optimal management of the locally available water resources, by intelligent management of the land. The local diversion of floods in rivers to spread the water in the plains and thus increase the infiltration, the building of small terraces on the slope of mountains to keep the soil and increase the infiltration, the storing of water in small ponds in valleys, to increase infiltration and permit some irrigation downstream, e.g. in India where they are called Ery, or the mulching of crops, as was practiced e.g. by the ancient North American Civilisations, are just a few examples of such ecological practices.

One may think that these different approaches are used on a case-by-case basis, according to the local needs and available technologies. While this is partly true, I believe that there also is a component of ethical and religious belief in these two approaches to water management. The Christian religions, for instance, imply a clear mandate to mankind to domesticate nature and turn it onto an obedient tool for the prosperity of man. Its modern successor, the Marxist philosophy, goes in exactly the same direction. Other religions, like the Buddhist, Hindu or Islam, or the ancient North American culture, have a much greater respect for nature and life, and do not aim at nature's domestication, they aim at establishing harmony between man and its environment.

We agreed today that we need to improve and deepen our teaching and continuous learning of water and environment. Many of us have emphasized the need to combine the "engineering" and "ecological" approach to water management, and to make our hydrologists aware of the global impact of their action on the water cycle. Is this to say that there is just one "scientific approach" to water management, which we should try to identify and to teach to our students ? I believe this is the view taken by many, and for instance by the group just presented to us a moment ago as the "World Water Vision", which tries to develop this unique "vision" of the world's water problem. While their efforts may be of value, I am very much concerned that their approach may be too narrow, too much influenced by the "Roman" or "Saxon" origin of their philosophy, and not compatible with other major "visions".

We all came here at this meeting because we believe that there is merit in exchanging experience and ideas for improving our understanding and teaching about water and environment. But we should always keep in mind that there is not just one approach to harnessing our water resources, and that we will only succeed in preventing the loss of our many fragile "Sarasvati Rivers" if each culture is given the freedom to make its own evaluation and decision.

Ghislain de Marsily
Professor, University Paris VI
Chairman of the Scientific Committee of the Symposium

Towards a water-education-training vision

During the FORUM at the end of the Symposium the Rapporteurs have highlighted the seven Themes of the Symposium. The rationale for a World Water Vision was presented by W.J. Cosgrove, Director of the World Water Vision (WWV)Unit at UNESCO's headquarters.
The WWV involves a process study, consultation and promotion which will :
*develop **knowledge** on what is happening in the world of water regionally and globally
*based on this knowledge, produce a consensus on a VISION for the 21st century
*raise awareness of water issues among the general population
*utilise the knowledge and support generated to provide input to the Global Water Partnership.
The VISION consultations include consultations on **sector** issues, **regional** consultations and **network** consultations.
Given the first objective of WWV -**develop knowledge**- the Symposium participants recommended a **sectoral consultation of educators and a educational component of the World Water Vision.**
A background document for the WATER-EDUCATION-TRAINING (WET) VISION was prepared by the International Hydrological Programme(IHP) of UNESCO and serves as a Discussion Forum on the website of WWV <http://watervision.org>

POSTERS

Linking educational tools to models, an integrated approach

Aschalew D.A.[1], Bauwens W.[1], Fuchs L.[2]
[1] Laboratory of Hydrology, Free University of Brussels, Belgium
[2] ITWH, Hannover

Abstract
In response to the large number of models developed during the past few decades, the need for educational tools has become more and more obvious. As a result, many have come up with such tools on various subjects in the water resources sector. However, most of these efforts do not reap the results they deserve mainly because they are too theoretical in the sense that they lack demonstrative insights into the subjects under consideration. An integrated approach, where educational tools are coupled with models is discussed and presented in this paper. As an example, an educational tool on urban drainage management linked to the sewer system simulation model, HYSTEM-EXTRAN is presented.

1. General Introduction

One of the main tasks of civil engineering in the field of urban storm drainage is the collection and disposal of stormwater runoff, to prevent flooding in urban areas. Equally important is the task of protecting the ecosystem. The actual problems regarding sewer systems in most of the industrialized countries do not concern the dimensioning of new sewer systems but the renovation of existing ones. The old sewer systems within the cities have to drain catchments, which have become bigger and bigger during the last decades and with a much higher percentage of imperviousness than they were designed for.

In response to these demands, numerous sewer flow simulators were developed around the world. The actual simulators are nearly all based on hydrograph methods, whereby the system is discretised spatially and temporally and whereby the rainfall-runoff process on the surface as well as the dynamic flow phenomena in the sewer system are accounted for. They are therefore more complex than the methods that were used in the past for dimensioning the systems. Moreover, the available simulators vary widely in their degree of complexity, on the concepts and assumptions they are built on and the algorithms that are used. As a result, their effective use depends on understanding their degree of complexity and assumptions.

Model developers have been focusing for years on improving the quality of their models while only a very small fraction of that effort is put towards improving the quality of model users. Although there is a help system accompanying most models, many have witnessed that most are far from being adequate. This is even more true not only in the lack of details but also in the way these help materials are presented. Consequently most of these efforts put into improving the models are not exploited by most users.

A simulation run without any error message doesn't mean that the results are suggestive or logical according to the actual problem. A computer simply helps the engineer with mathematical computations in order to investigate lots of different variants to find out the optimal solution for the actual problem. More than in the past the engineer must be able to analyze and work up the problem and perhaps to simplify and prepare the data for the simulation. The simulation results have to be checked and judged with a critical view in order to make safe and at the same time economical decisions.

In using such complex models, detailed knowledge about the basics, algorithms, limitations and usage of the model under consideration are more necessary than the knowledge in computer technology.

Since few years researchers in the field of water resources have come to realize the need for a comprehensive educational tool, which could give at least the general idea behind such complex methods. Efforts such as those shown by the Computer Aided Learning in Water Resources Engineering (CALWARE), and

In the Computer Aided Learning tool for Urban Drainage Management (UDMCAL), the basics of sewer system modeling is explained with the urban sewer system simulator HYSTEM-EXTRAN used as an example.

The tool provides detailed explanations to help novice users of the simulator to understand its background (tutorial) and assists the initial users of the package (user guide). It is linked to HYSTEM-EXTRAN and functions as an on-line help system for the simulators.

HYSTEM-EXTRAN is an urban drainage system simulation model developed by ITWH, Hanover. It is composed of the surface runoff simulator HYSTEM and a pipe flow simulation model ITWH-EXTRAN (Fuchs, 1990).

2. Introduction to UDMCAL

UDMCAL is an outcome of the combined efforts of the Laboratory of Hydrology of the Free University of Brussels (Belgium) and ITWH and was developed within the framework of the CALWARE project (Computer Aided Learning in Water Resources Engineering). CALWARE has brought together staff from across Europe who are leading experts in the field of water resources and who are also skilled teachers (J. Davis, 1993). These skills are utilized by computer aided learning specialists to provide effective learning tools.

UDMCAL is developed using the authoring tool Authorware® Professional™ for Windows™ 2.0 (Anonymous, 1991). Authorware® is a general, flowchart oriented authoring tool for educators and trainers who need to create computer-based interactive multimedia materials.

The creation of an interactive program with Authorware is based on the definition of a flowchart of the actions that need to be undertaken. This chart consists of icons that are dragged and dropped from a tool palette (shown on the left side of Fig.1). The tool provides about a dozen icons, each representing a command that triggers a generic action, such as playing a movie, running an animation, or performing a calculation. When an icon is dropped onto the flow line, a dialog box allows for the customization of the action. After an initial learning period, the process of building an interactive learning program becomes as easy as doing page layout or video editing, even though one is literally programming.

3. Some features of UDMCAL

The training material consists of texts, figures, photographs and animations. As the simulator is linked to UDMCAL, it may also be considered as an integral part of the training tool.

The text is organized in the form of a book, with chapters and subchapters following the structure of the original manual and largely consists of a condensed formulation of the texts in that manual.

Navigation through the tool can be done through a multitude of ways using menus, drop down menus, through flow charts, specific buttons, page sliders and using the hypertext features (Fig.2).

UDMCAL contains a number of animations that provide added value to the text, especially for the explanation of dynamic phenomena. One such example relates to the way off-line pumping stations are represented in the simulator (Fig.3). The functioning of this system is discussed step by step while the rotors of the pumps are animated and the in- and outflow hydrographs are gradually built up.

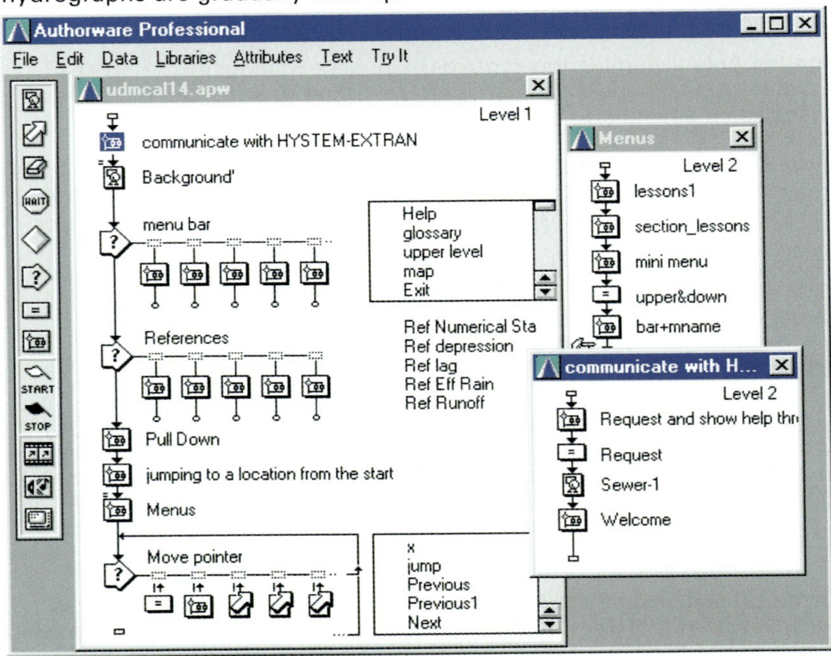

Figure 1. Authorware's authoring environment

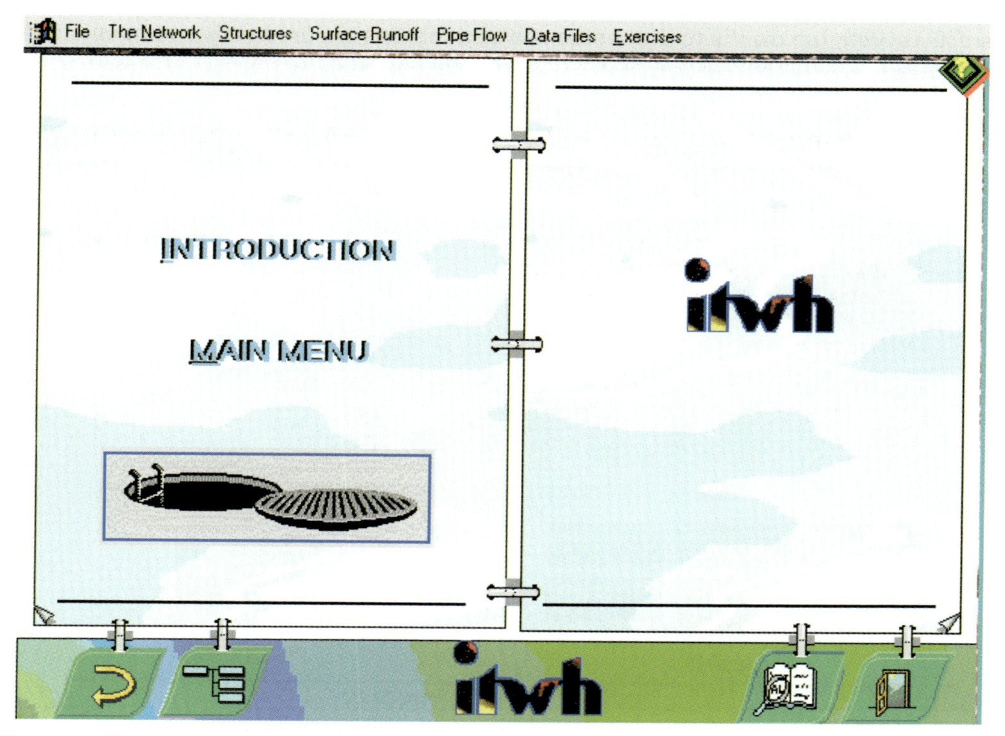

Figure 2. The main menu of UDMCAL

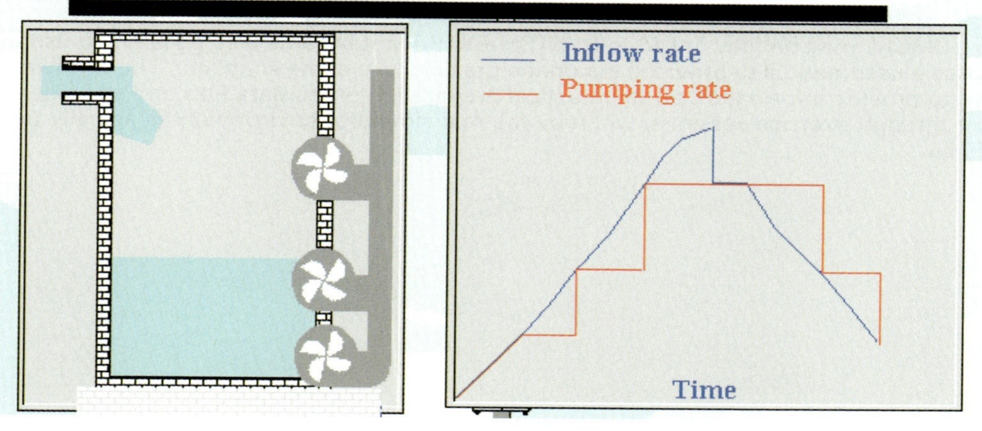

Figure 3. Demonstration of how an off-line pump is modeled in HYSTEM-EXTRAN

Interactivity with the user is further enhanced by exercises and interactive simulations. An example of the latter concerns the calculation of the net rainfall in HYSTEM (Fig.4).

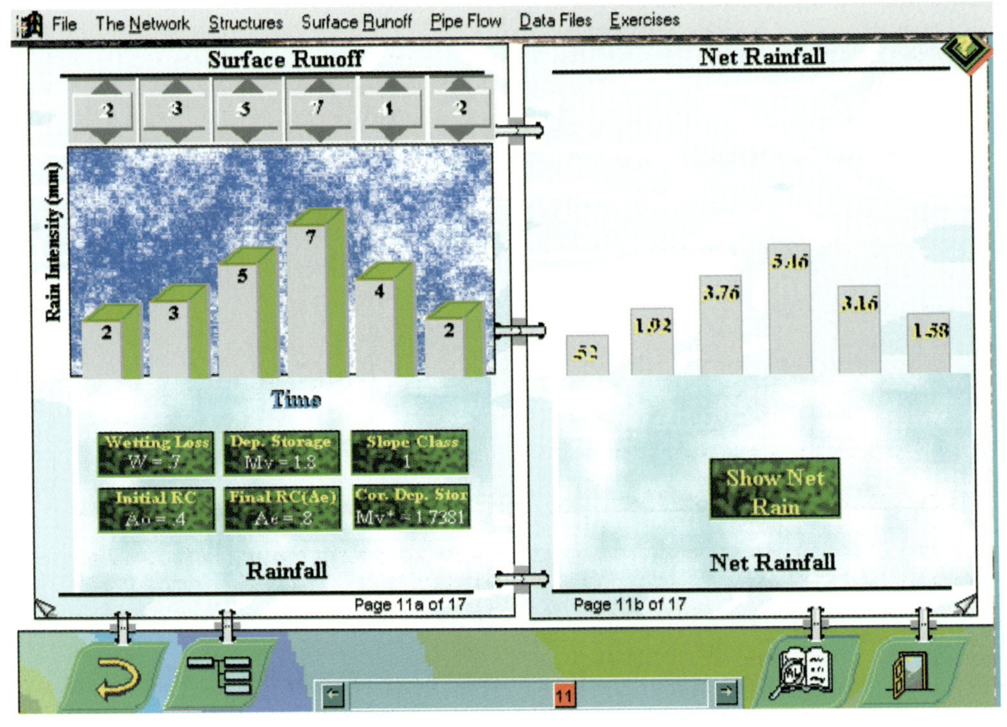

Figure 4. Training on rainfall abstractions

The user may define a simple hyetograph through dragging the rainfall bars (left side of the figure) and set the values of the model parameters (bottom left). This information is then automatically transferred to HYSTEM and the resulting net rain is visualized (right side of the figure).

As UDMCAL may be fully linked to HYSTEM-EXTRAN, it became also possible to use the tool as a user manual to provide help on the practical use of the simulator. The tool is thus able to provide information on the practical preparation of the data files and to guide the user through exercise sessions, whereby the real simulator environment is actually used (Fig.5).

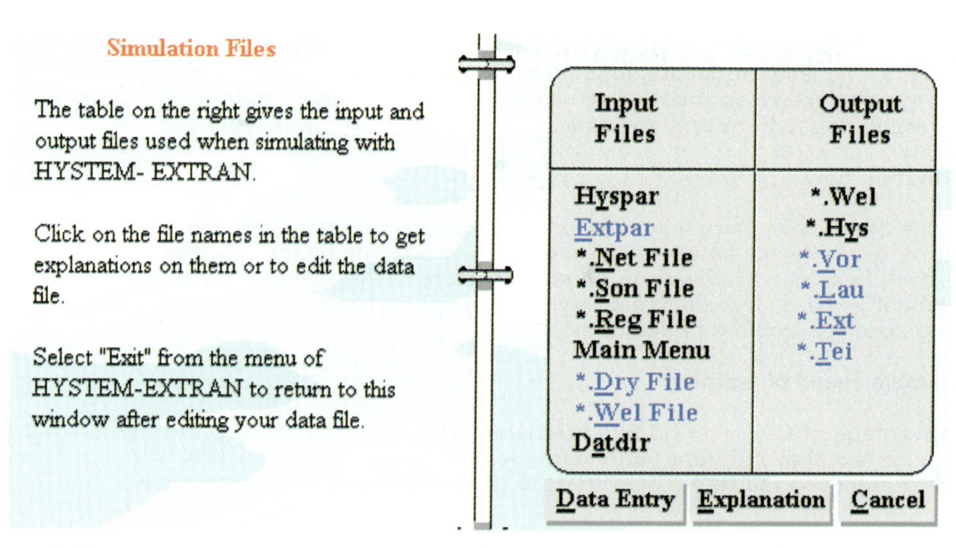

Figure 5. Data entry and explanation of data files

4. Advantages of UDMCAL

Computer aided learning technology offers several advantages when compared to the traditional manuals provided with computer simulators :

Full integration with the simulator

One of the many advantages of CAL tools is the possibility of full, user transparent communication among different software tools through the use of the Dynamic Data Exchange (DDE) and Object Linking and Embedding (OLE) possibilities provided by Windows™. Two way communication is possible between the training tool and the simulator : UDMCAL can be called from within the HYSTEM environment or the model can be started from within UDMCAL.

If called from within HYSTEM, UDMCAL serves as an on line help tool. The main advantage resides in the fact that the user is hereby sent automatically to the section of the 'manual' that is relevant to the problem (depending on the form or field from where the help was requested). Hereby, the help is not restricted to a specific help message as is often the case in more traditional on line help systems : the full CAL tool is accessible for further research if desired.

The fact that HYSTEM can be called and executed from within UDMCAL was also used for several purposes : to illustrate concepts used in the simulator - hereby using actual (sub-) components of the simulator to perform the calculations - or in the framework of the 'user manual', to guide the user through the actual pre- and post processors of the simulator.

Interactivity

Interactivity is very important in the learning process. Students listen to teachers and ask questions and get relevant answers. Teachers interrogate to test the level of their students and adjust their method of teaching depending on the level of their students. Although CAL tools are still far from providing these possibilities, encouraging developments in this area are emerging following the breakthrough with the speed and memories of computers.

Even if the possibilities of Authorware to evaluate the students and adapt the presented material to the level of the students have not been used within UDMCAL, the interactivy with regard to exercises and the link with the simulator represent an important advantage with respect to the book type manuals.

Information flow

Because of the flow chart type of programming approach and the different possibilities to organize relationships between different parts of documents through menus, buttons and hypertext, information flows can be well organized. The tool can therefore be consulted sequentially in a quite systematic way (beginners), randomly or - more difficult in a book - along specific, problem driven paths.

Adjustable speed of learning

An advantage of CAL over human teachers is the of adjustable speed of learning. It is a matter of fact that different students have different speeds with which they can learn. Every student can follow his own steps to go through the learning material

References

Fuchs, L. (1990) HYSTEM-EXTRAN, Integrating computer aided design tools for urban drainage problems. *Proceedings of the Fifth International Conference on Urban Storm Drainage, Osaka, Japan, pp. 1287-1292*

Authorware® Professional™ for Windows™ (1991). *Reference Manual*

Anonymous (1991)

Davis, J. (1994) Hydroinformatics'94. *Proceeding vol. 1, pp. 483-488*

An internationally distributed environmental information system – the Danube basin information network

Brilly M.
University of Ljubljana, Slovenija

Abstract
The Internet and WWW pages provide an excellent opportunity for the development of an International Environmental Information System with only proper linkage and the organization of national country home pages. A small team of scientists from Budapest, Hungary, Ljubljana, Slovenia, and Fort Collins, Colorado, USA, supported by a NATO ASI Linkage Grant, has developed an umbrella information system - the Danube Basin Information Network (DBIN). The ultimate goal was to produce the prototype of a cost-effective information-database system which would meet the needs of a maximum number of group and individual decision makers in the Danube Basin and elsewhere. The system was constructed as a connection of set linkages with international and national web sites. A National DBIN web site for Slovenia was also established.

1. Introduction

The development of an environmental information system is a necessity and a challenge for science today. This is an issue of multidisciplinary dynamics involving various sources and users from different backgrounds and with different areas and levels of understanding.

The Danube River Basin, 817,000 square kilometers in area, is shared by seventeen countries, more than any other river basin; the area has been affected by conflict and geopolitical changes from the two world wars and the cold war (Murphy, 1997). Cooperation between countries in the Danube River basin has a long history; cooperation in water management was first accelerated by an economic interest in navigation. The first agreement was made in the last century (1856): however, a more fundamental international act for navigation on the Danube River was the Belgrade Declaration of 1948 (http://ksh.fgg.uni-lj.si/danube/belgconv/).

Intergovernmental cooperation in the field of environmental science also has a long tradition. A hydrological monograph was published in 1988 (Stan_ik A., Jovanovi_, et al.), an IHP UNESCO Regional Program for the Danube River Basin was launched in 1987, and in 1985 the countries involved began the long term monitoring of water quality in the main stream with the signing of the Bucharest Declaration.

After several years of negotiation, the Danube River Protection Convention was signed in 1994. In the meantime, the Environmental Program for the Danube River Basin (EPDRB) had been in force as a framework for activities to improve the environment in the basin. GEF, the EU Phare Program, the World Bank, the European Bank and others all supported the program, and several activities are still being performed under the EPDRB umbrella. The most important of these is the Strategic Action Plan (www.infoterm.or.at/ceit/sap1.htm). The overall improvement of information management is an important part of the SAP and several projects in the region. The EPDRB has been given financial support for: providing public information, publishing the Danube Watch, providing textual information - the Danube Information System (DANIS), providing water quality data - the monitoring laboratory and information management agency (MLIM), providing warning in cases of environmental accidents - the Accident Warning System (AEWS), and several workshops for governmental and NGO institutional awareness.

The need to address issues about communications on international water resources was addressed at a NATO Science Committee Advanced Research Workshop in Budapest in 1996 (Murphy, 1997). This identified the goals of future activities which were successfully launched by a small team of experts from: the Colorado State University (Fort Collins, Colorado, USA), the Central European University (Budapest), the University of Ljubljana (Ljubljana, Slovenia) and VITUKI (Budapest). These developed a demonstration facility on the CEU web-server - the Danube Basin Information Network (DBIN).

Suggestions made by the leading team included the establishment of a management committee, to include Danube Basin country specialists, in order to help standardize the various categories of information and promote the development of accurate and compatible water quality and data-sharing. This committee could include web masters for the country home pages as well as the BIN web master. Glossaries could be expanded to include definitions in the languages of all the Danube River Basin countries.

2. Some rules for publishing on the Internet and the web

The Internet is a modern information tool which provides the possibility of establishing an open and transparent information network between various information and data bases (Figure 1).

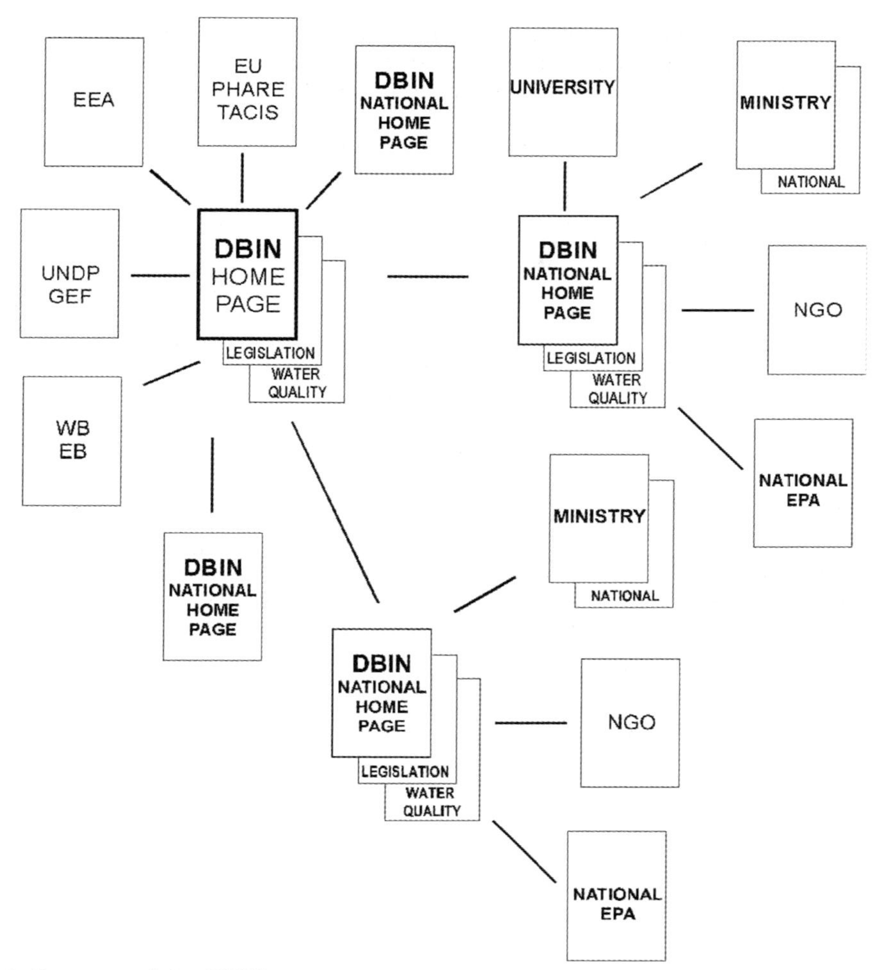

Figure 1. Concept of the DBIN.

Here the concern is how to link scientific data and find it without time-consuming surfing through the rapidly increasing number of web pages. These scientific data should be equipped with proper meta-data and provide source or officially-published data. The time-consuming surfing could be avoided by properly designed and daily-maintained web pages for navigation.

The Internet is a new challenge. It is the fastest developing new medium in the world (Hoffman, 1995), offering new possibilities almost every day. Thousands upon thousands of new web pages are being opened each day and offer a range of interesting information, rather like blooming algae in nutrient-rich water. The most frequent questions are: "How to provide a new web page to present information?" and "How to equip the web page with the scientific information which we need?"

The web is a new medium and adequate rules are still under development. The rules for book formatting and publishing (page numbering, indexes, tables of contents, title pages, etc.), took more than 100 years after Gutenberg to formulate; the development of web pages should follow a similar course, (Lynch, 1997). Some authors design web pages freely, without any rules at all, and the results can be either refreshingly interesting or chaotic. The rules for publishing which have been developed in more the five hundred years of publishing design are fairly relevant, but even the relatively short history of the WWW has provided us with some experience:

1. Design your pages in both text and graphic versions. Figures are attractive but time-consuming to download and some users without sophisticated equipment can not use them. You should give the user a choice (Wilson, 1995).
2. Navigation should be transparent with links in every section. The pages and links should be provided with identifiers for users who arrive directly from external links or for users who download your pages (Wilson, 1995).
3. Put content on every page; do not confuse the viewer with oversized pages that require scrolling; do not use slow graphics with a lot of colors and make everything as large as possible (Black, 1997).
4. Provide pages with the date the page was last updated and the contact person.
5. You are responsible for what you link to. Please link responsibly (CSU regulative).
6. Nobody reads anything- At least not everything. The only person that will read every word of what you've written is your mother (Black, 1997).

When we start a web design, we should decide what kind of information we want to provide and for whom. If we want to provide linkage with other significant pages either within or outside the WWW site, then clear and user-friendly navigation with transparent hierarchy, structured relationships, and logical functionality situated on compact one screen pages are all of utmost importance. The design of data- or information-related pages depends on the kind of information and the way we would like to provide the client therewith. Long multiple screen HTM files are useful if clients are happy with downloading the page contents and read them at home (Lynch, 1997).

If we want to establish an international information network on the Internet, there may be some problems with language. Obviously English is chosen for common communication and home pages are designed in both national (local language) and international (English) versions. The question is, how to homogenize the terminology used by developers or web page masters with different backgrounds for whom English is not the mother language, and how to mine through huge national databases of original documents and data in the local language. The solution lies in a glossary and a common multilingual thesaurus.

An international database may only represent a percentage of the national one; i.e.: it is only an abstract of the national one. The development of an international network is closely related to the development of a national database and a national information network. The development of an international database in English, without national support, is a very expensive, one-off business, but not as expensive as a centralized information system which collapses as soon as the international funds for its maintenance are removed.

3. The Danube basin information network - DBIN

Team of Scientists: Dr. Ed Bellinger and Andrey Semichaevsky (Central European University), Dr. Mitja Brilly, Dr. Damjana Drobne, MSc. Andrey Vidmar (University of Ljubljana), Dr. Darrell Fontane (Colorado State University) and Dr. Peter Bakonyi (VITUKI) led by Ms. Irene L. Murphy and financed by NATO ASI, created The Danube Basin Information Network (DBIN) for electronic access to scientific data and information about the Danube Basin.

The Danube Basin Information Network (DBIN) was constructed as an information system which includes sets of relational databases and linkages with international and national web sites (Figure 1). Sources of information in each country generally have their own web sites with relevant scientific information and take on the responsibility for updating and expanding such information. The primary DBIN page, programmed by Mr. A. Semichaevsky (Figure 2, http://www.syslab.ceu.hu/~semichae/danube), consists of connections with other pages, a map of all the Danube River Basin countries, a lower feedback portion, a description of the basin and a brief glossary. It is possible to navigate from the main page to other pages and to the country pages included in the DBIN. The Internet network was structured as a mesh of the main subjects and countries.

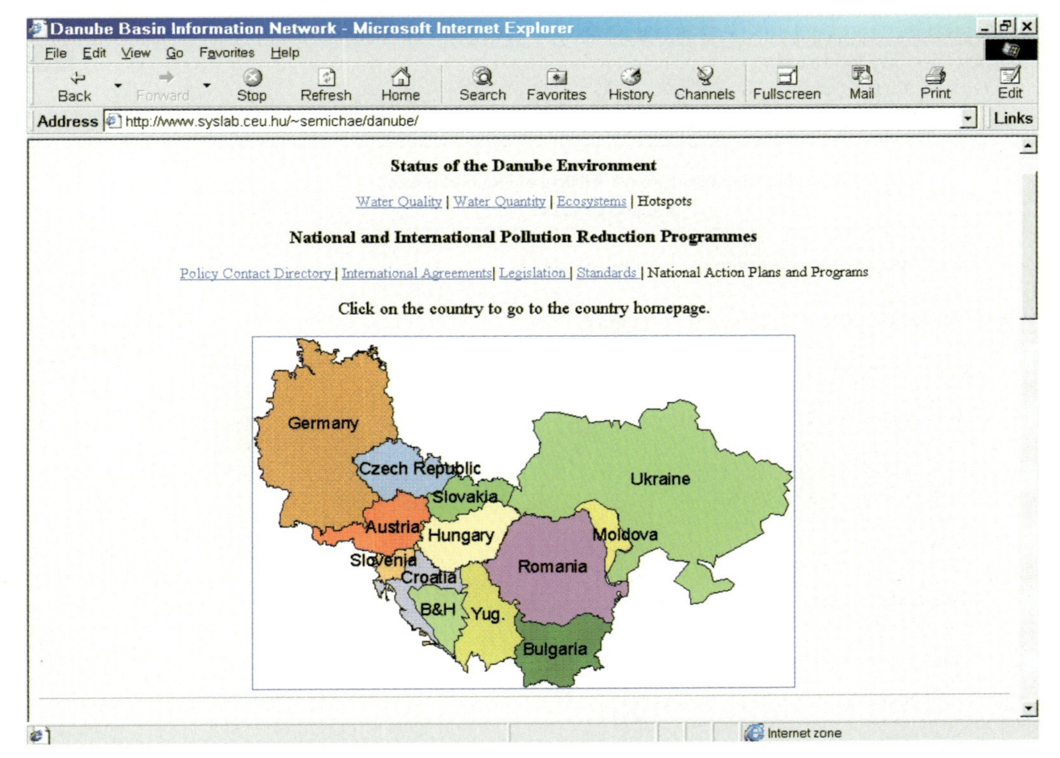

Figure 2. Danube Basin Information Centre - first page of the information system on the http://www.syslab.ceu.hu/~semichae/danube/

The categories of information provided include:
Directory of people and organizations involved in environmental activities in the Basin and connections with the home pages of the Ministries; Universities; Consultants, NGOs, etc. responsible for the environment, or lists with names and addresses.

Legislation: with connections to country web pages with environmental laws translated into English.
Standards: The water quality standards of the Danube Basin countries.
Conventions: The texts of the major conventions affecting the Danube Basin can be accessed. There is also a link with the US CIA list of environmental conventions.
Water Quality Data: Links to water quality data provided national and international sources.
Water Quantity Data: Links to water balance data.
Ecosystem: Data on flora, fauna and biota's or national parks accessible on country home pages.

The pages have been produced for scientific purposes and are reachable by password. A DBIN country page has also been established for Slovenia. The pages are presented in two languages: English, and Slovenian for domestic purposes (Figure 3). We could navigate throughout the national environmental pages or return to the main page of the DBIN and dig down to original documents such as the Danube Environmental Convention and the Danube Navigation Convention. We provide a lot of connection with NGO pages in Slovenia. The NGOs are very interested and ready to make data and information publicly available. The official services are more conservative about providing data on the Internet.

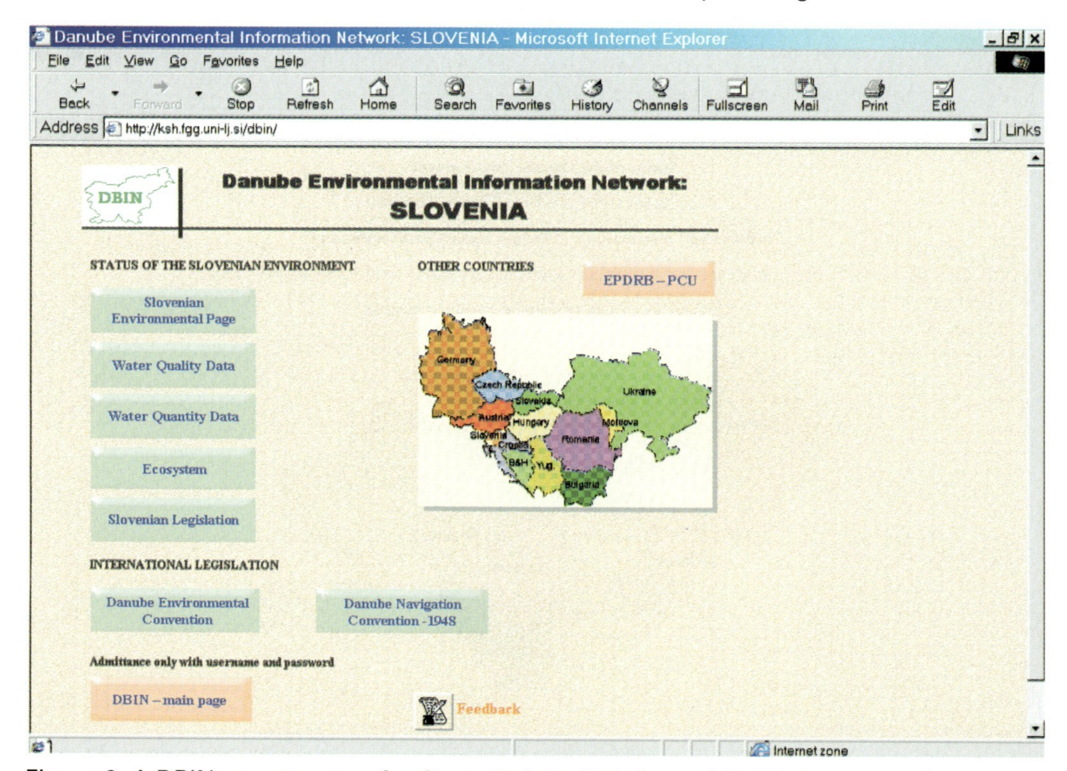

Figure 3. A DBIN country page for Slovenia http://ksh.fgg.uni-lj.si/dbin/

Environmental web pages were found for each country. Some of them provide data on mining facilities (www.ubavie.gv.at); others have only some environmental descriptions. Web pages for the countries of the Danube River Basin are increasing rapidly. There is no country in the Danube River Basin without a web page with information on environmental topics. During the two-year period of our investigation we found that access to such information on web sites has increased by more than ten times.

The DBIN has been found to be user friendly, cost-effective in operation and maintenance, and helpful to decision-makers in government and non-government organizations. The DBIN was found to be an excellent umbrella system at the GEF/UNDP Danube River Basin Information System Workshop on March 23-26 1998 in Baden, Austria. Several national and international projects for data management and public access to environmental information such as a Regional Center for Central and Eastern Europe in Budapest, a Catalogue of Data Sources with Environmental Information and Observation Network, a Central European Environmental Data Request Facility, a Danube Information System, and several UNEP projects (GRID, INFOTERRA) which provide web pages with scientific data and information were developed there.

The DBIN was oriented to promote the development and harmonization of environmental web pages in the countries of the Danube region. We would like to open information on source, in institutions which collect and maintain data or which are responsible for information.

4. Conclusions and recommendations

Participants in the demonstration and users of the Danube Basin Information Network (DBIN) reached the following conclusions:
1) The format and content of the DBIN could be used to satisfy a range of users from government officials to the general public who need access to data and information about the Basin.
2) Its concept, to use relational databases and links to Danube Basin country web sites, can provide quick access to a number of easily-updated databases, including legislative data and information, water quality standards for all the Danube countries, available water quality and quantity data, and data and information about Danube Basin ecosystems. Links to Danube Basin country home pages which provide more detailed back-up data and information, are an important feature of the DBIN.
3) A network of country home page web masters should be formed to facilitate the development of data and information relevant to those in the DBIN in each country. The Ministries of the Environment in several Danube Basin countries are now making general data and information about their countries' policies available on the Internet and may well be the vehicle for such a network.

References

Black, Roger: "Web Sites That Work", Adobe Press, San Jose, California,1997
Hoffman, P.: Destination Internet&World Wide Web, IDG Books Worldwide, Inc., Foster City, California, USA, 1995.
Lynch, P.: Horton S, Yale C/AIM Web Style Guide,1997
Murphy, I.L.: "Protecting Danube River Basin Resources", Kluwer Academic Publishers, 1997
Stančik, A., Jovanovič, S, et al.: "Hydrology of the River Danube", Priroda publishing house, Bratislava, 1988.
Wilson Stephen: "World wide WEB design guide", Hayden books, Macmillan Computer Publishing, 1995.

Proposals on IPTRID Network Development

Propositions aux IPTRID Reseau Developpement

Dukhovny V. A.
SIC ICWC, Uzbekistan

Abstract
Information exchange is important for professionals and scientists specializing on irrigation, drainage and water management for creating ability to use advanced experience in all directions of activity and, first of all, for raising efficiency of deficit natural resources use: water and irrigated lands. It has not only an economic, but nature protective meaning as allow mankind to survive and satisfy growing needs in these resources without increase of volumes of their use and through demands management and saving. Regional information exchange is organized both by regional and big national centers within countries or their different zones on a bilateral basis. On the one hand, they process information from central centers of IPTRID and disseminate it between their own information centers in different region's states or big country's zones according to their peculiarities, on the other hand, collect information by unified samples, formats and principles from these centers, aggregate it, translate and transfer to the IPTRID central centers in accordance with their technical orientation. Regional information exchange is organized both by regional and big national centers within countries or their different zones on a bilateral basis. On the one hand, they process information from central centers of IPTRID and disseminate it between their own information centers in different region's states or big country's zones according to their peculiarities, on the other hand, collect information by unified samples, formats and principles from these centers, aggregate it, translate and transfer to the IPTRID central centers in accordance with their technical orientation.

a) Introduction

Within the framework of IPTRID Program an effective and efficient mechanism, which aimed at establishment of the world information network on irrigation and drainage, research development and modern technologies introduction has been organized and developed. As it was noted at the meeting of July 10...11 in Wallingford and at the ICID conference of September 10, 1997 in Oxford, for the last years a huge work was done, particularly, establishment of a network of information centers and organization of information exchange. This has been done by organizers and initiators of the program (Guy le Moigne, Ashot Subramainan, Frank Hartwelt, M. Chitale, B. Swedema, etc) as well as by creators of IPTRID Central Network (HR Wallingford - Geoff Pearce; ILRI, Netherlands - Gerrit Naber; Cemagref, Frans - Allian Vidal; USBR, USA - Stanley Ponce). They initiated establishment of Network's regional and national centers and made possible bilateral and effective information exchange between central network and above centers.

At the same time, five years of its activity showed certain disadvantages, which are evident for the initial stage of such world network:

- information exchange is unsystematic and provided with different regularity, sometimes chance, and usually depends on responsibilities and abilities of central and regional Network's managers;
- the general rules of information exchange and distribution are not elaborated, especially due to stated in many countries laws about intellectual property and consideration of information as a type of business;
- direction towards impelling research in different developed countries, which provided quite good initial surveys by several countries (Morocco, Egypt, India, China, etc.) further transformed into additional sub-division of the World Bank, which prepared technical base for investment projects of this huge credit organization.

On the base of analysis of activity of SANIIRI, the regional research organization of Central Asian states, and then of Scientific Information Center (SIC) of the Interstate Coordination Water Commission (ICWC), established by five states, an attempt to show scheme of works organization and to elaborate proposals on strengthening and development of IPTRID networking has been made, particularly in connection with supposed transference of leadership and coordination of this Network to FAO. From our point of view this corresponds by directions and type of activity to tasks and objectives of such world recognized international organization.

Analysis of activities within the IPTRID Network between IFAS and ICWC organizations in Central Asia

1.1. Scientific Information Center of ICWC. According to decision of the Interstate Coordination Water Commission (ICWC), made on December 5, 1993, the Scientific Information Center by ICWC on water problems was established. SIC ICWC consists of two regional and 11 national research and design organizations of Central Asia. SIC ICWC's national branches in each republic are the centers of regional activity coordination at a national level.

Scientific Information Center (SIC) of ICWC prepares draft decisions in respect to perspective planning and water policies, perfection of water management and use and improvement of environmental situation in the region as well. SIC is the main analytical core of ICWC. It provides information exchange between organizations involved in ICWC and their subdivisions and with foreign institutions as well.

1.2. SIC ICWC is a regional center of IPTRID. The regional center of IPTRID was established in 1995 within the framework of «Contribution to the IPTRID information network» project, initially, by SANIIRI then it was transferred to SIC ICWC (Pic.1).

The main task of the regional center is to organize information exchange and promote establishment of contacts between professionals, who specializes in irrigation and drainage in Central Asia, and the world community, as well as inside the states.

The center's tasks also include the following:

- provision of access to IPTRID Central Network for local users;
- creation of national data base and register of on-going research projects on irrigation and drainage;
- keeping Central Asian professionals informed about newest achievements in irrigation and drainage;
- provision of bibliographic services for professionals;
- organization of technical tours abroad for professionals and training for young professionals in relevant foreign institutes and organizations.

1.3 Collection and dissemination of information. According to the tasks, the regional center provides the following:

- collection of information from the states involved in ICWC (research results, project drafting, publications on water management, land reclamation, irrigation and drainage problems);
- translation from English of papers from journals and brochures on relevant themes and their circulation;
- dissemination of knowledge through publication of ICWC bulletin, newsletters and juridical brochures, surveys of articles from journals, books and other specific literature.

Collection of information in sub-divisions and its transference to SIC ICWC have been organized for timely implementation of these tasks. Persons, who carry out this work, have necessary knowledge and abilities, which allow information to be appropriately selected and timely transmitted to SIC ICWC.

The list of collected and disseminated information include:

- proceedings and directives of ICWC and IFAS;
- proceedings, decisions made on meetings and information from work teams on implementation of the "Program of concrete actions", approved by Heads of five states in January 1994, and financing of these programs;
- analysis of hydrological and meteorological situation for the last year or period by main rivers and basins in connection with analysis of change in water consumption;
- state of the Aral Sea and its adjacent zone;
- all the new cut mines and introductions;
- world achievements, which are useful for the region;
- international projects in the region;
- newly developed and produced equipment;
- legal provision of joint water resources on transboundary rivers on the basis of the world analysis;
- opinions on principle issues of water management development.

1.4 SIC ICWC publications. At present time, the regional center publicates and disseminates the following:

- ICWC billetins, which contain official documents in Russian (15 issues) and in English (5 issues);
- abstract journals with foreign publication included on problems of land reclamation and water management (4 issues);
- publications arranged according to subject, which include the most interesting information about world experience, achievements in international water cooperation, issues of water and land resources management and use in the world (4 issues);
- juridical publications, which contain water laws of Central Asian states, international legal issues on water resources, transboundary rivers, etc. (2 issues).

ICWC bulletin in English is circulated among 19 foreign organizations, which provide information for our IPTRID center on a free basis. At the same time, SIC ICWC publications in Russian are delivered to ICWC organizations in Central Asian states via SIC's sub-divisions and ministries-members of ICWC. SIC's sub-divisions make contracts on informational provision with provincial organizations of the ministries of water resources and agriculture, Goskompriroda (State Committee of Nature) as well as with research and design organizations not involved in SIC ICWC. Prices in contract deliveries are minimal, they cover costs of reproduction and publication.

1.5 The main disadvantages in this work are:

- irregular receipt of information outside;
- weak equipping;
- communication difficulties, absence of connection to Internet;
- financial difficulties in respect to purchase of publications in Russia, abroad and inside the country.

2. Development of information exchange

2.1 Information exchange is important for professionals and scientists specializing on irrigation, drainage and water management for creating ability to use advanced experience in all directions of activity and, first of all, for raising efficiency of deficit natural resources use: water and irrigated lands. It has not only an economic, but nature protective meaning as allow mankind to survive and satisfy growing needs in these resources without increase of volumes of their use and through demands management and saving.

2.2 From this point of view in central IPTRID network centers subject and purposeful selection and dissemination of information between regional and national centers should be organized in the following main directions:

- analysis of macroecomonic tendencies in the world and regions connected with dynamics of water and land resources use; forecasts of water consumption, irrigation and drainage development;
- levels of advanced models achieved in water and land resources use in different regions, including water productivity by various water users and consumers, and in turn, by different branches and economic sectors;
- advanced technologies, directed towards effective water saving, efficient use of irrigated lands, reduction of prices on water use products through keeping requirements of water saving; reduction of costs on O&M of irrigation and water systems;
- new types of industrial products in the interests of water management and drainage;
- development of "Capacity building" in advanced countries, particularly directed towards rise of management sustainability, including legal, institutional and economic improvement;
- development of main research works in the same directions;
- express-information about international conferences, workshops, training to be held;
- information about donor's programs, which can be used in the interests of the sector's
- research development in the;
- work experience on transboundary water sources.

2.3 Regional information exchange is organized both by regional and big national centers within countries or their different zones on a bilateral basis. On the one hand, they process information from central centers of IPTRID and disseminate it between their own information centers in different region's states or big country's zones according to their peculiarities, on the other hand, collect information by unified samples, formats and principles from these centers, aggregate it, translate and transfer to the IPTRID central centers in accordance with their technical orientation.

Such approach determines following directions of IPTRID network's regional centers activity:

a) preparation and dissemination in the region of systematic bulletins, which highlight official documents and actions carried out by regional organizations and national governments in respect to management, development and improvement of water resources and irrigated lands use (treaties, proceedings, official surveys, communique, etc.);

b) collection, processing, analysis and dissemination of outputs of research and design works in the region, information on progress within projects, carried out in the region both at expense of international donors and governments' own funds;
c) organization of target analysis of different projects and measures implemented in the region before and representing interest for their further use;
d) preparation of technical surveys of world experience and publications related to the region or its analogues;
e) provision of abstracts on editions published and received from foreign countries and abstract journals on water management, irrigation and drainage of the region's countries including selection of information from periodical press (newspapers and journals);
f) data collection and dissemination of knowledge on new industrial models related to the sector and on results of advanced technologies introduction in the region, their systematization and analysis in respect to the main task of the region's water bodies - water saving and raise of productivity in water resources use.

2.4 In this system of information exchange an important role is laid on national and zonal (in big countries) IPTRID centers. Particularly, they must establish a wide system of information points on places, which will be, on the one hand, creator of information mentioned in para 2.3 b, c, f, consumer of information according to para 2.3 a, c, d, e, f and, finally, customer of requirements for information, by para 2.2, which regional organizations will request from central IPTRID centers.

2.5 Information exchange between all active centers can be established on a free base under condition, that all participants of information networking from the bottom to the top regularly carry out obligations on submitting documentation. Such exchange should be based on contracts between system's sections: central centers with regional on a multilateral basis, regional with national on bilateral one and national with the lowest sections as the last.

Interrelations between regional and local organizations should be based on refunding expenses according to agreed prices, including costs of translation, postal delivery and reproduction.

Other organizations, which do not enter into the system's structure, can use its services in last turn.

2.6 In order to improve system of information exchange implementation of a range of measures is needed, proceeding from rise of technical orientation of transmitted information. Particularly, this should be taken into account due to complexity of works connected with need of bilateral translation from English (French, German) into language of users and vice versa.

For this purpose it is expedient to:

- coordinate themes of exchange between regional and national centers and central IPTRID networks depending on spheres of their interests;
- establish definite periodicity of exchange by information types and submission of publications (journals, newspapers and abstracts from books);
- provide preparation of all regional and, desirably, national centers to Internet;
- develop uniform scheme of information exchange by its various types;
- organize on a constant base workshops, training on information exchange;
- provide information centers with computers, modems and equipment for reproduction.

2.7 Specific issue contain inclusion into IPTRID network of information on modern program, model and system products relating to directions, which are the base of sectoral development. Distribution of these products should be strictly commercial. Thus, catalogue of working software should include price-lists and brief description of inputs, outputs, contents, needed data, principles of solution.

3. Organization of works on research development and advanced technologies achievements introduction within the IPTRID network

3.1 Taking into account, that funds in all developing countries are limited for development of own research in water management, irrigation and drainage, IPTRID's activity in this direction should be focused on priority tasks of introduction of new technique and advanced institutional methods, which have specific importance for each country or zone and, simultaneously, on the basis of world knowledge's data base and attraction of analogues should promote reduction of costs both on research and directly on advanced experience introduction.

3.2 IPTRID's activity on research development and advanced experience application in different regions, countries and zones can be built in two main types:

- solution of concrete problems or aspects of the program on inquiry of regions, countries and zone;
- develop system of appraisal of the state of advanced technologies introduction in the region, country or zone.

3.3 Solution of certain thematic issues on inquiry is made on the base of joint work of the region, country or zone with such organization within IPTRID central network system, which has most experience in this aspect. An example of such purposeful work is organized by SIC ICWC jointly with FAO (G. Wolter, M. Smith) elaboration of common water consumption norms in the region for establishment of water use strict limitation. Taking into account previous research works on water consumption rationing in different institutes and unified CROPWAT methodical base, both organizations developed work program, TOR, methods of work and search of donors. This allowed in half a year after request of SIC ICWC to FAO to hold workshop with all interested organizations, to consider their methods and implement this work on practice.

Verifying, adaptation and development of such type of work in the country is laid, mainly, upon local institutions with consultations, technical assistance and supervision of central IPTRID centers. Their activity can be organized using funds of different joint programs of NATO, European Union, UNDP and other international organizations and individual projects as well, in implementation of which proposed project can be useful.

3.4 It is expedient to elaborate and develop combined research program in the region or country on the base of "Regional (or national) water strategy", "Master plan of integrated water resources development in the basin" or macroeconomic program of agricultural development. According to above, priority directions and weak points in store of resorts, which has the region or country for their solution proceeding from local restrictions, are determined. In this case, the work should be organized, as it showed at Pic. 2, through consistent search of these solutions in the world data base or search of analogues in other countries of the region or zones with the same conditions. Using unified IPTRID's methods the following are being determined more definitely:

- the state of research and advanced experience achievements in the region (country, zone) to the present moment;

- specifying priorities and weak points;
- program expertise from scientific point of view;
- ability to remove these weak points using previous (regional or national) or world experience;
- necessity to adapt these methods to local conditions and program of adaptation;
- rendering the systematic assistance and search of financing for target research and implementation, if necessary, taking into account approaches and methods recommended by IPTRID's methodical center, , search of partners probably at inter-regional level;
- possibility of development of institutional structure and economic methods through innovatory approach to the more modern institutional forms of "Capacity building".

For instance, it will be expedient to involve IPTRID with its scientific and informational potential in the Aral Sea Basin Program (ASBP) planned now under the aegis of IFAS (International Fund for the Aral Sea Saving) with participation of GEF, the Netherlands government, European Union (TACIS) and other donors. IPTRID can conduct expertise on correctness of the chosen directions in technological progress, capacity building and economic methods on the base of available experience, information volume and scientific methods with attraction of a range of IPTRID Central Networks.

3.5 Role and potentialities of IPTRID are increasing due to its inclusion in the Global Water Partnership (GWP).
As known, GWP enunciated its role, mostly, in mobilization of donors' funds to the most urgent problems of the world water development. Its networks involves, besides IPTRID, IIMI, IWRA, WWC, Water consulting group and other organizations. Such conglomeration makes a steep extension in IPTRID's technical and practical expertise potentialities, determination of research priorities. Simultaneously, GWP, taking into account high authority of its constitutors, could mobilize not only attraction of ordinary grants, funds, but involvement of donors in IPTRID's local sub-programs by agreement, for instance, with European Union (DG-12), NATO, USAID, UNDP about allotment of target grants for research development and implementation to support other combined donor programs. Today these actions are carried out separately - for instance, DG-1 allocates funds for ASBP, and DG-12 allocates funds for research on the Aral Sea basin, which has amateurish and impractical nature, on example of Copernicus program.
At GWP level IPTRID can obtain target financing in such programs as Copernicus, «Partnership for Peace» with appropriate programs of development of the region, country in part of water management, irrigation and drainage.

3.6 Under such development of the system IPTRID's activity and its aid to donor organizations - the World Bank, ADB, EDB, OESF, CIDA, etc. - will be increased and will have practical orientation and simultaneously guarantee of certain effectiveness.

3.7 One more direction in IPTRID's activity, which was highlighted in Mr. Pearce's and Mr. Ponce's report, is unification of the world standards in part of water management, irrigation and drainage. Undoubtedly, each country due to its peculiarities and economic potentialities can adopt standards, which differ from the world ones, but, at least, this country will be guided by them as a desirable target. Particularly, this requirement is important for us because after collapse of USSR and turn of the world space to the quite open one attraction of foreign specialists requires unified approaches all over the world and standardization.

3.8 Proceeding form para 3.3 and para 3.4, IPTRID central network, and may be its future headquarters (FAO), face the task of elaboration of a range of directives, which should be as a common instrument for all IPTRID network:

- methods for appraisal of the state of research and scientific achievements implementation in countries and regions;
- methods for appraisal of weak points in national sectoral development and obstacles in their overcoming;
- methods for analysis of possibilities to remove bottle-neck in national development using available and developed experience;
- adaptation of existed in other regions or countries methods to given conditions;
- methods for Capacity building improvement.

Groundwater: An interdisciplinary challenge for continuing education and training on an European scale

Les eaux souterraines: un défi interdisciplinaire dans le domaine formation continue à une échelle européene

Helmig R.[1], Hinkelmann R.1, Troisi S.[2]
[1] Inst. für ComputerAnwendungen im Bauingenieurwesen, Technical University of Braunschweig, Germany
[2] Depart. di Difesa del Suolo, University of Calabria, Italy

Abstract
In this paper continuing education and training in the field of groundwater hydraulic engineering is discussed. Within the European Graduate School of Hydraulics (EGH), which is embedded in the International Association for Hydraulic Research (IAHR), a European network of continuing education short courses at research institutions and universities has been developed within the past few years. The interdisciplinary challenges in groundwater hydraulic engineering lead to a wide variety of different tasks. Especially groundwater pollution is a currently evolving field. Interdisciplinary educational goals and requirements are formulated. The contents and the educational concepts of three IAHR-EGH courses, which aim at a high specialization in a partial field of groundwater hydraulics, are explained more detailled. Finally, perspectives of continuing education and training are given. IAHR-EGH intends to develop coherent sets of coordinated courses for main topics which are based on an interdisciplinary education concept and which will benefit from an increasing use of new media.

Résumé
Dans cette publication il est question de formation continue dans le domaine d'ingénierie hydraulique des eaux souterraines. Dans l'Ecole Européene d'Hydraulique (European Graduate School of Hydraulics, EGH) faisant partie de l'Association Internationale des Recherche Hydraulique (AIRH), on a établi dans les dernières années un réseau européen de cours de formation continue liés aux instituts de recherches et aux universités. Les défis interdisciplinaires dans le secteur des sciences d'hydraulique des eaux souterraines conduisent à une grande variété des différentes tâches. Avant tout la pollution des eaux souterraines est actuellement un domaine de développement. On formule des objectives et demandes de formation à un niveau interdisciplinaires. On explique plus en détail les contenus et les concepts de formation de trois cours d' EGH visant à une grande spécialisation dans un secteur de l'hydraulique des eaux souterraines. A la fin on expose les perspectives de la formation continue. L'EGH vise à coordonner différents groupes de cours à propos de sujets principaux basant sur un concept de formation interdisciplinaire profitant d'un accroîssement d'application des nouveaux médias.

1. Introduction

Groundwater is one of the main sources for drinking water supply in Europe. Therefore, groundwater management, protection, and remediation are important for a sustainable development of our society. For coping with groundwater related problems, an increasing number of well qualified professionals is required. The intensive connection of groundwater with adjacent compartments of the terrestric and aquatic environment requires interdisciplinary continuing education and training concepts, which are normally not adequately considered in curricula of engineers and naturals scientists.
Groundwater is one of the main topics of the International Association for Hydraulic Research / European Graduate School of Hydraulics (IAHR-EGH), which is part of the European Thematic Network of Education and Training for Environment-Water (ETNET.ENVIRONMENT-WATER) and subject of a major activity of the IAHR Committee on Education and Professional Development (CEPD).

In this paper continuing education and training which have been developed for groundwater related problems is discussed. In chapter two the IAHR-EGH is briefly introduced. Interdisciplinary challenges in groundwater hydraulic engineering are explained in chapter three while chapter four depicts interdisciplinary educational goals and requirements for groundwater hydraulic engineering. In chapter five three IAHR-EGH courses, which the authors are involved in, are described. The paper closes with perspectives for continuing education.

2. European graduate school of hydraulics

A specific need for specialized training for research students was foreseen for Europe the universities of which, in contrast to North-American universities, did not offer any formal graduate program. Since 1991 the IAHR Committee on Continuing Education and Training has been pursuing the idea of a European Graduate School of Hydraulics. A network of continuing education short courses at research institutions and universities within Europe has been developed during the period between 1997 and 1999. The network is a first step towards formalizing the basic level of training for research in Europe. It is intended to fill an important gap and to lead to a better basic qualification of doctoral students in Europe. Currently more than 50 European partners are involved in IAHR-EGH.

EGH functions as an umbrella for the benefit of all partners, channeling and coordinating individual activities with a minimum of administration while maintaining a maximum of individual freedom and responsibility for the various course organizers and at the same time providing quality assurance for the activities. Irrespective of the coordination by EGH, the organizers are fully responsible for the course as far as content, finances, arrangements, and special announcements are concerned.

Courses within the EGH framework will normally be taught in English, and they will be directed at graduate students, post-graduates and professionals. In order to classify the courses they are differentiated into two categories: design-oriented courses directed mainly at MSc students and professionals from industry and research-oriented courses directed mainly at Ph.D.-students and staff from universities and research institutions.

The courses are divided into the following main topic groups:
- rivers and sediment transport
- hydraulic structures and hydropower
- groundwater
- environmental and urban hydraulics
- hydroinformatics and computational hydraulics

Successful participation, established by a final examination, will be acknowledged by an IAHR certificate. The European Credit Transfer System (ECTS) is implemented for the academic recognition of the courses.

For promotion and development of the program, a steering committee has been appointed by the member institutions. In the long run the various courses should complement each other in an appropriate manner, thus forming specialized segments of higher education oriented at the actual water-related problems. Further details are given in Lensing & Kobus (1997), Kobus & Lensing (1999) and http://www.uni-stuttgart.de/UNIuser/iws/IAHR/home.html.

3. Interdisciplinary challenges in groundwater hydraulic engineering

Hydraulic engineering encompasses a wide variety of tasks in planning, designing, operating, and maintaining projects and water resources systems. Groundwater flow includes both water quantity and water quality. Groundwater systems vary considerably due to the geological formation of the aquifer and are characterized by low velocities and extremely long exchange times. Across its open boundaries, groundwater can be polluted by point sources of contaminants or by diffusive sources like agricultural fields. Transport processes take place at the scale of the pore and grain size and are simultaneously effected by large-scale local heterogeneities of the geological formation as well as by adsorption, chemical reactions and microbiological activities. Engineers are involved in groundwater resources management (e.g. over-exploitation effects such as land subsidence or saltwater intrusion), groundwater uses (e.g. planning, design, and operation of water supply systems), and groundwater protection (e.g. pollution control and remediation schemes).

The pollution of groundwater (see fig. 1) is a currently evolving field, as it is among the most critical problems facing groundwater hydraulic engineers today. Possible sources include accidental releases of hazardous substances into the environment, like oil spills, releases of chemicals, or agricultural runoffs. Groundwater pollution due to industrial and agricultural activities requires enormous restoration efforts. Some current issues are the transport characteristics of contaminants, effects of chemical reactions and microbiological processes upon water quality, the development of effective water quality monitoring programs for specific sites, conjunctive use of surface water and groundwater, water quality control, and reduction of contaminant inputs from industrial and agricultural sources. Further information is found in Kobus et al. (1996).

Ex- and Infiltration of Contaminations

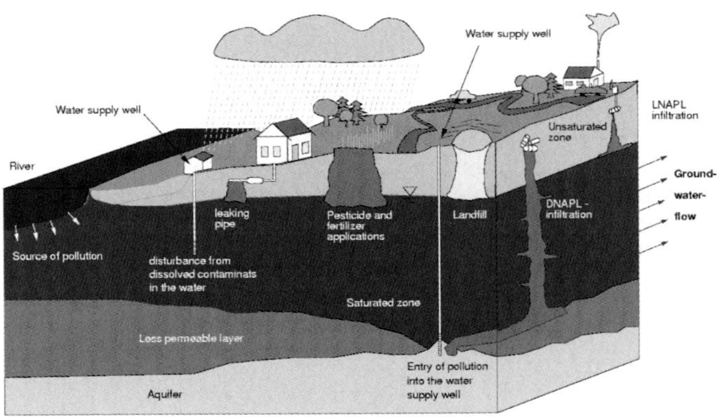

Figure 1: Infiltration and spreading processes in the subsurface

4. Interdisciplinary educational goals and requirements

A groundwater hydraulic engineer must be able to describe and quantify physical flow and transport processes in subsurface environmental systems, while considering chemical and biological processes. This technical-scientific field encompasses many basic and applied branches of science. His or her basic knowledge is fluid mechanics, especially hydromechanics, with a close relation to geology and soil mechanics. Computer sciences / hydroinformatics have gained more and more on importance in the past few years. In addition, the groundwater hydraulic engineer should be familiar with natural sciences (e.g. chemistry, biology) and mathematics / numerics. In a broader sense, there is a link to economics and social sciences.

Basic as well as continuing education in hydraulic engineering should be performed in three places: classroom, computer room and laboratory / field. The basic knowledge in fluid mechanics, geology, soil mechanics, numerics etc. is taught in the classroom. Computer exercises are essential for acquiring skills in the use of numerical models including pre- and postprocessing (e.g. mesh generation, CAD, visualization of results) as well as groundwater-related information systems (e.g. Geographical Information Systems (GIS)). Computers help to develop an understanding of physical processes, different algorithms and models. They are a good means to perform sensitivity analyses and to simulate variants of a given problem, i.e. to make predictions. In addition, they can be used to analyze experimental data. In combination with an information system, a comparison of physical experiments with numerical calculations is easily possible, and computers also enable to check the validity of the simplifying assumptions that are inherent in all numerical models. For learning and understanding fluid mechanics, some experience in the laboratory as well as in the field is required to become familiar with flow phenomena, experiments, measurements, and processing of experimental data. New developments in educational methodologies like teleteaching and telelearning open new horizons. Using e.g. video conferencing, a course program can easily be enlarged: international partners can be integrated giving lectures at their home departments. Moreover, laboratory / field experiments, which may be far away from the class or computer room, can be integrated. Telelearning is based on digital teaching material which can consist e.g. of movies showing numerical results and experiments or of recorded lectures. Telelearning is possible using the internet and can be done at home at any time. Using those new facilities is a challenge for the education of groundwater hydraulics and requires new concepts for education and training.

The working field of groundwater hydraulic engineers has fundamentally broadened in the past decades, and this development will continue in the future. Because of the wide range of tasks that fall within the interrelated fields of groundwater hydraulic engineering, no single curriculum can include all that may be required. Formal education must therefore include a variety of goals. Because the demands change as the working field develops, a comprehensive system of continuing education for lifelong learning is required to deepen, to widen, and to extend formal education. The universities should put emphasis on teaching basic and methodology knowledge. Lifelong learning can lead towards mainly becoming a generalist with a broad approach and background or towards mainly becoming a specialist with a detailed technical and engineering background.

5. Courses for continuing education and training on an international scale

Within the IAHR-EGH the authors are involved in three courses which have several features in common. They are research-oriented and deepen the knowledge in special fields of groundwater hydraulics (e.g. pollution) which have been gaining more and more importance in the past years. The courses focus on numerical simulation, teaching the physical and numerical basics in combination with computer exercises and a close link to laboratory and field measurements. The course teams consist of international experts not limited to a European scale.

5.1 Subsurface Modelling: Numerical Methods in Multiphase Flow and Transport Processes

Contamination of groundwater resources by hazardous substances that are immiscible with water (NAPLs - non aqueous phase liquids) has become an issue of increasing interest (see fig. 2). NAPL infiltration into subsurface systems (saturated and unsaturated zone) poses a serious long-term threat to groundwater systems. Complex multiphase systems exist where water, air, and NAPL phases occupy the pore space. Numerical models are the only means for quantifying processes such as infiltration, redistribution, and remobilization of NAPLs or for estimating the risk potential of contaminated sites. Knowledge of physical processes, mathematical formulations and assumptions is essential for the correct interpretation of the model's simulation results.

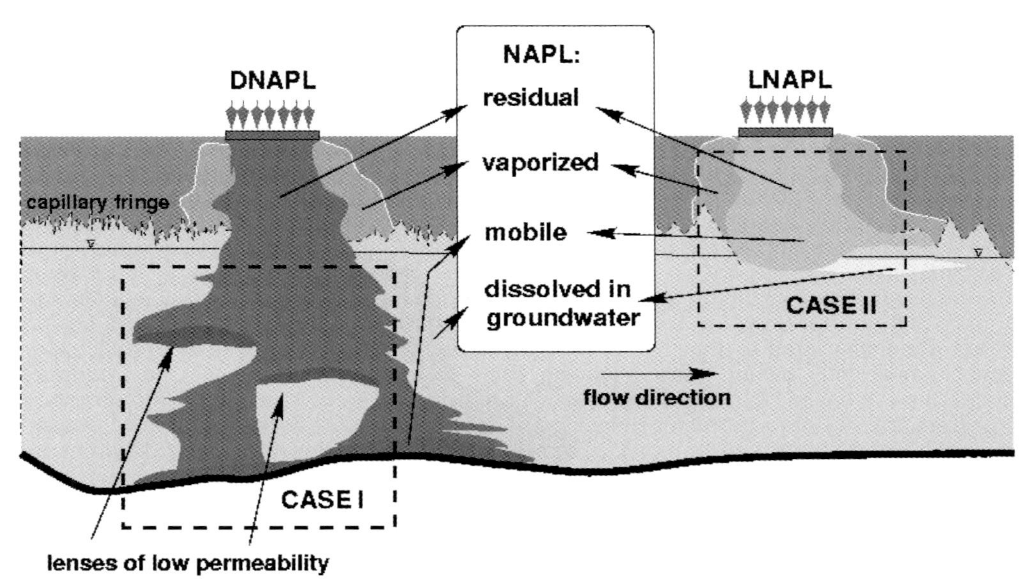

Figure 2: NAPL infiltration and remobilization in the subsurface

The short course provides an introduction to the physics of multiphase flow and transport processes in heterogeneous and fractured-porous media. The constitutive relationships of relative permeability and capillary pressure are discussed in detail, focussing on their influence on the behavior of multiphase flow systems. The governing flow and transport equations are derived, and their numerical solution by means of the Finite Element as well as Finite Volume Method is presented. The fundamentals of inverse modeling techniques for the determination of the hydrogeological properties are reviewed, providing additional insight into the interpretation and effect of the parameters used in numerical models. A number of illustrative sample problems are discussed and solved on computers. The participants are also invited to contribute problems from their own projects. The lecture notes (Helmig & Finsterle (1998)) as well as the book *Multiphase Flow and Transport Processes in the Subsurface* written by Helmig (1997) form the basis of the course and are handed out to the participants.

The short course is directed at Ph.D.-students and staff from research institutions. It has already been held six times (partially with a slightly different title). Since 1998 it has taken place at the Institut für Computer-Anwendungen im Bauingenieurwesen, Technical University of Braunschweig, Germany, in cooperation with the Versuchseinrichtung zur Grundwasser- und Altlastensanierung (VEGAS), Institut für Wasserbau, University of Stuttgart, Germany. In March 1998, about 30 participants from Europe, Asia, and Africa attended this short course.

5.2 Subsurface Modelling: Multiphase Flow, Transport and Bioremediation

Contamination of groundwater resources by hazardous substances that are immiscible with water (NAPLs - non aqueous phase liquids) has become an issue of increasing interest. NAPL infiltration into subsurface systems poses a serious long-term threat to groundwater systems. Since the remediation of contaminated sites can be expensive, a reliable capability for predicting the potential impact of a spill, for judging the prospect of passive remediation, and for evaluating the possibility of active remediation is essential. This capability requires knowledge of the physical, chemical, and biological processes (see fig. 3) as well as the

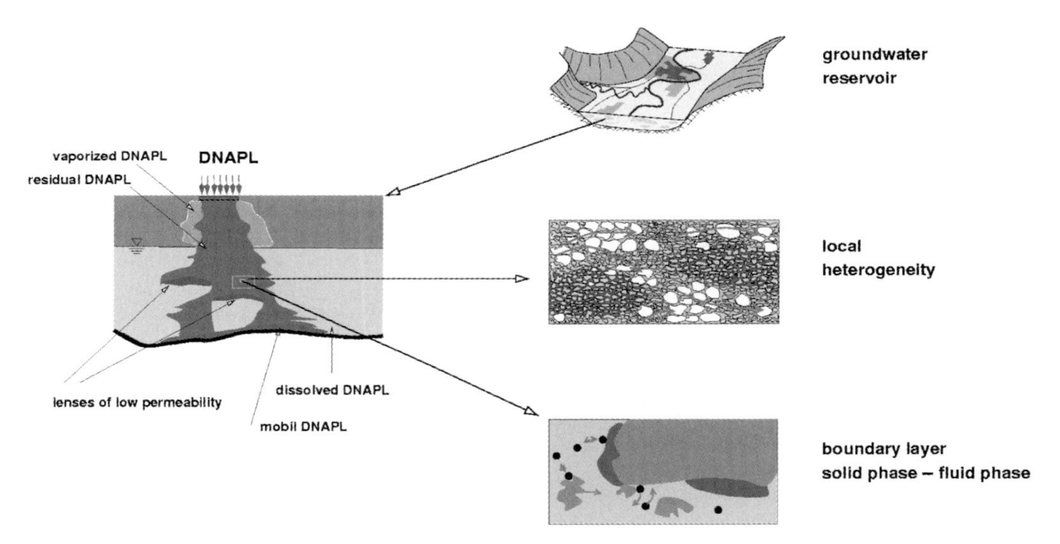

mathematical formulations and assumptions involved in a predictive modeling tool.
Figure 3: Scale dependent flow, transport and reaction processes in the subsurface

In the short course, fundamental principles governing the subsurface transport of dissolved and non aqueous phase organic contaminants are presented along with important microbial processes which govern the natural and enhanced attenuation of contaminant plumes via biotransformation. Transport / reaction concepts are illustrated using computational visualization methods with emphasis on how subsurface heterogeneities influence contaminant migration. Problem sets are assigned on key material (i.e. advection / dispersion, contaminant partitioning, surfactants, biotransformation kinetics, bioaugmentation, and natural attenuation) as well as discussed and solved on computers. The lecture notes (Cunningham et al. (1999)) form the basis of the course and are be handed out to the participants.

The short course is directed at Ph.D.-students and staff from research institutions. It has been given twice so far. Since 1998 it has been held at the Institut für ComputerAnwendungen im Bauingenieurwesen, Technical University of Braunschweig,

Germany, in cooperation with the Versuchseinrichtung zur Grundwasser- und Altlastensanierung (VEGAS), Institut für Wasserbau, University of Stuttgart, Germany. In March 1999, more than 30 participants from Germany, Austria, Italy, France, and some developing countries attended this short course.

5.3 Experimental data and validation of groundwater mathematical model for remediation of polluted sites

Contamination of groundwater systems has evolved as one of the major problems in environmental engineering in the last years. As remediation of polluted sites can be quite expensive, it is a practical use to develop special remediation techniques. The numerical simulation with groundwater mathematical models, which must be validated by experimental data, is

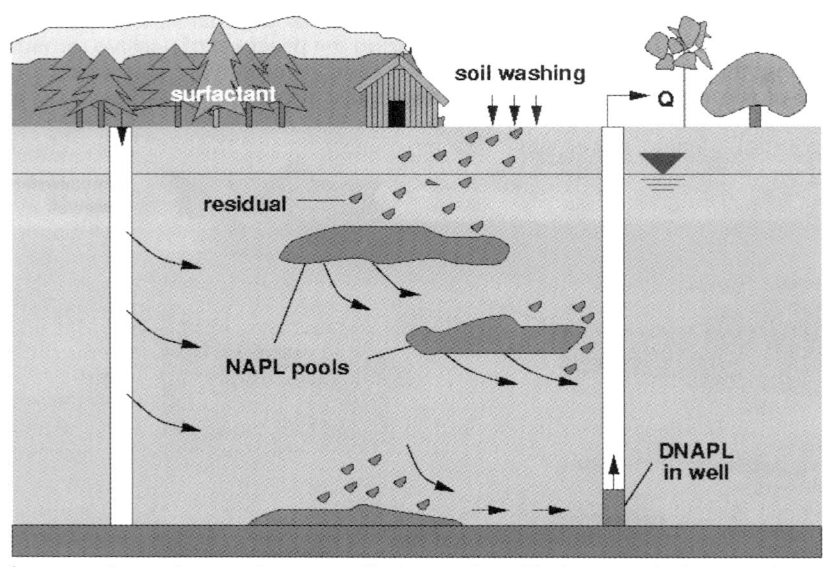

essential, because it can be used as a predictive tool and help to optimize certain measures.
Figure 4: A remediation technique for a polluted site in the subsurface

In this short course an introduction to modeling techniques for single and multiphase flow in the subsurface and in fractured media is given. A special focus is put on experimental data and model validation. Groundwater quality management and risk analysis are discussed in combination with advanced methodologies for the remediation of real aquifers, on the basis of the application of validated mathematical models and theoretical studies.

The lecture notes (Troisi et al. (1998)) form the basis of the course and are handed out to the participants. The short course is directed at Ph.D.-students and staff from research institutions. It was held for the first time in July 1999 at the Department of Difesa del Suolo, University of Calabria, Italy, in collaboration with EurAgEngSIG on Soil and Water and Ingegneria delle Risorse Idriche Sotterranee (IRIS). About 20 participants from Europe attended this short course.

6. Perspectives of continuing education and training

IAHR-EGH is going to integrate new courses within the next three-years' period and intends to develop coherent sets of coordinated courses for the main topics (see chapter

2) based on an interdisciplinary education concept. These different coordinated course programs should include MSc and PhD courses on a high academic level, fulfill the IAHR-EGH quality criteria, as well as consist of international teams of lecturers and an international audience. Additionally, the grouping of courses to summer schools and the integration of young scientist workshops – perhaps at annually changing locations - as well as the combination with teleteaching and telelearning methods can complete the programs. In the period posterior to the courses different groups of participants can continue their discussions by means of distributed learning, resulting in some positive influence on their own PhD and research work. For some of the (coordinated) courses it might be interesting to leave the European scale and cooperate with other international education programs in a worldwide and global sense. In the long run several segments of thematic linked and coordinated courses can serve as modules for an European or even an international degree of several universities and research institutions.

Acknowledgements
The authors would like to thank Corina Paland, Thomas Breiting and Gaby Hinkelmann for their fruitful discussions and the revision of the manuscript.

References

Cunningham, A., Ewing, R., Helmig, R. (1999) Subsurface Modelling: Multiphase Flow, Transport and Bioremediation, lecture notes of the IAHR-EGH short course, Insitut für ComputerAnwendungen im Bauingenieurwesen, Technische Universität Braunschweig, Germany.

Helmig, R. (1997) Multiphase Flow and Transport Processes in the Subsurface: A Contribution to the Modeling of Hydrosystems, Springer-Verlag, Berlin, Heidelberg, New York.

Helmig, R., Finsterle, S. (1998) Numerical Modelling: Multiphase Flow and Transport Processes in Subsurface Systems, lecture notes of the IAHR-EGH short course, Insitut für ComputerAnwendungen im Bauingenieurwesen, Technische Universität Braunschweig, Germany.

Kobus, H., Plate, E., Shen, H.W., Szöllösi-Nagy, A. (1996) Education of hydraulic engineers, Technical Documents in Hydrology, International Hydrological Programme, United Nations Educational, Scientific and Cultural Organization, UNESCO, Paris, France.

Kobus, H, Lensing, H.J. (1999) IAHR European Graduate School of Hydraulics: Towards a European Education and Training System, Proceedings of the International Symposium 'The Learning Society and the Water Environment', Paris, France.

Lensing, H.J., Kobus, H. (1997) IAHR European Graduate School of Hydraulics: Basic Network Concept, Development and Perspectives, Seminar on Continuing Education and Training, XXVII IAHR Congress, San Francisco, California.

Troisi,, S., Gambolati, G., Helmig, R., Giuliano, G., Di Luise, G., Tucciarelli, T., Vanderborght, J., Veselic, G., De Wrachien, D. (1998) Experimental data and validation of groundwater mathematical model for remediation of polluted sites, lecture notes of the IAHR-EGH short course, Department of 'Difesa del Suolo', University of Calabria, Italy.

Postgraduate education in European river engineering

Ervine D. A.[1], Armanini A.[2]
[1] University of Glasgow, Scotland
[2] Universita di Trento, Italy

Abstract
This paper advocates a pan-European structure for postgraduate education and training in the area of river engineering. Six key areas are proposed to form the core of the new structure. These are, (a) Catchment Hydrology and Flood warning (b) Analysis of floods and protection works, (c) Sediment behaviour in rivers, (d) River habitat and fisheries, (e) River water quality, and (f) The use of hydraulic structures in rivers. A co-ordinator will be appointed for each area above and will initiate a number of events and networks. Initiatives will include the development of distance learning courses, summer schools, as well as round table discussion groups.

In each case, courses and events will be targeted at two distinct audiences. The first of these is the Masters and PhD type at a research/academic level. The second is the more practically oriented continuing and life-long learning courses for those from industry. The scheme will form part of the European Graduate School of Hydraulics which is a network of more than 50 partners across Europe. The long-term ideal is the provision of a high quality, single European post-graduate experience in the field of rivers and their management.

1. Introduction

In view of the importance of rivers in the European context it is crucial that young researchers and designers in industry are exposed to the widest range of expertise available. For those at the start of a career, a summer school may provide a useful launchpad and a chance to meet leaders in the field. For those with busy lives, a distance learning course may provide a better option in terms of pace and convenience. Those at the cutting edge of a particular technology may benefit from a round table meeting with like-minded people. This applies to any branch of river engineering from flood estimation using principles of Hydrology, risk asessment associated with river floods, using 1-D, 2-D and 3-D numerical flood models, management of floodplain development, sediment transport in rivers with gravel beds or sand/silt beds, the latest numerical models for sediment transport, transport of pollutants in rivers, debris flows and torrents, water quality modelling in rivers, restoration of rivers, designing schemes to enhance the natural habitat conditions, as well as design of hydraulic structures to control rivers flows and to enhance the river environment.

The European Graduate School of Hydraulics (EGH) has laid the foundations for such a scheme. For the last 3 years it has delivered a number of short courses across Europe and has established a network of more than 50 active and passive partners in Universities and companies. It is now entering its second phase through the formation of a steering group and the establishment of five main areas of water education and training (Ref 1). One of the groups is concerned with European River Engineering the subject of this paper. The ideas expressed in this paper will also be the focus of an application for funding to the EU 5th Framework under the sub-heading of the Training and Mobility of Researchers (TMR). A number of trans-national partners is already in place, collaborating through the European Graduate School. In the river field this includes researchers from Portugal, Italy, UK, Sweden, Denmark, Netherlands and Germany.

2. Six sub-areas of the River Engineering programme

Technically and adminisratively it is convenient to sub-divide the river education and training programme into six areas. These are outlined in (a) to (f) below:

a) Catchment Hydrology and flood warning
River flooding is a natural and certain process to which mankind has responded in a number of ways, sometimes to the detriment of the river course. The prediction of the magnitude of river floods has exercised the minds of hydrologists for centuries and is now complicated by both urban development and the possibility of climate change. The latter predicts increased temperatures combined with higher wind speeds and more rainfall in many regions. Rising sea levels will exacerbate flooding problems in coastal regions. Predictions of flood discharge have traditionally concentrated on extreme value techniques or hydrograph analysis or by empirical estimates such as the Flood Studies Report in the UK. Tradition methods are questioned with new predictive methods proposed. Flood warning systems are making technological breakthroughs, and flood management strategies and hazard reduction systems need to keep pace. River Authorities require imaginative leadership particularly during flood crises when communication and information are at a premium. Significant human loss and economic damage are the result of poor systems. Improving the flood system at all levels is therefore the goal of the engineer. The level of post-graduate education across Europe therefore has an important role to play in ensuring that flood defence, flood warning and flood prediction are properly inderstood by the next generation. A co-ordinator will be appointed in this area to ensure this is a prime focus in the overall programme.

b) Analysis of floods and protection works
When flood magnitudes are known, the next step is the development of computer models to predict likely flood levels, as well as the influence of various flood defence works. Engineers in practice use one-dimensional steady and unsteady models such as HEC-RAS, MIKE 11 or ISIS. These are capable of being employed over the full catchment. Two and three dimensional models are also now being employed such as Telemac 2-d or 3-d but require enormous amounts of survey data and can be employed only over short river reaches. They are useful for investigating floodplain behaviour and in particular the influence of river defence measures on flood levels. In terms of flood defence, environmentally friendly strategies such as temporary storage ponds, distant floodbanks, two-stage channels and flood relief channels are in vogue.The importance of training engineers in the use of flood models as well as case studies on the best-practice for flood defence methods is clear. An understanding of the real role of floodplains and the effect of disconnecting them from a river by using levees and dikes is important. The specific problems of urban situations will also be addressed. The programmes intention is to have a co-ordinator for this area, instigating short courses, round tables and distance learning modules.

c) Sediment behaviour in rivers
Sediment transport in rivers is still a poorly understood subject. Numerical modelling of sediments in rivers is not well advanced. The first generation of sediment transport pre-

dictors for rivers has come to the end of its useful life. A new generation is in embryo state guided primarily by new breakthroughs in the physics of the flow phenomena, facilitated by the most recent instrumentation techniques. (Ref.3).

There are numerous schools of thought about how sediment transport rates should be calculated from flow parameters. The lack of agreement between researchers on this point is reflected in the poor predictions of transport rate and sediment entrainment in engineered and natural streams. Despite several decades of research by eminent workers, still no consensus on the best strategy for improving the accuracy of the equations is available to engineers.
At a general level there is an ongoing debate about which parameter has the strongest influence on sediment behaviour. While mean boundary shear stress is commonly used to predict sediment transport rate in rivers, stream power (Bagnold), unit stream power (Yang) and volumetric flow rate (Schocklitsch) have also been used. This debate will continue until there is more work conducted on the influence of turbulence on sediment transport. Yet intuitively it seems that turbulence and associated coherent structures must play a vital, perhaps dominant role in sediment transport. The main thrust in this area will therefore be based on understanding the role of turbulence in sediment transport, particularly related to non-uniform gravel bed rivers. The work will investigate detailed bed texture and armouring and its interaction with fluid turbulence, perhaps as a crucial stepping-stone to the next generation of transport equations and numerical models.

d) River habitat and fisheries
A key feature of this area will be its inter-disciplinary nature. River engineers of the future must have a good knowledge of the kinds of aquatic plants and vegetation on the river bed and banks. They must appreciate the kinds of invertebrate animals which exist on the river bed and the substrate, and the hydraulic conditions required for their existence. An appreciation of fish spawning processes in the river bed is essential, particularly for salmon and their spawning in gravel beds. The influence of pollutants on plant and animal life is a key part of this programme. Once these processes are better understood by engineers, then the means of real communicaion with environmental scientists, botanists and zoologists becomes a possibility. At present, communication between the disciplines is patchy particularly from engineers who are traditionally trained within the narrow confines of their discipline. Ideally the co-ordinator for this area would be an environmental scientist with an engineering background with capability for synthesising different disciplines and bringing together people of different backgrounds.

e) River water quality
One of the great benefits of the wider European project to date has been the EC directives on water quality which are being pursued by each country with some vigour. National efforts are important and are often achieved through River Environment Agencies with powers to prosecute polluters. As a result, rivers across Europe are becoming cleaner and fish are returning to once polluted rivers. The battle, however, is not yet won. The educational role of this programme through summer schools and distance learning modes will be to outline the latest EC directives and to promote skills related to river water quality measurements of oxygen demand, suspended solids, algae growth, the transport of pollutants such as heavy metals as well as the problem of agricultural runoff. Students should also be skilled in the use of river water quality models simulating the processes of diffusion and advection. The limitations of such models should be understood in the light of dead zones in rivers, sorption of pollutants into the bed material, and the lack of understanding of the physics of many of the processes associated with modelling pollutants

f) Hydraulic structures in rivers
Many rivers are regulated to control flood levels and flood magnitudes. This is usually achieved by sluice gates and weirs in the case of flood levels, and by temporary storage

basins in the case of flood peaks. There is a range of other structures including siphons, labyrinths, stepped weirs, fish passes, dropshafts, ogees, the latter related to dams in rivers which play an important role in the river environment. Other areas of interest include bridges, estimation of bridge afflux, structures designed to re-align secondary currents at bends, as well as low stone weirs and spur dykes to improve habitat conditions particularly for fish. It is important to teach the design principles for such structures also ensuring they are not harmful to the river environment. The passage of fish through fish pass structures is an important area as is the creation of pools and riffles.

3. A rational framework

Co-ordination of the six areas of river engineering noted above across the range of disciplines and across the continent of Europe (including Eastern Europe) requires both funding, and a team committment to this long term project. A possible framework is outlined in Fig.1.

The programme will be run by six co-ordinators, with each co-ordinator responsible for one of the six scientific programmes outlined in Fig. 1. Ideally it would be preferable to have each co-ordinator based in a different country to encourage a pan-European influence. Certainly no more than two co-ordinators would be allowed from any one country. The co-ordinators would form a steering group meeting twice per year, to ensure synthesis of the whole project and to promote activities in Journals and magazines.

An important role for each co-ordinator will be the organisation of the distance learning material for the sub-area. The material would be collected from invited experts, from the guest speakers at the summer schools and also round table events. Thus a major task will be selecting and collating all the material which would then be available on the World Wide Web for all participants.

A further role for each co-ordinator will be the organisation of a summer school each year which will be attended by those on distance learning mode and also students simply wishing to attend the summer school. Summer schools will alternate between academic and practical orientation, with academic courses essentially research based and practical courses based on application to practical problems. The latter therefore will focus on problem-based learning and case studies. In an ideal world it would be preferable to combine the two types of course, but experience has indicated a division is preferred by the majority of students.

4. Funding

A three year period is required to establish this programme in all six sub-areas. This is particularly true for distance learning modes which require preparation of study material and subsequent transfer to the Worldwide Web. Funding will be sought for a number of research assistants required for this work. A number of studentships are also required each year to allow attendance at the summer schools and to provide support to students on the distance learning mode. Funding will also be sought for staff travel and attendance at summer schools and for steering group meetings.

The intention is to submit an application to the EU 5th Framework during 1999. It is also intended to enlist the support of companies and environmental bodies in the river engineering field with a view to them providing delegates to the practice oriented courses which would then become self-supporting in terms of funding.

References

Van der Beken, A., Report of the second year of Etnet.Environment-Water, Brussels, 1998
Kobus, H., and Lensing, H.J. Minutes of the first steering group of the European Graduate School of Hydraulics. Stuttgart, Nov 1998
McEwan, I. " Thoughts on future sediment transport research" Private Communication, Dec 1998

IUPWARE : a postgraduate programme in Water Resources Engineering

IUPWARE : un programme de troisième cycle en Ingénierie des Ressources en Eau

Feyen J.[1], De Smedt F.[2], Batelaan O.[2], Raes D.[1], Berlamont J.[3], De Meester L.[4], Bauwens W.[2], Van der Beken A.[2]

[1] Institute for Land and Water Management, Katholieke Universiteit Leuven, Belgium

[2] Laboratory of Hydrology, Vrije Universiteit Brussel, Belgium

[3] Laboratory of Hydraulics, Katholieke Universiteit Leuven, Belgium

[4] Laboratory of Ecology and Aquaculture, Katholieke Universiteit Leuven, Belgium

Abstract

The Interuniversity Programme in Water Resources Engineering (IUPWARE) is a 2-year post-graduate programme leading to the degree of Master of Sciences in Water Resources Engineering. The programme was established in 1994 through the integration of two existing postgraduate programmes in Irrigation Engineering and Hydrology, respectively. The main objective of the post-graduate programme is to provide multidisciplinary and professional training in water resources engineering, and to equip future personnel with the necessary technical and managerial knowledge and skills that they will require to successfully design and operate water resources schemes. The programme mainly focuses on the training of young academicians and junior professional staff of the developing countries working in the field of water resources. IUPWARE is organised by the Katholieke Universiteit Leuven (K.U.Leuven) and the Vrije Universiteit Brussel (V.U.B.), in co-operation with lecturers from other universities in Belgium and overseas countries. The poster gives a summary of the aims, study programmes offered, admission requirement and application procedure, and a brief description of the organising institutes.

Résume

Le Programme Interuniversitaire en Ingénierie des Ressources en Eau est un programme de formation de troisième cycle sanctionné par le diplôme "Master of Sciences in Water Resources Engineering". Le programme a été créé en 1994 par l'intégration de deux programmes de troisième cycle existants, respectivement d'Ingénierie en Irrigation et d'Hydrologie. L'objectif principal du programme de troisième cycle est d'offrir une formation multidisciplinaire et professionnelle dans le domaine de l'ingénierie des ressources en eau, et d'équiper les futurs cadres avec les connaissances techniques et de gestion ainsi que les capacités indispensables pour concevoir et piloter des projets de ressources en eau. Le programme est orienté principalement vers la formation de jeunes universitaires et de jeunes professionnels issus de pays en développement et actifs dans le domaine des ressources en eau. IUPWARE est organisé conjointement par la "Katholieke Universiteit Leuven" (K.U.Leuven) et la "Vrije Universiteit Brussel" (V.U.B.) et en collaboration avec des intervenants d'autres universités belges et d'outre-mer. Le poster résume les objectifs, le programme d'études proposé, les conditions d'admission, la procédure de candidature et une brève description des institutions organisatrices.

1. History

IUPWARE arose in 1994-95 from the integration of two post-graduate programmes, namely Irrigation Engineering of the Katholieke Universiteit Leuven and Hydrology of the Vrije Universiteit Brussel. Both programmes have operated successfully since their formation in the early 80's. The advantages of integrating into a new interuniversity programme were: (i) to be able to offer an inter-disciplinary approach; (ii) to optimise the available training resources: and (iii) to offer more efficient and professional training. Through the amalgamation of these post-graduate programmes, a newly Interuniversity Programme for post-graduate training in Water Resources Engineering (IUPWARE) was born. Its role is not only to provide a long-term research oriented advanced study programme of MSc in Water Resources Engineering (WRE), but also to pool the existing expertise and to enhance parallel activities such as PhD and post-doctoral training, continuing education, overseas project development and consulting.

2. Aims and scope

IUPWARE aims at accomplishing the following two main goals: (i) the training of engineers and experts in specific disciplines of water resources such as surface and groundwater hydrology, irrigation, water quality management and aquatic ecology; and (ii) integration of technical multidisciplinary teams, efficiently co-ordinated by generalists with focus on management. IUPWARE puts emphasis on imparting knowledge and skills to trainees at three levels, namely, basic, specific and integrated. The first level includes disciplines such as hydrology, statistics, hydraulics, ecology, hydrogeology, economy and water quality, mathematics and computing. Specific disciplines refer to the various activities with water such as irrigation, water supplies, waste-water treatment, hydropower generation, hydraulic works, urban drainage, flood control, soil and water conservation and general use management. The last level relates to technical disciplines that treat water resources in an integrated way. In general these technical issues range from the operation research to the legal and institutional sectors. Subjects of social, institutional, political and communication areas form part of the study curriculum in anticipation to the growing tendency of public participation in the planning process. In addition, workshops on environmental and policy sciences are included to emphasis the close relationship between water quantity and water quality.

3. MSc study programme

The duration of the MSc programme is 2 academic years and the language of instruction English. The 1st year is called the Complementary Study Programme in WRE, and the 2nd year the Advanced Study Programme in WRE. In the following sections, details about the study programme of the 1st and 2nd year of the MSc in WRE are given.

1st year of the MSc in WRE
The 1st year programme 'Complementary Studies in Water Resources Engineering' offers basic courses and workshops. The programme has a total study load of 1,605 hours and consists of: a prerequisite course in Basic calculus; courses in Mathematical methods; Statistical methods; Land evaluation; Hydraulics; Surface hydrology; Groundwater hydrology; Irrigation agronomy; Aquatic ecology; and Water quality and treatment; and workshops in the field of Information technology; Hydrometry; Social, political and institution-

al aspects of water engineering; Environmental impact assessment, and Economic analysis of water resources projects. Table 1 gives a summary of the courses and workshops, including the total load in contact hours and credit points.

Table 1. Programme outline of the 1st year of the MSc in WRE

COURSES			
Subject	Th	Pr	Credit points
Basic calculus (pre-requisite)	-	30	-
Mathematical methods	30	30	5
Statistical methods	30	30	5
Land evaluation	30	30	5
Hydraulics	30	30	5
Surface hydrology	30	30	5
Groundwater hydrology	30	30	5
Irrigation agronomy	30	30	5
Aquatic ecology	30	30	5
Water quality and treatment	30	30	5
Subtotal	270	300	45

WORKSHOPS			
Subject	Th	Pr	Credit points
Information technology	-	60	3
Hydrometry	-	30	3
Social, political and institutional aspects of water engineering	-	30	3
Environmental impact assessment	-	30	3
Economic analysis of water resources projects	-	30	3
Subtotal	-	180	15

2nd year of the MSc in WRE

The 2nd year programme 'Advanced Studies in Water Resources Engineering' has a total study load of 1,665 hours and consists of a common core of 5 courses (Geographic information systems and remote sensing in water resources engineering; Advanced hydraulics; Water quality modelling; Systems approach to water management and Management of water use and re-use), an integrated project design and a Master

Table 2. Programme outline of the common core of the 2nd year of the MSc in WRE

Common core: COURSES/SEMINARS/ WORKSHOP/THESIS			
Subject	Th	Pr	Credit points
GIS & remote sensing in WRE	30	30	5
Advanced hydraulics	30	30	5
Water quality modelling	30	30	5
Systems approach to water management	30	30	5
Management of water use and re-use	30	30	5
Seminar	-	60	3
Integrated project design	-	90	5
Thesis research	-	120	17
Subtotal	150	420	50

of Science thesis. In addition, the student selects one of the 4 specific oriented options with 2 courses each in Hydrology (Surface water modelling, and Groundwater modelling); Irrigation (Irrigation engineering and technology, and Planning, operation and management of irrigation projects); Waste-water treatment (Hydraulics of waste-water collection and water supply, and Water and waste-water treatment); or Aquatic ecology (Monitoring of water quality, and Advanced aquatic ecology). The outline of the common core of courses is listed in Table 2.

Table 3 gives the outline of programme of the 4 options, hydrology, irrigation, water quality management and aquatic ecology, respectively.

Table 3. Programme outline of the options in the 2nd year of the MSc in WRE

Subject	Theory	Practice	Credit points
Option : HYDROLOGY			
Surface water modelling	30	30	5
Groundwater modelling	30	30	5
Subtotal	60	60	10
Option : IRRIGATION			
Irrigation engineering and technology	30	30	5
Planning, operation and management of irrigation projects	30	30	5
Subtotal	60	60	10
Option : WASTE-WATER TREATMENT			
Hydraulics of waste-water collection and water supply	30	30	5
Water and waste-water treatment	30	30	5
Subtotal	60	60	10
Option : AQUATIC ECOLOGY			
Monitoring of water quality	30	30	5
Advanced aquatic ecology	30	30	5
Subtotal	60	60	10

4. Admission requirements

The admission requirement for trainees enrolling in the 1st year are :
For overseas students : Bachelor of Science/Bachelor of Engineering/Bachelor of Technology in Agricultural or Applied Sciences of a 4-year university programme with minimum grade 2nd class honours, upper division and TOEFL (minimum score : 550) ; and
For students with a European or American university degree : Graduates from a 4 year university programme in sciences and graduates from a 4 year non-university degree in industrial engineering.

Students meeting the following requirements are allowed to enrol directly in the 2nd year:
For overseas students : Diploma of Complementary Studies in Water Resources Engineering, or Master of Science/Master of Engineering/Master of Technology degree in Agricultural or Civil Engineering (2-year programme) and TOEFL (minimum score : 550) ; and
For students with a European or American university degree : Graduates from a 5 year university programme in agriculture or engineering.

5. Certificates

After completion of the 1st year a diploma of Complementary Studies in Water Resources Engineering is awarded to successful candidates; and after successful completion of the 2nd year trainees obtain a diploma of Master of Science in Water Resources Engineering (option Hydrology, Irrigation, Waste-water treatment or Aquatic ecology).

6. Tuition fees

The tuition fee is function of the financial satus of the applicant. Foreign students having a grant pay an annual tuition fee of 80 EURO (equivalent to 3,200 BEF), whereas non-grantees pay 460 EURO (equivalent to 18,500 BEF).

7. Number of trainees

Figure 1 summarises the number of trainees that enrolled at the ICPs in Irrigation Engineering, Hydrology and Water Resources Engineering. As indicated above, the ICPs in Irrigation Engineering (K.U.Leuven) and Hydrology (V.U.B.) merged to the Interuniversity Programme in Water Resources Engineering (K.U.Leuven & V.U.B.) after the academic year 1993-1994. Therefore the graphical presentation of number of students/graduates was split into three sections, i.e., for the ICPs in Irrigation Engineering and Hydrology for the period 1981-1994 (top and middle section of Fig. 1), and the ICP in Water Resources Engineering for the period 1994-1999 (bottom section of Fig. 1). In Fig. 1, the light gray bars represent the number of trainees that enrolled, including also free students, doctoral students and the students enrolling within the frame of the European ERASMUS programme. The drak gray bars represent the number of graduates.

The first observation that can be made from Fig. 1 is the relative large difference in a given academic year between the number of students and the number of graduates. The foregoing can be explained as follows: the number of students for a given academic year is the total number of students registered at the registration office for the 1st and 2nd year, whereas the number of graduates refer only to the number of students that successfully completed the exams of the 2nd year of the MSc programme, i.e., the students receiving the degree of Master of Science in Water Resources Engineering. Further the free and doctoral students enrolling for a limited number of courses, with among them most of the ERASMUS students, are accounted as enrolling students but not as graduating MSc-students, because they do not finish the complete study programme. Over a period of 14 years (1981-1982 to 1993-1994) 441 students enrolled at K.U.Leuven for the ICP in Irrigation Engineering, of which 54% obtained the MSc-degree in Irrigation Engineering. Within the same period 448 students enrolled at the V.U.B. for the ICP in Hydrology, of whom 34% obtained the MSc-degree in Hydrology. The larger difference between the number of registered students and students obtaining the MSc-degree at the V.U.B. is due to the fact that the number of free and doctoral students at the V.U.B., following a limited number of courses, was considerably larger than at K.U.Leuven. Both programmes have provided training to 889 students in a period of 14 years, or an average of 63 students per year, and the degree of Master of Science in Irrigation Engineering (240) and Hydrology (152) was awarded to 392 trainees.

ICP in IRRIGATION ENGINEERING

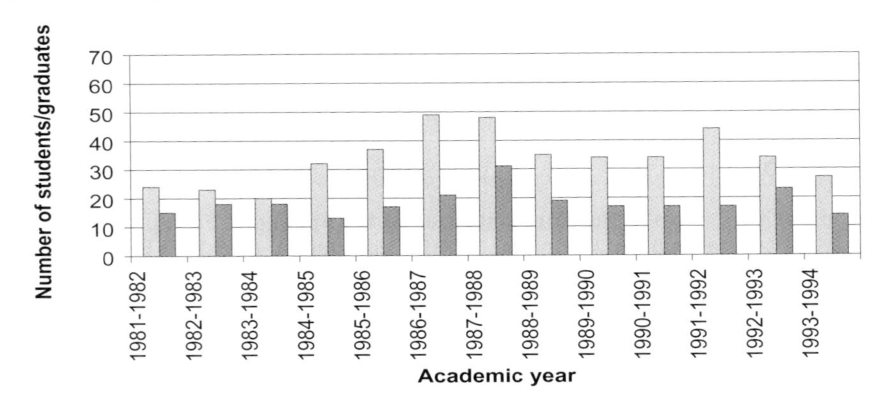

ICP in HYDROLOGY

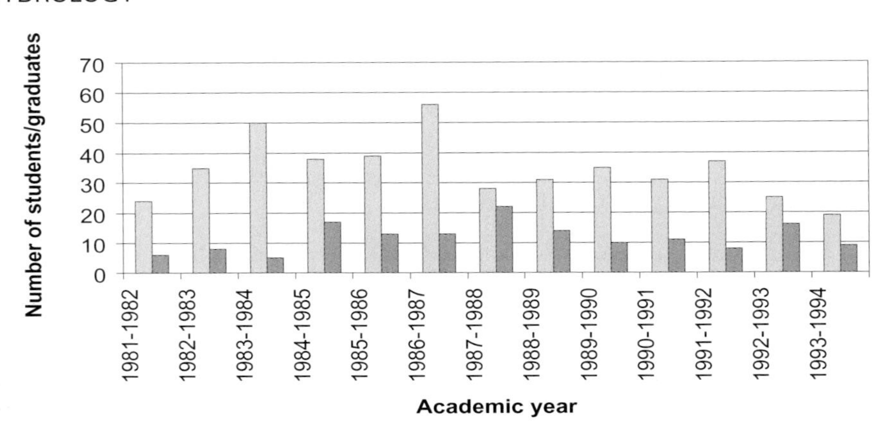

ICP in WATER RESOURCES ENGINEERING

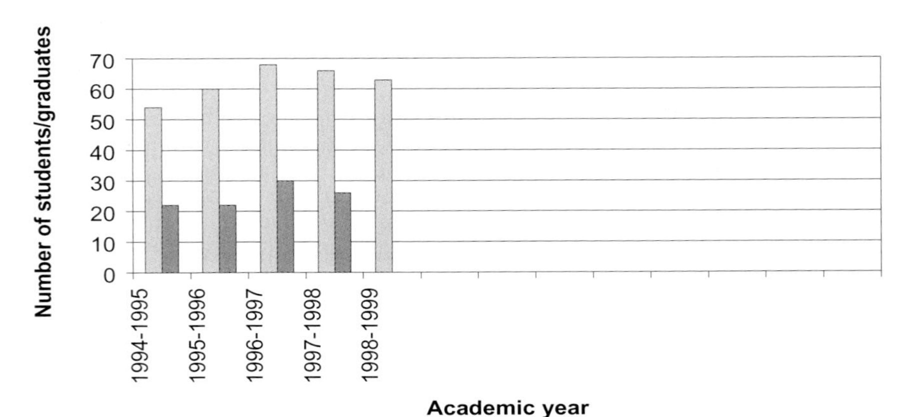

Figure 1. Evolution of the number of students/graduates in the ICPs in Irrigation Engineering, Hydrology and Water Resources Engineering

Figure 1 also reveals that the number of students can vary considerably, ranging for the ICP in Irrigation Engineering between 49 (academic year 1986-1987) and 23 (1982-1983), and for the ICP in Hydrology between 56 (academic year 1986-1987) and 19 (academic year 1993-1994). The data in the top and middle section of Fig. 1 clearly indicate the declining trend in number of students since the academic year 1991-1992. The foregoing was due to a reduction in grants for overseas students. This combined with the decline in the funding of both ICPs, and the need to anticipate to a changing environment and interest, urged the responsibles of both ICPs to integrate. The foregoing, as indicated in the bottom section of Fig. 1, resulted in a significant increase in number of students. The integration further resulted in a better functionality and professionalism, and a more cost effective exercise.

8. PhD-programme

Holders of a MSc/MEng/MTech degree in water related disciplines, with a minimum score of 70%, equivalent to 'Distinction' can apply for a PhD-programme at the K.U.Leuven or the V.U.B. The candidates are requested to follow the instructions for enrolment of each faculty/university and furthermore, they are subject to the by-laws of the PhD-programme of the faculty/university where they take enrolment. The registration fee is 200 EURO (7,800 BEF) for the first and last year of the PhD-programme, and 25 EURO (1,000 BEF) for all years in between.
The study programme of a doctoral student consists of course work and other study activities with a total load equal to at least 1,500 hours. The trainee's supervisor in consultation with the trainee and two other members of the academic staff design the programme. The latter form together with the supervisor or promoter, the supervising committee of the PhD-student. In general, the total length of the PhD-programme ranges from 3 to 4 years. Students have to submit yearly a progress report to the supervising committee for evaluation.

The main task of the PhD-programme is the conduct of a doctoral research project, the development of a PhD-dissertation and the public defence of the manuscript summarising the research findings. Upon the successful completion of the study programme and the public defence of the PhD-dissertation, the degree of Doctor in Applied Sciences, Doctor in Sciences or Doctor in Applied Biological Sciences is awarded. The doctoral degree is offered by the faculty/university where the doctoral student is enrolled. Participants in the PhD-programme have to enrol as students at the university where they carry out the research.

9. Continuing education

In addition to the long-term post-graduate course in Water Resources Engineering and the PhD-programme, IUPWARE and its supporting institutes/laboratories offer also tailor-made professional development and training programmes in the field of water resources. The content and the organisational costs of those training programmes are negotiated in direct co-operation with the client. Tailor-made training can, depending upon the request and the local capacity of the client, either be organised in Belgium or in the country of the requesting organisation/university.

IUPWARE is also active in several European Thematic Networks of Education and Training such as SOCRATES, ERASMUS, LEONARDO, TEMPUS, and the University Enterprise Training Partnership (UETP) entitled "TECHWARE" (TECHnology for Water Resources), and projects of distant-learning.

10. Teaching staff

Lecturing and assisting staff is mainly recruited from the Faculties of Engineering, Agricultural and Applied Biological Engineering and Sciences from the K.U.Leuven and from the Faculty of Engineering from the V.U.B. In addition lecturing staff is hired for specific fields from other Flemish Universities and from overseas leading institutions in the field of water resources (universities, research institutes, government agencies, consulting firms, etc.).

11. Management

IUPWARE is managed by a steering committee consisting of 4 members of each organising university, K.U.Leuven and V.U.B., respectively. Tasks of the steering committee comprise among others (i) definition of short, medium and long term objectives, (ii) organisation of the study programmes (1st and 2nd year) of the Master of Science in Water Resources Engineering, (iii) recruitment of visiting staff, (iv) financial management of the programme, (v) co-ordination of internal evaluation sessions, and (vi) development of national and international research projects and links with programmes of the European Community, international and national agencies, overseas universities and research institutions.

12. Supporting laboratories

Katholieke Universiteit Leuven

The *Institute for Land and Water Management* is a research unit of the Department of Land Management, and belongs to the Faculty of Agricultural and Applied Biological Sciences. Research in the Institute is focused on:

research in the field of soil geography, land evaluation, soil physics, modelling of the transfer and fate of nutrients and contaminants in soils and aquifers, soil and catchment hydrology, forestry, nature conservation and management, and landscape planning and ecology;

research in irrigation towards the development and evaluation of new knowledge required to efficiently manage water in irrigation schemes; and

research on image analysis, multi-scale integration of earth observation data and ground information (geographic information systems), operational monitoring tools (remote sensing), and decision support systems.

The *Laboratory of Hydraulics* is a research unit of the Department of Civil Engineering, of the Faculty of Engineering. Research is carried out in the following fields:

coastal and estuarine hydraulics: numerical modelling of tidal flow and storm surges, wave modelling;

urban hydrology: numerical modelling of flow in sewers, and of the impact of combined sewer overflows on the receiving waters;

pipe and canal networks; and

sediment mechanics: erosion sedimentation and transport of cohesive and non cohesive sediments in steady and unsteady flow conditions.

The *Laboratory of Ecology and Aquaculture* is a research unit of the Department of Biology, Faculty of Sciences. The research mainly focuses on fish and freshwater zooplankton and zoobenthos. The main lines of research are:

aquatic ecology: community analysis of fish and zooplankton;

population genetics of freshwater fishes, marine fishes and zooplankton;

assessment of water quality: the development of continuous biomonitors, including the development of rapid ecotoxicology tests; and

aquaculture: study of fish diseases, genetics of fish stocks and fish reproduction.

Vrije Universiteit Brussel

The *Laboratory of Hydrology* is a research unit of the Department of Civil Engineering of the Faculty of Applied Sciences. Main research activities can be summarised as:

the simulation and forecasting of the water balance of river basins;
the simulation and analysis of the processes controlling runoff and the quality of surface water;
the evaluation and design of storm drainage networks;
the assessment of the flow (quantity and quality) of regional groundwater reservoirs; and
the development of techniques to minimise the impact of waste disposal sites on the environment.

13. Conclusions

The poster highlights the different aspects (history, aims and scope of the programme, study programme, admission requirements, certificates, tuition fee, number of trainees, teaching staff, and management) of the Interuniversity Programme in Water Resources Engineering, being an international postgraduate programme primarily for trainees of the developing countries and organised by the Katholieke Universiteit Leuven (K.U.Leuven) and the Vrije Universiteit Brussel (V.U.B.). The present structure is the result of a continuous evolutionary process that started since the inception of the former ICPs in Irrigation Engineering and Hydrology. The programme is supported by the Belgian Administration for Development Co-operation (BADC) and the Flemish Interuniversity Council (VL.I.R.). In addition, the poster also briefly describes secondary activities, that are deployed within the frame of the MSc-programme, it are the possibilities for doctoral training in the field of water resources engineering at the K.U.Leuven and the V.U.B., and the activities of continuing education. In the final section of the poster, a short description is given of the main research activities of the supporting laboratories in the K.U.Leuven and the V.U.B. The latter carry not only the teaching load of the main bulk of courses in the 1st and 2nd year of the MSc in WRE, but provide also accommodation and assistance for MSc- and PhD-thesis research projects.

Acknowledgements
The authors like to thank the universities of K.U.Leuven and V.U.B. for providing the environment and infrastructure for the establishment of the Interuniversity Programme in Water Resources Engineering (IUPWARE). Also the financial support from the Belgian Administration for Development Co-operation (BADC), today provided through the Flemish Inrteruniversity Council (VL.I.R.), is very much appreciated, because without this financial support and the several external evaluations, IUPWARE would never have reached its present level of functionality and professionalism.

Report on the Course "A comparison among four different European biotic indexes (IBE, BBI, BMWP', RIVPACS) for river quality evaluation"

G. Flaim [*], G. Ziglio [**], M. Siligardi [*], F. Ciutti [*], C. Monauni [*], C. Cappelletti [*]

[*] Agricultural Institute, Trento, Italy

[**] Dept. Civil and Environmental Engineering, Univ. Trento, Italy

Abstract

As part of a European training course in water quality measurements, the biological quality of three rivers was evaluated using four different biotic indexes. The six-day course was structured to provide two days of refresher courses covering various aspects of river ecology, three days of sampling and laboratory work and a last day seminar. Every participant was able to utilize each of the four methods in the field and process samples in the laboratory. The thirty participants from 16 European countries were all associated with government, university or water authorities.

Résumé

Dans un Course Européen relatif à la qualité des eaux courantes, la qualité biologique de trois fleuves a été évalué avec quatre "biotic index ". Le course, duré six jours, prévoyait les deux premiers jours pour la mise au courant des différants aspects relatifs à l'écologie fluvial, trois jours pour l'échantillonnage en champ et l'analyse au laboratoire, le dernier jour pour le séminaire. Tous les participants ont eu la possibilité d'acquérir la pratique pour tous le quatre méthodes, sois au champ, sois au laboratoire.

Les vingt participantes, provenants de seize pays européens, étaient tous liés avec Organismes de l'Etat, Université, Autorités de Bassi.

1. Introduction

In the last decades, the importance of biological assessment methods for running water quality evaluation has been widely recognized (Woodiwiss, 1978; Ghetti, 1980). Twenty years ago in fact, the European Community organized three technical seminars for the intercalibration of several biotic assessment methods including Chandler's Biotic Score, the Saprobien System, the Verneaux-Tuffery method, and Woodiwiss's Extended Biotic Index. These seminars resulted in the publication *of Biological Methods for the Evaluation of Water Quality* (Ghetti & Bonazzi, 1980) and underlined in particular that:

a) a method for the biological classification of running waters was necessary to accurately define the ecological state of Europe's rivers;

b) biological assessment methods would have to become part of official monitoring procedures;

c) and methods based on the use of macroinvertebrates were the most suited to large scale application because they include numerous taxa with different pollution tolerance, are easily sampled and classified, are stable and are representative of a given tract of river.

Since then a vast array of biotic indexes all basically derived from the Trent Biotic Index (Woodiwiss, 1964) have been developed to assess the environmental quality of running waters. These methods are based on qualitative and semi-quantitative modifications in macroinvertebrate community composition brought on by pollution or significant alterations in the physical river environment.
TECHWARE, the main funding organization, is a non-competitive body whose mission is to be a bridge between national and European levels for universities, enterprises and public authorities. With this aim in mind, a comparison of four widely used biotic indexes was the subject of one of five courses in the "Water Quality Measurements" - European Training Courses co-ordinated by TECHWARE and approved by EC-DG XII Standards, Measurements & Testing. The methods (IBE – Italy, BBI - Belgium, BMWP' – Spain, RIVPACS – England) are the official biological assessment methods in their respective countries.
The aim of this course was:

a) to study the aquatic environment in a comprehensive way and to promote the use of biotic methods in evaluating the quality of surface waters.
b) to take the opportunity to stimulate a comparison among different procedures used by European countries, with a special emphasis towards those based on the qualitative and semi-qualitative examination of benthic communities.
c) to have an opportunity to confront some practical aspects and problems related to sampling and identification techniques.
d) to have an opportunity to exchange knowledge between teachers and participants of different European countries.
e) to provide an opportunity to discuss the similarities and contrasts between the methods and the applications of methods;
f) to provide a forum for potential standardisation of biological assessment methods of running waters on a European level.

2. Target audience

The course was targeted at the intermediate/advanced level, for personnel with monitoring responsibilities and/or working for water authority organisations in their countries. It was reasoned that these were the people who could most influence their local or national government on the use of biotic methods. Each participant was expected to have working experience in the use of macroinvertebrates for the quality assessment of running waters.
The large number of applications received (96) denoted the high esteem that the use of biotic indexes for water quality assessments have. Originally the course was geared towards 24 participants, but due to the high demand participants were increased to 31. They were chosen of the basis of their professional curriculum, affiliation and geographical origin. Personnel with monitoring responsibilities were given preference. Participants had the following geographical distribution: Belgium (2), Denmark (2), Finland (2), France (1), Germany (1), Greece (2), Hungary (1), Ireland (1), Italy (9), Netherlands (1), Romania (1), Russia (1), Poland (2), Scotland (1), Slovenia (2) and Spain (2). Most participants worked for government or research institutions and with a single exception, all were directly involved in water quality monitoring programs or were in positions to influence monitoring programs in their countries.

3. Course organization and structure

The course *Use of biotic indexes to evaluate the quality of freshwater streams: a comparison among four different European methods* (IBE, BBI, BMWP', RIVPACS) was organized locally by the University of Trento (Department of Civil and Environmental Engineering) and the Agricultural Institute of S. Michele all'Adige, Trento – I (Department of Natural and Environmental Research), were it was held (22-27 June 1998).

The course was entirely oriented towards exchanging knowledge on the use of biological indexes and acquiring confidence and skill capacity in the application of these biotic indexes to real situations (selected rivers). Course organization was based on the very successful IBE training course that is held at the Istituto Agrario of San Michele each year (Vittori et al. 1998). The 16th edition will be held in 1999 and has provided training to almost 200 technicians and researchers in the correct use of the IBE biotic index. Its success is based on the close contact of participants with instructors made possible by the very favourable instructor/student ratio. This type of course also encourages permanent professional contacts that are often difficult to develop for people who do not work in a research orientated environment.

Classroom work

The correct application of biological indices for the ecological surveillance of rivers requires an adequate formation in ecology, hydrobiology and taxonomy and a period of training with qualified personnel. With this in mind and in order to obtain a balanced learning process the course started with two days of refresher courses held by the instructors. The instructors, from major Universities and Research Institutes of Belgium, Finland, Italy, Spain and the UK, are all in the forefront in the field of biological quality evaluation in their respective countries. Each instructor provided reference material on methodologies and fluvial ecology. This aspect was especially appreciated by those participants from Eastern European countries.

The first day dealt with such topics as general characteristics of aquatic environments, river hydraulics and morphology, biological quality indexes and instream and bankside habitat assessment. The second day dealt more specifically with biological assessment methods considered in this course. Each instructor presented the method used in his country:

a) IBE - Extended Biotic Index – Italy,
b) BBI - Belgian Biotic Index,
c) BMWP' - Biological Monitoring Working Party – Spain,
d) RIVPACS - River Invertebrate Prediction and Classification System -UK.

The last day of the course was dedicated to European legislation on biological quality assessment of running waters and to a round table discussion of the methods utilized.

Field and laboratory work

The two days of classroom work was followed by three days of intensive "on the job training". Each participant was furnished with all of the necessary equipment for river sampling (with the exception of hip boots) and laboratory work. In addition participants were provided with access to taxonomic guides and a copy of the photographic macroinvertebrate guide used in Italy (Sansoni, 1988).

Three different rivers, one for each day, were chosen on the basis of varying ecological quality and physical habitat. Participants were divided into working groups each consisting of one instructor and tutor and five students. Each group was able to sample with two methods each day with instructors and tutors rotating in a round robin fashion. In this way participants were able to utilize four different methods. In particular, the practical application of the Extended Biotic Index used in Italy, based on sampling and identification of the macrobenthonic community and the subsequent classification of river quality, was developed and compared with the other three indexes (Table1).

Table 1. Rotation scheme for field and laboratory groups

Group	day 1	day 2	day 3
A	IBE - BBI	IBE- BMWP'	IBE – RIVPACS
B	BBI - IBE	BMWP' - IBE	RIVPACS - IBE
C	IBE - RIVPACS	IBE - BBI	IBE- BMWP'
D	RIVPACS - IBE	BBI - IBE	BMWP' - IBE
E	IBE- BMWP'	IBE - RIVPACS	IBE - BBI
F	BMWP' - IBE	RIVPACS - IBE	BBI - IBE

Mornings were dedicated to sampling and afternoons to laboratory processing of the samples and river assessment following the same order as morning sampling. Small group size and direct interaction both directly in the field and laboratory between participants and instructors offered the participants an unique opportunity to share ideas, problems and know how between experts and users.

Video
During the duration of the course, a 25-minute documentary was filmed. The video briefly describes the various phases of laboratory and fieldwork, in addition to a short overview of each method given by an instructor. This type of documentation is a very useful teaching instrument for those interested in biological water quality assessment methods currently used in Europe. Visual aids in fact, render more comprehensible the various phases of the assessment procedure both in the field and in the laboratory.

5. Course evaluation

Course evaluation forms distributed to both lecturers and participants, resulted in a very positive course assessment (Van den Berghe, 1998) The course was able to provide on the job training in the important area that is river quality assessment. It also produced a video useful for teaching biological assessment methods. An important achievement of the course were the professional contacts that were made between instructors and participants which were especially helpful for those who worked in isolated situations. Both instructors and participants felt that the course should be repeated. If this is possible, the organizers intend to include other European methods that rely more on quantitative methodologies, e.i, the Saprobien System, for a more ample comparison of biological quality assessment methods.
Most importantly however the course had the great advantage of underlining that all four biological methods studied came to essentially the same conclusion as to the biological quality of the three rivers sampled. This suggests that standarization of a single method for use in Europe is not necessary. The peculiarities in environmental differences and method development of each geographical area are to be respected. Where we feel that more effort is warranted is in how the results are presented. Number and significance of quality classes and their colour coding are different throughout Europe and readily lead to confusion. Standardizing the number of quality classes, their increasing or decreasing environmental value and their colour coding could lead in a very short time to a complete biological quality map of Europe's waterways.

References

Ghetti, P.F. & Bonazzi, G. (1980) Biological water assessment methods: 3rd Technical Seminar. Final Report. Commission of the European Communities.
Sansoni G. (1988) *Atlante per il riconoscimento dei macroinvertebrati dei corsi d'acqua italiani.* (Guide to the macroinvertebrates of Italian running waters) Provincia Autonoma di Trento. 191 pp.

Van den Berghe, W. (1995) *Achieving Quality in Training: European Guide for Collaborative Training Projects.* Tilkon bvba, Belgium.

Van den Berghe, W. (1998) Summary evaluation of the five courses of Techware Water Quality Measurement Course Series. Final Report Techware, Belgium 7 September 1998. 5 pp.

Vittori, A., Siligardi, M., Ciutti, F. (1998) L'attività formativa del corso sull'uso del metodo I.B.E. (Formative activity in the IBE method course) in G.N. Baldaccini & G.Sansoni (eds) *I biologi e l'ambiente oltre il duemila.* Venezia 22-23 nov. 1996 (in press)

Woodiwiss, F. S. (1964). The biological system of stream classification used by the Trent River Board. Chemistry and Industry, **14**, 443-447.

Woodiwiss, F. S. (1978) Comparative study of biological - ecological water quality assessment methods. Summary Report. Commission of the European communities.

Curriculum development of a traditional post-graduate course on hydrology

Gayer J., Jolánkai G, Bíró I.
VITUKI, Hungary

1. Introduction

Every country in the world aims at the sustainable development of its national economy. It is also generally recognised that aspects of the protection of the environment constitute a most significant limitation to that development, in fact the two must be harmonised in all their relations. The endeavour of achieving this goal has brought to life in recent decades the new discipline of Environmental Management, which in our days is inseparable from all plans for development. Water is - in one way or another - a decisive factor in all plans for development and at the same time a constituent of the environment. It is therefore of vital importance how we approach problems related to it. The abundance or scarcity and the quality of natural surface and subsurface waters affect almost all aspects of urbanisation, agriculture and industry, - in different ways under different climatic conditions and at different levels of development. Knowledge about the natural laws governing the occurrence and behaviour of these waters - enabling optimal, but environmental-friendly management of the same - is therefore of paramount importance for every nation in the world.

International co-operation for sharing experiences about the above mentioned subject has started some decades ago with special reference to hydrological sciences and with an emphasis on helping the developing nations. The International Hydrological Decade (IHD) launched by UNESCO in 1965 and the International Hydrological Programme (IHP) have been the driving forces of this co-operation. UNESCO's role of sponsoring international courses has gained world-wide significance, and it was within this framework that Hungary was among the first countries to start its courses under the title „Hydrological Methods for Developing Water Resources Management".

The other leading participant in this co-operation is the World Meteorological Organisation (WMO), the specialised agency of the United Nations for meteorology and operational hydrology. WMO has among its primary objectives to help member countries in developing human resources by providing professional and financial support in various forms for participants of training courses like the Budapest one.

2. The post-graduate course of VITUKI

Hungary, possessing great experiences - dating back to one and a half centuries - in carrying out extensive works of water resources development with due regard to both economic development and preserving the values of Nature, is ready to share these experiences, to disseminate the country's results and to contribute to the sustainable development world-wide. These courses are Hungary's contribution to the International Hydrological Programme (the continuation of IHD) by assisting developing nations and countries in transition by strengthening international co-operation in hydrology in general. From 1966 to 1974 the courses were held biannually, then from 1975 annually. During the 29 courses completed so far 452 participants from 76 countries have received training.

In the curriculum a three-level approach is followed to achieve this goal. First - as a brush-up - disciplines regarded as basic for the studies are reviewed. Second, a thorough analysis is presented of the hydrological processes encountered in the environment, how they are measured and monitored, and through which theoretical approaches they are described to achieve their practical evaluation. Third, the studies lead up to the investigation of the individual branches of water resources development, by applying the formerly gained knowledge to solving their problems as parts of sound environmental management. An assessment of the study results is provided for through tests, exams and an individually written Closing Paper. A Certificate of Attendance is finally awarded on the merit of all these. Subjects generally taught are: (1) Mathematics and computer applications, (2) Fluid mechanics, (3) Geosciences, (4) Hydrological processes, (5) Hydrological observations and primary data processing, (6) Hydrological analysis and modelling, (7) Water related environmental problems, (8) Applications of hydrology in water and environmental management.

In recent years the course has gained special importance from the viewpoint of the Central and Eastern European (CEE) countries because the water related environmental management has become of growing concern there and applications in an increasing number have been received from this region and especially from the associated countries to the European Union. These countries are at different stages of the accession process, but all of them need to adapt and adopt the EU legislation and requirements. One of the most important issues of this process is the environmental policy including water law. Taking into account the demands of these countries the curriculum of the 1999 course has been enlarged with a special workshop titled **Approximation of the EU environmental legislation on the field of water** presenting the EEC directives related to the aquatic environment and discussing the entailing tasks in the applicant countries.

Another special event of the 1999 course is the **International Conference on Participatory Processes in Water Management** to be organised by VITUKI with the attendance of the course's trainees. The basic legal framework for the exercise of rights to public participation has recently been introduced in Central and Eastern Europe and still missing in many developing countries. The limited experience of its use till now demonstrates that public participation in decisions and effective conflict resolution among those who have a stake in the decisions, are the a key components for efficient democratic process. The conference, therefore, will be an important occasion for the students to learn how to initiate a dialogue between public participation practitioners and the water management community, scientists and social scientists.

With the growing awareness of preserving the natural environment, the course's curriculum has recently gained environmental dimensions (complying with UNESCO's intentions). Water related environmental problems are taught in higher number of lecture hours, new topics have been inserted into the curriculum, the systems approach are emphasised in connection with environmental management. The title of the course has been changed accordingly to **Hydrological Methods of Environmental Management**. It is also envisaged that each year the curriculum will be updated and the programme will include special events, intensive short courses focussed on a special field or conferences dealing with certain aspects of water or environmental management.

In the new curriculum of the course a new main subject, the "**Water related environmental problems**" was introduced, consisting of six series of lectures on water quality and environmental management problems and the tools, procedures and strategies to solve them. The novel feature of the new lecture notes of nearly six hundred pages (edited by the authors of this poster) of this main subject is that it is associated with a **Computer aided learning** (CAL) software on the basics of river water quality modelling. The software efficiently aids teaching, with special regard to a series of practical exercises, in which latter the students are to solve water quality impact assessment problems. Main features and capabilities of this CAL programme are described briefly below.

3. Computer aided teaching of river water quality modelling

This computer aided learning software (CAL) has been prepared by the last two authors of this poster for UNESCO, as first part of an intended series, in the framework of the IHP-IV Project on the preparation of didactic materials in hydrology (CAL), to aid teachers and students both at university and post-graduate level in teaching and learning, respectively, the basis of river water quality modelling.

The software has been prepared on the basis of a booklet of the title "Hydrological, chemical and biological processes of contaminant transformation and transport in river and lake systems" written by the second author. It was published by UNESCO/IHP in the Series "Technical Documents in Hydrology" (1992, WS-93/WS.15). The users guide and the software was also published separately under the title "**Basic river water quality models. Computer aided learning (CAL) programme on water quality modelling; -WQMCAL version 1.1**" by UNESCO also in the Series "Technical Documents in Hydrology in 1997 (SC-97/WS/80) and distributed among potentially interested users. To the knowledge of the authors (on the basis of responses received) the programme is in fairly wide use at various universities.

The programme and software is fully user friendly and works in "windows" environment. The main feature of the software is that users can enter data (also by scroll bars) and see immediately the graphical and numerical results of altering inputs to or the properties of the aquatic system (see photograph). In this sense it is especially suited for problem solving exercises, simplified environmental impact assessments, such as the analysis of the impact of a sewage discharge and the design of the treatment efficiency needed to achieve water quality target levels in the recipient. Series of calculation examples to aid such practices were developed by the author (not published yet) and used for teaching the basics of designing control measures. The users can also set water quality class and alarm levels (criteria and standard values).

The software consists of the following main parts:

1. General theoretical background of modelling water quality. This module contains text, figures and equations and derives the basic differential equation of water quality modelling on the basis of the continuity and conservation of mass principles, deriving further certain simple model equations of coupled processes.

2. The second menu block is dealing with the models of the oxygen household. Coupled models of the Biochemical Oxygen Demand (BOD, the measure of biodegradable organic matter present in the water) and Dissolved Oxygen are presented here. They are called BOD-DO models of which three versions are included in the programme;
- The traditional "oxygen sag" curve model, which considers two major processes: the decomposition of organic matter and the reaeration of the river. Screens of the "practice" sub-menu of this block are shown in **Figure 1 and Figure 2**, also showing the oxygen curve of the recipient stream in the case of a sewage discharge (Figure 1.) and in that after sewage treatment (Figure 2.).
- A more sophisticated version which includes further three processes; 1. the sedimentation of biodegradable organic matter, 2. the benthic oxygen demand (diffuse source of BOD) and 3. an internal oxygen source (to account for photosynthesis).
- A second expanded BOD-DO model (sub-model of model system SENSMOD, developed by the same authors) which, in addition to the above processes includes diffuse loads of BOD and DO to the river, arriving via lateral inflow.

3. Dispersion-advection models, which include two sub-model menu blocks:
- A one-dimensional pollutant-spill model version which describes the propagation, downstream travelling, of pollution "waves". This model is especially suitable for analysing the data of accidental pollution events and forecasting the level of expectable pollution for downstream water users. Screen capture of **Figure 3** shows such an event in the form of concentration-time "waves" of different downstream sections. Animation (moving graphs) of the downstream travelling pollution wave can also be triggered on by the user.
- A two-dimensional (2D) transversal mixing model, which simulates the pollution plume downstream of effluent discharges with 10 concentration distribution curves across the river, each of which corresponds to a given downstream distance. The user can "highlight" the curve of interest (the distance of interest) and then the numerical value of the maximum calculated concentration also appears on the curve. The distance of the source from the riverbank can also be altered by the user. Effects of this latter (Y0) can be seen from the comparison of the two captured screens of Figures **4 and 5**.

4. This block contains written advises to teachers/users for creating numerical examples for the use of the above model blocks.

5. The fifth block is a test of 10 different questions related to basic knowledge on river water quality modelling. The user can select from 3-4 different answers and will receive information whether the answer was correct or not.

6. In this menu block the user can select (alter) water quality criteria for different (five) water quality classes of BOD and DO (while the values of the Hungarian standard are the default ones). The user can also alter the "alarm level" for the pollutant spill model and the criteria for the transversal mixing model. The respective screens of these model blocks will show this criteria for each example.

The authors wish to state that no existing, commercially available river water quality software has been utilised for writing this programme. The only exception is the BOD-DO submodel of the model system SENSMOD, which has also been developed by the same authors. This means, that the software is a genuine product, involving no copyright matters whatsoever and that all property rights of this material and programme stays with the authors and UNESCO. **Limited copies of the software will be freely available at the conference** and the authors expect the responses of the users.

It is to be emphasised that the software and the models are not intended for use in practical work (design, water pollution control planning, environmental impact assessment, etc), neither in the present nor in the final form, and serve solely for teaching purposes. Therefore the authors wish to state that they do not assume any responsibility for failures, faults or damages caused by such non-intended use of the software and the programme.

This programme and software is intended to be the first part of a series of similar CALs of which the likely next one will be dealing with the basics of lake water quality modelling, with special regard to plant nutrient budgets and eutrophication. The eco-hydrological approach, in which interactions of hydrology and ecology are handled with special care, will be emphasised in this second part of the software.

4. Conclusions

The world-widely-recognised necessity of utilising water resources in harmony with environmental conservation implies the importance of such courses and encourages the organisation of them. It can be stated that the continuous development of the curriculum by applying the latest achievements of the field is a must in order to manage water resources properly in the future. The dissemination of research results plays and important role in this endeavour. It is hoped that VITUKI's course can contribute to the transfer of knowledge required to maintain a sustainable development.

References

Jolánkai G. and Gayer J. (ed.) (1977) Water related environmental problems. Lecture notes, VITUKI

Jolánkai G. (1992) Hydrological, Chemical and Biological Processes of Contaminant Transformation and Transport in River and Lake Systems, UNESCO series; Technical Documents in Hydrology. WS-93/WS.15. UNESCO, Paris, p. 147.

Jolánkai G., Bíró I. (1997) Basic River Water Quality Models; Computer aided learning (CAL) programme on water quality modelling (WQMCAL version 1.1), UNESCO/IHP-V, Technical Documents in Hydrology, No. 13, SC-97/WS/80, UNESCO Paris p. 52 + software

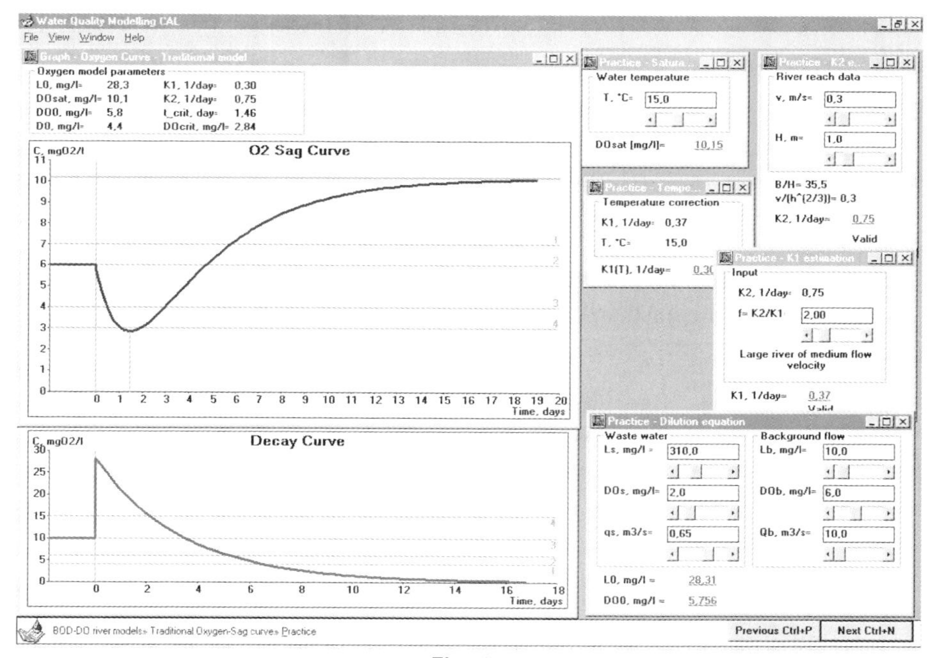

Fig. 1

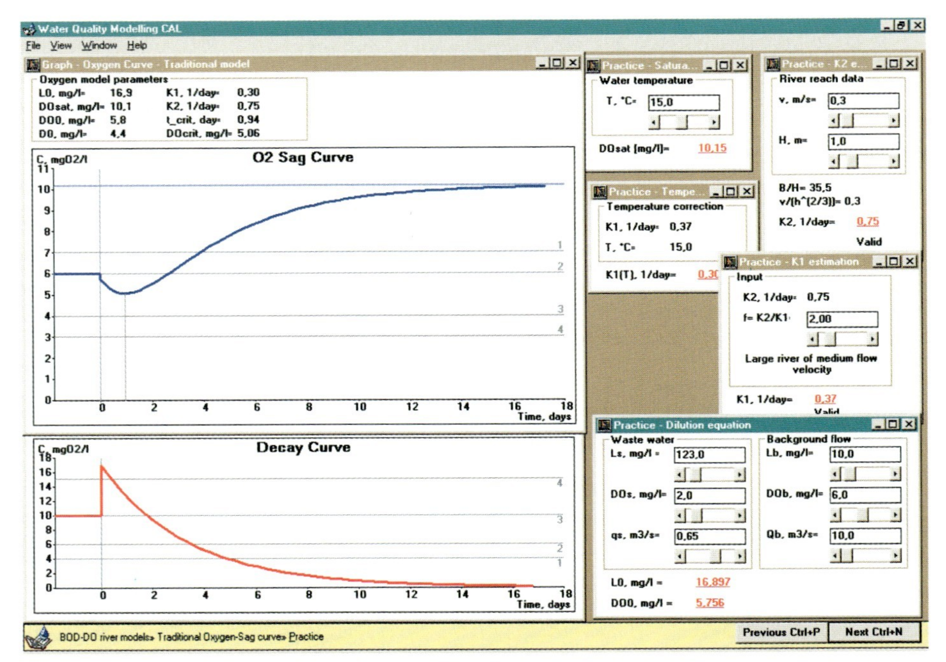

Water Quality Modelling CAL
File View Window Help
Graph - Oxygen Curve - Traditional model
Oxygen model parameters
L0, mg/l= 16,9
DOsat, mg/l= 10,1
DO0, mg/l= 5,8
D0, mg/l= 4,4
K1, 1/day= 0,30
K2, 1/day= 0,75
t_crit, day= 0,94
DOcrit, mg/l= 5,06
O2 Sag Curve
C, mgO2/l
Time, days
Decay Curve
C, mgO2/l
Time, days
Practice - Satura...
Water temperature
T, °C= 15,0
DOsat [mg/l]= 10,15
Practice - K2 e...
River reach data
v, m/s= 0,3
H, m= 1,0
B/H= 35,5
v/(h^(2/3))= 0,3
K2, 1/day= 0,75
Valid
Practice - Tempe...
Temperature correction
K1, 1/day= 0,37
T, °C= 15,0
K1(T), 1/day=
Practice - K1 estimation
Input
K2, 1/day= 0,75
f= K2/K1= 2,00
Large river of medium flow velocity
K1, 1/day= 0,37
Practice - Dilution equation
Waste water
Ls, mg/l = 123,0
DOs, mg/l= 2,0
qs, m3/s= 0,65
Background flow
Lb, mg/l= 10,0
DOb, mg/l= 6,0
Qb, m3/s= 10,0
L0, mg/l = 16,897
DO0, mg/l = 5,756
BOD-DO river models» Traditional Oxygen-Sag curve» Practice
Previous Ctrl+P
Next Ctrl+N

Fig. 2

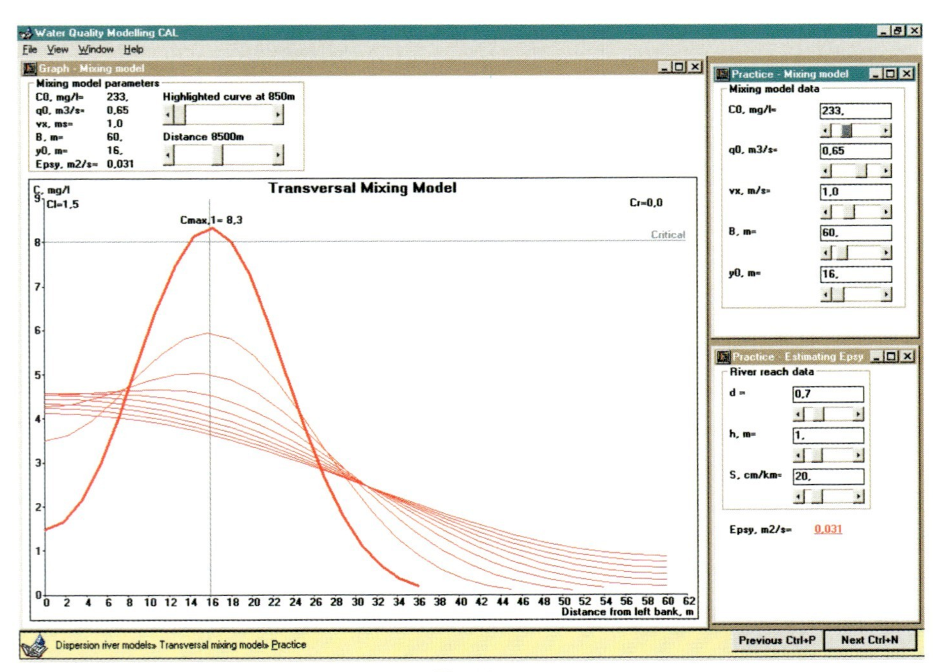

Water Quality Modelling CAL
File View Window Help
Graph - Mixing model
Mixing model parameters
C0, mg/l= 233,
q0, m3/s= 0,65
vx, ms= 1,0
B, m= 60,
y0, m= 16,
Epsy, m2/s= 0,031
Highlighted curve at 850m
Distance 8500m
Transversal Mixing Model
C, mg/l
Cl=1,5
Cmax,1= 8,3
Cr=0,0
Critical
Distance from left bank, m
Practice - Mixing model
Mixing model data
C0, mg/l= 233,
q0, m3/s= 0,65
vx, m/s= 1,0
B, m= 60,
y0, m= 16,
Practice - Estimating Epsy
River reach data
d = 0,7
h, m= 1,
S, cm/km= 20,
Epsy, m2/s= 0,031
Dispersion river models» Transversal mixing model» Practice
Previous Ctrl+P
Next Ctrl+N

Fig. 3

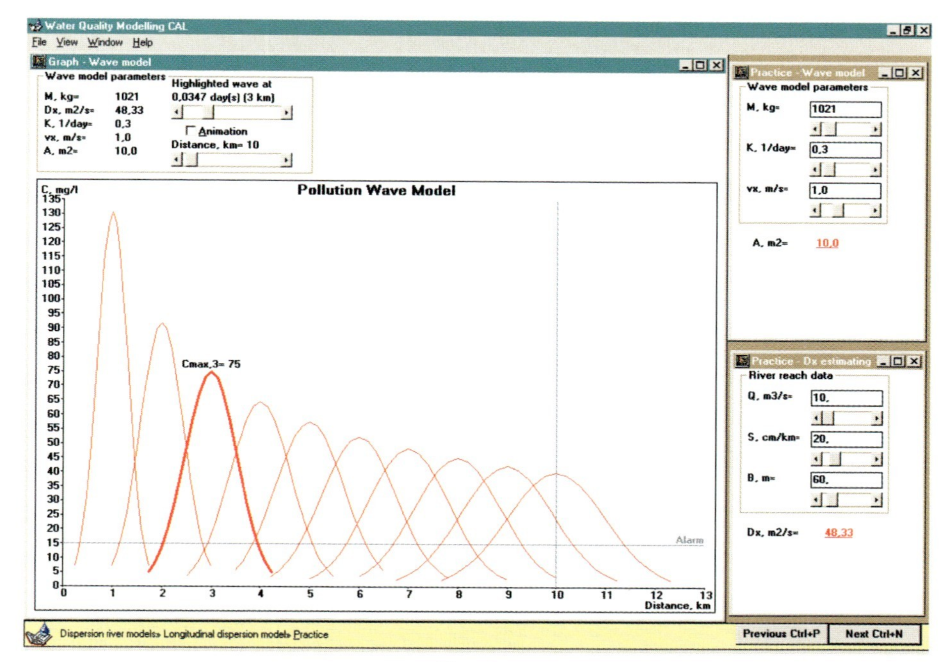
Water Quality Modelling CAL
File View Window Help
Graph - Wave model
Wave model parameters
M, kg= 1021
Dx, m2/s= 48,33
K, 1/day= 0,3
vx, m/s= 1,0
A, m2= 10,0
Highlighted wave at
0,0347 day(s) (3 km)
Animation
Distance, km= 10
Pollution Wave Model
C, mg/l
Cmax,3= 75
Alarm
Distance, km
Practice - Wave model
Wave model parameters
M, kg= 1021
K, 1/day= 0,3
vx, m/s= 1,0
A, m2= 10,0
Practice - Dx estimating
River reach data
Q, m3/s= 10,
S, cm/km= 20,
B, m= 60,
Dx, m2/s= 48,33
Dispersion river models» Longitudinal dispersion model» Practice
Previous Ctrl+P
Next Ctrl+N

Fig. 4

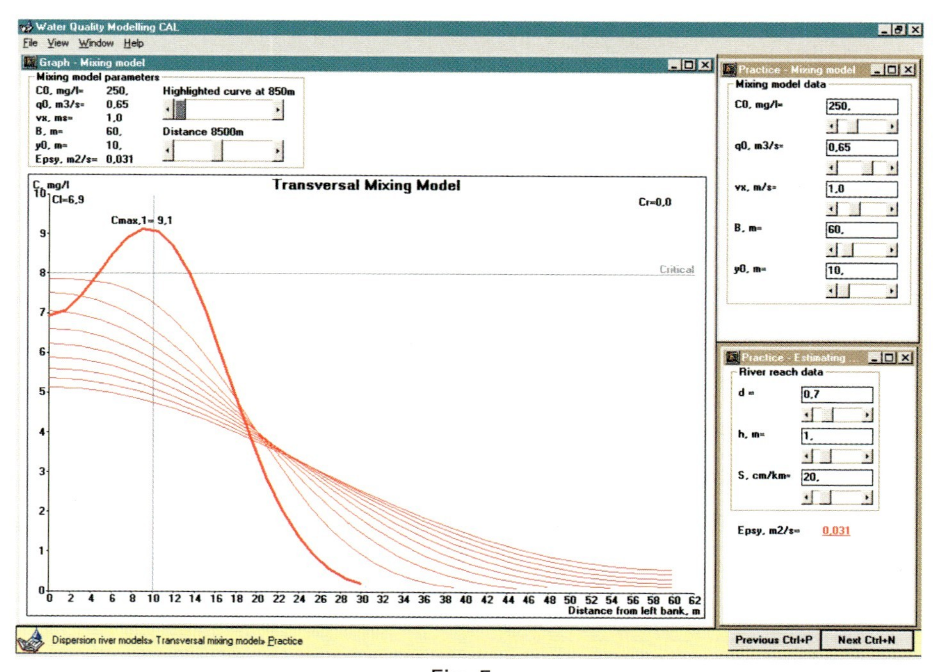
Water Quality Modelling CAL
File View Window Help
Graph - Mixing model
Mixing model parameters
C0, mg/l= 250,
q0, m3/s= 0,65
vx, ms= 1,0
B, m= 60,
y0, m= 10,
Epsy, m2/s= 0,031
Highlighted curve at 850m
Distance 8500m
Transversal Mixing Model
C, mg/l
Cl=6,9
Cr=0,0
Cmax,1= 9,1
Critical
Distance from left bank, m
Practice - Mixing model
Mixing model data
C0, mg/l= 250,
q0, m3/s= 0,65
vx, m/s= 1,0
B, m= 60,
y0, m= 10,
Practice - Estimating ...
River reach data
d = 0,7
h, m= 1,
S, cm/km= 20,
Epsy, m2/s= 0,031
Dispersion river models» Transversal mixing model» Practice
Previous Ctrl+P
Next Ctrl+N

Fig. 5

Interactive educational materials to popularize water issues at a global scale

Ladel J., Thibault J.
Association Graine de Chimiste -
Université Pierre et Marie Curie, France

Abstract
Consumers, trainers and decision makers request organized scientific bases which will allow them to interpret media information and to understand the field of the stakes related to water resources and their use. Educational materials realized by the 'Association Graine de Chimiste' with the aim to provide them with these bases, and in particular the itinerant interactive posters 'Planet Water at Home' are presented herein as well as remarks based on our field experience.
Key Words : Science Education, Water, Educational Materials

Résumé
Consommateurs, formateurs et décideurs ont besoin de bases organisées qui leur permettront d'interpréter les informations médiatiques et de comprendre l'ampleur des enjeux liés aux ressources en eau et à leur usage. Les matériels pédagogiques réalisés par l'Association Graine de Chimiste dans le but de leur fournir ces bases, et notamment l'exposition itinérante interactive "L'eau de la planète à la maison", sont présentés ici ainsi que des réflexions tirées de notre expérience de terrain.
***Mots Clés** : Education Scientifique, Eau, Matériels pédagogiques*

1. Objectives and means of education about water issues

The complexity of environment issues, and water in particular, requires an interdisciplinary and global approach. Most people have no means to decipher the abundant available information and to fulfil their responsibilities as planet citizen, regarding natural resources protection. There is a gap between specialists of water issues, information transfer professionals (mostly media and pedagogy professionals) and the public.

Only few scientists [6,13,24] succeeded in popularizing water issues as the interdisciplinarity required is large and still not usual in the scientific community. Water is not only the central interest for hydrologists, but also for hydrogeologists, climatologists, glaciologists, oceanologists, biologists, chemists, physicians, politicians, economists, etc. On the opposite, pedagogues do not always master specialized terms and complex notions used 'naturally' by specialists [26].

The challenge for education on the environment is to focus on that transfer of knowledge so that the Learning Society can provide mediator persons capable of understanding specialized works and papers and to talk with experts and of transmitting pre-requisites and essential information by adapting them to different public targets of consumers, trainers or even decision makers. Certain educational materials and books presently available provide such an overview on environmental issues [4,5,9,10,11,12,14,17,23,25].

2. An exhibition entitled 'L'eau de la planète à la maison'

An interactive and itinerant exhibition entitled '*Planet water at home*' was designed in 1997, for a French public older than 10 years by Dr Julie LADEL and Jérôme THIBAULT from the '*Association Graine de Chimiste*' [8,19] which received in 1995 the 'Scientific and Technical Culture Award' from the French Ministry of Higher Education and Research.

Both designers are scientific lecturers who have gained a large experience with the Association in transmitting knowledge and know-how to children since 5 years old, teenagers, students, adults, teachers and professionals. One of the designers is a water quality specialist, has experienced several fields of water sciences, the other one did not know anything about water. They decided to consider the theme of water by both its quantitative and qualitative aspects at a global scale as the management of natural resources is a world-wide shared responsibility and in order to initiate environmentally sound citizen behaviors [15].
Consistency, reliability and accuracy of certain scientific information were problematic, mostly for world water resources references. Finally, the international community publications [20,21(updated by 27)] were used to overcome this and experts from different fields kindly reviewed the draft.

The exhibition is composed of 10 posters in canvas (80cm x 120cm) with interactive elements. Its total weight does not exceed 30 kg which enable its transportation from one site to the other. Young people since 10 years old, non-initiated public, scientific lecturers and teachers in training are the target publics. A scientific lecturer trained by the '*Association Graine de Chimiste*' presents the exhibition, depending on the context, the audience level and the questions. In a scholarship context, a 1 hour-presentation is proposed and the exhibition is hired for a whole week to be deepened by teachers with the help of a guidebook specially aimed at them. This time allows the teaching team to exploit the information provided and to address more relevant questions formulated by pupils. It is also displayed during cultural, leisure and training contexts, for continuing teacher's training for example (Fig.).
Content and objectives of the exhibition

The global aim of the exhibition is to make the target population aware of the fragility of the Earth water resources. Limited by nature and threatened by human activities, the planet citizenship is mobilized so that anyone feels responsible for a sound management and an improved protection of water resources. By means of posters, a scientific lecturer transmit basic scientific knowledge to clarify the natural behavior of water resources and the stakes of their use by human society (Fig.1).

Fig.1. General view of the exhibition 'Planet Water at Home' with a scientific lecturer ('Planet Water at Home' by the 'Association Graine de Chimiste')

Except the first poster '*Planet Water at Home*' which serves to motivate the audience, the following '*Three States of Water*', '*Six Water 'Reservoirs'*' and '*Natural Water Resources*' are necessary steps to make understandable the poster '*Travel of Water on Earth*' which deals with the natural water cycle. Impacts of human activities on natural resources are studied in the '*Hydrological Cycle and Human Activities*' and the '*Aquatic Ecosystems Modification*'. The three last posters entitled '*Waste Water Purification*', '*Drinking Water Production*' and '*Drinking Water*' show the technological efforts realized to treat waters to safeguard the environment and human health. The content and objectives of posters are presented thereafter:

Poster 1 - Planet water at home

Any human being uses water. Source of pleasure, feeding, energy, water can also bring suffering and illness about. Few photographs display some extreme natural situations (drought, flood) and usual uses of water by human.

Objective: to show the importance of water resources for human activities and survival

Poster 2 - Three states of water

As any substance, water can exist under three states (solid, liquid and gas). A scale of temperature shows the existence fields of each state and their diverse transformations.

Objective: to prepare the further understanding of the global water cycle

Poster 3 - Six water "reservoirs"

In nature, water is stored in 'reservoirs'. By an interactive game, the vocabulary known by the public is used to identify the six natural water reservoirs on our planet.

Objective: to introduce the notion of 'reservoir' required for the water cycle

Poster 4 - Natural water resources

An illustration of the Earth viewed from space shows a high proportion of water. The diagram hidden by this view represents the distribution of the total volume of water among the six reservoirs. Anyone can then appreciate the quantity of freshwater available on our planet. Surface and volume are distinguished.

Objective: to make people aware of the small proportion of freshwater on the Earth

Remark: 'Are water resources evenly shared over the world?'

Poster 5 - Travel of water on Earth

Water moves from one water reservoir to another, globally from the atmosphere to the continents, then to the oceans and again to the atmosphere... The same water circulates on Earth since its formation. A natural landscape displayed enables the public to identify the natural reservoirs and some exchanges that link them. The public is made aware of the influence of climate on the natural cycle.

Objective: to understand the water cycle and its consequences on water resources

Remark: 'Climate is changing, earth is warming: what are the consequences for our environment?'

Poster 6 - Hydrological cycle and human activities

Human activities withdraw and/or pollute a part of the water present in natural reservoirs and linked by the hydrological cycle. By an attentive observation of a picture, everyone can list different perturbations of this natural cycle by human.

Objective: to sensitize to the impact of human activities on water resources and cycle

Remark: 'Which human activity requires the more water?'

Poster 7 - Aquatic ecosystems modification

Two illustrations show the impacts of human activities on aquatic ecosystems. Sometimes linked in a food-chain which starts in these environments, humans can be victims of their own pollution.

Objective: to show the fragility of aquatic ecosystems to human pressure

Remark: 'Reserves of life and regulating floods, wetlands are also natural filters of waters.'

Poster 8 - Waste water purification

In order to limit the degradation of natural watercourses during their release, water used by human can be treated. A diagram details the principle of the treatment steps.

Objective: to show that waste waters purification is still insufficient to avoid pollution

Poster 9 - Drinking water production
Natural water can sometimes be used to produce drinking water, freshwater disinfected to prevent the spreading of water-related diseases [1]. The principle of the treatment of drinking water production is explained on a diagram, which constitutes with the previous one, the domestic circuit of water.

Objective: to show the effort realized for the vital need of safe drinking water

Poster 10 - Drinking water
To drink is an essential need for human. Sometimes people have access to several types of drinking water, either treated natural waters, or natural waters supplied in bottles. But do tap waters, spring waters and mineral waters have the same characteristics [7]?

Objective: to help consumers to choose a suitable drinking water

Remark: 'When you will be 8,3 billion people, around 2025, will there be enough water for all?'

Interactivity of the exhibition
The interactivity of the exhibition is provided by a scientific lecturer trained by the '*Association Graine de Chimiste*' who presents the posters and interacts with the public depending on its level and reactions. This interactive character is emphasized by cards masking certain information that are removed as the public give the right answers. Major issues (defined before as 'remarks') raised on posters by a pelican and initially occulted by the speech, become predominating when basic notions have been understood. Thus, non-specialists have gained keys (notions, vocabulary) to decipher water issues and understand the challenges faced by our modern society.

After an interactive lecture on the exhibition, a schoolchild remarked: "What mostly impressed me is that we are going to fight for water. It's frightening!"

(2) Experimental development of the exhibition
In complement of the interactive exhibition '*Planet water at home*', hands-on workshops based on the '*Association Graine de Chimiste*' methodology were designed on pre-requisites (water states, properties and treatments), with the support of professionals such as drinking water producers and tasters. They have been adapted to the public level, since 5 years old (nursery garden) to professionals (Fig.2 and Fig.3).

Fig.2. Teacher producing drinking water, IUFM Teachers' Training College of Paris ('Drinking Water Production Treatment' by the 'Association Graine de Chimiste')

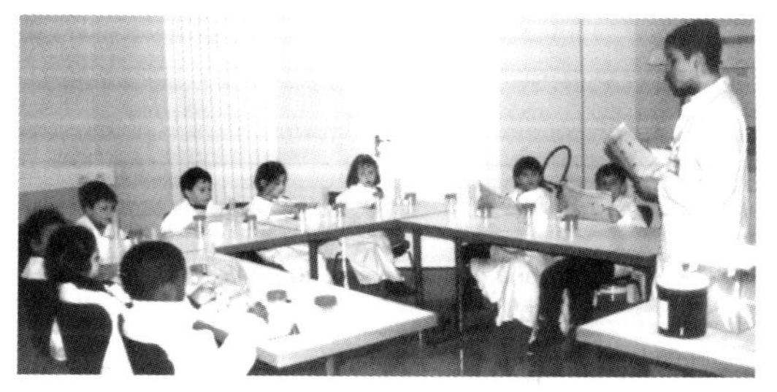

Fig.3. General view of a workshop aimed at 6-years old children during a 'water class' ('Drinking Water Production Treatment' by the 'Association Graine de Chimiste')

The experimental approach is a favorable support to the acquirement of necessary theoretical concepts and vocabulary as requested in official programs [16]. In order to help teachers to organize manipulation sequences in their classroom, the '*Association Graine de Chimiste*' has edited an educational booklet entitled *Pour une approche expérimentale de l'eau* (*For an experimental approach of water*) with the Research Group on Chemistry Didactics (GREDIC) from the University of Paris and voluntary primary school teachers [5].

The aim of these workshops is to transmit the basis to clarify, in particular, the processes intervening during the natural water cycle (*'Water fusion', 'Water boiling'*) [2], mineralization and pollution origins (*'Dissolution', 'Carbon dioxide dissolution in water', 'Experiments around mixtures', 'Taste of drinking waters', 'Water hardness',*), control and treatment possibilities (*'Hard-waters and detergents', 'Demineralized water', 'Drinking water production treatment'*) [3]. The understanding of the natural water cycle firstly requires the mastery of the state change notion and its related vocabulary. Other workshops demonstrate that natural waters are not pure substances because they can contain solids, liquids and gases. Their objective is to explain why natural waters have a different quality and why and how they can be polluted. When atom and molecule concepts are not yet acquired, these experiments can introduce to the vocabulary related to the concept of infinitely small, with the apprehension of the microscopic world.

(2) Evaluation of the interactive exhibition 'Planet water at home'
The evaluation of this exhibition contribution to an awareness of the value of water is in progress. Although it is incomplete, certain reactions seem relevant enough to be already underlined. In general, few people are aware that water resources are limited and the public is deeply impressed by this information. Young urban audience has difficulties to distinguish natural processes from anthropogenic ones. The public also discovers that water is a transboundary resource and that it can be a source of tensions, or even conflicts, between neighbor countries [18]. It seems that a real awareness of the extent of world stakes is raised as well as a wish to modify anyone's attitude concerning water.

3. Conclusion

The objective of the Learning Society is to provide any human with some scientific and technological information at global and local levels to promote the perception of global issues and to learn about each particular local environment. For this, many tools now exists such as the Internet. But other means could still be used too, with little funding and for other target public that have no access to the Internet.

Any initiative should provide any mediators or trainers with some interdisciplinary theoretical background, educational objective and activities related to the day-to-day life in order to help them to design their educational programs around one or several topics, for example, from the followings: water and nature; water and climate; water and life; water, hygiene and health; water and civilizations; water and human activities (with mention of the disparities related to gender); water and demography; water and conflicts.

To succeed in this challenge, the provision of training courses adapted to each population targets (consumers, trainers and decision-makers) and the production and distribution of adequate educational material are essential and could be initiated whether implemented by the Learning Society.

Acknowledgements
The exhibition "Planet water at home" was achieved with the support of the French Ministry of National Education, Research and Technology, the University of Paris '*Université Pierre et Marie Curie*', the Academic Inspection, the water agency '*Agence de l'Eau Seine-Normandie*' and the water suppliers '*Société Anonyme de Gestion des Eaux de Paris*' and '*Lyonnaise des Eaux*'. Graphic design and illustrations were made by Karine LABBAY. Photographs were provided by '*Agence de l'Eau Seine-Normandie*', '*Agence Spatiale Européenne*', '*Bombardier Inc.*', Olivier Faÿ, Frédérik Froument, Julie Ladel, '*Lyonnaise des Eaux*', '*La Photothèque d'EDF*', '*Société Anonyme de Gestion des Eaux de Paris*' and '*UNESCO*'.
The designers thank for their review and advice MM.Paul CARO (Cité des Sciences et de l'Industrie, Paris), Ghislain DE MARSILY (Laboratoire d'Hydrogéologie, Université Pierre et Marie Curie, Paris), Florent DOMINE (Laboratoire de Glaciologie et de Géophysique de l'Environnement, Université Grenoble I), François RAMADE (Laboratoire d'Ecologie aquatique et d'Ecotoxicologie, Université Paris Sud, Orsay) and Habib ZEBIDI (Division des Sciences de l'Eau, UNESCO, Paris).

References

Bontoux, Jean (1993) *Introduction à l'étude des eaux douces. Eaux naturelles, eaux usées, eaux de boisson*. Ed. Cébédoc, Tec&Doc-Lavoisier, 169 p.
Centre National de Documentation Pédagogique (1993) Le cycle de l'eau. *TDC*, 638, January 1993, 31 p.
Centre National de Documentation Pédagogique (1994) L'eau potable. TDC, 677, 31p.
Colas, René (1981) *Papa, dis-moi, l'eau, qu'est-ce que c'est ?* Ed. Orphys, Palais de la Découverte, 146 p.
CFES et C.I.eau (1996) *Léo et l'eau*. Coffret pédagogique. Collection "Les chemins de la santé", 16 p.
De Marsily, Ghislain (1995) *L'eau*. Ed. Flammarion, Coll.Dominos, 51, 126 p.
Evina, Emmanuelle (1997) *Le guide du buveur d'eau*. Ed. Solar, 2nd edition, 263 p.
Graine de Chimiste (1994) Dossier "Graine de Chimiste". *Journal des Instituteurs et des Institutrices*. October 1994, 2, 53-67.
Groupe de Recherche en Didactique de la Chimie, Graine de Chimiste et des enseignants (1997) *Pour une approche expérimentale de l'eau*. Ed. Centre Régional de Documentation Pédagogique de Poitou-Charentes, Poitiers, France, 129 p.
International Environmental Education Programme (1995) 8 posters. Ed. UNESCO-UNEP.
International Environmental Education Programme (1995) Freshwater resources. (ed.: Skofteland, Egil) Ed. UNESCO-UNEP, Environmental Education Module, Paris, 107 p.
Internationalization and Innovation of Teacher Education (ITE) and UNESCO & UNEP International Environmental Education Programme (IEEP) (1993) *Water in our life. Textbook for teachers in Environmental Education*. (ed.: Classen-Bauer, Ingrid). Ed. Verlag für Wissenschaft und Bildung, Berlin, 300 p.

Kandel, Robert (1998) *Les eaux du ciel*. Ed. Hachette Littérature. 329 p.
Kohler, Pierre (1997) *Voyage d'une goutte d'eau*. Ed. Ecole Active, Montreal, Canada. 44 p.
Ladel, J. & Thibault, J. (1998) Contribution with Interdisciplinary Scientific Notions to the Education for all to the Respect of Water. *In: Proceedings of the International Congress "Education for a Culture of a Shared and Preserved Water", Jounieh (Lebanon), 17-20 June 1998*. 10 p.
Ministère de l'Education Nationale, de la Recherche et de la Technologie, Direction des Lycées et des Collèges (1997) *Programmes du cycle central 5e et 4e*. B.O. n°5 du 30/01/97. Collection Collège. Ed. CNDP, 251 p.
Seuling, Barbara (1996) *Notre Planète Terre*. Ed.Flammarion, Coll.Castor Doc, D1, 88 p.
Sironneau, Jacques (1996) *L'eau. Nouvel enjeu stratégique mondial*. Ed. Economica, 111 p.
Thibault, J., Davous, D. & Masson, A. (1993) Une approche interactive de la chimie. *Didaskalia*. 2, 121-130.
UNESCO & OMM (1992) *Glossaire International d'Hydrologie*. Ed.UNESCO& OMM. 413 p.
UNESCO & UNEP (1994) Water: an educational and informative approach. *Connect*, Ed. UNESCO. Vol. XIX, 2, June 1994.
UNESCO (1998) *World Water Resources. A new appraisal and assessment for the 21st century*. (ed. Prof. I.A. Shiklomanov) Ed. UNESCO, Paris, 37 p.
US Geological Survey's Water Resources Education Initiative (1997) *Poster series on water-resources education*.
Villeneuve, Claude (1998) *Qui a peur de l'an 2000? Guide d'éducation relative à l'environnement pour le développement durable*. Ed. UNESCO/Multimondes. 303 p.
Wick, Walter (1997) *A Drop of Water*. Ed. Scholastic Press, New York, 40 p.
WMO & UNESCO (1997) *Is there enough water ?* 857. Ed. WMO, UNESCO, Paris. 22 p.
WMO (1997) *Comprehensive Assessment of the Freshwater Resources of the World*. (ed.: Prof. I.A. Shiklomanov). Ed. WMO, with the support of Stockholm Environment Institute, Paris. 88 p.

The life and the water vs. the life of the water

Macri M. V.

RGAB, Romania

Abstract

Like the energy, the water stands as an essential component of almost all the human occupations. Water supply is vital for the nourishment of the increasing world population, for the production of goods that result in the growth of the life level and for the maintenance of the natural systems integrity of what the life of the earth depends upon. The integrity of the water cycle makes downstream activities vulnerable to upstream land and water resources management. Upstream pollution may make accessible water unusable downstream. Recent trends towards ecorealism include focus on risk assessment on a catchment basis and on what particular ecological services primarily have to be protected. Long term planning will have to take into account a desired long-term future, mental anticipation, prevention of decreasing water usability downstream.

Résumé

Comme l'énergie, l'eau est une componente essentielle de presque tous les occupations humaines. Le fournissement avec d'eau est vitale pour l'alimentation de la population du monde en cours d'expansion, pour la production des bienes qui determine l'augmentation du standard de la vie et pour l'entretien de l'integrité des systèmes natureaux sur lesquelles depends la vie sur terre. L'integrité du cycle d'eau fait que les activités en aval soient vulnerables par rapport à la gestion des resources d'eau et de terre en amont. La polution en amont peut faire d'eau être non-usable en aval. Les tendances recentes vers l'ecorealism includent l'accent sur l'évolution des risque à la base du captage et sur les services écologists particulières qui doivent être protegées en première instance. Une planification à long terme devra considerer une future desirable à long terme, l'anticipation mentale, la prévention de décroissance d'utilisation d'eau en aval.

1. WATER - A VITAL RESOURCE -"Let's be the caretakers of the world"

Drinking water is a precious resource, but most people take it for granted. Many people assume that water will always come out of their kitchen tap and that it will always be wholesome. Like the energy, the water stands as an essential component of almost all the human occupations. Water supply is vital for nourishment of the increasing world population, for the production of goods that result in the growth of the life level and for the maintenance of the natural systems integrity of what the life of the earth depends upon. And besides all these, in most of the states it is amazingly little known about the used quantities of water and where exactly, when and by whom have they been consumed. Although almost any political leader may cite the price of the day for the oil barrel, not many of them are able to say which is the cost price for ensuring a plus of 1000 cubic meters of water.

Water, in spite of its apparent simplicity, is a highly complex substance with many parallel functions in landscape as well as society. As a consequence, the problems are not being solved by agreeing on temporary, sectorial action plans.

It is the job of the water system operator to get the water from the source to the consumer's tap. This may involve pumping water out of the ground or diverting a stream, removing harmful contaminants, and pumping the water through kilometers of pipes. All of this costs money. Water in the ground may be free, but getting the water from the source to the people's homes and making sure that it is safe costs money. An important part of the operator's job is to help people to understand why piped water to their homes is not free. If the operator can gain the support of the community, then his job will be easier, and he can better protect the precious resource which is the drinking water.

Water is now acknowledged as a major limiting factor in the socio - economic development of a world with a rapidly expanding population. Poverty eradication and food supply both contribute to growing per capita needs. This generates serious dilemmas in low income regions: How can life support and quality of life be balanced against preservation of downstream ecosystems? Is it preferable in dry climate regions to import water or import food? Is there an underground reserve of underutilized groundwater available? Or is it the groundwater already overexploited?

Unfortunately, both the time abundance as well as the water abundance may prove to be an illusion.

Romania, a country situated in the Eastern part of Europe, has a total surface of 230,340 sq. km. Of the country's population of 22,730,000 inhabitants, 54.7% live in urban areas (including 267 towns and conurbation) and 45.3% in rural areas (2686 communes including 13000 villages).

Romania has relatively modest resources of fresh water, accounting for about 9500 m3/person/year. The sources in the order of their use are: inland waterways, approx. 67%; subterranean sources, approx. 17%; the Danube, approx. 12%; and the lakes, approx. 4%. From a total length of 66000 km, 22000 km of the internal waterways are monitored for quality. The current situation shows that at present 2460 km, or approx. 11%, is depredated in quality and cannot be used. The Danube, subterranean waters and lakes are also in various states of pollution, but can be used with preliminary treatment.

The overall figures for the use of fresh water are shown in Table 1.

Table 1. Demand for water in billions of cubic meters

Sector	1989 (year of reference)	1995
Industry	10.6	4.2
Agriculture	9.1*	2.0**
Drinking water for areas within the industrial network	3.4	2.0
Total	23.1	8.2

* For the irrigation of approx. 2000000 ha

** For irrigation of approx. 400000 ha

The drastic decrease by about 35% in the demand for fresh water since 1989, when the radical changes in the political and economic system resulting from the "December revolution" began, can be accounted for as follows:

- a decline in industry which has reduced production by half, especially in the sector of the heavy industry of former communism; and
- a crisis in the use of the existing irrigation system on an area of approx. 3000000 ha, mainly through the reinstatement of private land of a total area of 6.5 Mha.

As an initial conclusion, pressure on resources of fresh water has fallen considerably leaving unresolved, however, the problem of the supply of drinking water to many areas and communities. Although all the cities have central water supply system, not all the inhabitants living in urban areas are connected to them.
In rural areas, only 18% of the villages have a water supply system.
Waste water networks are provided in all the cities and in 2.5% of the number of villages (346).

Due to various factors such as: non-uniform distribution of the sources, increasing pollution of both surface and underground sources, decrease of the available amount of water in summer (due to drought) and in winter (due to ice and snow) in a large number of major cities of Romania the water supply is not continuous.

The average drinking water consumption in Romania is around 600 l/inh./day, one of the largest in Europe. However, this figure includes not only all the losses in the distribution network but also a great amount of waste in the appartments due to the lack of metering systems.

As a general policy, until 1990, water production, treatment and distribution were ancillary activities of the "local commutes", entirely subsidized by the state. The price of the water was a symbolical one, such that the general feeling was that the water is cheap and available in an unlimited quantity. Every new investment was made by the state so, on one hand, it did not matter how much it will cost as long as you got the approval and, on the other hand, it did not change the price of water.
The consequences of that policy are visible in our transition period:
- strong reactions against a rise in price of water;
- demand for more investment money from the central government;
- very high water consumption, due to water waste.

In the appartments the water waste has also some essential causes:
- the very poor quality of the fittings which causes the taps and toilets to leak continuously ;
- the total absence of consumers awareness with respect to the scarceness of this resource;
- the total absence of individual reliable metering systems. Only as late as 1991 a complex action has been taken towards individual metering, but most of the high-rise appartment buildings have the interior facilities laid out in such a manner that one meter for cold water and one for hot water are not enough;
- the fact that, besides cold water supply, most of the cities have also a warm water distribution network which is distributing warm water according to a daily program, increasing the peak demand for water.

The actual policy concerning water has been changed in Romania since 1990; the status of the water companies has become a better one, the structures have been improved, their operation started to rely on a sounder financial basis, their short and long term policies are to improve the situation both in the technological field and in the institutional one.

2. Pollution - where it comes from

Life in the absence of water is not possible.

The seeming water abundance has blinded the society, making it ignore the necessity of conducting a viable water husbandry and to adapt within certain limited availability. No society can exploit all the resources it has at its disposal and, at the same time to keep the advantages offered by the water left in its natural state.

The need of protecting these natural functions thus stands for an essential meaningful element regarding the analysis of the manner how the society is using its water resources. An old British customary law, based on which the owner of a river plot was obliged not to diminish the water quantity and quality dedicated to the downstream consumers, it actually ensured the ecological protection for the habitats on the river stream.

A watercourse that is polluted can pose a serious threat to the natural environment, endangering human life and aquatic flora and fauna, as well as becoming a potential hazard via industrial and domestic water supplies.

We believe that in order to achieve and maintain an acceptable standard of living, the access to a safe, reliable source of clean drinking water is essential. This will become a viable objective only by keeping our valuable water resources as pollution - free as possible.

Pollution is a catchword, which describes the process of contamination of the natural environment such that it becomes offensive or harmful to human, animal or plant life.

Although pollution may occur naturally, it is largely man-made.

Pollution may be separated into point and diffuse source pollution. Point source pollution includes pollution originating from discrete measurable sources such as discharges from sewerage works and industrial discharges. Diffuse sources pollution occurs when pollution enters the rivers and streams either from the atmosphere or from water draining the land anywhere in the catchment, and it's thus difficult to identify and control. Industrial discharges, discharges from farms including run-off of agricultural wastes and fertilizers, seepage of leachate from waste disposal sites and pollution from domestic sewage effluent are only some of the sources of point and diffuse pollution. Each type of pollution contains characteristic pollutants, which in high enough concentrations, may have a detrimental effect on the receiving water quality.

Examples of potentially dangerous pollutants are:

Industrial waste - oils, solvents, acids, alkalis, metals.
Agricultural waste - nutrients from fertilizer run-off, pesticides, suspended solids from soil run-off.
Domestic effluent - disease bearing fecal bacteria and nutrients, organic material.
Local water allocation has to fit into a larger framework of upstream/downstream water sharing where due attention is paid to ecological risk assessments of downstream wetlands and biodiversity. The slogan "Think globally, act locally" is misleading for freshwater - it originates from a time when the worlds' attention was focused on atmospheric pollutants. Water related problems have to be addressed in terms of "Think regionally - Act locally".

As long as human attitude will be inadequate with respect to the environment protection it will be necessary all the time to find a guard who will be charged with discovering and fighting against that negative behavior. This guard is one of us who has to be paid for his job (wasteful in absolute value, useful just because of the negligence of some of the members of the society that we belong at) and the payment is done from the product made by all the members of the society

3. The effects of pollution

Litter, oil scrums and foam patches are visually unpleasant and may detract from recreational enjoyment. Raw or partially treated sewage escaping from inadequate systems may lead to foul-smelling unpleasant-looking rivers and streams.
The pollutants in the form of organic substances, acids, heavy metals nutrients such as nitrogen and phosphorous, and disease - bearing organisms may not be visible, but are nonetheless great potential sources of pollution.
Chemical pollutants such pesticides or heavy metals reaching a water body may have a catastrophic effect on the natural ecological balance of the system, and may even cause the water to lose its capacity for self-purification. Eventually the water body will stagnate and become unsuitable for all but the most pollution-tolerant species.
Nutrient enrichment of a water body, originating mainly from treated and untreated sewage and agricultural run-off, may cause excessive plant and alga growth. The death and decay of plant growth causes a reduction in the dissolved oxygen levels in the water column.
Once the dissolved oxygen has been removed from the water, oxygen is drawn from nutrients in the bottom sediments. This chemical change results in the characteristic "rotten egg smell" of deoxygenated waters. As the dissolved oxygen in the water drops and the quality of the water deteriorates, the diversity of species is reduced.
Contamination of water by raw or partially treated sewage carries an associated threat of disease. Some of these diseases are life-threatening to humans.

4. Pollution - Causes - *"Let's meet the future more resourcefully"*

There is general agreement worldwide that gross contamination of rivers, lakes, and shorelines is unsustainable and exacts a heavy price on the health of population and the aquatic ecosystems. Many large cities around the world discharge almost all of their wastewater into the environment, virtually untreated. As a consequence, downstream communities suffer since untreated effluent discharges incur steadily rising costs to make the water potable. This unhealthy and unsustainable situation has largely resulted from the low priority given to wastewater treatment with devastating consequences for the environment.

This form of pollution presents an ever-increasing problem in the catchment as the population grows and the trend towards urbanization continues.

The error of not giving attention to these stress signs and at not viability basing the water use is threatening the existence itself of the basic resources and of the economic systems that are based upon these resources.

A large proportion of the population lives in settlements (either formal or informal), where sanitation facilities are non-existent or at the best inadequate. In some of the formal settlements, the available facilities were not planned to deal with the vast amounts of raw sewage they are currently presented with, and the systems are subject to overloading and failure. Many of the informal settlements have sanitation systems, which consist of inadequately constructed pit latrines. During periods of rain or in areas where ground water table is near the surface, sewage from these latrines may contaminate the ground water and the rivers.

It rarely happens that people pay the real price for the water they use. The lack of concern as to maintain these constitutes a proof of the short - sight and it's a mistake that the future generations will not be able to forgive and is reasonable so.

5. Implications of this pollution

Large amounts of untreated sewage carry proportionately large loads of disease-causing bacteria. This places the informal communities at risk of contracting the diseases carried in the water, since the inhabitants rely on this same water for washing, bathing, watering animals and, in some cases, for potable use.

High nutrient loads in these waters create ideal conditions for algal growth in dams, resulting in certain dams containing high numbers of algal cells. Problems arise since certain algae may be toxic and may produce chemicals which, upon chlorination, make the water taste and smell unpleasant. Expensive treatment processes may be required to produce water of acceptable quality.

6. Remedial Measures - "Man's concern for his environment will save the world"

The beneficiaries should start paying also for the treatment of the water they pollute.
The industrialization must go forward with the same speed as the capacity of paying for the fighting the pollution it determines. On a long-term, the sacrification of water quality for the industrialization's sake entails more damages than advantages.
Rural and urban informal and formal settlements must be supplied with adequate sanitation facilities without further undue delay.
The Romanian Ministry of Public Works, has designed to supply potable water through centralized systems for 3137 villages until year 2000. Adequate sanitation facilities will go hand in hand with this objective.
Rural and urban developments should be planned and sited so as to minimize their effects on our valuable water resources.
By education, people can be motivated to improve the quality of their lifestyles - and ultimately the quality of their environments. In particular industrialists and agriculturalists must become more aware of their responsibilities for the catchment. It is well known that preventing contamination is a more economical and safer measure than correcting the damage after rivers and lakes have been polluted. In this context, environmental considerations gain prime consideration for the Romanian's water and sewerage systems.
The environmental Agency personnel has to visit potential polluters and provide free advice and assistance where necessary. Effluent re-use and reduction must be encouraged.
Within many communities throughout the world, approaching or reaching the limits of their available water supplies, water reclamation and reuse has become an attractive option for conserving and extending the available water resources. Water reuse may also present to the communities an opportunity for pollution abatement when it replaces effluent discharge to sensitive surface waters.
It is far more cost-effective to minimize pollution loads entering our streams and rivers than to channel unnecessary costs into expensive water treatment processes.
The World Business Council is presently disseminating the concept of eco-efficiency within companies (including life cycle analysis) which includes concern both for ecology and the economics of business operations. The transnational corporations, have recognized that pollution prevention is more cost - effective than end-of-pipe clean up. Avoiding environmental disasters is much better than paying for them after the fact. This important business message has not yet reached many of the smaller and medium sized companies or even some of the larger national companies still sheltered behind protective legislation in some countries.
Public opinion may be utilized to make it uncomfortable for polluters discharging inadequately treated effluent into our streams and rivers.

Legal action should be taken against those irresponsible industrialists who persist in their pollution activities, with penalties sufficiently severe to make it preferable for them to deal with their wastes responsibly.
It is in the interest of every water user in the catchment to play a part in maintaining or recovering the quality of our water. The users need to lobby for reforms such as the need for adequate safe disposal of chemical residues, and the urgent need for provision of basic services to disadvantaged settlements.
The causes and solutions to water pollution are not to be found in the water, but in the catchment. The catchment includes the entire drainage basin of a river system from mountains to sea.
Mismanagement of the catchment harms our water resources, whilst harm inflicted upstream has an influence on the downstream water quality. The system cannot function in isolation. Management of water on a catchment basis - from source to tap and from sewer to sea - where the costs are shared by all consumers and founded by water tariffs, will facilitate the protection of our water sources from pollution.

7. What you can do to prevent water pollution - "Let's work together for a cleaner, greener world"

Each of us can play a vital role in protecting our water resources from pollution.
Tell-tale signs of pollution include changes in the appearance of the water such as colour, smell, cloudiness or foaming. The presence of a film on the water surface, or distressed fish are also good indicators of pollution. You can help by keeping a watchful eye on your local rivers and dams and reporting any signs of pollution to the local environmental agencies.

We have to be the caretakers of the world!

References

Crook, J., Ammerman, D. K., Okun, D. A., Matthews, R. L. - Guidelines for Water Reuse. Cambridge, Massachusetts, in October 1992.
Gazdaru, A. - National Report Romania. 21st International Water Supply Congress and Exhibition. Madrid, September 1997.
Damian, R., Anton, A., Macri, M.V. - Rehabilitation of Water Supply Systems in Romania, Techware Assambly, Rome, 1997.
Postel, S. - (1985) State of the World

The laws affecting water policy and water – environment issues in Albania

Selenica A.
Institute of Hydrometeorology, Albania

Abstract
The water resources constitute an important natural resource for Albania, which compared with other European Countries is considered as one of the richest. Thus, the mean annual precipitation are 1485 mm and the mean annual runoff 891 mm.
Several sectors, organisations, institutions and legal structures have been developed over the years dealing with water from its own perspective (drinking water, sewerage, irrigation, nature protection, fishing, etc.).
Currently there are three laws affecting water policy and protection in Albania. The aim of this paper is to propagate the current assignment of responsibilities under these laws and to describe some of water-environment issues coming of the overlapping competencies defined by the three laws. Parliamentary actions are recommended in order to enforce the current water legislation in Albania.

Résumé
Les ressources en eau constituent des ressources naturelles importantes pour Albanie, laquelle comparée avec les autres pays de l'Europe est considérée comme une des plus riches. Ainsi, les précipitations et l'écoulement moyennes annuelles ont respectivement les valeurs 1485 mm et 891 mm.
Certains secteurs, organismes, institutions et structures légales sont développées au cours des années concernant l'eau et leur perspectives (l'eau potable, canalisation, irrigation, protection de la nature, la pêche etc.).
Actuellement en Albanie il y a trois lois concernant la politique de l'eau. L'intention de notre matériel est de faire connaître les responsabilités de ses lois et de décrire les problèmes liés aux eaux et environnement, en conséquence de superposition des compétences déterminées par les différentes lois. Des actions parlementaires ce sont recommander pour renforcer la législation courante.

Keywords
Water law, water policy, overlapping competencies, institutional improvements, National Water Strategy, parliamentary action.

1. Legislation : current situation

Currently there are three laws affecting water policy and protection in Albania. The oldest is the Law on Environmental Protection of 1993 (the LEP), which is a framework law establishing a basic structure for environmental impact assessment, permitting of land development and industrial operations, nature protection and environmental monitoring under the authority of the Ministry of Health and Environment (MHE), through the Committee for Environmental Protection (CEP).

The other two laws were adopted on 28 March 1996. The Water Resources Law (WRL) establishes a framework for the regulation of all water resources in Albania under the direction of the National Water Council (NWC) - a committee of ministers of the national government. The NWC was set up by a decision of the Council of Ministers in 1994.

The Water Supply and Sanitation Regulation[1] (WSSR) establishes a control structure for the soon-to-be privatised sector of waterworks, sewerage, and waste water treatment facilities under the direction of an independent National Water Supply and Sanitation Regulatory Commission (the Commission)[v]. At this time, the water works and sewerage systems are run by the municipalities. There is no wastewater treatment, but at least nine projects to build waste water treatment plants are known to have started.
Each law assigns competency to manage or to protect Albanian water resources to its set of institutions. So far, the only one to have a functioning implementing structure is the LEP, but CEP and the Environmental Inspectorates are weak in terms of staffing, resources, experience, and regulatory structure. Permitting and control actions are taken on an ad hoc basis, (in theory) after consideration of relevant European Union and western national standards and requirements[3].

It should be noted that the Public Health Directorate of MHE also has authority over drinking water quality, and the Ministry of Agriculture and Food has authority over irrigation waters and activities. In contrast to the new water legislation, the offices responsible for these activities have been in existence for a considerable time and are functioning with a reasonable number of staff. However, perhaps because they feel secure in their long-standing positions of sole competency in their areas of authority, they have not yet become involved in the discussion over competencies for water use and protection under the newer legislation.

The WRL's claim to automatically abrogate any previous laws which contradict its text[4] is unlikely to be enforceable, as it is too vague and can inadvertently put too many holes in other regulatory systems without providing a replacement.

The WRL and the WSSR will set up new institutions independent of Government ministries to make and implement water policy. The WRL follows a traditional approach of setting up a "political committee" of ministers, chaired by the Prime Minister, to be jointly responsible for the definition and implementation of policy. In practice, implementation would probably be delegated to the Technical Secretariat.

Broad water policy would be adopted by the NWC, upon a proposal of the Technical Secretariat.

The WSSR sets up an independent, professionally qualified regulatory Commission following the American model. The Commission would both make and implement policy within the framework of the management structure and basic regulations which are to be adopted by the Council of Ministers.

---

[1] Although translated as "regulation", this is actually a normal law adopted by the Parliament.
[2] This is according to the American model, where the professional qualifications and independence from influence is guaranteed by the selection procedure and conditions against conflict of interest placed on the individual Commission members.
[3] A project has recently been proposed to Phare to bring national permitting officials from the Member States to Albania to advise their colleagues on specific licensing issues.
[4] Art. 73.

The LEP was adopted earlier, and uses the normal ministry structure to define policy which would be implemented by the hierarchy of the MHE, CEP and the regional environmental Inspectorates, or by the local authorities, depending on the issue. Broad environmental policy would be adopted by the Council of Ministers, with more specific action plans and environmental policies adopted by the MHE.

WRL	LEP	Min. Agriculture	WSSL
WC Is the only institution with overall authority to decide water protection and management strategy Is the only institution with authority to control well-drilling, land management for water protection, banks and shorelines Competing authority for "permitting" sewerage and treatment works, discharges to water and land, water use Competing authority to regulate irrigation Lacks staff, budget and political authority to carry out these responsibilities	MHE/CEP has overall authority to protect the environment Has authority to compel EIA for activities "having a strong impact on the environment and which are particularly dangerous to human health" Competing authority for regulating and issuing permits for activities which "have an impact on the environment" (e.g. discharges to water, air and land) Has authority to supervise environmental monitoring, collect and process data Has structure and staff, but lacks training and equipment to carry out these responsibilities	Min.Agr. has competing authority to regulate irrigation Has structure, staff and enthusiastic co-operation of water users groups in carrying out its responsibilities	Commission has focused authority to regulate water supply and sanitation services to the public Competing authority (with NWC) to regulate water supply and sanitation services to the public Potentially competing authority with MHE/CEP to regulate discharges from WW plants Currently lacks staff, budget and political authority, but this is expected to change within a year

Table 1. Institutional Responsibilities defined by the three Laws

2. Overlapping of the competencies and its implications

The positive side of the equation is that Albania has three laws which give the Government adequate and extensive authority to regulate and protect water resources, and has created a structure to define and implement a comprehensive national water management and protection strategy. Theoretically, each system could operate independently within the framework of the National Water Strategy, with conflicts between competencies sorted out through an agreement of the responsible Ministers and institutions, all of whom are on the NWC.

Regarding the regulation of private water uses and discharges, an applicant would need one or multiple "permits"[1] to carry out their intentions, depending on the type of activity. Well-drilling requires only a permit from the NWC; construction and operation of a waste water treatment plant would need at least three permits: operation permits under the WSSR; discharge permits (and concession?) under the WRL and LEP. A leather processing plant would need an environmental permit under the LEP and under the WRL for its discharges to surface water and for any discharges to land which might endanger the groundwater.
Each institution would be responsible for monitoring compliance with, controlling and enforcing its own permits. Each would establish a fee structure which would support the administration necessary to implement the permitting system.

At a very general level, the critical overlapping competencies may be summarised as follows:
- All three laws authorise their institutions to establish regulatory structures for water: the WRL for all aspects of water management, the LEP for water protection; the WSSR for water supply and treatment systems.
- The WSSR's scope is entirely within the scope of the WRL.
- The LEP's water permitting authority overlaps substantially with the scope of the WRL, and also overlaps with the WSSR's regulation of waste water treatment facilities.

At the moment, two regulatory structures exist:
- CEP and its regional environmental inspectorates, and the local authorities (municipalities and districts) who are responsible for permitting land development and industrial facilities and discharges;
- and the (embryonic) NWC, which has a Technical Secretariat staff.
- WSSR will receive a staff of 30 experts. A USAID team is currently preparing implementing regulations and developing a pricing structure for the WSSR and is prepared to do the same for the WRL. The WSSR is being driven forward by the twin forces of privatisation of the water industry and the existing nine or more major foreign aid projects in the water sector, plus further projects in the pipeline[2].

The WRL is designed to create a new, powerful, centralised structure at the highest political level to decide policy and manage Albania's largely pure and abundant water resources. The policy of the management had been developed through Phare Project 95-1145.00 *National Water Strategy for Albania.*

The implications of the present situation for the future are the following:
- The WRL and the National Water Strategy remain empty shells, due to the lack of commitment and action by the Prime Minister, the Minister of Public Works and Transport, and the Minister of Health and Environment to convene the NWC and to provide the intended staff and budget.
- The occasional permit for mining or dredging operations might be issued in the name of the NWC, but the TS will not be in a position to develop or implement policies.
- The WSSR will receive the necessary staff and funding because it is linked to privatisation of the water sector - an important priority of the Government and of foreign aid institutions.
- The Ministry of Agriculture will continue to regulate irrigation, and may become active in prosecuting polluters of irrigation networks. However, it has only administrative, not civil

---

[1] The WRL uses several terms for different types of permissions for different activities; we use the generic term "permit" to cover all of these permissions (permits, authorisations, concessions, licences, etc.)

[2] E.g. Phare investment funds of ECU 20 million in the water sector are foreseen for 1997-99.

or criminal prosecution authority.
- The EIA and industrial permitting systems set up under the LEP will continue to hobble on, due to a combination of a lack of personnel, funds, equipment, training, information and no clear commitment in the Government to strengthen the permitting systems[3].
- There will be no clear water management policy. The quality of Albania's waters will decline as new industrial and agricultural activities are introduced[4]. Today, no clear knowledge about the present water quality (surface or groundwater) exist, due to the "collapse" of the monitoring institutes.
- There will be a very preliminary de facto division of regulatory authority between four bodies:
- The Ministry of Agriculture will continue to manage irrigation works and water distribution, using the considerable budget and institutions retained from the past, and now revitalised through the World Bank project for seven districts.
- The Commission will regulate water works and waste water treatment facilities. (Pressure from donors and the impetus of privatisation will ensure that it receives a budget and staff.)
- The NWC's Technical Secretariat will have some influence on mining, hydropower and other developments affecting water resources, but this influence will be minimal, as no further staff are likely to be appointed. The Water Basin Authorities will not be established until 1999, at the earliest, because initially it will be necessary and practical to centralise control in the national Government.
- The overburdened LEP industrial permitting system will capture a few industrial establishments, but most will escape regulation and can be expected to produce discharges which will damage the quality of Albania's surface and groundwater. A further threat could come from the growing prosperity of farmers, who become able to buy pesticides and fertilisers, thus producing run-off which can infiltrate groundwater and pollute surface waters and would not be subject to any permit regime.

3. Conclusions

Analysing the three laws affecting water policy and protection it is clear that some confusions and conflicts exist, mainly due to the overlapping of the competencies of water institutions defined by different laws.

For this reason, two alternative parliamentary actions are recommended :
- Review the three laws in order to avoid the overlaps and also the conflicts through an agreement of the responsible Ministers and institutions, all of whom are on National Water Council.
- Adopt a new integral law concerning water policy and protection, which also should improve the water institutions.

References

Law of the Environment Protection(LEP), Official Newspaper, Tirana 1993.
Water Supply and Sanity Regulation(WSSR), Official Newspaper, Tirana 1996.
Water Resources Law(WRL), Official Newspaper, Tirana 1996.
National Water Strategy for Albania - Final Report, BCEOM French Engineering Consultant, Tirana 1996.

---

[3] E.g., the Phare environment projects were completed and no subsequent Phare environment programmes is planned.
[4] Tanneries are operating near Gjirokaster above the unpolluted Drinos River, and near Tirana above agricultural fields; it is not clear whether or not they have applied for a permit and obtained it.

Electronic Information on the Environment

Heljä Tarmo, Raija Lappäjärvi and Eeva-Liisa Hallanaro

Finnish Environment Institute, Finland

Abstract

The Finnish Environment Institute (FEI) is a national research and development centre which plays a key role in producing and distributing information on the environment. During 1997 two extensive projects on electronic environmental information were started: producing regularly updated information on the environment for the web pages and producing a multimedia CD-ROM on the state of the Finnish environment. The popularity of the web pages produced by the environmental adminis-tration has already indicated that dissemination of environmental information electroni-cally is a very effective way to respond to the growing public demand for environmental information.

Résumé

L'Institut Finlandais pour l'Environnement (IFE), centre national de recherche et développement, joue un rôle clé en matière de production et de diffusion d'information sur l'environnement. Deux grands projets sur l'information environnementale sous forme électronique ont été lancés en 1997: production d'une information régulièrement mise à jour sur l'environnement à diffuser sur pages Internet; et production d'un CD-ROM multimédia sur l'état de l'environnement finlandais. L'ampleur des pages WWW de l'administration de l'Environnement et leur très grande popularité démontrent le bien-fondé de l'orientation du développement.

The Finnish Environment Institute as a producer of electronic environmental information: role and goals

The Finnish Environment Institute has a key role in producing, compiling, assessing and disseminating information on the environment. The results of monitoring, assessment, research and development are collected, compiled and distributed to meet the needs of different target groups such as politicians, businessmen, journalists, teachers, farmers and the society at large.

The primary aim is to ensure that decisions concerning the environment are based on reliable and objective information. The main aims of the communication process at the FEI are that the information is reliable and up-to-date, the distribution of information is performed openly and actively, and that there is cooperation with users of the information and with the producers of environmental information.

In response to growing public demand for environmental information FEI is making more environmental information electronically available and doing so in ways that make it easy for people to understand and use. During 1997 two extensive projects on electronic environmental information were started: producing regularly updated information on the Finnish environment for the web pages and producing a multimedia CD-ROM on the state of the Finnish environment.

Environmental information on the web pages

The Environment Administration's (Ministry of the Environment, the 13 Regional Environment Centres, the Finnish Environment Institute and the Housing Fund of

Finland) home pages have been produced and developed intensively since 1995. The information on the web pages is mostly in Finnish and serves different target groups in Finland. The development of an English-language service was also started during 1998. Maintenance and coordination of the environmental administration's web site and the related expertise in managing the site are largely the concern of the FEI.

Although the service was initially very inadequate, the site was well used. The regularly updated information given on water conditions was especially popular. The site was laid out according to the work of the different organization units, which was not the best possible arrangement from the users' point of view. The largest groups of users at the early stage were the universities and other institutions of higher education, which together accounted for about one third of all visits to the site.

The good response and rising public demand for environmental information motivated further improvements in the service, and a reorganized site was opened in March 1998. The principle of the site is that the information itself is the overriding factor, rather than the organizational unit within the environmental administration that produced the information. The following 14 thematic units were thus created:

Current consisting of all kinds of actual information and press releases
Services which includes information for special groups e.g. companies and communities, and ordinary citizens, EU-information, information about how to manage when an environmental accident happens, information about environmental databases, library services, publications, education and the link list of other environmental web sites.
State of the environment. State of the environment is based on the monitoring and research data produced by the FEI and the regional environment centres and on the information provided by the data systems of Environmental Administration. A large entity consists of water issues (surface and ground waters), which describes both hydrological situation and watershed forecasting, water quality, eutrophication and algae situation in inland waters and coastal waters. Both the hydrological situation and the watershed forecasting are realtime processing information automatically updated from registers. In addition, in summer the web pages report weekly about the algae situation in inland and coastal waters. This unit also contains information about the soil, air, biodiversity of nature and change of climate and other environmental problems.
Environmental protection contains information about the main tasks of environmental protection, about the environmental legislation and permissions and about the control systems of environmental issues.
Nature conservation unit provides information about the Finnish Natura 2000-environmental protection network, protection of species, biodiversity of nature, nature conservation law and of the LUMONET information dissemination system related to the biodiversity of the Finnish nature. This unit also deals with protection of species and sustainable use of living nature resources.
Use, management and restoration of the environment contains information about water resources, such as regulation, flood protection, restoration of waters and natural water construction. The thematic unit also contains information about water and wastewater management including both distribution and treatment; as well as information related to soil and nature .
Aims of environmental policies - this unit reports about sustainable development and environmental programs. It also contains data of the assessment of environmental impacts and related permits and decisions.
Land use contains data related to planning, land use and cultural environments.
Housing offers information related to housing and housing policy.
Building contains issues related to the constructed environment and building.
Environmental damage provides instructions for the management of different environmental accidents.

Research contains the information about research and development programs, monitoring programs and EU-research projects. The research concentrates on environmental changes and their causes and on methods of solving environmental problems.
International cooperation gives information about the Arctic and the Antarctic, Barents area, cooperation between Finland and countries in Central and Eastern Europe, and the international consulting services of FEI.
Organization and contacts provides data on the organization of environmental administration and about the different organization units and their contact information

The information produced by the FEI (Finnish Environment Institute) mainly concerns the subject areas of services, the state of the environment, environmental damage and nature conservation.

As mentioned above, the popularity of the service has continued to grow, and in 1998 a total of over 2 million visits were registered. The major users are now companies and communities (34 %) and other public sector bodies (21 %). Universities and institutions of higher education now account for only 8 per cent of all visits.

Further development of the service will be based on the considerable feedback received and monitoring of the types of visitors to the site. Target group-specific material, for example, for students and school children would promote dissemination of environ-mental information and improve people's environmental awareness.

A multimedia publication on the state of the Finnish environment

FEI and its former organization, the National Board of Waters and the Environment has issued volumes on the State of the Finnish environment at regular intervals (The State of the Finnish Environment, 1992; The Future of the Finnish Environment, 1996). The next report on the state of the Finnish environment will be published in the form of a multimedia in 2000. It is like the previous publications, aimed at everybody with an interest in these issues: ordinary citizens, students, decision-makers in commerce and industry, politicians and public sector employees involved in issues relating to the environment.

It contains basic facts on environmental processes and trends and their causes and consequences, and shows how environmental problems can be avoided or alleviated. The CD-ROM is intended to be used for reference and as an interesting source of environmental information. It is designed to encourage users to get a better understanding of environmental issues and to help them find solutions to problems.

The CD-ROM presents important environmental information in an attractive way:
- Texts are compact and easy to read.
- Many photos, maps, graphs and tables are included.
- Stimulating videos, animations, music and the sounds of nature feature prominently.
- The pathways through the material are interactive.
- The search function helps users to find data easily.

The environmental information is presented through four main perspectives:
The Earth examines environmental issues from a geographical perspective, on a global or European scale, and looks at the regional variations and dimensions of environmental processes and problems. The Baltic Sea and the whole Baltic region feature prominently here.
Ecosystems deals with the state of the Finnish environment from the point of view of nature. Topics are analysed according to the relevant habitats, such as forests, mires and urban areas. The central theme here is how human activity has changed these habitats, and how these changes have affected the flora and fauna found there, and indeed people themselves.

Environmental concerns focuses on problems, looking at their causes and consequences, as well as indicating possible solutions. Many of the themes included here are familiar ones: climate change, acidification and declining biodiversity, for example. **People and society** concentrates on how human activity has led to changes in the environment, how we suffer or benefit from these changes, and how we can repair any damage done. Environmental issues are examined through the perspective of different aspects of society, such as consumption, agriculture and transport. Finnish values, our everyday way of life and social welfare are central themes here.

The pathways through the information form a structured hierarchy shaped like the branches of a tree. The first choice users face is to select one of the four main perspectives described above. From then on, the branches lead users ever deeper into this fount of valuable environmental data. Internal links allow users to leap from branch to branch, or change to another main perspective. It is intended that users move freely around the information according to their own interests.

The CD-ROM will contain about 1200 factual features, with a total of 2000-2500 compact pages of text, 1500 photos, 1000 graphs, maps or tables, along with animated features, videos, backed with plenty of the sounds of nature, other sound effects and music.

Aquatic environments

One of the most interesting and essential topics of the ecosystems perspective is Aquatic Environment. Finland's aquatic environments, from the Baltic Sea through lakes and rivers to the smallest water features, will feature prominently in a new multimedia CD-ROM publication on the state of the Finnish environment to be published in 2000. Aquatic environments come under the section 'Ecosystems', where both topical and eternal issues related to water are discussed. Other habitats included in this section of the CD-ROM include shores, forests, mires, arctic fells, farmland and urban areas.

Throughout the CD-ROM the pathways through the information form a structured hierarchy shaped like the branches of a tree. Users can find the information they need by choosing their own route according to their interests. Each alternative route leads to concise environmental data in the form of texts, pictures and diagrams, and most likely further alternatives offering more detailed information on related topics.

The aquatic environments section has five main headings. Under the first heading, **water and water resources**, many of the special and vital characteristics of water are dealt with, and Finland's water resources are quantified and considered. Aquatic environments are particularly important in Finland, since a tenth of the country's land area is covered by lakes, and the waters of the Baltic Sea form a lengthy coastline in the south and west. There are also hundreds of thousands of ponds, streams, springs and other small water features in Finland.
The second subsection covers Finland's **aquatic ecosystems** and their biodiversity, examining the plants and animals which are dependent on the continued purity of the water. One special feature deals with the Saimaa seal (Phoca hispida saimensis), a subspecies of the ringed seal only found in Lake Saimaa in eastern Finland.

Information about the current condition of Finland's aquatic environments according to various parameters can be found under the next heading, **the state of lakes, rivers and the sea**. Fortunately the concentrations of many toxic chemicals in aquatic environments are declining nowadays. Contrastingly, eutrophication is an ever increasing problem in Finnish lakes and the Baltic Sea.

Along the branch **use and misuse** the environmental loads and risks related to our use of water resources are catalogued. Waste water effluent and nutrients leaching into water from farmland are often considered to put the most serious pressure on aquatic environments, but boating, fishing and energy production all also have potentially harmful effects.

The fourth and final main heading under aquatic environments is **protection and restoration**. This subsection covers what has been or could be done in Finland to help preserve aquatic ecosystems. It can be seen here that there is always room for improve-ment, in terms of reducing environmental loads as well as designating more protected areas, for example. Finland is actively involved in the management of aquatic environ-ments covered by many international agreements and joint cooperation projects described here.

Many of the vital issues related to aquatic environments also crop up in other sections of the new multimedia CD-ROM. Special features in the 'Environmental Concerns' section deal with eutrophication and acidification, for example, while the state of the world's oceans is covered in the section 'The Earth'.

Further information
Environmental Administration's web site http://www.vyh.fi or http://www.vyh.fi/eng and multimedia on the state of the Finnish environment. http://www.vyh.fi/environ/state/mm2000.

A new book on water management of Russia

Tchernyaev A., Prohorova N., Dalkov M.
RosNIIVKh, Russia

The development of the efficient water management strategy in Russia can not do without studying the knowledge of the real condition of water reserves, social-economic problems at water bodies areas, nor can it do without developing ecosystem approach to water management. Thus, currently leading experts of Russia started working on a fundamental monograph entitled "Water Resources and Water Management in Russia" at the Russian Research Institute for Integrated Water Management (RosNIIVKh) of the Ministry of Natural Resources of the Russian Federation.

Russia possesses vast water resources. However, continuous technological impact on them is the problem of the deepest concern. A detailed study of the data on the condition of natural water shows that over last decades no sufficient improvements have taken place. Even under the circumstances of the transition period crisis resulting in the considerable decrease of industrial output the condition of water is still worsening as to some parameters.

That is why it is very important, first of all, to evaluate natural water resources as the basis for the optimum water use and, secondly, to pay close attention to engineering and technologies as tools for increasing water potential.

The monograph is going to consist of several volumes, which are separate books. The first one has chapters touching on the following problems:
1.Natural climatic water resources potential
2.Geological water resources potential
3.Hydro-geological water resources potential
4.Hydrological water resources potential
5.Engineering and technical potential
6.Resources and quality of natural water

The book is well illustrated. It was the first attempt at assessing the importance of engineering-technological measures to be taken to create conditions for sustainable water use. It also tried to evaluate fresh water reserves, to distinguish among mineral, thermal and industrial waters. These data are presented in a non-traditional way, understandable in terms of water management practice. Tentative circulation of the book (300 copies) has caused great interest on behalf of experts and has enabled to make some amendments for the final edition.

The second book entitled "Current Water Management" was also at first published tentatively, it includes the following chapters:
Structure and features of water sector
Usage of water by different industries
Water bodies protection from pollution
Water management systems at residential areas and industrial centers

The second part of the book studies social-economic problems of water bodies, including territorial and inland seas, big lakes, water reservoirs, small rivers. Of a special interest is, in our opinion, the third book, dealing with the description of water management systems of the biggest Russian rivers catchments, such as the Volga, the Ob, the Northern Dvina, the Dnieper, the Don, the Kuban, the Ural, the Terek, the Yenisey, the Angara, the Irtysh, the Lena, the Kolyma, the Amur. Also given are the characteristics of water resources of catchments, their use and enumeration of problems, examples of the biggest municipal water management systems in the catchments are cited.

The fourth book comments on the principles and methods of water management of Russia. Methodological approaches are touched on in detail, they being based on studying global biosphere laws of water formation and ecological approach to water use, the experience of Russian experts in rehabilitating geological ecosystems in a river catchment.

One of the sections is devoted to the national water policy aimed at sustainable water use and conceptual approach to solving main social ecological water problems, the latter being those of drinking water, floods, water conservation, water reservoirs and small rivers.

Of an utmost interest is the section characterizing all the levels of water use management – from the federal up to local (municipal) ones. The legislative and normative base to support water use management is touched on, scientific foundation of the system is dealt with.

The fifth book tells about computerizing water management sector, automated management systems, mathematical models used as a decision support system.

The authors are convinced that the book is going to be interesting to a wide audience of experts in water management and to all those who would like to learn more about one of the strategic resources of Russia – its pure water.

Un nouveau livre sur le management d´eau en Russie

Le développement d'une stratégie effective de l'exploitation d'eau en Russie est impossible sans analyse des connaissances de l'état réel du potentiel de ressources en eau, des problèmes socio-économiques des pièces d'eau, du développement des méthodes du management d'écosystème de l'exploitation d'eau. C'est pourquoi dans le centre de recherche de l'utilisation complexe et de la protection des ressources en eau (RosNIIVKh) du Ministère des ressources naturelles de la Fédération Russe avec la participation des spécialistes principaux de la Russie on commence la publication d'une monographie fondamentale «Des ressources en eau et l'exploitation d'eau de la Russie».
La publication de la monographie se réalise en forme des livres.

Dans le premier livre on a essayé pour la première fois d'estimer l'importance des mesures d'ingénieur techniques pour la création des conditions de l'usage d'eau stable. On a estimé des ressource en eau fraîche, on a fait attention aux eaux minérales, termales et industrielles.

Dans le deuxième livre il y a une analyse des problèmes socio-écologiques des pièces d'eau, y compris des mers méditerranéennes et territoriales, de grands lacs, réservoirs, petites rivières.

Le troisième livre est consacré à la description des systèmes d'exploitation d'eau des bassins de grands fleuves de la Russie: Volga, Ob, Sévernaya Dvina, Dnepr, Don, Koubagne, Ural, Térec, Enisseil, Angara, Irtich, Léna, Kolima, Amur.

Le quatrième livre donne l'idée des principes et méthodes de la gestion de l'exploitation d'eau en Russie. Des approches méthodologiques sont décrites en détail, basées sur l'étude des régularités globales biosphériques de la formation des eaux et de l'usage d'eau, considérant les intérêts de l'environment, on cosidère l'expérience des savants russes dans la restoration des géosystèmes des bassins des rivières.

Le cinqième livre nous parle du système de l'utilisation des ordinateurs dans l'industrie de l'exploitation d'eau, des systèmes automatiques du management, des models mathématiques, utilisés pour apporter un appui à la prise des décisions de management.
Les autheurs sont sûrs, que le livre sera utile aux milieux les plus grands de spécialistes qui s'occupent de l'exploitation d'eau, et pour tous les autres qui veulent connaître mieux l'un des ressources stratégiques de la Russie - c'est l'eau claire.

DIGITAL EDUCATIONAL OUTCOMES OF THE SYMPOSIUM

1. CD - ROMs

- Basic river water quality modelling (WQMCAL version 1.1).
  Mail to: jbogardi@unesco.org
- Fluid Mechanics
  Mail to: info@techware.org
- L'OR BLEU
  Mail to: publishing.promotion@unesco.org
- Urban Drainage Management (UDMCAL)
  Mail to: wbauwens@vub.ac.be
- Water distribution systems
  http://water.fce.vutbr.cz
- Environmental information of Finland
  http://www.vgh.fi/environ/state/mm2000

2. Internet courses

- GIS and watershed systems design
  http://danpatch.ecn.purdue.edu/
- Global Change for decision-makers
  http://www.litap.iastate.edu/gcp/gcp.html
- Hydroinformatics
  http://www.bauinf.tu-cottbus.de/EGH
- Hydrologie urbaine
  http://www.enpc.fr/cergrene/HomePages/tassin/hydurb99/index.html
- Urban water quality management
  http://www.imt.dtu.dk/uwrem/

3. Internet tests

- TEST-EAU
  http://socrates.civil.auth.gr/test-eau
- TEST-EAUPRO
  http://www2.oieau.fr/testeaupro/

Organising Committee & Scientific Committee

Comité d'organisation / Organising Committee

- P. HUBERT (IHP-France) Président/Chairman
- J. BOGARDI (UNESCO)
- G. NEVEU (OIE/IOW)
- W. GILBRICH (TECHWARE)
- A. VAN DER BEKEN (ETNET/TECHWARE)

Comité scientifique / Scientific Committee

- G. de MARSILY (France) Président/Chairman
- M.E. ALMEIDA-TEIXEIRA (UE/EU)
- G. ARDUINO (OMM/WMO)
- R. N. ATHAVALE (Inde/India)
- M. BESBES (Tunisie/Tunisia)
- J. BOGARDI (UNESCO)
- B. DIENG (Burkina Faso)
- P. EFREMOV (Russie/Russia)
- Y.J. LIU (Chine/China)
- R.A. LOPARDO (Argentine/Argentina)
- K. O'CONNOR (Irlande/Ireland)
- I. RAMALOHLANYE (Botswana)
- N. TAMAI (Japon/Japan)
- A. VAN DER BEKEN (TECHWARE)
- J.P. VILLENEUVE (Canada)

Organisers

European Thematic Network of Education and Training
ETNET. Environment-Water
http://etnet.vub.ac.be/

International Office for Water / *Office International de l'Eau*
IOW -OIE
http://www.oieau.fr

International Hydrological Programme
IHP - UNESCO
http://www.pangea.org/orgs/unesco/
And the French National Committee for IHP-UNESCO

TECHnology for WAter REsources
TECHWARE
http://keywater.euro.net/

Co-organisers

WMO World Meteorological Organisation
Hydrology and Water Resources Programme (HWRP)
http://www.wmo.ch/
UNEP United Nations Environment Programme
http://www.unep.org

International Association of Hydrological Sciences IAHS
http://www.wlu.ca/~wwwiahs/

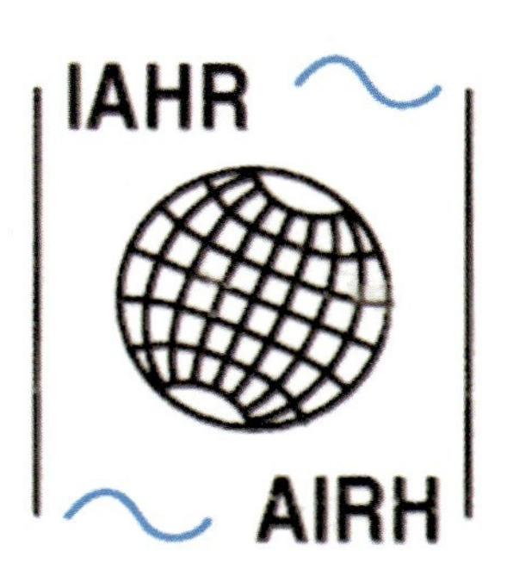

International Association for Hydraulic Research IAHR
http://www.iahr.nl

List of participants

Salisu ABDULMUMIN
National Water Resources Institute
Mando Road, P.M.B. 2309
Kaduna
Nigéria / Nigeria

ADALI
12, rue du Bac
75007 Paris
France

Ari AKOUVI
CRG de Thonon
47 Avenue de Corzent
74200 Thonon-Les-Bains
France

Ismail AL BAZ
Carl Duisberg Gesellschaff, Lutzowufer 6-9
10785 Berlin
Allemagne / Germany

Judith ALBERTI
American Water Resources Inc.
1606 Hermosa Place
CO 80906 Colorado Springs
USA

M. - E. ALMEIDA - TEIXEIRA
European Commission, DG XXII
Rue de la Loi 200, Office B7 06/07
B-1049 Brussels
CE / EC

Debebe Ayele ASCHALEW
Laboratory of Hydrology, Pleinlaan-2
1050 Brussels
Belgique / Belgium

R.N. ATHAVALE
National Geophysical Institute
Uppal Road
500 007 Hyderabad
Inde / India

Panagiotis BALABANIS
200 rue de la Loi (SOME 7/34)
1049 Bruxelles
CE / EC

Zoubeida BARGAOUI
ENIT
BP 37
1002 Tunis
Tunisie / Tunisia

Giovanni BARROCU
Department of Territorial Engineering, University of Cagliari
09123 Cagliari
Italie / Italy

Hocine BENDJOUDI
Université Pierre et Marie Curie
LGA, 4, place Jussieu, Case 123
75252 Paris Cedex 5
France

Mustapha BESBES
ENIT
BP 37 Le Belvédère
Tunis
Tunisie / Tunisia

Ernest R. BLATCHLEY
School of Civil Engineering/Purdue University
West Lafayette, Indiana
DN 47907-1284
USA

Petru BOERIU
IHE Delft
Waestvest 7, PO BOX 3015
2601DA Delft
Pays - Bas / The Netherlands

Janos BOGARDI
UNESCO
Division of Water Sciences, 1 rue Miolis
F-75732 Paris
UNESCO

Marie - Line BOUILLON
OIEAU
France

Gabrielle BOULEAU
ENGREF
648 rue J.F. Breton, BP 5093
34033 Montpellier cedex 1
France

Mitja BRILLY
Hajdrihova 28
1000 Ljubljana
Slovénie / Slovenia

Alan BRUCE
Regional TECHWARE Bureau UK and Ireland
c/o FWR, Allen House, The Listons Marlow
SL7 1FD Bucks
Royaume-Uni / UK

Ion BUSUIOC
Splaiul Independentei nr. 202 A, etage 9, secteur 6
Bucarest
Roumanie / Romania

Enrique CABRERA
Universidad Politecnica de Valencia
Gropo Mecanica de Fluidos,
Apartado de Correros 22012
46071 Valencia
Espagne / Spain

Giuseppe CANE
Italie / Italy

Roger CANS
WHAT Water Commission
6 rue de l'église
72510 Saint-Jean de la Motte
France

Jean - Pierre CARBONNEL
Université Pierre et Marie Curie
LGA, case 123, 4 place Jussieu
75252 Paris
France

Christine CARBONNEL
STRASS
190 rue de Vaugirard
75015 Paris
France

Raoul CARUBA
Réseau Méditérranée et Chaire UNESCO sur les Ressources en eau, Université de Nice - Sophia Antipolis, Parc Valrose
06108 Nice Cedex 02
France

Christiane CARUBA
Université de Nice
France

Nick CHAPPELL
Lancaster University
Environmental Science
LA1 41Q Lancaster
UK

Luisa COLLA
CIMA
Universita degli studi di Genova, Via Cadorna 7
17100 Savona
Italie / Italy

W. COSGROVE
World Water Vision

René COULOMB
94 rue de Provence
75009 Paris
France

Maria DA CONCEICAO CUNHA
ISEC
Quinta da Nora
3000 Coimbra
Portugal

Luis Veiga DA CUNHA
Universidade Nova de Lisboa
FCT/DCEA, Quinta de Torre
2825-114 Caparica
Portugal

Hédi DAGHARI
Institut National Agronomique de Tunisie
43, avenue Charles Nicole
1082 Tunis
Tunisie / Tunisia

Denis DAKOURE
CRG
BP 510, 47 Avenue de Corzent
F-74203 Thonon Cedex
France

Radu - Mircea DAMIAN
Technical Univ. of Civil Engineering
Bd. Lacul Tei 124, Sector 2
Bucarest
Roumanie / Romania

Victor DE KOSINSKY
Université de Liège
AUEF
Liège
Belgique / Belgium

J. M. DE LA FUENTE GONZALEZ
9, Allée du Château
78600 Le-Mesnil-le-Roi
France

P. J. M. DE LAAT
IHE
PO box 3015
2601DA Delft
Pays - Bas / The Netherlands

Ghislain DE MARSILY
Université Pierre et Marie Curie
LGA, Case 123, 4 place Jussieu
75252 Paris
France

André DELISLE
16, rue du Cherche - Midi
75006 Paris
France

Joseph DELLAPENNA
Villanova University School of Law
299 N. Spring Mill Road
PA 19085 Villanova
USA

Jacques W. DELLEUR
School of Civil Engineering/Purdue University
West Lafayette, Indiana
47907-1284
USA

Jean - Claude DEUTSCH
CERGRENE/Labam
ENPC-ENGREF-UPVM, 6-8 rue Blaise Pascal, Cité Descartes Champs sur Marne
France

Laura DIACO
Federgasaqua, via Corona 179
00181 Roma
Italie / Italy

Babacar DIENG
EIER
BP 7023
Ouagadougou
Burkina Faso

Cyrill DIRSCHERL
European Commission, DG XII
200 rue de la Loi
B-1049 Bruxelles
CE / EC

Francesca DODA
AMGA S.p.A.
Staff Direzione Generale, V. SS. Giacomo e Filippo 7
16112 Genova
Italie / Italy

François J. DONZIER
OIEAU
France

Ion DRAGHICI
World Meteorological Organisation
7 bis Avenue de la Paix
Genève
OMM / WMO

Radu DROBOT
UTCB
124 Lacul tei, S2
72302 Bucarest
Roumanie / Romania

Viktor DUKHOVNY
SIC Interstate Coordination Water Commission Aral Sea Basin
11 Massiv Karasu 4
700187 Tashkent
Ouzbékistan / Uzbekistan

Ovidiu DUMITRESCU
Splaiul Independentei 202 A, etaj 9, sector 6
Bucarest
Roumanie / Romania

Pavel EFREMOV
Moscow State University
Hydrology Division, Faculty of Geography
119899 Moscow
Russie / Russia

Hagen EINAR
Norwegian University of Science and Technology
Department of Hydraulic and Environmental Engineering
N-7491 Trondheim
Norvège / Norway

Mohamed ELFLEET
Environment Department, 1 Park Drive
G3 6LP Glasgow
Royaume-Uni / UK

Alain ERVINE
University of Glasgow
Department of Civil Engineering
G12 8LT Glasgow
Ecosse / Scotland

Jean - Pierre FAILLAT
ISAMOR Technopôle Brest - Iroise
29280 Plouzané
France

Guozhang FENG
Northwestern Agricultural University
College of Water Ressource and Architectural Engineering
712100 Yangling, Shaanxi
Chine / China

Jan FEYEN
KU Leuven
Institute for Land and Water management,
Vital Decosterstraat 102
3000 Leuven
Belgique / Belgium

Douglas FILS
Iowa State University
International Institute of Theoretical and Applied
Physics
IA 50011 Ames
USA

Mongi FOURY
Solidarité Franco-Maghrébine
45, rue Grabriel Péri
94270 Le Kremlin Bicêtre
France

J. GANOULIS
Aristotle University of Thessaloniki
Laboratory of Hydraulics
54006 Thessaloniki
Grèce / Greece

Christopher GEORGE
Executive Director IAHR
Rotterdamsweg 185
2629 MH Delft
The Netherlands

Wilfried H. GILBRICH
2, Avenue du Vert Bois
92410 Ville d'Avray
France

M. GIUDICI
Universita' degli Studi di Milano
Dipartimento di scienze della Terra, via Cicognara 7
I - 20129 Milano
Italie / Italy

Peter GLADISH
Akamedia GmbH
Silberstr. 22
D - 44137 Dortmund
Allemagne / Germany

François GODLEWSKI
DDE des Yvelines
BP 1115
78011 Versailles cedex
France

Philippe GOURBESVILLE
Université de Nice-Sophia Antipolis
UMR 5651, 98 bvd Edouard Herriot
06000 Nice
France

Albert GREGOIRE
CIBE (Compagnie Intercommunale
Bruxelloise des Eaux)
Rue aux Laines 70
B-1000 Bruxelles
Belgique / Belgium

Benny HAGELSKJAER
Vejlsovej 51
DK-8600 Silkerborg
Danemark / Denmark

Philippe HARANG
10, rue Dauzats
33000 Bordeaux
France

Poul HARREMOES
Technical University of Denmark
Institute of Environmental Science and Engineering,
Building 115
DK-2800 Lyngby
Danemark / Denmark

Thomas HAVENS
American Water Resources Inc.
1606 Hermosa Place
CO 80906 Colorado Springs
USA

Tobias HOFMANN
Bauhaus - Universitaet Weimar
Coudraystr. 7, Zimmer 206
99423 Weimar
Allemagne / Germany

Pierre HUBERT
Ecoles des Mines de Paris
CIG, 35 rue Saint-Honoré
77305 Fontainebleau
France

Marc HUYGENS
Hydraulics Laboratory University Gent,
Sint - Pietersnieuwstraat, 41
B-9000 Gent
Belgique / Belgium

George ICONOMI
Albanie / Albania

Stefan IGNAR
Warsaw Agricultural University
Department of Hydraulic Structures,
ut. Novorwsyowska 166
02-787 Warsav
Pologne / Poland

Istvan IJJAS
Muegyetem rakpart 3
1111 Budapest
Hongrie / Hungary

Sandrine IRACE
Université Pierre et Marie Curie
4 place Jussieu, Case 134
75005 Paris
France

Harouna KARAMBIRI
Université P. et M. Curie
Labo de Géologie Appliquée, case 123, 4 place Jussieu
75252 Paris cedex 05
France

Dimitris KARPOUZOS
Fondation Helenique (CIUP), 47 Bd Jourdan
75690 Paris
France

René KERSAUZE
O.M.S.
Bureau de projets, 149 rue Gabriel Péri
54500 Vandoeuvre
France

Vivek S. KHADPEKAR
Centre for Environment Education
Thaltei Tekra
380 051 Ahmedabad
Inde / India

Helmut KOBUS
Universitat Stuttgart
Institut fur Wasserbau, Pfaffenwaldring 61
70550 Stuttgart
Allemagne / Germany

Istvan KONTUR
Technical University of Budapest
Dep. of Water Ressources Management,
Muegyetem rkp. 3
H-1111 Budapest
Hongrie / Hungary

Zdenek KOS
Université Technique de Prague
Faculté de Génie Civil, Thakurova 7
166 29 Praha 6
République Tchèque / Czech Republik

Pavel KOVAR
Czech University of Agriculture
Dept. of Land and Water Engineering,
Kamycka 129
CZ-16521 Prague
République Tchèque / Czech Republik

Martine LE BEC
6 rue Vauvouleurs
75011 Paris
France

Risto LEMMELA
HUT/Water Res. Eng.
Huhtatie 12
04300 Tuusula
Finlande / Finland

Hoerby LENSING
Universitaet Stuttgart
Institut fuer Wasserbau, Pfaffenwaldring 61
70550 Stuttgart
Allemagne / Germany

Raija LEPPAJARVI
Finnish Environment Institute
P.O. Box 140
FIN - 00251 Helsinki
Finlande / Finland

Yu Jie LIU
China Meteorological Administration
No 46 Baishiqiaolu
100081 Beijing
Chine / China

Willibald LOISKANDL
Universitaet fuer Bodenkultur Wien
IHLWW, Muthgasse 18
A-1190 Vienne
Autriche / Austria

Raul Antonio LOPARDO
National Institute of Water and Environment
Cassila de Correo 46
1802 Aeropuerto Ezeiza
Argentine / Argentina

Jan LUIJENDIJK
IHE Delft
Waestvest 7, PO BOX 3015
2601DA Delft
Pays - Bas / The Netherlands

Khlifa MAALEL
Ecole Nationale d'Ingénieurs de Tunis
BP 37 Le Belvédère
Tunis
Tunisie / Tunisia

Mircea - Valentin MACRI
Bucharest General Water Company
18 Bd. Mircea Eliade , sector 1
Bucharest
Roumanie / Romania

Ramiz MAMEDOV
Institute of Geography of Academy of Sciences of Azerbaidjan Republic
H. Javid ave. 31
Baku
Azerbaidjan

Anne MARLEIX
STRASS
190 rue de Vaugirard
75015 Paris
France

Jean-Marc MATHIEU
34 Rue de l'Abbée GROULT
75015 Paris
France

Marta MAZURO
Gdansk Water Foundation
9 Rycerska Street
80-882 Gdansk
Pologne / Poland

Margareta MIHAILESCU
Université Pierre et Marie Curie
LGA, Case 123, 4 place Jussieu
75252 Paris

Annie MOISSET
ENGEES
1 quai Koch, BP 1039
F-67070 Strasbourg cedex
France

Frank MOLKENTHIN
Brandenburg University of Technology at Cottbus
Institut fur Bauinformatik, Universitaets-platz 3-4
D - 03044 Cottbus
Allemagne / Germany

André MUSY
Ecole Polytechnique Fédérale de Lausanne
I.A. Terre et Eaux
CH 1015 Lausanne
Suisse / Switzerland

Stefania NASCIMBEN
Research and Training Center for Water Systems Control
Strada 52, Poggio dei pini
09012 Capoterra
Italie / Italy

Gilles NEVEU
Office International de l'Eau
Rue E. Chamberland
87065 Limoges
France

Tuong - Vi NGUYEN
Académie de l'Eau
51 rue Salvador Allende
92027 Nanterre Cedex
France

Markus NIERODA
Bildunsgzentrum fuer die Entsorgungs und Wasserwirtschaft GmbH
Dr. Detlev-Carsten-Rohwedder-Str. 70
47228 Duisburg-Rheinhausen
Allemagne / Germany

Valeriu NISTREANU
University Politechnica of Bucharest
Splaiul Independentei no 313, sector 6
Bucharest
Roumanie / Romania

Ralf OSINSKI
IKV, Kaiserstr. 46
40429 Dusseldorf
Allemagne / Germany

Ada PANDELE
INMH
Sos. Bucuresti - Ploiesti 97, sector 1
71552 Bucarest
Roumanie / Romania

Ulderica PARODI
CIMA
Universita degli studi di Genova, Via Cadorna 7
17100 Savona
Italie / Italy

R. PEREZ
Universidad Politecnica de Valencia
Gropo Mecanica de Fluidos, Apartado de Correros 22012
46071 Valencia
Espagne / Spain

Sara C. PEREZ DE VARGAS
Université Nationale de La Plata
Faculté des Sciences de l'Education
Argentine / Argentina

Gustavo PERRUSQUIA
Chalmers University
Dept. of Hydraulics
SE-412 96 Goteborg
Suède / Sweden

Timm PETERS
Akamedia GmbH
Silberstr. 22
D - 44137 Dortmund
Allemagne / Germany

Iulian POPA
Université de Bucarest
Faculté de Géologie et Géophysique, 6 rue Traian Vuia, sector 1
70139 Bucarest
Roumanie Romania

Anne POWELL
The Forge, Denton Green, Cuddesdon
OX44 9JE Oxon
Royaume-Uni / UK

Eva PRENOSILOVA
Université Technique de Prague
Faculté de Génie Civil, Thakurova 7
166 29 Praha 6
République Tchèque / Czech Republik

Emil PRODAN
Splaiul Independentei 202 A, étage 9, ch. 21 - 22, sector 6
Bucarest
Roumanie / Romania

Markku PUUPPONEN
Finnish Environment Institute
Hydrology and Water Management Division, Kesakatu 6, P.O. BOX 140
FIN - 00251 Helsinki
Finlande / Finland

Pierre - André ROCHE
Directeur de l'Agence de l'Eau Seine-Normandie
France

John C. RODDA
IAHS President
Institute of Hydrology, Wallingford, OXON
OX10 8BB Wallingford
Royaume-Uni / UK

Albert Louis ROUX
Président du Conseil d'Administration de l'Agence Rhone Méditerranée Corse
2-4 Allée de Lodz
69363 Lyon cedex 07
France

M. B. A. SAAD
Hydraulics Research Institute
PO BOX 13621, Delta Barrage
Cairo
Egypte / Egypt

Claude SALVETTI
Agence de l'eau Seine-Normandie
51 rue Salvador Allende
92027 Nanterre cedex
France

Rajinder SAXENA
Inst. for Geovetenskaper
Programmet for Hydrologi, Villavagen - 16
S-752 36 Uppsala
Suède / Sweden

T. SCHMIDT - TJARKSEN
Universitaet Weimar
Bauhaus
Weimar
Allemagne / Germany

Paul P. SCHOT
Utrecht University
Environmental Sciences, Faculty of Geographical Sciences, P.O. Box 80.115
3508TC Utrecht
Pays - Bas / The Netherlands

Martin SEIDL
6, rue Filles du calvaire
75003 Paris
France

Motshabi Miriam SEKATI
Department of Water Affairs and Forestry
Private Bag X 313
0001 Pretoria
Afrique du Sud / South Africa

Robert SELLIN
The Cottage, 28 West Hill,
BUDLEIGH SALTERTON
EX9 6BU Devon
Royaume-Uni / UK

Petru SERBAN
Str. Edgar Quinet nr 6
70106 Bucarest
Roumanie / Romania

Muhammad SHATANAWI
University of Jordan
Water and Environment Research and Study Center
Amman
Jordanie / Jordan

Vladimir Yu. SMAKHTIN
Rhodes University
Institute for Water Research, PO Box 94
6140 Grahamstown
Afrique du Sud / South Africa

Olga SMAKHTINA
Univ. of Fort Hare
Dept. of Geography, Private Bag X1314
5700 Alice
Afrique du Sud / South Africa

Zbigniew SOBOCINSKI
Gdansk Water Foundation
9 Rycerska Street
80-882 Gdansk
Pologne / Poland

Viorel Al. STANESCU
National Institute of Meteorology and Hydrology
Sos. Bucuresti - Ploiesti 97
71552 Bucarest
Roumanie / Romania

Vlastimil STARA
Technical University of Brno
Institute of Municipal Water Management, Zizkova 17
602 00 Brno
République Tchèque / Czech Republic

Eugene S. TAKLE
Iowa State University
International Institute of Theoretical and Applied Physics, 3010 Agronomy Hall
IA 50011 Ames
USA

Amal TALBI
Université Pierre et Marie Curie
LGA, Case 123, 4 place Jussieu
75252 Paris
France

N. TAMAI
University of Tokyo
Department of Civil Engineering, Bunkyo-ku
Tokyo
Japon / Japan

Helja TARMO
Finnish Environment Institute
P.O. Box 140
FIN-00251 Helsinki
Finlande / Finland

Bruno TASSIN
CERGRENE/Labam
ENPC-ENGREF-UPVM, 6-8 rue Blaise Pascal,
Cité Descartes Champs sur Marne
F 77 455 Marne la Vallée
France

P. - F. TENIERE - BUCHOT
9, place de Bretend
75007 Paris
France

Daniel THEVENOT
CERGRENE - LABAM
Université Paris XII - Val de Marne,
Faculte des Sciences et Techniques,
61 Avenue du général de Gaulle,
94010 Creteil
France

Janine THIBAULT
Association "Graine de chimiste"
Université Pierre et Marie Curie, boîte 67, 4 place Jussieu
75252 Paris cedex 05
France

Claude THIRRIOT
ENSEEIHT
Departement d'Hydraulique, 2 rue Camichel, BP7122
31071 Toulouse
France

Arne TOLLAN
Norwegian Water Ressources and Energy Directorate
P.O. Box 5091, Majorstua
0301 Oslo
Norvège / Norway

Claude TOUTANT
Office International de l'Eau
rue E. Chamberland
87065 Limoges
France

Maria Cristina TRIFU
INMH
Sos. Bucuresti - Ploiesti 97, sector 1
71552 Bucarest
Roumanie Romania

Ladislav TUHOVCAK
Technical University of Brno
Institute of Municipal Water Management,
Zizkova 17
602 00 Brno
République Tchèque / Czech Republik

E. M. VALENTINE
University of Newcastle
Dept. of Civil Engineering
NE1 7RU Newcastle upon Tyne
Royaume-Uni / UK

Wouter VAN DEN BERGHE
Deloite&Touche Management Solutions
Berkenlaan 1
B-1831 Diegem
Belgique / Belgium

André VAN DER BEKEN
Vrije Universiteit Brussel (VUB),
Pleinlaan 2, B-1050 Brussel
Belgique / Belgium

Ruud VAN DER HELM
UNESCO SC/HYD
World Water Vision

M. P. M. VAN DER PLOEG
Delft University of Technology
SEPA, Jaffalaan 5
2628 BX Delft
Pays -Bas / The Netherlands

Cèlia VENDRELL
Ministeri de Salut i Benestar
Ed. Clara Rabassa, Av. Princep Benlloch,
30 Andorra la Vella.
Principat d'Andorra

A. VERHALLEN
Wageningen Agricultural University
Nieuwe Kanaal 11
6709 PA Wageningen
Pays - Bas / The Netherlands

Ronny VERHOEVEN
University of Gent
Hydraulics Laboratory, Sint - Pietersnieuwstraat, 41
B-9000 Gent
Belgique / Belgium

Laurent VERNAY
BRL Inginérie
Département Eau, Av. P. Mendès, BP 4001
30001 Nimes
France

Isabel VILCHES DUENAS
Centro de Estudios Y Experimentation de Obras Publicas (CEDEX)
c/ Alfonso XII, n 73
28014 Madrid
Espagne / Spain

H. J. VINKERS
Kiwa NV, BP 1072
3430 BB Nieuwegein
Pays -Bas / The Netherlands

Paul VLASE
87 Avenue Gallieni
93 800 Epinay sur Seine
France

Aileen VOGEL
Department of Chemistry, Private bag X07
0116 Pretoria north
South Africa

Eric WARNAARS
Technical University of Denmark
Department of Environmental Science and Engineerig,
Building 115
DK-2800 Lyngby
Danemark / Denmark

Peter WARREN
34 Plough Lane, Purvy, Surrey
Royaume-Uni / UK

Johannes WESSEL
Adelaert 18
2202 PM Noordwijk
Pays - Bas / The Netherlands

Ewa WIETSMA - LACKA
Nieuve kanaal 11
6709PA Wageningen
Pays - Bas / The Netherlands

Herman G. WIND
P.O.Box 217
7500 Enschede
Pays - Bas / The Netherlands

G. ZIGLIO
University of Trento
Dept. Civil Enginnering and Environment,
Via Mesiano 77
I-38050 Trento
Italie / Italy

Floris ZUIDEMA
Netherlands National Committee IHP-OHP
c/o RNMI, Postbox 201
3730 AE DcBilt
Pays Bas / The Netherlands

European Commission
Commission européenne

The learning society and the water-environment
La société cognitive et les problèmes de l'eau

Luxembourg: Office for Official Publications of the European Communities

2000 — XIII, 512 pp. — 17.6 x 25 cm

ISBN 92-828-8308-6

Price (excluding VAT) in Luxembourg: EUR 20
Prix au Luxembourg (TVA exclue): EUR 20